Conceptual
PHYSICS

Third Edition

The High School Physics Program

Written and illustrated by

Paul G. Hewitt

City College of San Francisco
San Francisco, California

▲▼ **Addison-Wesley Publishing Company**

Menlo Park, California • Reading, Massachusetts • New York
Don Mills, Ontario • Wokingham, England • Amsterdam • Bonn
Paris • Milan • Madrid • Sydney • Singapore • Tokyo
Seoul • Taipei • Mexico City • San Juan

Pilot Teachers

Marshall Ellenstein
Maine West High School
Des Plaines, Illinois

Paul Robinson
Woodcreek High School
Roseville, California

Nathan A. Unterman
Glenbrook North High School
Northbrook, Illinois

Nancy T. Watson
Burris Laboratory School
Muncie, Indiana

Consultants

Clarence Bakken
Gunn High School
Palo Alto, California

Art Farmer
Gunn High School
Palo Alto, California

Kenneth W. Ford
American Institute of Physics (retired),
Germantown Academy
Fort Washington, Pennsylvania

Sheron Snyder
Mason High School
Mason, Michigan

Charles A. Spiegel
California Academy of Mathematics and Science
Carson, California

ISBN 0-201-46697-X
4 5 6 7 8 9 10—VH—00 99 98 97

To Kenneth W. Ford
Eminent physicist and great human being

Contents

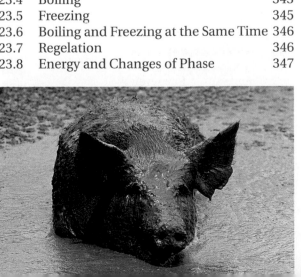

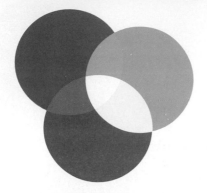

Unit V
Electricity and Magnetism 499

Unit VI
Atomic and Nuclear Physics   595

URANIUM-238 — URANIUM-239 — NEPTUNIUM-239 — PLUTONIUM-239

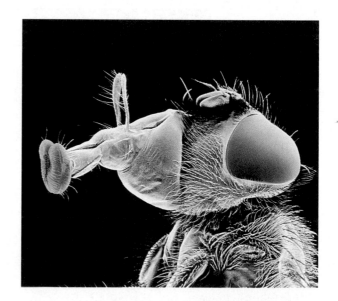

To the Student

You can't fully enjoy a game unless you know its rules. Whether it's a ball game, computer game, or party game—if you don't know the rules, it can be boring. You miss out on what others enjoy. Just as a musician hears what untrained ears can't, and just as a cook tastes in food what others miss, a person who knows nature's rules can better appreciate nature.

Learning that satellites follow the same rules as tossed baseballs changes the way you see orbiting astronauts on TV. Learning the rules of light changes the way you see blue skies, white clouds, and rainbows. Richness in life is not only seeing the world with wide open eyes, but knowing what to look for. We begin by looking at some of nature's basic rules—physics.

We treat physics conceptually in this book, which means concepts are presented in familiar English, with equations as "guides to thinking." Comprehension of concepts before calculation is the key to understanding.

Enjoy your physics!

PAUL G. HEWITT

1 About Science

Spaceship Earth in outer space.

What would it be like to live in outer space? At first thought we might think this question is for astronauts, particularly those who have experienced space walks. But on second thought, we realize this question is for everybody, for all of us *are* in outer space. At every moment we are riding on our home planet Earth, which has been in outer space from day one, hurtling completely out of human control around and around the sun. Photos of Earth taken by astronauts have greatly influenced the way we see our home in space—a relatively small life-supporting garden planet.

We can't control Earth's motion, but we have learned the rules by which it moves—rules that were painstakingly discovered by investigators throughout much of human history. The study of nature's rules is what science is about. These rules, which are surprisingly few in number, explain such things as why Earth is round, why its oceans and sky are blue, and why its sunsets are red. The richness of life is not only seeing the world with wide open eyes, but knowing about the connections between things. To know nature's rules is to add richness to the way we see our world.

The roots of science go back to before recorded history, when humans first discovered regularities and relationships in nature. One regularity was the appearance of the star patterns in the night sky. Another was the weather patterns during the year—when the rainy season started or the days grew longer. People learned to make predictions based on these regularities and to make connections between things that at first seemed to have no relationship. More and more they learned about the workings of nature. That body of knowledge, growing all the time, is part of science. The greater part of science, however, is the methods used to produce that knowledge. Science is a way of thinking, as well as a body of knowledge.

1.1 The Basic Science—Physics

Science is the present-day equivalent of what used to be called natural philosophy. Natural philosophy was the study of unanswered questions about nature. As the answers were found, they became part of what is now called science.

The study of science today branches into the study of living things and nonliving things—the life sciences and the physical sciences. The life sciences branch into areas such as biology, zoology, and botany. The physical sciences branch into areas such as geology, astronomy, chemistry, and physics.

Physics is more than a part of the physical sciences, it is the most basic of all the sciences. It's about the nature of basic things such as motion, forces, energy, matter, heat, sound, light, and the composition of atoms. Chemistry is about how matter is put together, how atoms combine to form molecules, and how the molecules combine to make up the many kinds of matter around us. Biology is still more complex and involves matter that is alive. So physics supports chemistry, which in turn supports biology. The ideas of physics are fundamental to these more complicated sciences. That's why physics is the most basic science. You can understand other sciences much better if you first understand physics.

1.2 Mathematics—The Language of Science

Science was transformed in the seventeenth century when it was learned that nature can be analyzed and described mathematically. When the ideas of science are expressed in mathematical terms, they are unambiguous. They don't have the double meanings that so often confuse the discussion of ideas expressed in common language. When the findings in nature are expressed mathematically, they are easier to verify or to disprove by experiment.* The methods of mathematics and experimentation led to the enormous success of science.

---

* Although mathematics is very important to scientific mastery, it will not be the focus of attention in this book. This book focuses instead upon what should come first: the basic ideas and concepts of physics—in English. You'll note that at the end of most chapters problems requiring elementary algebra are outnumbered by *Think and Explain* exercises. In this book, you will learn physics primarily through word descriptions that help you to visualize ideas and concepts, and only secondary emphasis will be placed on mathematical descriptions. By postponing emphasis on algebraic problem solving (which often tends to obscure the physics) to a follow-up course, you will gain a better comprehension of the conceptual foundation of physics.

Figure 1.1 ▶
Galileo (left) and Francis Bacon
(right) have been credited as the
founders of the scientific method.

1.3 The Scientific Method

The Italian physicist Galileo Galilei (1564–1642) and the English
philosopher Francis Bacon (1561–1626) are usually credited as
the principal founders of the **scientific method**—a method that is
extremely effective in gaining, organizing, and applying new knowledge. This method is essentially as follows:

1. Recognize a problem.

2. Make an educated guess—a **hypothesis**—about the answer.

3. Predict the consequences of the hypothesis.

4. Perform experiments to test predictions.

5. Formulate the simplest general rule that organizes the three main
   ingredients: hypothesis, prediction, and experimental outcome.

   Although this cookbook method has a certain appeal, it is not
the universal key to the discoveries and advances in science. Trial
and error, experimentation without guessing, or just plain accidental
discovery accounts for much of the progress in science. The success
of science has more to do with an attitude common to scientists
than with a particular method. This attitude is one of inquiry, experimentation, and humility before the facts.

1.4 The Scientific Attitude

In science, a **fact** is a close agreement by competent observers who
make a series of observations of the same phenomenon. A scientific
hypothesis, on the other hand, is an educated guess that is only presumed to be factual until demonstrated by experiment. When
hypotheses are tested over and over again and not contradicted,
they may become known as **laws** or **principles.**

If a scientist finds evidence that contradicts a hypothesis, law, or principle, then in the scientific spirit the hypothesis, law, or principle must be changed or abandoned (unless the contradicting evidence turns out to be wrong, which sometimes happens). A scientist must be prepared to change or abandon an idea. As an example, the greatly respected Greek philosopher Aristotle (384–322 B.C.) claimed that an object twice as heavy as another falls twice as fast. This false idea was held to be true for nearly 2000 years because of Aristotle's compelling authority. In the scientific spirit, however, a single verifiable experiment to the contrary outweighs any authority, regardless of reputation or the number of followers or advocates. In modern science, argument by appeal to authority is of little value.

Scientists must accept their findings even when they would like them to be different. They must strive to distinguish between what they see and what they wish to see. Scientists, like most people, have a vast capacity for fooling themselves.* People have always tended to adopt general rules, beliefs, creeds, ideas, and hypotheses without thoroughly questioning their validity and to retain them long after they have been shown to be false or at least questionable. The most widespread assumptions are often the least questioned. Most often when an idea is adopted, particular attention is given to cases that seem to support it, while cases that seem to refute it are distorted, belittled, or ignored.

Scientists use the word *theory* differently from the way it is used in everyday speech. In everyday speech a theory is the same as a hypothesis—a supposition that has not been verified. A scientific **theory**, on the other hand, is a synthesis of a large body of information that encompasses well-tested and verified hypotheses about certain aspects of the natural world. For example, physicists speak of atomic theory; biologists speak of cell theory.

The theories of science are not fixed, but rather they undergo change. Scientific theories evolve as they go through stages of redefinition and refinement. During the past hundred years, the theory of the atom has been refined as new evidence was gathered. Similarly, biologists have refined the cell theory.

The refinement of theories is a strength of science, not a weakness. Many people feel that it is a sign of weakness to "change your mind." Yet competent scientists must be experts at changing their minds. They change their minds, however, only when confronted with solid experimental evidence to the contrary, or when a conceptually simpler hypothesis forces them to a new point of view. More important than defending beliefs is improving them. Better hypotheses are made by those who are honest in the face of fact.

The scientific attitude accompanies a search for order, for uniformities, and for lawful relations among the events of nature. These enable prediction. By better understanding nature, we can better control our destinies.

Figure 1.2 ▲
Scientific theories rely on scientific facts.

---

*   In your education it is not enough to be aware that other people may try to fool you, but mainly to be aware of your own tendency to fool yourself.

Figure 1.3 ▲
Experiments test scientific hypotheses.

1.5 Scientific Hypotheses Must Be Testable

Before a hypothesis can be classified as scientific, it must link to a general understanding of nature and conform to a cardinal rule. The rule is that the hypothesis must be testable. It is more important that there be a means of proving it *wrong* than that there be a means of proving it correct. On first consideration this may seem strange, for usually we concern ourselves with verifying that something is true. Scientific hypotheses are different. In fact, if you want to determine whether a hypothesis is scientific or not, look to see if there is a test for proving it wrong. If there is no test for its possible wrongness, then it is not scientific. Albert Einstein put it well when he stated, "No number of experiments can prove me right; a single experiment can prove me wrong."

Consider the hypothesis "The alignment of planets in the sky determines the best time for making decisions." Many people believe it, but this hypothesis is not scientific. It cannot be proven wrong, nor can it be proven right. It is *speculation.* Likewise, the hypothesis "Intelligent life exists on other planets somewhere in the universe" is not scientific. Although it can be proven correct by the verification of a single instance of intelligent life existing elsewhere in the universe, there is no way to prove it wrong if no life is ever found. If we searched the far reaches of the universe for eons and found no life, we would not prove that it doesn't exist "around the next corner." The hypothesis "Most people stop for red lights" is also outside of science, but for a different reason. Although it can easily

■ Question

Which of these is a scientific hypothesis?

a. Atoms are the smallest particles of matter.

b. The universe is surrounded by a second universe, the existence of which cannot be detected by scientists.

c. Albert Einstein is the greatest physicist of the twentieth century.

■ Answer

Only (a) is scientific, because there is a test for its wrongness. The statement is not only *capable* of being proven wrong, but in fact it *has* been proven wrong. Statement (b) has no test for possible wrongness and is therefore unscientific. Some pseudoscientists and other pretenders of knowledge will not even consider a test for the possible wrongness of their statements. Statement (c) is an assertion that has no test for possible wrongness. If Einstein was not the greatest physicist, how would we know? It is important to note that because the name Einstein is generally held in high esteem, it is a favorite of pseudoscientists. So we should not be surprised that the name of Einstein, like the names of other prominent and respected figures, is often cited by charlatans who wish to bring respect to themselves and their points of view.

be tested and shown to be right or wrong, the hypothesis doesn't link up to our general understanding of nature. It doesn't fit into the structure of science.

Here is a hypothesis that *is* scientific: "No material object can travel faster than light." Even if it were supported by a thousand other experiments, this hypothesis could be proven wrong by a single experiment. (So far, we believe it to be true.) A hypothesis that has no test for its possible wrongness lies outside the domain of science.

1.6 Science, Technology, and Society

Science and technology are different. Science is a method of answering theoretical questions; technology is a method of solving practical problems. Science has to do with discovering facts and relationships between observable phenomena in nature and with establishing theories that organize and make sense of these facts and relationships. Technology has to do with tools, techniques, and procedures for putting the findings of science to use.

Science and technology are human enterprises, but in different ways. In deciding what problems to work on, scientists are guided by their own interests, and sometimes by a desire to help other people or to serve their nation. Most often scientists are driven primarily by curiosity, the simple urge to know. They pursue knowledge, insofar as is possible, that is free of current fashion, beliefs, and value judgments. What scientists discover may shock or anger some people, as did Darwin's theory of evolution. But science by itself does not intrude on human life—technology does. Once developed, it can hardly be ignored. Technologists specifically set out to design, create, or build something for the use and enjoyment of humans, often for the betterment of human life. Yet some technology can have adverse side effects or create other problems that must be solved. Although technology derives from science, it has to be judged on how it affects human life.

We are all familiar with the abuses of technology. Many people blame technology itself for widespread pollution, resource depletion, and even social decay. The blame placed on technology often obscures its promise. That promise is a cleaner and healthier world. It is much wiser to combat the dangers of technology with knowledge than with ignorance. Wise applications of science and technology can lead to a better world.

Science and technology make up a larger part of our everyday lives than ever before. Humans now have much influence over nature's delicate balance. With that power comes a responsibility to maintain that balance, and to do that, we must understand nature's basic rules. Citizens must be knowledgeable about how the world works in order to deal with issues such as acid rain, global warming, and toxic wastes. The scientific way of thinking becomes vital to society as new facts are discovered and new ideas for caring for the planet are needed.

Figure 1.4 ▲
Science complements technology.

1.7 Science, Art, and Religion

The search for order and meaning in the world takes different forms; one is science, another is art, and another is religion. Although the roots of all three go back thousands of years, the traditions of science are relatively recent. More important, the domains of science, art, and religion are different, even though they overlap. Science is mostly concerned with discovering and recording natural phenomena, the arts are concerned with the value of human interactions as they pertain to the senses, and religion is concerned with the source, purpose, and meaning of everything.

The principal values of science and the arts are comparable. Literature describes the human experience. It allows us to learn about emotions, even if we haven't yet experienced them. The arts do not necessarily give us those experiences, but they describe them to us and suggest what may be in store for us. Similarly, science tells us what is possible in nature. Scientific knowledge helps us to predict possibilities in nature even before these possibilities have been experienced. It provides us with a way of connecting things, of seeing relationships between and among them, and of making sense of the many natural events we find around us. Science widens our perspective of nature. A truly educated person is knowledgeable in both the arts and science.

Science and religion are different. The domain of science is natural order; the domain of religion is nature's purpose. Religious beliefs and practices usually involve faith in and worship of a supreme being and the creation of human community—not the practices of science. In this respect, science and religion are as different as apples and oranges and do not contradict each other. The two complement rather than contradict each other.

When we study the nature of light later in this book, we will treat light first as a wave and then as a particle. To the person who knows

Figure 1.5 ▲
Science and religion have different domains.

■ Question

Which of the following involves great amounts of human passion, talent, and intelligence?

a. art   b. literature   c. music   d. science

■ Answer

All of them! In this book we focus on science, which is an enchanting human activity shared by a wide variety of people. With present-day tools and know-how, we are reaching farther and finding out more about ourselves and our environment than people in the past were ever able to do. The more we know about science, the more passionate we feel toward our surroundings. There is physics in everything we see, hear, smell, taste, and touch!

only a little physics, waves and particles are contradictory. Light can be only one or the other, and we have to choose between them. But to the enlightened physicist, waves and particles complement each other and provide a deeper understanding of light. Similarly, people who are either uninformed or misinformed about the deeper nature of both science and religion often feel they must choose between them. But if we have an understanding of science and religion, we can embrace both without contradiction.

1.8 In Perspective

More than 3000 years ago enormous human effort went into the construction of great pyramids in Egypt. They were the world's greatest monuments to a vision of the universe. The Pyramids testify to human genius, endurance, and thirst for deeper understanding. A few centuries ago the then-modern world directed its brilliance to the building of great stone and marble structures. Cathedrals, synagogues, temples, and mosques were manifestations of people's vision. Some of these structures took more than a century to build, which means that nobody witnessed both the beginning and the end of construction. Even the architects and early builders who lived to a ripe old age never saw the finished results of their labors. Entire lifetimes were spent in the shadows of construction that must have seemed without beginning or end. This enormous focus of human energy was inspired by a vision that went beyond world concerns—a vision of the cosmos. To the people of that time, the structures they erected were their "spaceships of faith"—firmly anchored but pointing to the cosmos.

Today the efforts of many of our most skilled scientists, engineers, and artisans are directed toward building the spaceships that orbit the earth, and others that will voyage beyond. The time required to build these spaceships is extremely brief compared with the time spent building the stone and marble structures of the past. Many people working on today's spaceships were alive before the first jetliner carried passengers. Where will younger lives lead in a comparable time?

We seem to be at the dawn of a major change in human growth, not unlike the stage of a chicken embryo before it fully matures. When the chicken embryo exhausts the last of its inner-egg resources and before it pokes its way out of its shell, it may seem to be at its last moments. But what seems like an end is really only a beginning. Are we like the hatching chicks ready to poke through to a whole new range of possibilities? Are our space-faring efforts the early signs of a new human era?

The earth is our cradle and has served us well. But cradles, however comfortable, are outgrown one day. With inspiration similar to the inspiration of those who built the early cathedrals, synagogues, temples, and mosques, we aim for the cosmos.

We live at an exciting time!

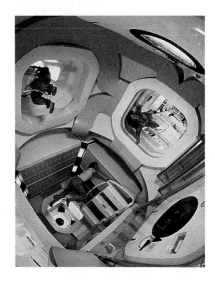

Figure 1.6 ▲
A NASA mock-up of a spaceship of the future. New discoveries await the people who will venture out into and beyond our solar system.

1 Chapter Review

Concept Summary

Science is the study of nature's rules.

Science is a way of thinking as well as a body of knowledge.

- Physics is the most basic of all the sciences.
- The use of mathematics helps make ideas in science unambiguous.

The scientific method is a procedure for answering questions about the world by testing educated guesses (hypotheses) and formulating general rules.

- Hypotheses in science must be testable. They are changed or abandoned if they are contradicted by experimental evidence.

A scientific theory is a body of knowledge and well-tested hypotheses about some aspect of the natural world.

- Theories are modified as new evidence is gathered.

Science deals with knowledge for its own sake, while technology is an application of scientific knowledge.

- Science deals with theoretical questions, while technology deals with practical problems.

Important Terms

fact (1.4)
hypothesis (1.3)
law (1.4)
principle (1.4)
scientific method (1.3)
theory (1.4)

Review Questions

1. Why is physics the most basic science? (1.1)

2. Why is mathematics important to science? Why is the use of mathematics minimized in this book? (1.2)

3. What is the scientific method? (1.3)

4. Is a scientific fact something that is absolute and unchanging? Explain. (1.4)

5. Scientific theories undergo change. Is this a strength or a weakness of science? Explain. (1.4)

6. What does it mean to say that if a hypothesis is scientific, then there must be a means of proving it wrong? (1.5)

7. How do science and technology differ? (1.6)

8. How are science and the arts similar? (1.7)

9. How do science and religion differ? (1.7)

10. Why do citizens have a responsibility to have some basic understanding of nature's rules? (1.7)

Think and Explain

1. Why does science tend to be a "self-correcting" way of knowing about things?

2. What is likely being misunderstood by someone who says, "But that's *only* a scientific theory"?

3. **a.** Make an argument for halting the advances of technology.

   b. Make an argument for continuing advances in technology.

   c. Contrast your two arguments.

Mechanics

There's a lot of physics in simply holding an apple. Not only is the apple pulled downward by the earth, the earth is pulled upward by the apple, with just as much force! Both the earth and the apple pull on each other with equal forces in opposite directions. The pair of forces, called **action** and **reaction**, comprise a **single gravitational interaction**. Read on, discover the rules of mechanics, and you'll do more than pass a physics course— you'll sharpen your intuition of nature!

2 Linear Motion

We experience the concepts of speed, velocity, and acceleration when riding in automobiles.

Motion occurs all around us. We see it in the everyday activity of people, of cars on the highway, in trees that sway in the wind, and with patience, we see it in the nighttime stars. There is motion at the microscopic level that we cannot see directly: jostling atoms make heat and sound, flowing electrons make electricity, and vibrating electrons produce radio and television. Even the light that lets us see motion is caused by the motion of electrons in atoms. Truly, motion is everywhere.

Motion is easy to recognize, but it is hard to describe. Even the Greek scientists of more than 2000 years ago, who had a very good understanding of many of the ideas of physics we study today, had great difficulty describing motion. They failed because they did not understand the idea of *rate.* A quantity divided by *time* is a **rate.** It tells how fast something happens, or how much something changes in a certain amount of time. In this chapter we will describe motion by the rates known as *speed, velocity,* and *acceleration.* It would be very nice if this chapter helped you to master these concepts, but it will be enough for you to become familiar with them and to be able to distinguish among them. So we'll consider only the simplest form of motion, motion along a straight-line path—*linear motion.* In Chapter 3 we'll extend these concepts to include motion along curved paths. The following chapters will sharpen your understanding of the concepts of motion.

2.1 Motion Is Relative

Everything moves. Even things that appear to be at rest move. They move with respect to, or **relative** to, the sun and stars. A book that is at rest, relative to the table it lies on, is moving at about 30 kilometers per second relative to the sun. The book moves even faster relative to the center of our galaxy. When we discuss the motion of something, we describe its motion relative to something else. When we say that a space shuttle moves at 8 kilometers per second, we mean its movement relative to the earth below. When we say a racing car in the Indy 500 reaches a speed of 300 kilometers per hour, of course we mean relative to the track. Unless stated otherwise, when we discuss the speeds of things in our environment, we mean speed with respect to the surface of the earth. Motion is relative.

2.2 Speed

A moving object travels a certain distance in a given time. A car, for example, travels so many kilometers in an hour. **Speed** is a measure of how fast something is moving. It is the rate at which distance is covered. Remember, the word *rate* is a clue that something is being *divided by time*. Speed is always measured in terms of a unit of distance divided by a unit of time. Speed is defined as the distance covered *per* unit of time. The word *per* means "divided by."

◀ **Figure 2.1**
A cheetah is the fastest land animal over distances less than 500 meters and can achieve peak speeds of 100 km/h.

Any combination of units for distance and time that are useful and convenient are legitimate for describing speed. Miles per hour (mi/h), kilometers per hour (km/h), centimeters per day (the speed of a sick snail?), or light-years per century are all legitimate units for speed. The slash symbol (/) is read as "per." Throughout this book

Table 2.1 Approximate Speeds in Different Units

20 km/h =	12 mi/h =	6 m/s
40 km/h =	25 mi/h =	11 m/s
60 km/h =	37 mi/h =	17 m/s
80 km/h =	50 mi/h =	22 m/s
100 km/h =	62 mi/h =	28 m/s
120 km/h =	75 mi/h =	33 m/s

we'll primarily use the units *meters per second* (m/s) for speed. Table 2.1 shows some comparative speeds in different units.

Instantaneous Speed

A car does not always move at the same speed. A car may travel down a street at 50 km/h, slow to 0 km/h at a red light, and speed up to only 30 km/h because of traffic. You can tell the speed of the car at any instant by looking at the car's speedometer. The speed at any instant is called the **instantaneous speed.** A car traveling at 50 km/h may go at that speed for only one minute. If the car continued at that speed for a full hour, it would cover 50 km. If it continued at that speed for only half an hour, it would cover only half that distance, or 25 km. In one minute the car would cover less than 1 km.

Average Speed

In planning a trip by car, the driver often wants to know how long it will take to cover a certain distance. The car will certainly not travel at the same speed all during the trip. The driver cares only about the **average speed** for the trip as a whole. The average speed is defined as follows:

$$\text{average speed} = \frac{\text{total distance covered}}{\text{time interval}}$$

Average speed can be calculated rather easily. For example, if we drive a distance of 60 kilometers during a time of 1 hour, we say our average speed is 60 kilometers per hour (60 km/h). Or, if we travel 240 kilometers in 4 hours,

$$\text{average speed} = \frac{\text{total distance covered}}{\text{time interval}} = \frac{240 \text{ km}}{4 \text{ h}} = 60 \text{ km/h}$$

Note that when a distance in kilometers (km) is divided by a time in hours (h), the answer is in kilometers per hour (km/h).

Since average speed is the distance covered divided by the time of travel, it does not indicate variations in the speed that may take place during the trip. In practice, we experience a variety of speeds on most trips, so the average speed is often quite different from the instantaneous speed. Whether we talk about average speed or instantaneous speed, we are talking about the rates at which distance is traveled.

Figure 2.2 ▲
The speedometer for a North American car gives readings of instantaneous speed in both mi/h and km/h. Odometers for the U.S. market give readings in miles; those for the Canadian market give readings in kilometers.

1. The speedometer in every car also has an odometer that records the distance traveled.

   a. If the odometer reads zero at the beginning of a trip and 35 km a half hour later, what is the average speed?

   b. Would it be possible to attain this average speed and never exceed a reading of 70 km/h on the speedometer?

2. If a cheetah can maintain a constant speed of 25 m/s, it will cover 25 meters every second. At this rate, how far will it travel in 10 seconds? In 1 minute?

2.3 Velocity

In everyday language, we can use the words *speed* and *velocity* interchangeably. In physics, we make a distinction between the two. Very simply, the difference is that **velocity** is speed in a given direction. When we say a car travels at 60 km/h, we are specifying its speed. But if we say a car moves at 60 km/h to the north, we are specifying its velocity. Speed is a description of how fast an object moves; velocity is how fast and in what direction it moves. We will see in the next section that there are good reasons for the distinction between speed and velocity.

■ **Answers**

(Are you reading this before you have formulated a reasoned answer in your mind? If so, do you also exercise your body by watching others do push-ups? Exercise your thinking! When you encounter the many questions throughout this book, *think* before you read the footnoted answers. You'll not only learn more, you'll enjoy learning more.)

1. a. average speed = $\dfrac{\text{total distance covered}}{\text{time interval}} = \dfrac{35 \text{ km}}{0.5 \text{ h}} = 70$ km/h

   b. No, not if the trip started from rest and ended at rest, because any intervals with an instantaneous speed less than 70 km/h would have to be compensated with instantaneous speeds greater than 70 km/h to yield an average of 70 km/h. In practice, average speeds are usually appreciably less than peak instantaneous speeds.

2. In 10 s the cheetah will cover 250 m, and in 1 minute (or 60 s) it will cover 1500 m, more than 15 football fields! If we know the average speed and the time of travel, then the distance covered is

$$\text{distance} = \text{average speed} \times \text{time interval}$$

$$\text{distance} = (25 \text{ m/s}) \times (10 \text{ s}) = 250 \text{ m}$$

$$\text{distance} = (25 \text{ m/s}) \times (60 \text{ s}) = 1500 \text{ m}$$

A little thought will show that this relationship is simply a rearrangement of

$$\text{average speed} = \dfrac{\text{distance}}{\text{time interval}}$$

■ Question

The speedometer of a car moving northward reads 60 km/h. It passes another car that travels southward at 60 km/h. Do both cars have the same speed? Do they have the same velocity?

Constant Velocity

From the definition of velocity, it follows that to have a constant velocity requires both constant speed *and* constant direction. Constant speed means that the motion remains at the same speed—the object does not move faster or slower. Constant direction means that the motion is in a straight line—the object's path does not curve at all. Motion at constant velocity is motion in a straight line at constant speed.

Figure 2.3 ▶
The car on the circular track may have a constant speed but not a constant velocity, because its direction of motion is changing every instant.

Changing Velocity

If *either* the speed *or* the direction (or both) is changing, then the velocity is changing. Constant speed and constant velocity are not the same. A body may move at constant speed along a curved path, for example, but it does not move with constant velocity, because its direction is changing every instant.

In a car there are three controls that are used to change the velocity. One is the gas pedal, which is used to maintain or increase the speed. The second is the brake, which is used to decrease the speed. The third is the steering wheel, which is used to change the direction.

■ Answer

Both cars have the same speed, but they have opposite velocities because they are moving in opposite directions.

2.4 Acceleration

We can change the state of motion of an object by changing its speed, its direction of motion, or both. Any of these changes is a change in velocity. Sometimes we are interested in how fast the velocity is changing. A driver on a two-lane road who wants to pass another car would like to be able to speed up and pass in the shortest possible time. The rate at which the velocity is changing is called **acceleration.** Because acceleration is a rate, it is a measure of how the velocity is changing with respect to time.

$$\text{acceleration} = \frac{\text{change of velocity}}{\text{time interval}}$$

We are familiar with acceleration in an automobile. The driver depresses the gas pedal, which is appropriately called the accelerator. The passengers then experience acceleration, or "pickup" as it is sometimes called, as they are pressed into their seats. The key idea that defines acceleration is *change.* Whenever we change our state of motion, we are accelerating. A car that can accelerate well has the ability to change its velocity rapidly. A car that can go from zero to 60 km/h in 5 seconds has a greater acceleration than another car that can go from zero to 80 km/h in 10 seconds. So having good acceleration means being able to change velocity quickly and does not necessarily refer to how fast something is moving.

In physics, the term *acceleration* applies to decreases as well as increases in speed. The brakes of a car can produce large retarding accelerations, that is, they can produce a large decrease per second in the speed. This is often called *deceleration,* or *negative acceleration.* We experience deceleration when the driver of a bus or car slams on the brakes and we tend to hurtle forward.

Acceleration applies to changes in *direction* as well as changes in speed. If you ride around a curve at a constant speed of 50 km/h, you feel the effects of acceleration as your body tends to move outward toward the outside of the curve. You may round the curve at constant speed, but your velocity is not constant, because your direction is changing every instant. Your state of motion is changing:

Figure 2.4 ▲
A car is accelerating whenever there is a *change* in its state of motion.

you are accelerating. Now you can see why it is important to distinguish between speed and velocity and why acceleration is defined as the rate of change in *velocity*, rather than *speed*. Acceleration, like velocity, is directional. If we change either speed or direction, or both, we change velocity and we accelerate.

In much of this book we will be concerned only with motion along a straight line. When straight-line motion is considered, it is common to use speed and velocity interchangeably. When the direction is not changing, acceleration may be expressed as the rate at which *speed* changes.

$$\text{acceleration (along a straight line)} = \frac{\text{change in speed}}{\text{time interval}}$$

Speed and velocity are measured in units of distance per time. The units of acceleration are a bit more complicated. Since acceleration is the change in velocity or speed per time interval, its units are those of speed per time. If we speed up, without changing direction, from zero to 10 km/h in 1 second, our change in speed is 10 km/h in a time interval of 1 s. Our acceleration along a straight line is

$$\text{acceleration} = \frac{\text{change in speed}}{\text{time interval}} = \frac{10 \text{ km/h}}{1 \text{ s}} = 10 \text{ km/h·s}$$

The acceleration is 10 km/h·s, which is read as "10 kilometers per hour-second." Note that a unit for time enters twice: once for the unit of speed and again for the interval of time in which the speed is changing. If you understand this, you can answer the following questions. If you don't, maybe the answers to the questions will help.

■ Questions

1. Suppose a car moving in a straight line steadily increases its speed each second, first from 35 to 40 km/h, then from 40 to 45 km/h, then from 45 to 50 km/h. What is its acceleration?

2. In 5 seconds a car moving in a straight line increases its speed from 50 km/h to 65 km/h, while a truck goes from rest to 15 km/h in a straight line. Which undergoes greater acceleration? What is the acceleration of each vehicle?

■ Answers

1. We see that the speed increases by 5 km/h during each 1-s interval. The acceleration is therefore 5 km/h·s during each interval.

2. The car and truck both increase their speed by 15 km/h during the same time interval, so their acceleration is the same. If you realized this without first calculating the accelerations, you're thinking conceptually. The acceleration of each vehicle is

$$\text{acceleration} = \frac{\text{change in speed}}{\text{time interval}} = \frac{15 \text{ km/h}}{5 \text{ s}} = 3 \text{ km/h·s}$$

Although the speeds involved are quite different, the rates of change in speed are the same. Hence, the accelerations are equal.

2.5 Free Fall: How Fast

An apple falls from a tree. Does it accelerate while falling? We know it starts from a rest position and gains speed as it falls. We know this because it would be safe to catch if it fell a meter or two, but not if it fell from a high-flying balloon. Thus, the apple must gain more speed during the time it drops from a great height than during the shorter time it takes to drop a meter. This gain in speed indicates that the apple does accelerate as it falls.

Gravity causes the apple to accelerate downward once it begins falling. In real life, air resistance affects the acceleration of a falling object. Let's imagine there is no air resistance and that gravity is the only thing affecting a falling object. Such an object would then be in **free fall.** Freely falling objects are affected only by gravity. Table 2.2 shows the instantaneous speed at the end of each second of fall of a freely falling object dropped from rest. The **elapsed time** is the time that has elapsed, or passed, since the beginning of the fall.

Figure 2.5 ▲
If a falling rock were somehow equipped with a speedometer, in each succeeding second of fall its reading would increase by 10 m/s. Table 2.2 shows the speed we would read at various seconds of fall.

Table 2.2	
Free Fall Speeds of Objects Dropped from Rest	
Elapsed Time (seconds)	**Instantaneous Speed (meters/second)**
0	0
1	10
2	20
3	30
4	40
5	50
•	•
•	•
•	•
t	$10t$

Note in Table 2.2 the way the speed changes. During each second of fall the instantaneous speed of the object increases by an additional 10 meters per second. This gain in speed per second is the acceleration.

$$\text{acceleration} = \frac{\text{change in speed}}{\text{time interval}} = \frac{10 \text{ m/s}}{1 \text{ s}} = 10 \text{ m/s}^2$$

Note that when the change in speed is in m/s and the time interval is in s, the acceleration is in m/s^2, which is read as "meters per second squared." The unit of time, the second, occurs twice—once for the unit of speed and again for the time interval during which the speed changes.

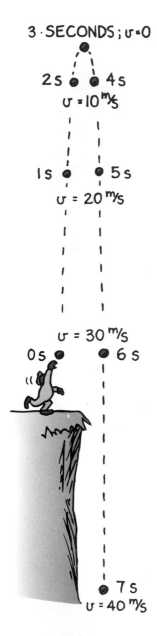

3 · SECONDS ; $v = 0$

2s   4s
$v = 10 \, {}^m\!/\!s$

1s   5s
$v = 20 \, {}^m\!/\!s$

$v = 30 \, {}^m\!/\!s$
0s   6s

7s
$v = 40 \, {}^m\!/\!s$

Figure 2.6 ▲
The rate at which the speed changes each second is the same whether the ball is going upward or downward.

The acceleration of an object falling under conditions where air resistance is negligible is about 10 meters per second squared (10 m/s²). For free fall, it is customary to use the letter g to represent the acceleration because the acceleration is due to gravity. Although g varies slightly in different parts of the world, its average value is nearly 10 m/s². More accurately, g is 9.8 m/s², but it is easier to see the ideas involved when it is rounded off to 10 m/s². Where accuracy is important, the value of 9.8 m/s² should be used for the acceleration during free fall. Note in Table 2.2 that the instantaneous speed of an object falling from rest is equal to the acceleration multiplied by the amount of time it falls, the elapsed time.

$$\text{instantaneous speed} = \text{acceleration} \times \text{elapsed time}$$

The instantaneous speed v of an object falling from rest after an elapsed time t can be expressed in equation form*

$$v = gt$$

The letter v symbolizes both speed and velocity. Take a moment to check this equation with Table 2.2. You will see that whenever the acceleration $g = 10$ m/s² is multiplied by the elapsed time in seconds, the result is the instantaneous speed in meters per second.

■ Question

What would the speedometer reading on the falling rock shown in Figure 2.5 be 4.5 seconds after it drops from rest? How about 8 seconds after it is dropped? 15 seconds?

So far, we have been looking at objects moving straight downward due to gravity. Now consider an object thrown straight up. It continues to move upward for a while, then it comes back down. At the highest point, when the object is changing its direction of motion from upward to downward, its instantaneous speed is zero. It then starts downward just as if it had been dropped from rest at that height.

During the upward part of this motion, the object slows from its initial upward velocity to zero velocity. We know the object is accelerating because its velocity is changing. How much does its speed decrease each second? It should come as no surprise that the speed decreases at the same rate it increases when moving downward—at 10 meters per second each second. So as Figure 2.6 shows, the

■ Answer

The speedometer readings would be 45 m/s, 80 m/s, and 150 m/s, respectively. You can reason this from Table 2.2 or use the equation $v = gt$, where g is replaced by 10 m/s².

* This relationship follows from the definition of acceleration when the acceleration is g and the initial speed is zero. If the object is initially moving downward at speed v_o, the speed v after any elapsed time t is $v = v_o + gt$. This book will not focus on such added complications. You can learn a lot from even the simplest cases!

instantaneous speed at points of equal elevation in the path is the same whether the object is moving upward or downward. The *velocities* are different of course, because they are in opposite directions. During each second, the speed or the velocity changes by 10 m/s. The acceleration is 10 m/s² the entire time, whether the object is moving upward or downward.

2.6 Free Fall: How Far

How *fast* something moves is entirely different from how *far* it moves—speed and distance are not the same thing. To understand the difference, return to Table 2.2. At the end of the first second, the falling object has an instantaneous speed of 10 m/s. Does this mean it falls a distance of 10 meters during this first second? No. Here's where the difference between instantaneous speed and average speed comes in. If the object falls 10 meters the first second, its *average* speed is 10 m/s. But we know the speed began at zero and took a full second to get to 10 m/s. So the average speed is between zero and 10 m/s. For any object moving in a straight line with constant acceleration, we find the average speed the way we find the average of any two numbers: add them and divide by 2. So adding the initial speed of zero and the final speed of 10 m/s and then dividing by 2, we get 5 m/s. During the first second, the object has an average speed of 5 m/s. So it falls a distance of 5 meters. To check your understanding of this, carefully consider the following question before going further.

■ Question

During the span of the second time interval in Table 2.2, the object begins at 10 m/s and ends at 20 m/s. What is the average speed of the object during this 1-second interval? What is its acceleration?

Table 2.3 shows the total distance moved by a freely falling object dropped from rest. At the end of one second, it has fallen 5 meters. At the end of 2 seconds, it has dropped a total distance of 20 meters. At the end of 3 seconds, it has dropped 45 meters altogether. These distances form a mathematical pattern: at the

■ Answer

The average speed will be

$$\frac{\text{beginning speed} + \text{final speed}}{2} = \frac{10 \text{ m/s} + 20 \text{ m/s}}{2} = \frac{30 \text{ m/s}}{2} = 15 \text{ m/s}$$

The acceleration will be

$$\frac{\text{change in speed}}{\text{time interval}} = \frac{20 \text{ m/s} - 10 \text{ m/s}}{1 \text{ s}} = \frac{10 \text{ m/s}}{1 \text{ s}} = 10 \text{ m/s}^2$$

Table 2.3	
Free Fall Distances of an Object Dropped from Rest	
Elapsed Time (seconds)	**Distances Fallen (meters)**
0	0
1	5
2	20
3	45
4	80
5	125
·	·
·	·
·	·
t	$\frac{1}{2}gt^2$

Figure 2.7 ▲
Pretend that a falling rock is somehow equipped with an *odometer*. The readings of distance fallen increase with time and are shown in Table 2.3.

end of time t, the object has fallen a distance d of $\frac{1}{2}gt^2$.* Try using $g = 10 \text{ m/s}^2$ to calculate the distance fallen for some of the times shown in Table 2.3.

■ Question

An apple drops from a tree and hits the ground in one second. What is its speed upon striking the ground? What is its average speed during the one second? How high above ground was the apple when it first dropped?

■ Answer

Using 10 m/s² for g we find

speed $v = gt = (10 \text{ m/s}^2)(1 \text{ s}) = 10 \text{ m/s}$

average speed $\overline{v} = \dfrac{\text{beginning } v + \text{final } v}{2} = \dfrac{0 \text{ m/s} + 10 \text{ m/s}}{2} = 5 \text{ m/s}$

(The bar over the symbol v denotes average speed $\overline{v}$.)

distance d = average speed × time interval = (5 m/s)(1 s) = 5 m or equivalently,

distance $d = \dfrac{1}{2}gt^2 = (\dfrac{1}{2})(10 \text{ m/s}^2)(1 \text{ s})^2 = 5 \text{ m}$

Notice that the distance can be found by either of these equivalent relationships.

$$
\begin{aligned}
\text{* distance} &= \text{average speed} \times \text{time interval} \\
&= \frac{\text{beginning speed} + \text{final speed}}{2} \times \text{time} \\
&= \frac{0 + gt}{2} \times t \\
&= \frac{1}{2}gt^2
\end{aligned}
$$

Reaction Time

Try this with your friends. Hold a dollar bill so that the midpoint hangs between a friend's fingers. Challenge your friend to catch it by snapping his or her fingers shut when you release it. The bill won't be caught!

Explanation: It takes at least 1/7 second for nerve impulses to travel from the eye to the brain to the fingers. But, according to the equation $d = (1/2)gt^2$, in only 1/8 second the bill falls 8 centimeters—half the length of the bill.

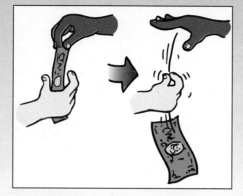

Activity

We used freely falling objects to describe the relationship between distance traveled, acceleration, and velocity acquired. In our examples, we used the acceleration of gravity, $g = 10$ m/s^2. But accelerating objects need not be freely falling objects. A car accelerates when we step on the gas or the brake pedal. Whenever an object's initial speed is zero and the acceleration a is constant, that is, steady and "nonjerky," the equations* for the velocity and distance traveled are

$$v = at \text{ and } d = \frac{1}{2}at^2$$

2.7 Graphs of Motion

Equations and tables are not the only way to describe relationships such as velocity and acceleration. Another way is to use graphs that visually describe relationships. Since you'll develop basic graphing skills in the laboratory, we won't make a big deal of graphs. Here we'll simply show the graphs for Tables 2.2 and 2.3.

Figure 2.9 is a graph of the speed-versus-time data in Table 2.2. Note that speed v is plotted on the vertical axis and time t is plotted on the horizontal axis. In this case, the "curve" that best fits the points forms a straight line. The straightness of the curve indicates a "linear"

---

* If the object has an initial speed v_o, some thought will show that the equations for velocity and distance traveled become $v = v_0 + at$ and $d = v_0 t + \frac{1}{2}at^2$.

Hang Time

Some people, such as basketball players and ballet dancers, are gifted with great jumping ability. Leaping straight up, they seem to hang in the air in defiance of gravity. Ask your friends to estimate the "hang time" of the great jumpers—the amount of time a jumper is airborne (feet off the ground). One or two seconds? Several seconds? Nope. Surprisingly, the hang time of the greatest jumpers is almost always less than 1 second! Our perception of a longer hang time is one of many illusions we have about nature.

Jumping ability is best measured by a standing vertical jump. Stand facing a wall, and with feet flat on the floor and arms extended upward, make a mark on the wall at the top of your reach. Then make your jump, and at the peak, make another mark. The distance between these two marks measures your vertical leap. If it's more than 2 feet (0.6 meters), you're exceptional.

Here's the physics. When you leap upward, jumping force is applied only as long as your feet are still in contact with the ground. The greater the force, the greater your launch speed and the higher the jump is. It is important to note that as soon as your feet leave the ground, whatever upward speed you attain immediately decreases at the steady rate of g, 10 m/s². Maximum height is attained when your upward speed decreases to zero. You then begin falling, gaining speed at exactly the same rate, g. If you land as you took off, upright with legs extended, then time rising equals time falling. Hang time is the sum of rising and falling times. While airborne, no amount of leg or arm pumping or other bodily motions will change your hang time.

The relationship between rising or falling time and vertical height is given by

$$d = \frac{1}{2}gt^2$$

▲ **Figure 2.8** How high can you jump?

If we know the vertical height, we can rearrange this expression to read

$$t = \sqrt{\frac{2d}{g}}$$

Basketball player Spud Webb's record jumping height is 1.25 meters (4 feet). Setting d* equal to 1.25 m, and using the more precise 9.8 m/s² for g, we find that t, half the hang time, is

$$t = \sqrt{\frac{2d}{g}} = \sqrt{\frac{2(1.25 \text{ m})}{9.8 \text{ m/s}^2}} = 0.50 \text{ s}$$

Doubling this, we see Spud's record hang time is 1 second!

We've only been talking about vertical motion. How about running jumps? We'll learn in Chapter 3 that hang time depends only on the jumper's vertical speed at launch; it does not depend on horizontal speed. While airborne, the jumper's horizontal speed remains constant but the vertical speed undergoes acceleration. Interesting physics!

---

* The value of 1.25 m for d represents the maximum height of the jumper's center of gravity. The height gained by the jumper's center of gravity is what's important in determining jumping ability. You will learn about center of gravity in Chapter 10.

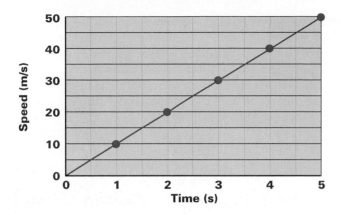

◀ **Figure 2.9**
A speed-versus-time graph of the data from Table 2.2.

relationship between speed and time. For every increase of 1 s, there is the same 10 m/s increase in speed. Mathematicians call this *linearity,* and the graph shows why—the curve is a straight line. Since the object is dropped from rest, the line starts at the origin, where both v and t are zero. If we double $t,$ we double $v;$ if we triple $t,$ we triple $v;$ and so on. This particular linearity is called a *direct proportion.*

The curve is a straight line, so its slope is constant—like a flight of stairs. *Slope* is the vertical change divided by the horizontal change for any part of the line. On this graph the slope measures speed per time, or acceleration. Note that for each 10 m/s of vertical change there is a corresponding horizontal change of 1 s. We see the slope is 10 m/s divided by 1 s, or 10 m/s^2. The straight line shows the acceleration is constant. If the acceleration were greater, the slope of the graph would be steeper. For more information about slope, see Appendix C.

Figure 2.10 is a graph of the distance-versus-time data in Table 2.3. Distance d is plotted on the vertical axis, and time t is on the horizontal axis. The result is a curved line. The curve shows that the relationship between distance traveled and time is not linear. The relationship shown here is *parabolic*—when we double $t,$ we do not double $d;$ we quadruple it. Distance depends on time *squared!*

A curved line also has a slope. If you look at the graph in Figure 2.10 you can see that the curve has a certain slant or "steepness" at

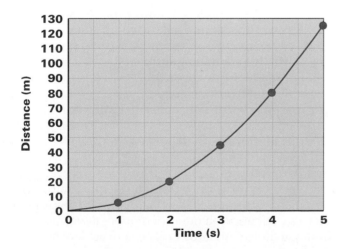

◀ **Figure 2.10**
A distance-versus-time graph of the data from Table 2.3.

every point. This slope changes from one point to the next. The slope of the curve on a distance-versus-time graph is very significant. It is speed, the *rate* at which distance is covered per unit of time. In this graph the slope steepens (becomes greater) as time passes. This shows that speed increases as time passes. In fact, if the slope could be measured accurately, you would find it increases by 10 meters per second each second.

2.8 Air Resistance and Falling Objects

Drop a feather and a coin and we notice the coin reaches the floor far ahead of the feather. Air resistance is responsible for these different accelerations. This fact can be shown quite nicely with a closed glass tube connected to a vacuum pump. The feather and coin are placed inside. When the tube is inverted with air inside, the coin falls much more rapidly than the feather. The feather flutters through the air. But if the air is removed with a vacuum pump and the tube is quickly inverted, the feather and coin fall side by side with the same acceleration, *g*, as shown in Figure 2.11.

Air resistance noticeably alters the motion of things like falling feathers or pieces of paper. But air resistance less noticeably affects the motion of more compact objects like stones and baseballs. In many cases the effect of air resistance is small enough to be neglected. With negligible air resistance, falling objects can be considered to be falling freely. Air resistance will be covered in more detail in Chapter 5.

Figure 2.11 ▲
A feather and a coin accelerate equally when there is no air around them.

2.9 How Fast, How Far, How Quickly How Fast Changes

Much of the confusion that occurs in analyzing the motion of falling objects comes about from mixing up "how fast" with "how far." When we wish to specify how fast something freely falls from rest after a certain elapsed time, we are talking about speed or velocity. The appropriate equation is $v = gt$. When we wish to specify how far that object has fallen, we are talking about distance. The appropriate equation is $d = \frac{1}{2}gt^2$. Velocity or speed (how fast) and distance (how far) are entirely different from each other.

One of the most confusing concepts encountered in this book is acceleration, or "how quickly does speed or velocity change." What makes acceleration so complex is that it is *a rate of a rate*. It is often confused with velocity, which is itself a rate (the rate at which distance is covered). Acceleration is not velocity, nor is it even a change in velocity; acceleration is the rate at which velocity itself changes.

Please be patient with yourself if you find that you require a few hours to achieve a clear understanding of motion. It took people nearly 2000 years from the time of Aristotle to Galileo to achieve as much!

Concept Summary

Motion is described relative to something.

Speed is a measure of how fast something is moving.

- Speed is the rate at which distance is covered, and it is measured in units of distance divided by time.

- Instantaneous speed is the speed at any instant.

- Average speed is the total distance covered divided by the time interval.

Velocity is speed together with direction.

- Velocity is constant only when speed and direction are both constant.

Acceleration is the rate at which velocity is changing with respect to time.

- An object accelerates when its speed is increasing, when its speed is decreasing, and/or when its direction is changing.

- Acceleration is measured in units of speed divided by time.

An object in free fall is falling under the influence of gravity alone when air resistance does not affect its motion.

- An object in free fall has a constant acceleration of about 10 m/s^2.

Important Terms

acceleration (2.4)	rate (2.0)
average speed (2.2)	relative (2.1)
elapsed time (2.5)	speed (2.2)
free fall (2.5)	velocity (2.3)
instantaneous speed (2.2)	

Review Questions

1. What do we mean when we say that motion is relative? What is everyday motion usually relative to? (2.1)

2. Speed is the rate at which what happens? (2.2)

3. You walk across the room at 2 kilometers per hour. Express this speed using abbreviated units. (2.2)

4. What is the difference between instantaneous speed and average speed? (2.2)

5. Does the speedometer of a car read instantaneous speed or average speed? (2.2)

6. What is the difference between speed and velocity? (2.3)

7. If the speedometer of a car reads a constant speed of 40 km/h, can you say that the car has a constant velocity? Why or why not? (2.3)

8. What two controls on a car cause a change in speed? What control causes a change in velocity? (2.3)

9. What quantity describes how quickly you change how fast you're traveling, or how quickly you change your direction? (2.4)

10. Acceleration is the rate at which what happens? (2.4)

11. What is the acceleration of a car that travels in a straight line at a constant speed of 100 km/h? (2.4)

12. What is the acceleration of a car moving along a straight-line path that increases its speed from zero to 100 km/h in 10 s? (2.4)

13. By how much does the speed of a vehicle moving in a straight line change each second when it is accelerating at 2 km/h·s? At 4 km/h·s? At 10 km/h·s? (2.4)

14. Why does the unit of time enter twice in the unit of acceleration? (2.4)

15. What is the meaning of *free fall?* (2.5)

16. For a freely falling object dropped from rest, what is the instantaneous speed at the end of the fifth second of fall? The sixth second? (2.5)

17. For a freely falling object dropped from rest, what is the *acceleration* at the end of the fifth second of fall? The sixth second? At the end of any elapsed time t? (2.5)

18. Toss a ball upward. What is the change in speed each second on the way up? On the way down? (2.5)

19. How far will a freely falling object fall from rest in five seconds? Six seconds? (2.6)

20. How far will an object move in one second if its average speed is 5 m/s? (2.6)

21. How far will a freely falling object have fallen from a position of rest when its instantaneous speed is 10 m/s? (2.6)

22. What does the slope of the curve on a distance-versus-time graph represent? (2.7)

23. What does the slope of the curve on a velocity-versus-time graph represent? (2.7)

24. Does air resistance increase or decrease the acceleration of a falling object? (2.8)

25. What is the appropriate equation for how fast an object freely falls from a position of rest? For how far that object falls? (2.9)

Activities

1. By any method you choose, determine your average speed of walking.

2. You can compare your reaction time with that of a friend by catching a ruler that is dropped between your fingers. Let your friend hold the ruler as shown in the figure. Snap your fingers shut as soon as you see the ruler released. The number of centimeters that pass through your fingers depends on your reaction time.

You can find your reaction time in seconds by solving $d = \frac{1}{2}gt^2$; for time, $t = \sqrt{2d/g}$. If you express d in meters (likely a fraction of a meter), then $t = 0.45\sqrt{d}$; if you express d in centimeters, then $t = 0.045\sqrt{d}$.

3. Calculate your personal "hang time," the time your feet are off the ground during a vertical jump.

Plug and Chug

1. Calculate the average speed (in m/s) of a cheetah that runs 140 meters in 5 seconds.

2. **a.** Calculate the average speed (in km/h) of Charlie, who runs to the store 4 kilometers away in 30 minutes.

    b. Calculate the distance (in km) that Charlie runs if he maintains this average speed for 1 hour.

3. Calculate the acceleration of a car (in km/h·s) that can go from rest to 100 km/h in 10 s.

4. Calculate the instantaneous speed (in m/s) at the 10-second mark for a car that accelerates at 2 m/s^2 from a position of rest.

5. Calculate the speed (in m/s) of a skateboarder who accelerates from rest for 3 seconds down a ramp at an acceleration of 5 m/s^2.

6. Calculate the instantaneous speed of an apple that falls freely from a rest position and accelerates at 10 m/s^2 for 1.5 seconds.

7. An object is dropped from rest and falls freely. After 6 seconds, calculate its instantaneous speed, average speed, and distance fallen.

8. Calculate the instantaneous speed and distance fallen for an object that falls freely from rest for 8 seconds.

Think and Explain

1. Why is it that an object can accelerate while traveling at constant speed, but not at constant velocity?

2. Light travels in a straight line at a constant speed of 300 000 km/s. What is the light's acceleration?

3. Which has more acceleration when moving in a straight line—a car increasing its speed from 50 to 60 km/h, or a bicycle that goes from zero to 10 km/h in the same time? Defend your answer.

4. **a.** If a freely falling rock were equipped with a speedometer, by how much would its speed readings increase with each second of fall?

   b. Suppose the freely falling rock were dropped near the surface of a planet where $g = 20$ m/s². By how much would its speed readings change each second?

5. If a freely falling rock were equipped with an odometer, would the readings for distance fallen each second stay the same, increase with time, or decrease with time?

6. **a.** When a ball is thrown straight up, by how much does the speed decrease each second? Neglect air resistance.

   b. After the ball reaches the top and begins its return back down, by how much does its speed increase each second?

   c. Compare the times going up and coming down.

7. Table 2.2 shows that the instantaneous speed of an object dropped from rest is 10 m/s after 1 second of fall. Table 2.3 shows that the object has fallen only 5 meters during this time. Your friend says this is incorrect, because distance traveled equals speed times time, so the object should fall 10 meters. What do you say?

8. A ball is thrown straight up. What will be the instantaneous velocity at the top of its path? What will be its acceleration at the top? Why are your answers different?

Think and Solve

1. If humans originated in Africa and migrated to other parts of the world, some time would be required for this to occur. At the modest rate of a mere one kilometer per year, how many centuries would it take for humans originating in Africa to migrate to China, some 10 000 kilometers away?

2. **a.** Find the speed required to throw a ball straight up and have it return 6 seconds later. Neglect air resistance.

   b. How high does the ball go?

3. Calculate the hang time of an athlete who jumps a vertical distance of 0.75 meter.

4. If a salmon swims straight upward in the water fast enough to break through the surface at a speed of 5 meters per second, how high can it jump above water?

3 Projectile Motion

Once airborne, the path of a projectile is determined only by initial velocity, gravity and air resistance.

In the previous chapter we studied simple straight-line motion—*linear motion*. We distinguished between motion with constant velocity, such as a bowling ball rolling horizontally, and accelerated motion, such as an object falling vertically under the influence of gravity. Now we extend these ideas to *nonlinear motion*—motion along a *curved path*. Throw a baseball and the path it follows is a curve. This curve is a combination of constant-velocity horizontal motion and accelerated vertical motion. We'll see that the velocity of a thrown ball at any instant has two *independent* "components" of motion—what happens horizontally is not affected by what happens vertically. To understand this idea we must first learn about *vectors*. We will learn that vectors are used to represent quantities that have both magnitude (how much) and direction (which way).

3.1 Vector and Scalar Quantities

It is often said that a picture is worth a thousand words. Sometimes a picture explains a physics concept better than an equation does. Physicists love sketching doodles and equations to explain ideas. Their doodles often include arrows, where each arrow represents the magnitude and the direction of a certain quantity. The quantity might be the electric current in a power line, the orbital velocity of a communication satellite, or the enormous force lifting the space shuttle off the ground.

A quantity that requires both magnitude and direction for a complete description is a **vector quantity.** Recall from the previous chapter that velocity differs from speed in that velocity includes direction in its description. Velocity is a vector quantity, as is acceleration. In later chapters we'll see that other quantities, such as force, are also vector quantities. For now we'll focus on the vector nature of velocity.

Many quantities in physics, such as mass, volume, and time, are completely specified by their magnitude only—they do not have direction. A quantity that is completely described by magnitude only is a **scalar quantity.** Scalars can be added, subtracted, multiplied, and divided like ordinary numbers. When 3 kg of sand is added to 1 kg of cement, the resulting mixture has a mass of 4 kg. When 5 liters of water are poured from a pail that has 8 liters of water in it, the resulting volume is 3 liters. If a scheduled 60-minute trip has a 15-minute delay, the trip takes 75 minutes. In each of these cases, no direction is involved. We see that descriptions such as 10 kilograms north, 5 liters east, or 15 minutes south have no meaning.

3.2 Velocity Vectors

An arrow is used to represent the magnitude and direction of a vector quantity. The length of the arrow, drawn to scale, indicates the magnitude of the vector quantity. The direction of the arrow indicates the direction of the vector quantity. Rules for combining arrows on a piece of paper are the same as the rules for combining vector quantities in nature. We call the arrow a **vector.** The vector in Figure 3.1 is scaled so that 1 centimeter represents 20 kilometers per hour. It is 3 centimeters long and points to the right; therefore it represents a velocity of 60 kilometers per hour to the right, or 60 km/h east.

A velocity is sometimes the result of combining two or more other velocities. For example, an airplane's velocity is a combination of the velocity of the airplane relative to the air and the velocity of the air relative to the ground, or the wind velocity. Consider a small airplane slowly flying north at 100 km/h relative to the surrounding air. Suppose there is a tailwind blowing north at a velocity of 20 km/h. This example is represented with vectors in Figure 3.2. Here the velocity vectors are scaled so that 1 cm represents 20 km/h. Thus, the 100-km/h velocity of the airplane is shown by the 5-cm vector, and the 20-km/h tailwind is shown by the 1-cm vector. With or without vectors we can see that the resulting velocity is going to be 120 km/h. Without the tailwind, the airplane travels 100 kilometers in one hour relative to the ground below. With the tailwind, it travels 120 kilometers in one hour relative to the ground below.

Now suppose the airplane makes a U-turn and flies into the wind rather than with the wind. The velocity vectors are now in opposite directions. The resulting speed of the airplane is 100 km/h − 20 km/h = 80 km/h. Flying against a 20-km/h wind, the airplane travels only 80 kilometers in one hour relative to the ground.

We didn't have to use vectors to answer questions about tailwinds and headwinds, but we'll now see that vectors are useful for combining velocities that are not parallel.

Figure 3.1 ▲
This vector, scaled so that 1 cm = 20 km/h, represents 60 km/h to the right.

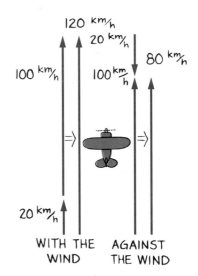

Figure 3.2 ▲
The airplane's velocity relative to the ground depends on the airplane's velocity relative to the air and on the wind's velocity.

■ Question

Suppose that the airplane discussed encounters wind at a right angle to its forward motion—a crosswind. Will the airplane fly faster or slower than 80 km/h?

Consider our 80-km/h airplane flying north caught in a strong crosswind of 60 km/h blowing from west to east. Figure 3.3 shows vectors for the airplane velocity and wind velocity. The scale is 1 cm = 20 km/h. The result of adding these two vectors, called the **resultant,** is the diagonal of the rectangle described by the two vectors.

Figure 3.3 ▶
An 80-km/h airplane flying in a 60-km/h crosswind has a resultant speed of 100 km/h relative to the ground.

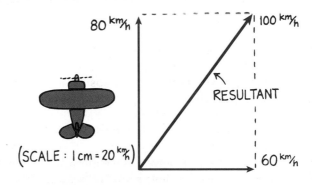

The resultant of two perpendicular vectors is the diagonal of a rectangle constructed with the two vectors as sides.

The diagonal of the constructed rectangle measures 5 cm, which represents 100 km/h. So relative to the ground, the airplane moves 100 km/h northeasterly.*

A simple three-step technique is used to find the resultant of a pair of vectors that are at right angles to each other. First, draw the

■ Answer

A crosswind would increase the speed of the airplane and blow it off course by a predictable amount.

---

*   Whenever a pair of vectors are at right angles (90°), their resultant can be found by the Pythagorean Theorem, a well-known tool of geometry. It states that the square of the hypotenuse of a right-angle triangle is equal to the sum of the squares of the other two sides. Note that two right triangles are present in the rectangle in Figure 3.3. From either one of these triangles we get

$$resultant^2 = (60 \text{ km/h})^2 + (80 \text{ km/h})^2$$

$$= 3600 \text{ (km/h)}^2 + 6400 \text{ (km/h)}^2$$

$$= 10\ 000 \text{ (km/h)}^2$$

The square root of 10 000 (km/h)^2 is 100 km/h, as expected.

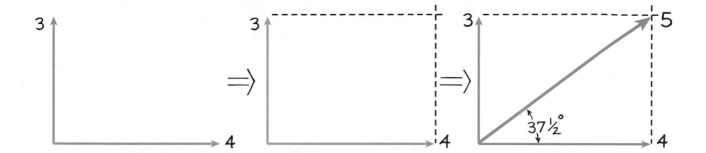

two vectors with their tails touching as shown in Figure 3.4. Second, draw a parallel projection of each vector with dashed lines to form a rectangle. Third, draw the diagonal from the point where the two tails are touching.

For the more general case when the vectors are not at right angles to each other, the resultant is found by constructing a *parallelogram.* A parallelogram is a four-sided figure with opposite sides that are parallel and of equal length. The parallelogram is constructed with the two vectors as sides. The resultant of the parallelogram is its diagonal. A rectangle is a special case of a parallelogram—one in which all the angles are right angles. We'll concern ourselves only with rectangles in this chapter and leave parallelograms for Chapter 4.

Another special case of the parallelogram is a square. To add two vectors that are equal in magnitude and at right angles to each other, we construct a square. For any square, the length of its diagonal is $\sqrt{2}$, or 1.414, times either of the sides. Thus, the resultant is $\sqrt{2}$ times either of the vectors. For instance, the resultant of two equal vectors of magnitude 100 acting at right angles to each other is 141.4.*

Figure 3.4 ▲
The 3-unit and 4-unit vectors add to produce a resultant vector of 5 units, at 37.5° from the horizontal.

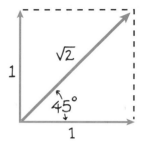

Figure 3.5 ▲
The diagonal of a square is $\sqrt{2}$, or 1.414, times the length of one of its sides.

3.3 Components of Vectors

Often we will need to change a single vector into an equivalent set of two *component* vectors at right angles to each other. Any vector can be "resolved" into two component vectors at right angles to each other. This pair of vectors are known as the **components** of the given vector they replace. The process of determining the components of a vector is called **resolution.** Any vector drawn on a piece of paper can be resolved into vertical and horizontal components.

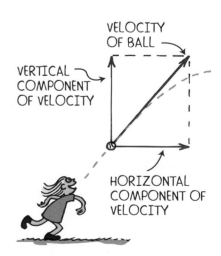

Figure 3.6 ▲
The horizontal and vertical components of a ball's velocity.

---

* An important property of vectors is that they can be moved around as long as their length and direction are not changed. Vectors can be rearranged into a chain, tail-to-head in any order. A vector drawn from the tail of the first vector to the head of the last vector represents the resultant of the entire chain of vectors.

Surfing

Surfing nicely illustrates component and resultant vectors. (1) When surfing in the same direction as the wave, our velocity is the same as the wave's velocity, $v_\perp$. This velocity is called $v_\perp$ because we are moving perpendicular to the wave front. (2) To go faster, we surf at an angle to the wave front. Now we have a component of velocity parallel to the wave front, $v_\parallel$, as well as the perpendicular component $v_\perp$. We can vary $v_\parallel$, but $v_\perp$ stays relatively constant as long as we ride the wave. Adding components, we see that when surfing at an angle to the wave front our resultant velocity, v_r, exceeds $v_\perp$. (3) As we increase our angle relative to the wave front, the resultant velocity also increases.

Vector resolution is illustrated in Figure 3.7. Vector V represents a vector quantity. First, vertical and horizontal lines are drawn from the tail of the vector (top). Second, a rectangle is drawn that encloses the vector V as its diagonal (bottom). The sides of this rectangle are the desired components, vectors X and Y.

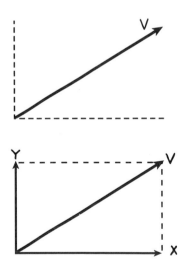

Figure 3.7 ▲
Construction of the vertical and horizontal components of a vector.

■ Exercise

With a ruler, draw the horizontal and vertical components of the two vectors shown. Measure the components and compare your findings with the answers given at the bottom of the page.

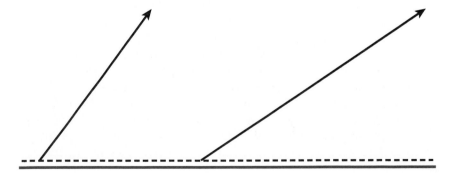

■ Answer

Left vector: the horizontal component is 3 cm; the vertical component is 4 cm. Right vector: the horizontal component is 6 cm; the vertical component is 4 cm.

3.4 Projectile Motion

A cannonball shot from a cannon, a stone thrown into the air, a ball rolling off the edge of a table, a spacecraft circling the earth—all of these are examples of **projectiles.** Projectiles near the earth follow a curved path that at first seems rather complicated. However, these paths are surprisingly simple when we look at the horizontal and vertical components of motion separately.

The horizontal component of motion for a projectile is just like the horizontal motion of a ball rolling freely along a level surface. When friction is negligible, a rolling ball moves at constant velocity. The ball covers equal distances in equal intervals of time as shown in Figure 3.8 (top). With no horizontal force acting on the ball there is no horizontal acceleration. The same is true for the projectile—when no horizontal force acts on the projectile, the horizontal velocity remains constant.

Figure 3.8 (Top) ▲
Roll a ball along a horizontal surface, and its velocity is constant because no component of gravitational force acts horizontally. (Right) ▶ Drop it, and it accelerates downward and covers greater vertical distances each second.

The vertical component of a projectile's velocity is like the motion described in Chapter 2 for a freely falling object. In the vertical direction, there is a force due to gravity. Like a ball dropped in midair, a projectile accelerates downward as shown on the right in Figure 3.8. Its vertical component of velocity changes with time. The increasing speed in the vertical direction causes greater distances to be covered in each successive equal time interval.

Most important, the horizontal component of motion for a projectile is completely independent of the vertical component of motion. Each component is independent of the other. Their combined effects produce the variety of curved paths that projectiles follow.

The multiple-flash exposure in Figure 3.9 shows equally timed successive positions for two balls. One ball is projected horizontally while the other is simply dropped. Study the photo carefully, for there's a lot of good physics here. Analyze the curved path of the ball by considering the horizontal and vertical velocity components separately. There are two important things to notice. The first is that the ball's horizontal component of motion remains constant. The ball moves the same horizontal distance in the equal time intervals between each flash, because no horizontal component of force is acting on it. Gravity acts only downward, so the only acceleration of

Figure 3.9 ▶
A strobe-light photo of two balls released simultaneously from a mechanism that allows one ball to drop freely while the other is projected horizontally. Notice that in equal times both balls fall the same vertical distance.

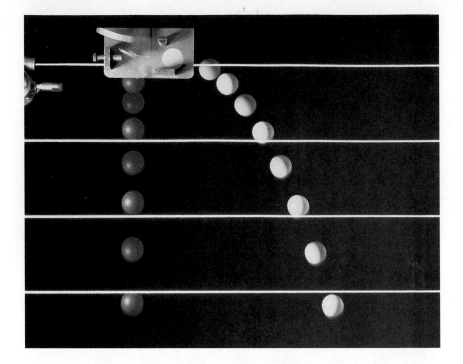

the ball is downward. The second thing to note is that both balls fall the same vertical distance in the same time. The vertical distance fallen has nothing to do with the horizontal component of motion. The downward motion of the horizontally projected ball is the same as that of free fall.

The path traced by a projectile accelerating only in the vertical direction while moving at constant horizontal velocity is a *parabola*. When air resistance is small enough to neglect—usually for slow-moving or very heavy projectiles—the curved paths are parabolic.

■ Question

At the instant a horizontally held rifle is fired, a bullet held at the rifle's side is released and drops to the ground. Which bullet strikes the ground first, the one fired from the rifle or the one dropped?

■ Answer

Both bullets fall the same vertical distance with the same acceleration g and therefore strike the ground at the same time. Do you see that this is consistent with our analysis of Figure 3.9? Ask which bullet strikes the ground first when the rifle is pointed at an upward angle. In this case, the bullet that is simply dropped hits the ground first. Now consider the case when the rifle is pointed downward. The fired bullet hits first. So upward, the dropped bullet hits first; downward, the fired bullet hits first. There must be some angle where both hit at the same time. Do you see it would be when the rifle is pointing neither upward nor downward, that is, when it is pointing horizontally?

Projectiles and Free Fall

Place a coin at the edge of a smooth table so that it hangs over slightly. Then place a second coin on the tabletop some distance from the overhanging coin. Slide the second coin across the table (such as by flicking it with your finger) so that it strikes the over-hanging coin and both coins fall to the floor below. See for yourself which coin, if either, hits the floor first. Does your answer depend on the speed of the sliding coin?

3.5 Upwardly Launched Projectiles

Consider a cannonball shot at an upward angle. Because of gravity, the cannonball follows a curved path and finally hits the ground. Pretend for a moment that there is no gravity and the cannonball follows the straight-line path shown by the dashed line in Figure 3.10. What really happens is quite fascinating. Get this: The cannonball continually falls beneath this imaginary line until it hits the ground. The vertical distance it falls *beneath any point on the dashed line* is the same vertical distance it would fall if it were dropped from rest and had been falling for the same amount of time. This dis-tance, introduced in Chapter 2, is given by $d = \frac{1}{2}gt^2$, where t is the elapsed time. Using the value of 10 m/s^2 for g in the equation yields $d = 5t^2$ meters.

We can put this another way. Shoot a pro-jectile skyward at some angle and pretend there is no gravity. After so many sec-onds t, it should be at a certain point along a straight-line path. But due to gravity, it isn't. Where is it? The answer is, it's directly below that point. How far below?

5 m

20 m

45 m

1 s

2 s

3 s

◀ **Figure 3.10**
With no gravity the projectile would follow the straight-line path (dashed line). But because of gravity it falls beneath this line the same vertical distance it would fall if it were released from rest. Compare the distances fallen with those in Table 2.3.

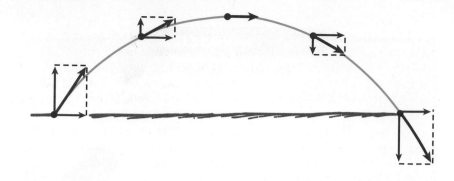

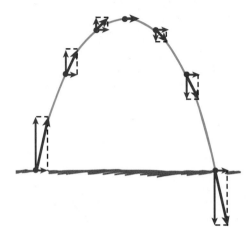

Figure 3.12 ▲
The path for a projectile fired at a steep angle.

The answer is $5t^2$ meters below that point. How about that? The vertical distance the cannonball falls below the imaginary straight-line path increases continually with time and is equal to $5t^2$ meters.

Note also from Figure 3.10 that since there is no horizontal acceleration, the cannonball moves equal horizontal distances in equal time intervals. That's because there is no horizontal acceleration. The only acceleration is vertical, in the direction of the earth's gravity.

Figure 3.11 shows vectors representing both the horizontal and vertical components of velocity for a projectile on a parabolic path. Notice that the horizontal component is always the same and that only the vertical component changes. Note also that the actual resultant velocity vector is represented by the diagonal of the rectangle formed by the vector components. At the top of the path the vertical component shrinks to zero, so the velocity there *is* the same as the horizontal component of velocity at all other points. Everywhere else the magnitude of velocity is greater, just as the diagonal of a rectangle is greater than either of its sides.

Figure 3.12 shows the path traced by a projectile with the same launching speed but at a steeper angle. Notice that the initial velocity vector has a greater vertical component than when the projection angle is less. This greater component results in a higher path. However, since the horizontal component is less, the range is less.

Figure 3.13 shows the paths of several projectiles all having the same initial speed but different projection angles. The figure neglects the effects of air resistance, so the paths are all parabolas. Notice that these projectiles reach different heights above the ground. They also travel different horizontal distances, that is, they have different *horizontal ranges.*

Figure 3.13 ▶
The paths of projectiles launched at the same speed but at different angles. The paths neglect air resistance.

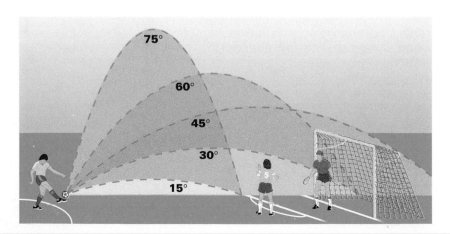

The remarkable thing to note from Figure 3.13 is that the same range is obtained for two different projection angles—angles that add up to 90 degrees! For example, an object thrown into the air at an angle of 60 degrees will have the same range as if it were thrown at 30 degrees with the same speed. Of course, for the smaller angle the object remains in the air for a shorter time.

We have emphasized the special case of projectile motion for negligible air resistance. As we can see in Figure 3.15, when the effect of air resistance is significant, the range of a projectile is diminished and the path is not a true parabola.

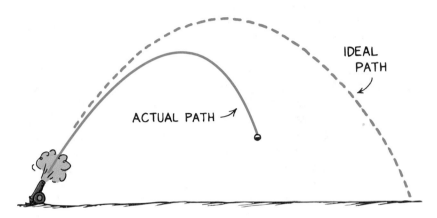

Figure 3.15 ▲
In the presence of air resistance, the path of a high-speed projectile falls below the idealized parabola and follows the dashed curve.

Figure 3.14 ▲
Maximum range is attained when the ball is batted at an angle of nearly 45°. For a thrown javelin, on the other hand, maximum range is achieved for an angle quite a bit less than 45°, because the force of gravity on the relatively heavy javelin is significant during launch. Just as you can't throw a heavy rock as fast upward as sideways, so it is that the javelin's launch speed is reduced when thrown upward.

■ Questions

1. A projectile is shot at an angle into the air. Neglecting air resistance, what is its vertical acceleration? Its horizontal acceleration?

2. At what point in its path does a projectile have minimum speed?

If air resistance is negligible, a projectile will rise to its maximum height in the same time it takes to fall from that height to the ground. This is due to the constant effect of gravity. The deceleration due to gravity going up is the same as the acceleration due to gravity

■ Answers

1. Its vertical acceleration is *g* because the force of gravity is downward. Its horizontal acceleration is zero because no horizontal force acts on it.

2. The minimum speed of a projectile occurs at the top of its path. If it is launched vertically, its speed at the top is zero. If it is projected at an angle, the vertical component of velocity is still zero at the top, leaving only the horizontal component. So the speed at the top is equal to the horizontal component of the projectile's velocity at any point. How about that?

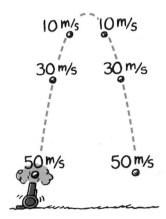

Figure 3.16 ▲
Without air resistance, the speed lost while the cannonball is going up equals the speed gained while it is coming down. The time to go up equals the time to come down.

Hang Time Revisited

Recall our discussion of hang time in Chapter 2. We stated that the time one is airborne during a jump is independent of horizontal speed. Now we see why this is so—horizontal and vertical components of motion are independent of each other. The rules of projectile motion apply to jumping. Once the feet are off the ground, if we neglect air resistance, the only force acting on the

jumper is gravity. Hang time depends only on the vertical component of liftoff velocity. It turns out that jumping force can be somewhat increased by the action of running, so hang time for a running jump usually exceeds that for a standing jump. However, once the feet are off the ground, *only* the vertical component of liftoff velocity determines hang time.

coming down. The speed it loses going up is therefore the same as the speed it gains coming down. So the projectile hits the ground with the same speed it had originally when it was projected upward from the ground.

For short-range projectile motion such as a batted ball in a baseball game, we usually assume the ground is flat. However, for very long range projectiles the curvature of the earth's surface must be taken into account. We'll see that if an object is projected fast enough, it will fall all the way around the earth and become an earth satellite!

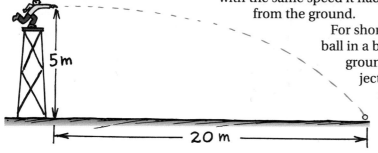

Figure 3.17 ▲
How fast is the ball thrown?

■ Question

The boy on the tower in Figure 3.17 throws a ball a distance of 20 m. At what speed is the ball thrown?

■ Answer

The ball is thrown horizontally, so its speed equals the horizontal distance divided by the time, that is, $v = d/t$. If we let the terms in the equation guide our thinking, we have the horizontal distance of 20 m but not the time. The equation guides us to look for the time. This we can do, for due to gravity the ball falls a vertical distance of 5 m—which takes 1 s. So the time of flight is 1 s. That means the ball travels horizontally a distance of 20 m in 1 s. The ball's horizontal component of velocity, the speed at which it is thrown, must be 20 m/s. This procedure works even when the numbers aren't so simple. Equations guide thinking!

3.6 Fast-Moving Projectiles—Satellites

Consider the ball thrown from the tower in Figure 3.17. If gravity did not act on the ball, the ball's path would be a straight horizontal line. But there is gravity and the ball falls below this straight-line path. One second after the ball is thrown it has fallen 5 meters below the dashed line. The ball falls 5 meters no matter how fast it is thrown. If the ball is thrown twice as fast, it will go twice as far in the same time. If thrown three times as fast, it will go three times as far in the same time.

What if the ball were thrown so fast that the curvature of the earth came into play? We know the ball follows a curved downward path. We also know the earth's surface curves. What if the ball were thrown fast enough so that its curved path matched the curve of the earth's surface? If there were no air resistance to slow it down, the ball would orbit the earth! An earth **satellite**, such as the space shuttle, is simply a projectile traveling fast enough to fall around the earth rather than into it.* We'll see in Chapter 14 that this speed is 8 km/s, or 18 000 mi/h.

At 8 km/s, atmospheric friction would burn a baseball to a crisp or melt a piece of iron. This is the fate of bits of rock and other meteorites that graze the earth's atmosphere and burn up, appearing as "falling stars." That is why satellites such as the space shuttle are launched to altitudes above 150 kilometers. A common misconception is that satellites orbiting at high altitudes are free from the earth's gravity. Nothing could be further from the truth. We will see later that the force of gravity on a satellite 150 kilometers above the earth's surface is almost as strong as it is at the surface. The high altitude puts the satellite beyond the earth's atmosphere, where air resistance is almost totally absent, but not beyond the earth's gravity. We'll return to gravity and satellite motion in Chapters 12–14.

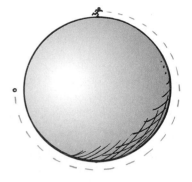

Figure 3.18 ▲
If Superman were to throw a ball so fast that it cleared the horizon, the ball would then orbit the earth and become an earth satellite.

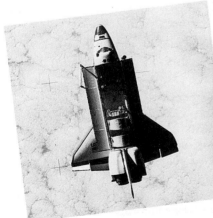

◄ **Figure 3.19**
The space shuttle and other satellites are projectiles in a constant state of free fall. Because their velocity is parallel to the earth's surface, they fall around the earth rather than vertically into it.

---

* The conventional definition of *to fall* is "to get closer to the earth." Satellites in circular orbit, such as the moon and the space shuttle, do not do this. In science, the technical definition often differs from the conventional one. For example, we say "the sun rises" and "the moon sets," but technically they do not.

Concept Summary

Vector quantities have both magnitude and direction.

A vector is represented by an arrow whose length represents the magnitude of the vector quantity and whose direction represents the direction of the vector quantity.

The resultant of two velocities can be determined from a vector diagram drawn to scale.

Any single vector can be replaced by two components that add by vector rules to form the original vector.

When gravity is the only force acting on a projectile near the earth, the horizontal component of its velocity does not change.

A satellite is continually falling around the earth.

Important Terms

component (3.3)   satellite (3.6)
projectile (3.4)   scalar quantity (3.1)
resolution (3.3)   vector (3.2)
resultant (3.2)   vector quantity (3.1)

Review Questions

1. How does a vector quantity differ from a scalar quantity? (3.1)

2. Why is speed classified as a scalar quantity and velocity classified as a vector quantity? (3.1)

3. If a vector that is 1 cm long represents a velocity of 10 km/h, what velocity does a vector 2 cm long drawn to the same scale represent? (3.2)

4. When a rectangle is constructed in order to add velocities, what represents the resultant of the velocities? (3.2)

5. Why do we say a rectangle is a special case of a parallelogram? (3.2)

6. Will the horizontal and vertical components of a vector at 45° to the horizontal be larger or smaller than the vector? By how much? (3.3)

7. Why does a bowling ball move without acceleration when it rolls along a bowling alley? (3.4)

8. In the absence of air resistance, why does the horizontal component of velocity for a projectile remain constant while the vertical component changes? (3.4)

9. How does the downward component of the motion of a projectile compare with the motion of free fall? (3.4)

10. At the instant a ball is thrown horizontally over a level range, a ball held at the side of the first is released and drops to the ground. If air resistance is neglected, which ball strikes the ground first? (3.4)

11. a. How far below an initial straight-line path will a projectile fall in one second? (3.5)

    b. Does your answer depend on the angle of launch or on the initial speed of the projectile? Defend your answer. (3.5)

12. At what angle should a slingshot be oriented for maximum altitude? For maximum horizontal range? (3.5)

13. Neglecting air resistance, if you throw a ball straight up with a speed of 20 m/s, how fast will it be moving when you catch it? (3.5)

14. a. Neglecting air resistance, if you throw a baseball at 20 m/s to your friend who is on first base, will the catching speed be greater than, equal to, or less than 20 m/s?

    b. Does the speed change if air resistance is a factor? (3.5)

15. What do we call a projectile that continually "falls" around the earth? (3.6)

16. How fast must a projectile moving horizontally travel so that the curve it follows matches the curve of the earth? (3.6)

17. Why is it important that such a satellite be above the earth's atmosphere? (3.6)

18. What force acts on a satellite that is above the earth's atmosphere? (3.6)

Plug and Chug

1. Calculate the resultant velocity of an airplane that normally flies at 200 km/h if it encounters a 50-km/h tailwind. If it encounters a 50-km/h headwind?

2. Calculate the resultant of the pair of velocities 100 km/h north and 75 km/h south. Calculate the resultant if both of the velocities are directed north.

3. Calculate the magnitude of the resultant of a pair of 100-km/h velocity vectors that are at right angles to each other.

4. Calculate the magnitude of the horizontal and vertical components of a vector that is 100 units long and is oriented at 45°.

5. Calculate the resultant of a horizontal vector with a magnitude of 4 units and a vertical vector with a magnitude of 3 units. How does the resultant compare with the hypotenuse of a right triangle having sides of 3 and 4 units?

6. A right triangle with sides of 3, 4, and 5 units has angles that are 37°, 53°, and 90°, respectively. Which of its sides is the hypotenuse? Which side is opposite the 37° angle? Which side is opposite the 53° angle?

7. What are the horizontal and vertical components of a 10-unit vector that is oriented 37° above the horizontal?

8. What are the horizontal and vertical components of a 10-unit vector that is oriented 53° above the horizontal?

9. The launching velocity of a projectile is 20 m/s at 53° above the horizontal. What is the vertical component of its velocity at launch? Its horizontal component of velocity? Neglecting air friction, which of these components remains constant throughout the flight path? Which of these components determines the projectile's time in the air?

10. Satellites in a circular, low earth-orbit move at 8 km/s. Convert this speed to miles per hour. (There are about 1.6 km in 1 mile and 3600 s in 1 h.)

Think and Explain

1. What is the maximum possible resultant of a pair of vectors with magnitudes of 4 and 5 units? What is the minimum possible resultant?

2. If you swim in a direction directly across a river and you end up downstream due to the flow of water, do you move faster than you would if the water didn't flow? Explain.

3. The speed of falling rain is the same 10 m above ground as it is just before it hits the ground. What does this tell you about whether or not the rain encounters air resistance?

4. Rain falling vertically will make vertical streaks on a car window. However, if the car is moving, the streaks are slanted. If the streaks from a vertically falling rain make 45° streaks, how fast is the car moving compared with the speed of the falling rain?

5. You're driving behind a car and wish to pass, so you turn to the left and pull into the passing lane without changing speed. Why does the distance increase between you and the car you're following?

6. A projectile is fired straight up at 141 m/s. How fast is it moving at the top of its trajectory? Suppose it is fired upward at 45° above the horizontal plane. How fast is it moving at the top of its curved trajectory?

7. When you jump up, your hang time is the time your feet are off the ground. Does hang time depend on your vertical component of velocity when you jump, your horizontal component of velocity, or both? Defend your answer.

8. The hang time of a basketball player who jumps a vertical distance of 2 feet (0.6 m) is 2/3 second. What will be the hang time if the player reaches the same height while jumping a horizontal distance of 4 feet (1.2 m)?

9. Assuming no air resistance, why does a projectile fired horizontally at 8 km/s not strike the earth's surface?

10. We think of something falling if it gets closer to the ground. Yet a satellite in circular orbit does not get closer to the ground, because the earth curves as much as the satellite's trajectory does. So how can we say it falls? (*Hint*: Compare the position of the satellite with the imaginary line it would follow if there were no gravity. Does it fall beneath this line?)

Think and Solve

1. A boat is rowed at 8 km/h directly across a river that flows at 6 km/h, as shown in the figure.

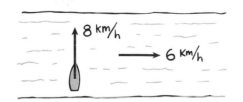

a. What is the resultant speed of the boat?

b. How fast and in what direction can the boat be rowed to reach a destination directly across the river?

2. Harry accidentally falls out of a helicopter that is traveling at 100 m/s. He plunges into a swimming pool 2 seconds later. Assuming no air resistance, what was the horizontal distance between Harry and the swimming pool when he fell from the helicopter?

3. Harry and Angela look from their balcony to a swimming pool below that is 15 m from the bottom of their building. They estimate the balcony is 45 m high and wonder how fast they would have to jump horizontally to succeed in reaching the pool. What is your answer?

4. A kid fires a slingshot pellet directly at a target that is far enough away to take one-half second to reach. How far below the target does the pellet hit? How high above the target should she aim?

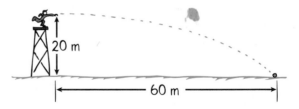

5. The boy on the tower in the figure throws a ball a distance of 60 m, as shown. At what speed, in m/s, is the ball thrown?

6. A shiny new sports car sits in the parking lot of a car dealership. Above is a cargo plane, flying horizontally at 50 m/s. At the exact moment the plane is 125 m directly above the car, a heavy crate accidentally falls from its cargo doors. Relative to the car, where will the crate hit?

4 Newton's First Law of Motion—Inertia

If you saw a boulder in the middle of a flat field suddenly begin to move across the ground, you'd look for the cause of its motion. You might look to see if someone was pulling the boulder with a rope or pushing it with a stick. You would reason that something was causing it to move. We don't believe that motion occurs without cause. In general, we say the cause of the boulder's motion is a force of some kind. We know that something forces the boulder to begin moving.

Because of its huge inertia, the ship is not easily pushed off course.

4.1 Aristotle on Motion

The idea that a force causes motion goes back to the fourth century B.C., when the Greeks were developing some of the ideas of science. Aristotle, the foremost Greek scientist, studied motion and divided it into two types: *natural motion* and *violent motion*.

Natural motion on the earth was thought to be either straight up or straight down, such as a boulder falling toward the ground or a puff of smoke rising in the air. Objects would seek their natural resting places: boulders on the ground and smoke high in the air like the clouds. It was "natural" for heavy things to fall and for very light things to rise. Aristotle proclaimed circular motion was natural for the heavens, for he saw both circular motion and the heavens as without beginning or end. Thus, the planets and stars moved in perfect circles about the earth. Since these motions were considered natural, they were not thought to be caused by forces.

Violent motion, on the other hand, was imposed motion. It was the result of forces that pushed or pulled. A cart moved because it was pulled by a horse; a tug-of-war was won by pulling on a rope; a ship was pushed by the force of the wind. The important thing about defining violent motion was that it had an external cause. Violent motion

Figure 4.1 ▲
Do boulders move without cause?

was imparted to objects. Objects in their natural resting places could not move by themselves; they had to be pushed or pulled.

It was commonly thought for nearly 2000 years that if an object was moving "against its nature," then a force of some kind was responsible. Such motion was possible only because of an outside force. If there were no force there would be no motion (except in the vertical direction). So the proper state of objects was one of rest, unless they were being pushed or pulled, or were moving toward their natural resting place. Since most thinkers up to the sixteenth century considered it obvious that the earth must be in its natural resting place and assumed that a force large enough to move it was unthinkable, it was clear that the earth did not move.

4.2 Copernicus and the Moving Earth

It was in this intellectual climate that the astronomer Nicolaus Copernicus (1473–1543) formulated his theory of the moving earth. Copernicus reasoned that the simplest way to interpret astronomical observations was to assume that the earth (and other planets) move around the sun. This idea was extremely controversial at the time. People preferred to believe that the earth was at the center of the universe. Copernicus worked on his ideas in secret to escape perse-cution. In the last days of his life and at the urging of close friends, he sent his ideas to the printer. The first copy of his work, *De Revolutionibus,* reached him on the day of his death, May 24, 1543.

4.3 Galileo on Motion

Galileo, the foremost scientist of late-Renaissance Italy, was outspoken in his support of Copernicus. As a result, he suffered a trial and house arrest. One of Galileo's great contributions to physics was demolishing the notion that a force is necessary to keep an object moving.

A **force** is any push or pull. **Friction** is the name given to the force that acts between materials that touch as they move past each other. Friction is caused by the irregularities in the surfaces of objects that are touching. Even very smooth surfaces have micro-scopic irregularities that obstruct motion. If friction were absent, a moving object would need no force whatever to remain in motion.

Galileo argued that only when friction is present—as it usually is—is a force needed to keep an object moving. He tested his idea by rolling balls along plane surfaces tilted at different angles. He noted that a ball rolling down such an inclined plane picks up speed, as shown on the left in Figure 4.2. The ball is rolling partly in the direc-tion of the pull of the earth's gravity. He also noted that a ball rolling up an inclined plane—in a direction opposed by gravity—slows down, as shown in the center in Figure 4.2. What about a ball rolling

on a level surface, as shown on the right in Figure 4.2? While rolling
level, the ball does not roll with or against gravity. Galileo found that
a ball rolling on a smooth horizontal plane has almost constant
velocity. He stated that if friction were entirely absent, a ball moving
horizontally would move forever. No push or pull would be required
to keep it moving once it is set in motion.

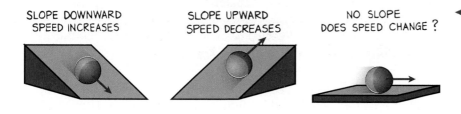

SLOPE DOWNWARD
SPEED INCREASES

SLOPE UPWARD
SPEED DECREASES

NO SLOPE
DOES SPEED CHANGE ?

◀ **Figure 4.2**
(Left) When the ball rolls downward, it moves with the earth's
gravity and its speed increases.
(Center) When the ball rolls
upward, it moves against gravity
and loses speed. (Right) When the
ball rolls on a level plane, it does
not move with or against gravity.
Does the ball's speed change
while rolling horizontally?

    Galileo's conclusion was supported by another line of reasoning.
He described two inclined planes facing each other, as in Figure 4.3.
A ball released to roll down one plane would roll up the other to
reach nearly the same height. The smoother the planes were, the
more nearly equal would be the initial and final heights. He noted
that the ball tended to attain the same height, even when the second
plane was longer and inclined at a smaller angle than the first plane.
In rolling to the same height, the ball had to roll farther. Additional
reductions of the angle of the upward plane gave the same results.
Always, the ball went farther and tended to reach the same height.
    What if the angle of incline of the second plane were reduced to
zero so that the plane was perfectly horizontal? How far would the
ball roll? He realized that only friction would keep it from rolling forever. It was not the nature of the ball to come to rest as Aristotle had
claimed. In the absence of friction, the moving ball would naturally
keep moving. Galileo stated that every material object resists change
to its state of motion. We call this resistance **inertia.**

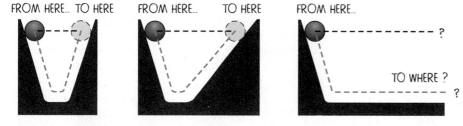

FROM HERE... TO HERE

FROM HERE... TO HERE

FROM HERE...

?

TO WHERE ?

?

Figure 4.3 ▲
(Left) The ball rolling down the incline rolls up the opposite incline and
reaches its initial height. (Center) As the angle of the upward incline is
reduced, the ball rolls a greater distance before reaching its initial height.
(Right) How far will the ball roll along the horizontal?

    Galileo was concerned with *how* things move rather than *why*
they move. He showed that experiment, not logic, is the best test of
knowledge. Galileo's findings about motion and his concept of inertia discredited Aristotle's theory of motion. The way was open for
Isaac Newton (1642–1727) to synthesize a new vision of the universe.

A ball is rolled across a counter top and rolls slowly to a stop. How would Aristotle interpret this behavior? How would Galileo interpret it? How would you interpret it?

Figure 4.4 ▲
Objects at rest tend to remain at rest.

4.4 Newton's Law of Inertia

On Christmas day in the year Galileo died, Isaac Newton was born. By age 24, he had developed his famous laws of motion. They replaced the Aristotelian ideas that dominated the thinking of the best minds for most of the previous 2000 years. This chapter covers the first of Newton's three laws of motion. Newton's two other laws of motion are covered in the following chapters.

Newton's first law, usually called the **law of inertia,** is a restatement of Galileo's idea.

> Every object continues in a state of rest, or of motion in a straight line at constant speed, unless it is compelled to change that state by forces exerted upon it.

Simply put, things tend to keep on doing what they're already doing. Dishes on a tabletop, for example, are in a state of rest. They tend to remain at rest, as is evidenced if you snap a tablecloth from beneath them. Try this at first with some unbreakable dishes. If you do it properly, you'll find the brief and small forces of friction are not significant enough to appreciably move the dishes (close inspection

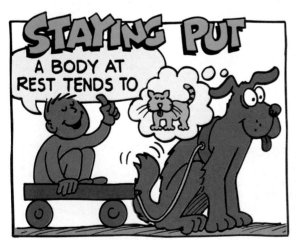

■ **Answer**

Aristotle would probably say that the ball stops because it seeks its natural state of rest. Galileo would probably say that the friction between the ball and the table overcomes the ball's natural tendency to continue rolling—overcomes the ball's inertia—and brings it to a stop. Only you can answer the last question!

will show that brief friction between the dishes and the fast-moving tablecloth starts the dishes moving, but immediately after the tablecloth is removed friction between the dishes and table stops them). Objects in a state of rest tend to remain at rest. Only a force will change that state.

Now consider an object in motion. If you slide a hockey puck along the surface of a city street, the puck quite soon comes to rest. If you slide it along ice, it slides for a longer distance. This is because the friction force is very small. If you slide it along an air table where friction is practically absent, it slides with no apparent loss in speed. We see that in the absence of forces, a moving object tends to move in a straight line indefinitely. Toss an object from a space station located in the vacuum of outer space, and the object will move forever. It will move by virtue of its own inertia.

◀ **Figure 4.5**
An air table: Blasts of air from many tiny holes provide a nearly friction-free surface.

We see that the law of inertia provides a completely different way of viewing motion. Whereas the ancients thought continual forces were needed to maintain motion, we now know that objects

47

Figure 4.6 ▲
You can tell how much matter is in a can when you kick it.

continue to move by themselves. Forces are needed to overcome any friction that may be present and to set objects in motion initially. Once the object is moving in a force-free environment, it will move in a straight line indefinitely. In the next chapter we'll see that forces are needed to accelerate objects, but not to maintain motion if there is no friction.

■ Questions

1. If the force of gravity between the sun and planets suddenly disappeared, what type of path would the planets follow?
2. Is it correct to say that the *reason* an object resists change and persists in its state of motion is that it has inertia?

4.5 Mass—A Measure of Inertia

Kick an empty can and it moves. Kick a can filled with sand and it doesn't move as much. Kick a can filled with steel nails and you'll hurt your foot. The nail-filled can has more inertia than the sand-filled can, which in turn has more inertia than the empty can. The amount of inertia an object has depends on its mass—which is roughly the amount of material present in the object. The more mass an object has, the greater its inertia and the more force it takes to change its state of motion. Mass is a measure of the inertia of an object.

Mass Is Not Volume

Do not confuse mass and volume. They are entirely different concepts. Volume is a measure of space and is measured in units such as cubic centimeters, cubic meters, and liters. Mass is measured in **kilograms.** If an object has a large mass, it may or may not have a large volume. For example, equal-size bags of cotton and nails may have equal volumes, but very unequal masses. How many kilograms of matter an object contains and how much space the object occupies

■ Answers

1. The planets would move in straight lines at constant speed—at constant velocity.

2. In a strict sense, no. We don't know the reason why objects persist in their motion when nothing acts on them, but we call this property *inertia*. We understand many things, and we have labels for these things. There are also many things we do not understand, and we have labels for these things too. Education consists not so much in acquiring new labels, but in learning what is understood, what is not—and why.

are two different things. (A liter of milk, juice, or soda—anything that is mainly water—has a mass of about one kilogram.)

Which has more mass, a feather pillow or a common automobile battery? Clearly an automobile battery is more difficult to set into motion. This is evidence of the battery's greater inertia and hence its greater mass. The pillow may be bigger, that is, it may have a larger volume, but it has less mass. Mass is different from volume.

Mass Is Not Weight

Mass is often confused with weight. We say a heavy object contains a lot of matter. We often determine the amount of matter in an object by measuring its gravitational attraction to the earth. However, mass is more fundamental than weight. Mass is a measure of the amount of material in an object and depends only on the number of and kind of atoms that compose it. Weight on the other hand is a measure of the gravitational force acting on the object. Weight depends on an object's location.

The amount of material in a particular stone is the same whether the stone is located on the earth, on the moon, or in outer space. Hence, the stone's mass is the same in all of these locations. This could be demonstrated by shaking the stone back and forth in these three locations. The same force would be required to shake the stone with the same rhythm whether the stone was on the earth, on the moon, or in a force-free region of outer space. The stone's inertia, or mass, is solely a property of the stone and not its location.

But the weight of the stone would be very different on the earth and on the moon, and still different in outer space. On the surface of the moon, the stone would have only one-sixth the weight it has on the earth. This is because the force of gravity on the moon is only one-sixth as strong as it is on the earth. If the stone were in a gravity-free region of space, its weight would be zero. Its mass, on the other hand, would not be zero. Mass is different from weight.

We can define mass and weight as follows:

> **Mass** is the quantity of matter in an object. More specifically, mass is a measure of the inertia, or "laziness," that an object exhibits in response to any effort made to start it, stop it, or otherwise change its state of motion.

> **Weight** is the force of gravity on an object.

While mass and weight are not the same thing, they are proportional to each other in a given place. Objects with great mass have great weight; objects with little mass have little weight. In the same location, twice the mass weighs twice as much. Mass and weight are proportional to each other, but they are not equal to each other. Remember that mass has to do with the amount of matter in the object, while weight has to do with how strongly that matter is attracted by gravity.

Figure 4.7 ▲
The pillow has a larger size (volume) but a smaller mass than the battery.

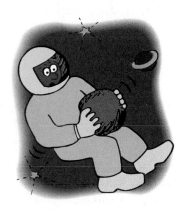

Figure 4.8 ▲
It's just as difficult to shake a stone in its weightless state in space as it is in its weighted state on earth.

1. Does a 2-kilogram iron block have twice as much *inertia* as a 1-kilogram block of iron? Twice as much *mass?* Twice as much *volume?* Twice as much *weight,* when weighed in the same location?

2. Does a 2-kilogram bunch of bananas have twice as much *inertia* as a 1-kilogram loaf of bread? Twice as much *mass?* Twice as much *volume?* Twice as much *weight,* when weighed in the same location?

Figure 4.9 ▲
One kilogram of nails weighs 9.8 newtons, which is equal to 2.2 pounds.

One Kilogram Weighs 9.8 Newtons

In the United States it is common to describe the amount of matter in an object by its gravitational pull to the earth, that is, by its weight. In English-speaking countries, the traditional unit of weight is the pound. In most parts of the world, however, the measure of matter is commonly expressed in units of mass. The SI* unit of mass is the kilogram; its symbol is kg. At the earth's surface, a 1-kg bag of nails has a weight of 2.2 pounds.

The SI unit of *force* is the **newton** (named after guess who?). One newton is equal to slightly less than a quarter pound, about the weight of a quarter-pound burger *after* it is cooked. The SI symbol for the newton is N and is written with a capital letter because it is named after a person. A 1-kg bag of nails weighs 9.8 N in SI units. Away from the earth's surface, where the force of gravity is less, the bag of nails weighs less.

If you know the mass of something in kilograms and want its weight in newtons, multiply the number of kilograms by 9.8. Or, if you know the weight in newtons, divide by 9.8 and you'll have the mass in kilograms. Weight and mass are proportional to each other. *Weight = mass × acceleration due to gravity*, or simply, *weight = mg.*

■ **Answers**

1. The answer is yes to all questions. A 2-kilogram block of iron has twice as many iron atoms, and therefore twice the amount of matter, mass, and weight. The blocks are made of the same material, so the 2-kilogram block also has twice the volume.

2. Two kilograms of *anything* has twice the inertia and twice the mass of one kilogram of anything else. In the same location, where mass and weight are proportional, two kilograms of anything will weigh twice as much as one kilogram of anything. Except for volume, the answer to all the questions is yes. Volume and mass are proportional only when the materials are the same or when equal masses occupy the same volume, that is, when they have the same *density.* Bananas are much more dense than bread, so two kilograms of bananas must occupy less volume than one kilogram of bread.

* The metric system was originally established in France in the 1790s. The International System of Units (abbreviated SI, after the French name *Le Système International d'Unités*) is a revised version of the metric system. The short forms of the SI units are called *symbols* rather than *abbreviations.*

■ **Question**

The text states that a 1-kg bag of nails weighs 9.8 N at the earth's surface. Does 1 kg of yogurt also weigh 9.8 N at the earth's surface?

APPLIED FORCES	NET FORCE
5 N 10 N	15 N
5 N 10 N	5 N
5 N 5 N	0 N

4.6 Net Force

In the absence of force, objects at rest stay at rest and objects in motion continue in motion. More specifically, in the absence of a *net force,* objects do not change their state of motion. For example, if you push with equal and opposite forces on opposite sides of an object at rest, it will remain at rest. The forces cancel each other out and there is no net force. The combination of all forces acting on an object is called the **net force.** It is the net force that changes an object's state of motion.

Figure 4.10 shows how forces combine to produce a net force. When you pull horizontally with a force of 10 N on an object resting on a frictionless surface, the net force acting on the object is 10 N. If a friend assists you and also pulls in the same direction with a force of 5 N, then the net force is the sum of these forces, or 15 N (Figure 4.10, top). The object moves as if it were pulled with a single 15-N force. However, if your friend pulls with a force of 5 N in the opposite direction, then the net force is the difference of these forces, or 5 N (Figure 4.10, center). The resulting motion of the object is the same as if it were pulled with a single 5-N force.

Figure 4.10 ▲
When more than one force acts on an object, the net force is the sum of the forces. When forces act in the same direction, the net force is the sum of the forces. When forces act in opposite directions, the net force is the difference of the forces.

4.7 Equilibrium—When Net Force Equals Zero

What forces act on your book while it is motionless on a table? Don't say just its weight. If only the force of gravity were acting on the book, it would be in free fall. The fact that the book is at rest is evidence that another force must be acting on it. This other force exactly balances the book's weight and produces a net force of zero. The other force is the **support force** of the table. Support force is often called **normal force.*** As shown in Figure 4.11, the table pushes up on the book with a force equal to the book's weight. When an

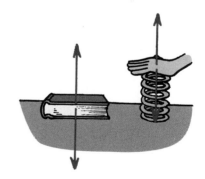

Figure 4.11 ▲
(Left) The table pushes up on the book with a force equal to the downward weight of the book. (Right) The spring pushes up on your hand with the same force you use to push down on the spring.

■ **Answer**

Yes, at the earth's surface 1 kg of anything weighs 9.8 N.

---

*   This force acts at right angles to the surface. *Normal to* means "at right angles to," which is why the force is called a normal force.

object is at rest, with the net force on it being zero, we say the object is in a state of **equilibrium.** The resting book is in equilibrium.

Does the table really push up on the book? Yes, just as a spring pushes up on your hand when you compress it. Look at Figure 4.11. You can feel the spring pushing back on your hand. Similarly, a book lying on a table compresses the atoms in the table. Behaving like microscopic springs, the atoms in contact with the book push back on it. Since the book is in equilibrium, the net force on it must be zero. The table actually pushes up on the book with the same force that the book presses down. An ant trapped between the book and the table would feel itself being squashed from both sides—the top and the bottom.

When you hang from a rope, the atoms in the rope are *not* compressed. Instead, the atoms are stretched apart. A tension force is established in the rope. A rope under tension "twangs" if you pluck it. How much tension is in the rope when you hang from it? If you are in equilibrium, then the tension must equal your weight. The rope pulls you up and the force of gravity pulls you down. Since the equal and opposite forces cancel each other, you hang motionless. Suppose you hang from a bar supported by two ropes, as in Figure 4.12. Neglecting the weight of the bar, the tension in each rope is one-half your weight. The total tension force acting upward ($\frac{1}{2}$ your weight + $\frac{1}{2}$ your weight) balances your weight, which acts downward. When doing pull-ups using both arms, each arm supports half of your weight. Have you ever tried pull-ups with just one arm? Why is it twice as difficult?

A spring scale can also be used to measure tension. When we weigh a bag of apples by suspending it on a spring, the scale reading tells us the weight. Consider a 1-kg bag of 10 apples weighing 10 N.

Figure 4.12 ▲
The sum of the rope tensions must equal your weight.

■ Question

When you step on a bathroom scale, the downward force supplied by your feet and the upward force supplied by the floor compress a calibrated spring. The compression of the spring gives your weight. In effect, the scale measures the floor's support force. What will each scale read if you stand on two scales with your weight divided equally between them? What happens if you stand with more of your weight on one foot than the other?

■ Answer

Since you are in equilibrium, the two scale readings must add up to your weight. The sum of the scale readings, which equals the support force of the floor, counteracts your weight so that the net force is zero. If you stand with your weight divided equally between the two scales, each scale will read half your weight. If you lean more on one scale, it will read more than half your weight. However, the two readings must still add up to your weight. For example, if one scale reads two-thirds of your weight, the other scale will read one-third of your weight. Get it?

When suspended by a single vertical scale as shown on the left in Figure 4.13, the scale reads 10 N. If instead we hang the 10-N bag from a pair of vertical scales as shown on the right in Figure 4.13, each scale reads 5 N. The two scales pull up with a combined force equaling the bag's weight, 10 N. Since the bag hangs at rest, the net force on it must be zero. The key concept is this: If a 10-N bag is in equilibrium, the *resultant of the forces* applied by the pair of springs must equal 10 N. If the spring scales are oriented vertically, this is easy: 5 N + 5 N = 10 N.

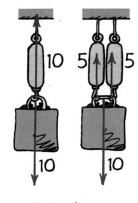

Figure 4.13 ▲
(Left) When a 10-N load hangs vertically from a single spring scale, the scale pulls upward with a force of 10 N. (Right) When the load hangs vertically from two spring scales, each scale pulls upward with a force equal to half of the load's weight, or 5 N.

4.8 Vector Addition of Forces

Let's now consider nonvertically oriented spring scales. The tension is greater in a pair of nonvertical spring scales and depends on their angle from the vertical. This more complicated situation is most easily understood using vectors. Recall our treatment of velocity vectors in Chapter 3. We'll now use the same vector addition techniques with forces. Force, like velocity, has magnitude and direction and is a vector quantity.

Look at the left of Figure 4.14. Notice that when the spring scales hang at 60° from the vertical, their readings are 10 N each—double the tension of the vertically hanging scales! Why? Because the sum of the two tension vectors must support the downward-acting 10-N weight. The sum of the vectors must be 10 N, directed vertically upward as shown by the diagonal of the parallelogram formed from the two vectors. For an angle 60° from the vertical, or 120° between the scales, 10 N of tension force is required in each scale. For angles greater than 60°, the scale readings would increase even more.

Notice that on the right of Figure 4.14, the angle from the vertical has increased to 75.5°. At this angle, each spring must pull with 20 N to produce the required 10-N vertically upward resultant. As the angle between the scales increases, the tension in the scales must increase for the resultant to remain 10 N. In terms of the parallelogram, as the scale's angle from the vertical increases, the magnitude of the tension force must also increase for the diagonal to remain the

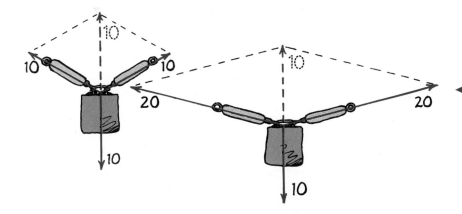

◀ **Figure 4.14**
As the angle between the spring scales increases, the scale readings increase to maintain the 10-N upward resultant. The 10-N resultant, shown as the dashed-line vector, is needed to support the 10-N load.

same, 10 N. If you understand this, you understand why a vertical clothesline can support your weight but a horizontal clothesline cannot. The tension in the nearly horizontal clothesline is much greater than the tension in the vertical clothesline, and therefore it breaks.

For any pair of scales, ropes, or wires supporting a load, the greater their angle from the vertical, the larger the tension force in them. The resultant of the tension forces, or the diagonal of the parallelogram they describe, must be equal and opposite to the load being supported. As you will see when you answer the following questions, the parallelogram technique of adding vectors yields some very interesting results.

Figure 4.15 ▶
You can safely hang from a vertically hanging clothesline, but you'll break the clothesline if it is strung horizontally.

■ Questions

1. If the kids on the swings are of equal weight, which swing is more likely to break?

2. Consider what would happen if you suspended a 10-N object midway along a very tight, horizontally stretched guitar string. Is it possible for the string to remain horizontal without a slight sag at the point of suspension?

■ Answers

1. The stretching force, or tension, is greater in the ropes hanging at an angle. The angled ropes are more likely to break than the vertical ropes.

2. No way! If the 10-N load is to hang in equilibrium, there must be a supporting 10-N upward resultant. The tension in each half of the guitar string must form a parallelogram with a vertically upward 10-N resultant. For a slight sag, the sides of the parallelogram are very, very long and the tension force is very large. To approach no sag is to approach an infinite tension. Turning this idea around, a little thought shows that pulling a tightly stretched string slightly to one side increases the tension in the string enormously. That's why a small sideways force can break a very strong guitar string!

4.9 The Moving Earth Again

Copernicus announced the idea of a moving earth in the sixteenth century. This controversial idea stimulated much argument and debate. One of the arguments against a moving earth was as follows. Consider a bird sitting at rest in the top of a tall tree. On the ground below is a fat, juicy worm. The bird sees the worm, drops down vertically, and catches it. It was argued that this would not be possible if the earth moved as Copernicus suggested. If Copernicus were correct, the earth would have to travel at a speed of 107 000 km/h to circle the sun in one year. Convert this speed to kilometers per second and you'll get 30 km/s. Even if the bird could descend from its branch in one second, the worm would have been swept away by the moving earth for a distance of 30 kilometers. For the bird to catch the worm under this circumstance would be an impossible task. The fact that birds do catch worms from high tree branches seemed to be clear evidence that the earth must be at rest.

Can you refute this argument? You can if you invoke the idea of inertia. You see, not only is the earth moving at 30 km/s, but so are the tree, the branch of the tree, the bird that sits on it, the worm below, and even the air in between. All are moving at 30 km/s. Things in motion remain in motion if no unbalanced forces act on them. So when the bird drops from the branch, its initial sideways motion of 30 km/s remains unchanged. It catches the worm and is quite unaffected by the motion of its total environment.

Stand next to a wall. Jump up so that your feet no longer touch the floor. Does the 30-km/s wall slam into you? Why not? Because you are also traveling at 30 km/s, before, during, and after your jump. The 30 km/s is the speed of the earth relative to the sun, not the speed of the wall relative to you.

Four hundred years ago, people had difficulty with ideas like these, not only because they failed to acknowledge the concept of inertia, but also because they were not accustomed to moving in high-speed vehicles. Slow, bumpy rides in horse-drawn carriages do not lend themselves to experiments that reveal inertia. Today we flip a coin in a high-speed car, bus, or plane and catch the vertically moving coin as we would if the vehicle were at rest. We see evidence for the law of inertia when the horizontal motion of the coin before, during, and after the catch is the same. The coin keeps up with us. The vertical force of gravity affects only the vertical motion of the coin.

Our notions of motion today are very different from those of our distant ancestors. Aristotle did not recognize the idea of inertia, because he did not see that all moving things follow the same rules. He imagined different rules for motion in the heavens and motion on the earth. He saw horizontal motion as "unnatural," requiring a sustained force. Galileo and Newton, on the other hand, saw that all moving things follow the same rules. To them, moving things required *no* force to keep moving if friction was not present. We can only wonder how differently science might have progressed if Aristotle had recognized the unity of all kinds of motion and friction's effect on motion.

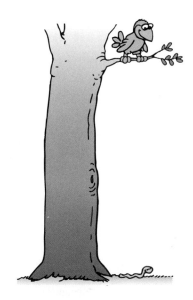

Figure 4.16 ▲
Must the earth be at rest for the bird to catch the worm?

Figure 4.17 ▲
Flip a coin in a high-speed airplane, and it behaves as if the plane were at rest. The coin keeps up with you—inertia in action!

4 Chapter Review

Concept Summary

Galileo concluded that if it were not for friction, an object in motion would keep moving forever.

Newton's first law of motion—the law of inertia.

> Every object continues in a state of rest, or in a state of motion in a straight line at constant speed, unless it is compelled to change that state by forces exerted upon it.

Inertia is the resistance an object has to a change in its state of motion.

- Mass is a measure of inertia.
- Mass is not the same as volume.
- Mass is not the same as weight.
- The mass of an object depends only on the number and kind of atoms in it. Mass does not depend on the location of the object.
- The weight of an object is the gravitational force acting on it. Weight depends on the location of the object.

The net force, which is the vector sum of all forces acting on an object, affects the object's state of motion.

- When an object is at rest, its weight is balanced by an equal and opposite support force.
- An object is in equilibrium when it is at rest, with zero net force acting on it.

Important Terms

equilibrium (4.7)
force (4.3)
friction (4.3)
inertia (4.3)
kilogram (4.5)
law of inertia (4.4)
mass (4.5)

net force (4.6)
newton (4.5)
Newton's first law (4.4)
normal force (4.7)
support force (4.7)
weight (4.5)

Review Questions

1. What distinction did Aristotle make between natural motion and violent motion? (4.1)

2. Why was Copernicus reluctant to publish his ideas? (4.2)

3. What is the effect of friction on a moving object? (4.3)

4. The speed of a ball increases as it rolls down an incline and decreases as it rolls up an incline. What happens to its speed on a smooth, horizontal surface? (4.3)

5. Galileo found that a ball rolling down one incline will pick up enough speed to roll up another. How high will it roll compared with its initial height? (4.3)

6. Does the law of inertia pertain to moving objects, objects at rest, or both? Support your answer with examples. (4.4)

7. The law of inertia states that no force is required to maintain motion. Why, then, do you have to keep peddling your bicycle to maintain motion? (4.4)

8. If you were in a spaceship and fired a cannonball into frictionless space, how much force would have to be exerted on the ball to keep it going? (4.4)

9. Does a 2-kilogram rock have twice the mass of a 1-kilogram rock? Twice the inertia? Twice the weight (when weighed in the same location)? (4.5)

10. Does a liter of molten lead have the same volume as a liter of apple juice? Does it have the same mass? (4.5)

11. Why do physicists say mass is more fundamental than weight? (4.5)

12. An elephant and a mouse would both have zero weight in gravity-free space. If they were moving toward you with the same speed, would they bump into you with the same effect? Explain. (4.5)

13. What is the weight of 2 kilograms of yogurt? (4.5)

14. What is the net force or, equivalently, the resultant force acting on an object in equilibrium? (4.6)

15. Forces of 10 N and 15 N in the same direction act on an object. What is the net force on the object? (4.6)

16. If forces of 10 N and 15 N act in opposite directions on an object, what is the net force? (4.6)

17. How does the tension in your arms compare when you let yourself dangle motionless by both arms and by one arm? (4.7)

18. A clothesline is under tension when you hang from it. Why is the tension greater when the clothesline is strung horizontally than when it hangs vertically? (4.8)

19. If you hold a coin above your head while in a bus that is not moving, the coin will land at your feet when you drop it. Where will it land if the bus is moving in a straight line at constant speed? Explain. (4.9)

20. In the cabin of a jetliner that cruises at 600 km/h, a pillow drops from an overhead rack into your lap below. Since the jetliner is moving so fast, why doesn't the pillow slam into the rear of the compartment when it drops? What is the horizontal speed of the pillow relative to the ground? Relative to you inside the jetliner? (4.9)

Plug and Chug

1. If a woman has a mass of 50 kg, calculate her weight in newtons.

2. Calculate in newtons the weight of a 2000-kg elephant.

3. Calculate in newtons the weight of a 2.5-kg melon. What is its weight in pounds?

4. An apple weighs about 1 N. What is its mass in kilograms? What is its weight in pounds?

5. Susie Small finds she weighs 300 N. Calculate her mass.

Think and Explain

1. Many automobile passengers suffer neck injuries when struck by cars from behind. How does Newton's law of inertia apply here? How do headrests help to guard against this type of injury?

2. Suppose you place a ball in the middle of a wagon and then accelerate the wagon forward. Describe the motion of the ball relative to (a) the ground and (b) the wagon.

3. When a junked car is crushed into a compact cube, does its mass change? Its volume? Its weight?

4. If an elephant were chasing you, its enormous mass would be very threatening. But if you zigzagged, the elephant's mass would be to your advantage. Why?

5. When you compress a sponge, which quantity changes: mass, inertia, volume, or weight?

6. A massive ball is suspended on a string and slowly pulled by another string attached to it from below, as shown in Figure A below.

   a. Is the string tension greater in the upper or the lower string? Which string is more likely to break? Which property, mass or weight, is important here?

   b. If the string is instead snapped downward, which string is more likely to break? Is mass or weight important this time?

Figure A

Figure B

7. The head of a hammer is loose and you wish to tighten it by banging it against the top of a workbench. Why is it better to hold the hammer with the handle down, as shown in Figure B above, rather than with the head down? Explain in terms of inertia.

8. The little girl in the figure hangs at rest from the ends of the rope. How does the reading on the scale compare with her weight?

9. Harry the painter swings year after year from his bosun's chair. His weight is 500 N and the rope, unknown to him, has a breaking point of 300 N. Why doesn't the rope break when he is supported as shown on the left? One day Harry is painting near a flagpole, and for a change, he ties the free end of the rope to the flagpole instead of to his chair (right). Why did Harry end up taking his vacation early?

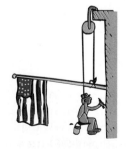

10. As the earth rotates about its axis, it takes three hours for the United States to pass beneath a point above the earth that is stationary relative to the sun. What is wrong with the following scheme? To travel from Washington D.C. to San Francisco using very little fuel, simply ascend in a helicopter high over Washington D.C. and wait three hours until San Francisco passes below.

11. In which position is the tension the least in the arms of the weightlifter shown? The most?

12. Why can't the strong man pull hard enough to make the chain straight?

Think and Solve

1. A medium-size American automobile has a weight of about 3000 pounds. What is its mass in kilograms?

2. If a woman weighs 500 N on Earth, what would she weigh on Jupiter, where the acceleration of gravity is 26 m/s^2?

5 Newton's Second Law of Motion—Force and Acceleration

The thrill of a roller coaster ride comes from the accelerations experienced.

Kick a football and it moves. Its path through the air is not a straight line—it curves downward due to gravity. Catch the ball and it stops. Most of the motion we see undergoes change. Most things start up, slow down, or curve as they move. The previous chapter covered objects at rest or moving at constant velocity. There was no net force acting on these objects. This chapter covers the more common cases in which there is a *change* in motion—that is, accelerated motion.

Recall from Chapter 2 that acceleration describes how quickly motion changes. Specifically, it is the change in velocity per certain time interval. In shorthand notation,

$$\text{acceleration} = \frac{\text{change in velocity}}{\text{time interval}}$$

This is the definition of acceleration.* This chapter focuses on the *cause* of acceleration: *force*.

Figure 5.1 ▲
Kick a football and it neither remains at rest nor moves in a straight line.

5.1 Force Causes Acceleration

Consider an object at rest, such as a hockey puck on ice. Apply a force and it starts to move. Since the puck was not moving before, it has accelerated—it has changed its motion. When the hockey stick is no longer pushing it, the puck moves at constant velocity. Apply another force by striking the puck again, and again the motion changes. The puck accelerates—force causes acceleration.

---

* The Greek letter Δ (delta) is often used as a symbol for "change in" or "difference in." In delta notation, $a = \Delta v/\Delta t$, where Δv is the change in velocity and Δt is the change in time (the time interval).

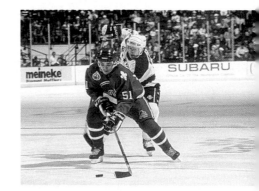

Figure 5.2 ▲
Puck about to be hit.

Most often, the force we apply is not the only force acting on an object. Other forces may act as well. Recall from the previous chapter that the combination of forces acting on an object is the *net force*. Acceleration depends on the *net force*. To increase the acceleration of an object, you must increase the net force acting on it. This makes good sense. Double the force on an object and its acceleration doubles. If you triple the force, its acceleration triples, and so on. We say an object's acceleration is directly proportional to the net force acting on it. We write

$$\text{acceleration} \sim \text{net force}$$

The symbol ~ stands for "is directly proportional to."

5.2 Mass Resists Acceleration

Push on an empty shopping cart. Then push equally hard on a heavily loaded shopping cart. The loaded shopping cart will accelerate much less than the empty cart. This shows that acceleration depends on the mass being pushed. The same force applied to twice as much mass results in only half the acceleration. For three times the mass, one-third the acceleration results. In other words, for a given force, the acceleration produced is *inversely proportional* to the mass. We write

$$\text{acceleration} \sim \frac{1}{\text{mass}}$$

By **inversely** we mean that the two values change in opposite directions. Mathematically we see that as the denominator increases, the whole quantity decreases. The quantity 1/100 is less than the quantity 1/10, for example.

Figure 5.3 ▲
The acceleration produced depends on the mass that is pushed.

5.3 Newton's Second Law

Newton was the first to realize that the acceleration produced when we move something depends not only on how hard we push or pull, but also on the object's mass. He came up with one of the most important rules of nature ever proposed, his second law of motion. **Newton's second law** states

> The acceleration produced by a net force on an object is directly proportional to the magnitude of the net force, is in the same direction as the net force, and is inversely proportional to the mass of the object.

Or, in equation form,

$$\text{acceleration} \sim \frac{\text{net force}}{\text{mass}}$$

By using consistent units, such as newtons (N) for force, kilograms (kg) for mass, and meters per second squared (m/s^2) for acceleration, we get the exact equation

$$\text{acceleration} = \frac{\text{net force}}{\text{mass}}$$

In briefest form, where a is acceleration, F is net force, and m is mass,

$$a = \frac{F}{m}$$

The acceleration is equal to the net force divided by the mass. From this relationship we see that doubling the net force acting on an object doubles its acceleration. Suppose instead that the mass is doubled. Then acceleration will be halved. If both the net force and the mass are doubled, the acceleration will be unchanged.

> ▶ **DOING PHYSICS**
>
> ### Acceleration, Which Way?
>
> The net force acting on an object and its resulting acceleration are always in the same direction. You can demonstrate this with a spool of thread. If the spool is pulled horizontally by the thread to the right, in which direction will it roll? Does it make a difference whether the string is on the bottom or the top? Try it and you might be surprised.
>
>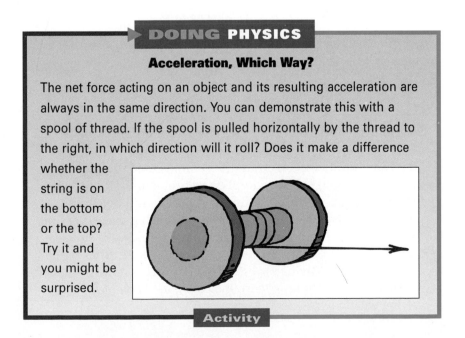
>
> **Activity**

61

If we know the mass of an object in kilograms (kg) and its acceleration in meters per second per second (m/s²), then the force will be expressed in newtons (N). One newton is the force needed to give a mass of one kilogram an acceleration of one meter per second squared. We can arrange Newton's second law to read

$$\text{force} = \text{mass} \times \text{acceleration}$$

$$1 \text{ N} = (1 \text{ kg})(1 \text{ m/s}^2)$$

We can see,

$$1 \text{ N} = 1 \text{ kg·m/s}^2$$

The dot between kg and m/s² means that the units are multiplied together.

If we know two of the quantities in Newton's second law, we can calculate the third. For example, how much force, or thrust, must a 30 000-kg jet plane develop to achieve an acceleration of 1.5 m/s²? We calculate

$$F = ma$$

$$= (30\ 000 \text{ kg})(1.5 \text{ m/s}^2)$$

$$= 45\ 000 \text{ kg·m/s}^2$$

$$= 45\ 000 \text{ N}$$

Suppose we know the force and the mass, and we want to find the acceleration. For example, what acceleration is produced by a force of 2000 N applied to a 1000-kg car? Using Newton's second law we find

$$a = \frac{F}{m} = \frac{2000 \text{ N}}{1000 \text{ kg}} = \frac{2000 \text{ kg·m/s}^2}{1000 \text{ kg}} = 2 \text{ m/s}^2$$

If the force is 4000 N, what is the acceleration?

$$a = \frac{F}{m} = \frac{4000 \text{ N}}{1000 \text{ kg}} = \frac{4000 \text{ kg·m/s}^2}{1000 \text{ kg}} = 4 \text{ m/s}^2$$

Doubling the force on the same mass simply doubles the acceleration.

Physics problems are often more complicated than these. We don't focus on solving complicated problems in this book. Instead we emphasize equations as guides to thinking about the relationships of basic physics concepts. The Plug and Chug section at the end of many chapters familiarizes you with equations, and the Problems sections go a step or two further for more challenge. Solving problems is a big part of more advanced physics courses. For now, learn the concepts! Then problem solving will be more meaningful.

■ Questions

1. If a car can accelerate at 2 m/s², what acceleration can it attain if it is towing another car of equal mass?

2. What kind of motion does a constant force produce on an object of fixed mass?

5.4 Friction

We discussed friction in Chapter 4. Friction is a force like any other force and affects motion. Friction acts on materials that are in contact with each other, and it always acts in a direction to oppose motion. When two solid objects come into contact, the friction is mainly due to irregularities in the two surfaces. When one object slides against another, it must either rise over the irregular bumps or else scrape them off. Either way requires force.

The force of friction between the surfaces depends on the kinds of material in contact and how much the surfaces are pressed together. For example, rubber against concrete produces more friction than steel against steel. That's why concrete road dividers have replaced steel rails; see Figure 5.5. The friction produced by a tire rubbing against the concrete is more effective in slowing the car than the friction produced by a steel car body sliding against a steel rail. Notice that the concrete divider is wider at the bottom to ensure that the tire of a sideswiping car will make contact with the divider before the steel car body does.

Friction is not restricted to solids sliding or tending to slide over one another. Friction also occurs in liquids and gases, both of which are called **fluids** (because they flow). Fluid friction occurs as an object pushes aside the fluid it is moving through. Have you ever tried running a 100-m dash through waist-deep water? The friction of liquids is appreciable, even at low speeds. **Air resistance,** which is the friction acting on something moving through air, is a very common form of fluid friction. You usually don't notice air resistance when walking or jogging, but you do notice it at the higher speeds that occur when riding a bicycle or skiing downhill.

When friction is present, an object may move with a constant velocity even when an outside force is applied to it. In such a case, the friction force just balances the applied force. The net force is zero, so there is no acceleration. For example, in Figure 5.6 the

Figure 5.5 ▲
Cross-section view of a concrete road divider and a steel road divider. Which design is best for slowing an out-of-control, side-swiping car?

■ Answers

1. The same force on twice the mass produces half the acceleration, or 1 m/s².

2. A constant force produces motion at a constant acceleration, in accordance with Newton's second law.

Figure 5.6 ▶

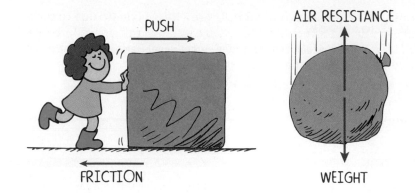

The direction of the force of friction always opposes the direction of motion. (Left) Push the crate to the right and friction acts toward the left. (Right) The sack falls downward and air friction acts upward.

crate moves with a constant velocity when the force pushing it just balances the force of friction. The sack will also fall with a constant velocity once the force due to air resistance balances the sack's weight.

■ Questions

1. Two forces act on a book resting on a table: its weight and the support force from the table. Does a force of friction act as well?

2. Suppose a high-flying jet cruises with a constant velocity when the thrust from its engines is a constant 80 000 N. What is the *acceleration* of the jet? What is the force of air resistance acting on the jet?

5.5 Applying Force—Pressure

No matter how you place a book on a table—on its back, upright, or even balancing on a single corner—the force of the book on the table is the same. You can check this by placing a book in any position on a bathroom scale. You'll read the same weight in all cases. Balance a book in different positions on the palm of your hand. Although the force is always the same, you'll notice differences in the way the

■ Answers

1. No, not unless the book tends to slide or does slide across the table. For example, if it is pushed toward the left by another force, then friction between the book and table will act toward the right. Friction forces occur only when an object tends to slide or is sliding.

2. The acceleration must be zero because the velocity is not changing—velocity is constant. Since the acceleration is zero, it follows from $a = F/m$ that the net force is zero. This means the force of air resistance must equal the engine's thrust. The air resistance is 80 000 N, and it acts in the direction opposite to the jet's motion.

book presses against your palm. These differences are due to differences in the area of contact for each case. The amount of force *per unit of area* is called **pressure.** More precisely, when the force is perpendicular to the surface area,

$$\text{pressure} = \frac{\text{force}}{\text{area of application}}$$

In equation form,

$$P = \frac{F}{A}$$

where *P* is the pressure and *A* is the area over which the force acts. Force, which is measured in newtons, is different from pressure, which is measured in newtons per square meter, or **pascals** (Pa). One newton per square meter is equal to one pascal.

You exert more pressure against the ground when you stand on one foot than when you stand on both feet. This is due to the decreased area of contact. Stand on one toe like a ballerina and the pressure is huge. The smaller the area supporting a given force, the greater the pressure on that surface.

You can calculate the pressure you exert on the ground when you are standing. One way is to moisten the bottom of your foot with water and step on a clean sheet of graph paper. Count the number of squares on the graph paper contained within your footprint. Divide your weight by this area and you have the average pressure you exert on the ground when standing on one foot. How will this pressure compare with the pressure you exert when you stand on two feet?

A dramatic illustration of pressure is shown in Figure 5.8. The author applies appreciable force when he breaks the cement block with the sledge hammer. Yet his friend (the author of the lab manual) sandwiched between two beds of sharp nails is unharmed. The friend is unharmed because much of the force is distributed over the more than 200 nails that make contact with his body. The combined surface area of this many nails results in a tolerable pressure that does not puncture the skin. CAUTION: This demonstration is quite dangerous. Do not attempt it on your own.

Figure 5.7 ▲
The upright book exerts the same force, but greater pressure, against the supporting surface.

Figure 5.8 ▼
The author applies a force to fellow physics teacher Paul Robinson, who is bravely sandwiched between beds of sharp nails. The driving force per nail is not enough to puncture the skin. CAUTION: Do not attempt this on your own!

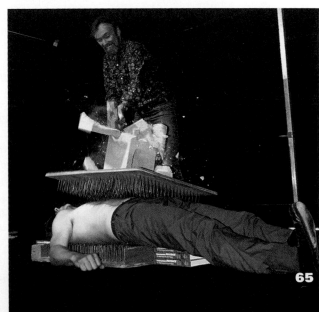

1. In attempting to do the demonstration shown in Figure 5.8, would it be wise to begin with a few nails and work upward to more nails?

2. The massiveness of the cement block plays an important role in this demonstration. Which provides more safety, a less massive block or a more massive one?

5.6 Free Fall Explained

Figure 5.9 ▲
Galileo's famous demonstration.

Galileo showed that falling objects accelerate equally, regardless of their masses. This is *strictly* true if air resistance is negligible, that is, if the objects are in free fall. It is *approximately* true when air resistance is very small compared with the weight of the falling object. For example, a 10-kg cannonball and a 1-kg stone dropped from an elevated position at the same time will fall together and strike the ground at practically the same time. This experiment, said to be done by Galileo from the Leaning Tower of Pisa, demolished the Aristotelian idea that an object that weighs ten times as much as another should fall ten times faster than the lighter object. Galileo's experiment and many others that showed the same result were convincing. But Galileo couldn't say *why* the accelerations were equal. The explanation is a straightforward application of Newton's second law and is the topic of the cartoon "Backyard Physics." Let's treat it separately here.

Recall that mass (a quantity of matter) and weight (the force due to gravity) are proportional. A 2-kg bag of nails weighs twice as much as a 1-kg bag of nails. So a 10-kg cannonball experiences 10 times as much gravitational force (weight) as a 1-kg stone. The followers of Aristotle believed that the cannonball should accelerate at a rate ten times that of the stone, because they considered only the cannonball's ten-times-greater weight. However, Newton's second law tells us to consider the mass as well. A little thought will show that ten times as much force acting on ten times as much mass produces the same acceleration as the smaller force acting on the smaller mass. In symbolic notation,

$$\frac{F}{m} = \frac{F}{m}$$

■ **Answers**

1. No, no, no! There would be one less physics teacher if the demonstration were performed with fewer nails. The resulting greater pressure would cause harm.

2. The greater the mass of the block, the smaller the acceleration of the block and bed of nails toward the friend. Much of the force wielded by the hammer goes into *breaking* the block. It is important that the block be massive and that it break upon impact.

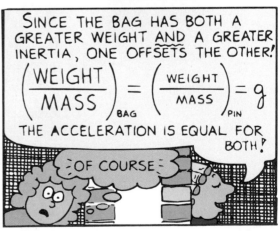

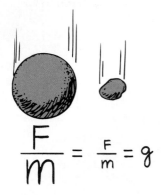

$$\frac{F}{m} = \frac{F}{m} = g$$

Figure 5.10 ▲
The ratio of weight (F) to mass (m) is the same for the 10-kg cannonball and the 1-kg stone.

where F stands for the force (weight) acting on the cannonball, and m stands for the correspondingly large mass of the cannonball. The small F and m stand for the smaller weight and smaller mass of the stone. We see that the *ratio* of weight to mass is the same for these or any objects. All freely falling objects undergo the same acceleration at the same place on the earth. In Chapter 2 we introduced the symbol g for the acceleration.

We can show the same result with numerical values. The weight of a 1-kg stone is 9.8 N at the earth's surface. The weight of a 10-kg cannonball is 98 N at the earth's surface. The force acting on a falling object is the force due to gravity—the object's weight. Using Newton's second law, the acceleration of the stone is

$$a = \frac{F}{m} = \frac{\text{weight}}{m} = \frac{9.8\ \text{N}}{1\ \text{kg}} = \frac{9.8\ \text{kg·m/s}^2}{1\ \text{kg}} = 9.8\ \text{m/s}^2 = g$$

and the acceleration of the cannonball is

$$a = \frac{F}{m} = \frac{\text{weight}}{m} = \frac{98\ \text{N}}{10\ \text{kg}} = \frac{98\ \text{kg·m/s}^2}{10\ \text{kg}} = 9.8\ \text{m/s}^2 = g$$

In the famous coin-and-feather-in-a-vacuum-tube demonstration discussed in Chapter 2, the reason for the equal accelerations was not discussed. Now we know that both freely falling objects fall with the same acceleration because the net force on each object is only its weight, and the ratio of weight to mass is the same for both.

■ Question

If you were on the moon and dropped a hammer and a feather from the same elevation at the same time, would they strike the surface of the moon at the same instant?

5.7 Falling and Air Resistance

The feather and coin fall with equal accelerations in a vacuum, but quite unequally in the presence of air. When air is let into the glass tube and it is again inverted and held upright, the coin falls quickly while the feather flutters to the bottom. Air resistance diminishes the net forces acting on the objects. The net force decreases a tiny bit for the coin and a lot for the feather. The downward acceleration for the feather is very brief because the air resistance builds up quickly and counteracts its tiny weight. The feather does not have to fall very long or very fast for this to happen. When the air resistance on the feather equals the weight of the feather, the net force is zero and no further

■ Answer

Yes. Astronaut David Scott did this exact experiment! On the moon, the hammer and feather weigh only one-sixth of their earth weight, and there is no air to provide friction. The ratio of moon-weight to mass for each object is the same, and they both accelerate at (1/6)g.

acceleration occurs. Acceleration terminates: the feather has reached its **terminal speed.** If we are concerned with direction, which is down for falling objects, we say the feather has reached its **terminal velocity.**

Air resistance does not have as much effect on the coin. At slow speeds, the force of air resistance is very small compared with the weight of the coin, and its acceleration is only slightly less than the acceleration of free fall, *g*. The coin might have to fall for several seconds before its speed is great enough for air resistance to equal and cancel its weight. Its speed at this point, perhaps as much as 200 km/h, would no longer increase. The coin reaches its terminal speed.

The terminal speed for a sky diver varies from about 150 to 200 km/h, depending on the weight and orientation of the body. A heavier person will attain a greater terminal speed than a lighter person. The greater weight is more effective in "plowing through" air. Body orientation also makes a difference. More air is encountered when the body is spread out and surface area is increased, like that of the flying squirrel in Figure 5.12. Terminal speed can be controlled by variations in body orientation. A heavy sky diver and a light sky diver can remain in close proximity to each other if the heavy person spreads out like a flying squirrel while the light person falls head or feet first. A parachute greatly increases air resistance, and cuts the terminal speed down to 15 to 25 km/h, slow enough for a safe landing.

If you hold a baseball and tennis ball at arm's length and release them at the same time, you'll see them strike the floor at the same time. But if you drop them from the top of a building, you'll notice

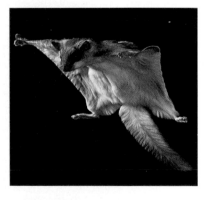

Figure 5.12 ▲
The flying squirrel increases its area by spreading out. This increases air resistance and decreases the speed of its fall.

■ Question

If a heavy person and a light person open their parachutes together at the same altitude and each wears the same size parachute, who will reach the ground first?

■ Answer

The heavy person will reach the ground first. Like a feather, the light person reaches terminal speed sooner, while the heavy person continues to accelerate until a greater terminal speed is reached. The heavy person moves ahead of the light person, and the separation continues to increase as they descend.

the heavier baseball strikes the ground first. This is due to the buildup of air resistance at higher speeds (like the parachutists in the question). At low speeds, air resistance is often negligible, but at high speeds, it can make quite a difference. The effect of air resistance is more pronounced on the lighter tennis ball than on the heavier baseball, so the acceleration of the fall is less for the tennis ball. The tennis ball behaves more like a parachute than the baseball does.

When Galileo reportedly dropped the objects of different weights from the Leaning Tower of Pisa, the heavier object *did* get to the ground first. However, the time difference was only a split second, rather than the pronounced time difference expected by the followers of Aristotle. The behavior of falling objects was never really understood until Newton announced his second law of motion.

Isaac Newton truly changed our way of seeing the world.

Figure 5.13 ▲
A stroboscopic photo of a golf ball and a Styrofoam® ball falling in air. The weight of the heavier golf ball is more effective in overcoming air resistance, so its acceleration is greater. Will both ultimately reach a terminal speed? Which will do so first? Why?

■ Question

If the force of air resistance is the same for a falling baseball and a falling tennis ball, which will have the greater acceleration?

■ Answer

Don't say "the same"! It's true the air resistance is the same for each, but that doesn't mean the net force is the same for each, or that the ratio of net force to mass is the same for each. The heavier baseball has the greater net force, and greater net force per mass, just as the heavier parachutist previously considered. Convince yourself of this by considering the upper limit of air resistance, that is, when air resistance is equal to the weight of the tennis ball. What would the tennis ball's acceleration be? It would be zero. But when this amount of air resistance acts on the heavier baseball, do you see that the baseball still accelerates? And with more thought, do you see that the baseball has the greater acceleration even when the air resistance is less than the weight of the tennis ball?

5 Chapter Review

Concept Summary

An object accelerates—changes speed and/or direction—when a net force acts on it.

- The acceleration of an object is directly proportional to the net force acting on it.
- The acceleration of an object is inversely proportional to the mass of the object.
- Acceleration equals net force divided by mass.
- Acceleration is in the same direction as the net force.

When an object moves with constant velocity while an applied force acts on it, an equal and opposite force, usually friction, must also act to balance the applied force.

The application of a force over an area produces pressure.

- When the force is perpendicular to the surface area, the pressure equals the force divided by the area over which it acts.

The acceleration of all objects in free fall is the same, regardless of their mass.

- When air resistance is present, a falling object accelerates only until it reaches its terminal speed.
- At terminal speed, the force of air resistance balances the force of gravity.

Important Terms

air resistance (5.4)
fluid (5.4)
inversely (5.2)
Newton's second law (5.3)
pascal (5.5)
pressure (5.5)
terminal speed (5.7)
terminal velocity (5.7)

Review Questions

1. Distinguish between the relationship that defines acceleration and the relationship that states how it is produced. (5.1)

2. What is meant by the net force that acts on an object? (5.1)

3. Suppose a cart is being moved by a certain net force. If the net force is doubled, by how much does the cart's acceleration change? (5.1)

4. Suppose a cart is being moved by a certain net force. If a load is dumped into the cart so its mass is doubled, by how much does the acceleration change? (5.2)

5. Distinguish between the concepts *directly proportional* and *inversely proportional.* Support your statement with examples. (5.1–5.2)

6. State Newton's second law in words and then in the form of an equation. (5.3)

7. How much force does a 20 000-kg rocket develop to accelerate 1 m/s^2? (5.3)

8. What is the cause of friction, and in what direction does it act with respect to the motion of a sliding object? (5.4)

9. If the force of friction acting on a sliding crate is 100 N, how much force must be applied to maintain a constant velocity? What will be the net force acting on the crate? What will be the acceleration? (5.4)

10. Distinguish between force and pressure. (5.5)

11. Which produces more pressure on the ground, a person standing up or the same person lying down? (5.5)

12. The force of gravity is twice as great on a 2-kg rock as on a 1-kg rock. Why does the 2-kg rock not fall with twice the acceleration? (5.6)

13. Why do a coin and a feather in a vacuum tube fall with the same acceleration? (5.7)

14. Why do a coin and a feather fall with different accelerations in the presence of air? (5.7)

15. How much air resistance acts on a 100-N bag of nails that falls at its terminal speed? (5.7)

16. How do the air resistance and the weight of a falling object compare when terminal speed is reached? (5.7)

17. All other things being equal, why does a heavy sky diver have a terminal speed greater than a light sky diver? What can be done so that both terminal speeds are equal? (5.7)

18. What is the net force acting on a 25-N freely falling object? What is the net force when the object encounters 15 N of air resistance? When it falls fast enough to encounter 25 N of air resistance? (5.6–5.7)

Activity

1. Drop two balls of different weights from the same height, and for low speeds, they practically fall together. Will they roll together down the same inclined plane? If each is suspended from the same length string, made into a pendulum, and then displaced through the same angle, will they swing back and forth in unison? Try it and see.

Plug and Chug

1. Calculate the acceleration of a 2000-kg, single-engine airplane just before takeoff when the thrust of its engine is 500 N.

2. Calculate the acceleration of a 300 000-kg jumbo jet just before takeoff when the thrust for each of its four engines is 30 000 N.

3. **a.** Calculate the acceleration if you push with a 20-N horizontal force on a 2-kg block on a horizontal friction-free air table.

   b. What acceleration occurs if the friction force is 4 N?

4. Calculate the horizontal force that must be applied to produce an acceleration of 1 g for a 1-kg puck on a horizontal friction-free air table.

5. Calculate the horizontal force that must be applied to produce an acceleration of 1.8 g for a 1.2-kg puck on a horizontal friction-free air table.

Think and Explain

1. What is the difference between saying that one quantity is proportional to another and saying it is equal to another?

2. If an object has no acceleration, can you conclude that no forces are exerted on it? Explain.

3. What is the acceleration of a rock at the top of its trajectory when thrown straight upward? Explain whether or not the answer is zero by using the equation $a = F/m$ as a guide to your thinking.

4. A rocket fired from its launching pad not only picks up speed, but its acceleration also increases significantly as firing continues. Why is this so? (*Hint:* About 90% of the mass of a newly launched rocket is fuel.)

5. When blocking in football, why does a defending lineman often attempt to get his body under that of his opponent and push upward? What effect does this have on the friction force between the opposing lineman's feet and the ground?

6. Why does a sharp knife cut better than a dull knife?

7. An aircraft gains speed during takeoff due to the constant thrust of its engines. When is the acceleration during takeoff greatest—at the beginning of the run along the runway or just before the aircraft lifts into the air? Think, then explain.

8. As a sky diver falls faster and faster through the air (before reaching terminal speed), does the net force on her increase, decrease, or remain unchanged? Does her acceleration increase, decrease, or remain unchanged? Defend your answers.

9. After she jumps, a sky diver reaches terminal speed after 10 seconds. Does she gain more speed during the first second of fall or the ninth second of fall? Compared with the first second of fall, does she fall a greater or a lesser distance during the ninth second?

10. A regular tennis ball and another one filled with heavy sand are dropped at the same time from the top of a high building. Your friend says that even though air resistance is present, both balls should hit the ground at the same time because they are the same size and "plow through" the same amount of air. What do you say?

Think and Solve

1. A horizontal force of 100 N is required to push a crate across a factory floor at a constant speed. What is the net force acting on the crate? What is the force of friction acting on the crate?

2. If a four-engine jet accelerates down the runway at 2 m/s^2 and one of the jet engines fails, how much acceleration will the other three produce?

3. What will be the acceleration of a sky diver when air resistance is half the weight of the sky diver?

4. If a loaded truck that can accelerate at 1 m/s^2 loses its load and has three-fourths of the original mass, what acceleration can it attain from the same driving force?

5. A 10-kg mass on a horizontal friction-free air track is accelerated by a string attached to another 10-kg mass hanging vertically from a pulley as shown. What is the force due to gravity in newtons of the hanging 10-kg mass? What is the acceleration of the system of both masses?

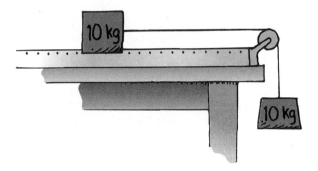

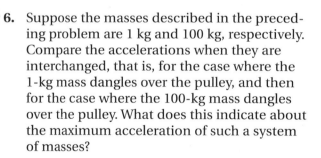

6. Suppose the masses described in the preceding problem are 1 kg and 100 kg, respectively. Compare the accelerations when they are interchanged, that is, for the case where the 1-kg mass dangles over the pulley, and then for the case where the 100-kg mass dangles over the pulley. What does this indicate about the maximum acceleration of such a system of masses?

6 Newton's Third Law of Motion—Action and Reaction

Every foot that pushes downward on the earth is at the same time pushed upward by the earth—that's the law!

Figure 6.1 ▲
When you push on the wall, the wall pushes on you.

Figure 6.2 ▲
The interaction that drives the nail is the same as the one that halts the hammer.

If you lean over too far, you'll fall. But if you lean over with your hand outstretched and make contact with a wall, you can do so without falling. When you push against the wall, it pushes back on you. That's why you are supported. Ask your friends why you don't topple over. How many will answer, "Because the wall is pushing on you and holding you in place"? Probably not very many people, unless they're physics types, realize that walls can push on us every bit as much as we push on them.*

6.1 Forces and Interactions

In the simplest sense, a force is a push or a pull. Looking closer, however, Newton realized that a force is not a thing in itself but part of a mutual action, an **interaction**, between one thing and another. For example, consider the interaction between a hammer and a nail. A hammer exerts a force on the nail and drives it into a board. But this force is only half the story, for there must also be a force exerted on the hammer to halt it in the process. What exerts this force? The nail does! Newton reasoned that while the hammer exerts a force on the nail, the nail exerts a force on the hammer. So, in the interaction between the hammer and the nail, there are a pair of forces, one acting on the nail and the other acting on the hammer. Such observations led Newton to his third law: the law of action and reaction.

---

* The terms *push* and *pull* usually invoke the idea of a living thing exerting a force. So, strictly speaking, to say "the wall pushes on you" is to say "the wall exerts a force as though it were pushing on you." As far as these mutual forces are concerned, there is no observable difference between the force exerted by you, a living being, and the force exerted by a wall, a nonliving object.

6.2 Newton's Third Law

Newton's third law states

> Whenever one object exerts a force on a second object, the second object exerts an equal and opposite force on the first object.

One force is called the **action force.** The other force is called the **reaction force.** It doesn't matter which force we call *action* and which we call *reaction.* The important thing is that they are coparts of a single interaction and that neither force exists without the other. They are equal in strength and opposite in direction. Newton's third law is often stated: "To every action there is always an equal opposing reaction."

In every interaction, the forces always occur in pairs. For example, you interact with the floor when you walk on it. You push against the floor, and the floor simultaneously pushes against you. Likewise, the tires of a car interact with the road to produce the car's motion. The tires push against the road, and the road simultaneously pushes back on the tires. When swimming, you interact with the water. You push the water backward, and the water pushes you forward. There are a pair of forces acting in each interaction. Notice that the interactions in these examples depend on friction. For example, a person trying to walk on ice, where friction is minimal, may not be able to exert an action force against the ice. Without the action force there cannot be a reaction force, and thus there is no resulting forward motion.

Figure 6.3 ▲
What happens to the boat when she jumps to shore?

■ Questions

1. Does a stick of dynamite contain force?

2. A car accelerates along a road. Strictly speaking, what is the force that moves the car?

6.3 Identifying Action and Reaction

Sometimes the identity of the pair of action and reaction forces in an interaction is not immediately obvious. For example, what are the

Figure 6.4 ▲
Does the dog wag the tail or does the tail wag the dog? Or both?

■ Answers

1. No. Force is not something an object has, like mass. Force is an interaction between one object and another. An object may possess the capability of exerting a force on another object, but it cannot possess force as a thing in itself. Later we will see that something like a stick of dynamite possesses *energy.*

2. It is the road that pushes the car along. Really! Except for air resistance, only the road provides a horizontal force on the car. How? The rotating tires push back on the road (action). The road simultaneously pushes forward on the tires (reaction). The next time you see a car moving along a road, tell your friends that the road pushes the car along. If at first they don't believe you, convince them that there is more to the physical world than meets the eye of the casual observer. Turn them on to some physics.

Figure 6.5 ▶
Force-pair between object A and object B. Note that when action is *A exerts force on B,* the reaction is simply *B exerts force on A.*

action and reaction forces in the case of a falling boulder? You might say that the earth's gravitational force on the boulder is the action force, but can you identify the reaction force? Is it the weight of the boulder? No, weight is simply another name for the force of gravity. Is it caused by the ground where the boulder lands? No, the ground does not act on the boulder until the boulder hits it.

It turns out that there is a simple recipe for treating action and reaction forces. First identify the interaction. Let's say one object, A, interacts with another object, B. The action and reaction forces can then be stated in the form

Action: Object A exerts a force on object B.

Reaction: Object B exerts a force on object A.

This is easy to remember. Just identify interacting objects A and B, and if the action is A on B, the reaction is simply B on A. So, in the case of the falling boulder, the interaction during the fall is the gravitational attraction between the boulder and the earth. If we call the *action* the earth exerting a force on the boulder, then the *reaction* is the boulder simultaneously exerting a force on the earth.

■ Question

We know that the earth pulls on the moon. Does the moon also pull on the earth? If so, which pull is stronger?

ACTION: TIRE PUSHES ROAD   REACTION: ROAD PUSHES TIRE

ACTION: ROCKET PUSHES GAS   REACTION: GAS PUSHES ROCKET

ACTION: EARTH PULLS BALL

REACTION: BALL PULLS EARTH

■ Answer

Yes, in the interaction between the earth and the moon, the earth and moon pull simultaneously on each other. The earth pulls on the moon while the moon pulls on the earth. Both pulls make up an action-reaction pair, are opposite in direction to each other, and have equal strength.

6.4 Action and Reaction on Different Masses

Interestingly enough, in the interaction between the boulder and the earth, the boulder pulls up on the earth with as much force as the earth pulls down on the boulder. The forces are equal in strength and opposite in direction. We say the boulder falls to the earth. Could we also say the earth falls to the boulder? The answer is yes, but the distance the earth falls is much less. Although the pair of forces between the boulder and the earth are the same, the masses are quite unequal. Recall that Newton's second law states that acceleration is not only proportional to the net force, but it is also inversely proportional to the mass. Because the earth has a huge mass, we don't sense its infinitesimally small acceleration. Although the earth's acceleration is negligible, strictly speaking it does move up toward the falling boulder. So when you step off a curb, the street actually comes up a tiny bit to meet you!

A similar, but less exaggerated, example occurs during the firing of a rifle. When the rifle is fired, there is an interaction between the rifle and the bullet. The force the rifle exerts on the bullet is exactly equal and opposite to the force the bullet exerts on the rifle, so the rifle "kicks." On first consideration, you might expect the rifle to kick more than it does, or you might wonder why the bullet moves so fast compared with the rifle. According to Newton's second law, we must also consider the masses.

Figure 6.6 ▲
The earth is pulled up by the boulder with just as much force as the boulder is pulled down by the earth.

◀ **Figure 6.7**
The force exerted against the recoiling rifle is just as great as the force that drives the bullet along the barrel. Why, then, does the bullet undergo more acceleration than the rifle?

Let F represent both the action and reaction forces; m, the mass of the rifle; and m, the mass of the bullet. Different size symbols are used to indicate the differences in relative masses and the resulting accelerations. The acceleration of the bullet and rifle are

$$\text{Bullet: } \frac{F}{m} = a \qquad \text{Rifle: } \frac{F}{m} = a$$

Do you see why the change in the velocity of the bullet is great compared with the change in velocity of the rifle? A given force exerted on a small mass produces a greater acceleration than the same force exerted on a large mass.

If we extend the basic idea of a rifle recoiling from the bullet it fires, we can understand rocket propulsion. Consider a machine gun recoiling each time a bullet is fired. If the machine gun is fastened on a vertical wire so that it is free to slide as shown in Figure 6.8, it accelerates upward as bullets are fired downward. A rocket accelerates in much the same way—it continually recoils from the exhaust gases ejected from its engine. Each molecule of exhaust gas acts like a tiny molecular bullet shot downward from the rocket.

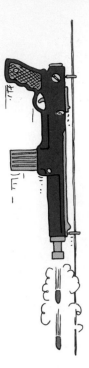

Figure 6.8 ▲
The machine gun recoils from the bullets it fires and climbs upward.

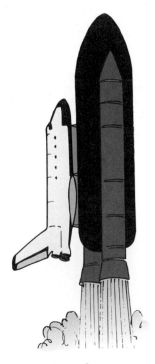

Figure 6.9 ▲
The rocket recoils from the "molecular bullets" it fires and climbs upward.

A common misconception is that a rocket is propelled by the impact of exhaust gases against the atmosphere. In fact, before the advent of rockets, it was commonly thought that sending a rocket to the moon was impossible because of the absence of an atmosphere for the rocket to push against. This is like saying a gun won't kick unless the bullet has air to push against. This is not true! Both the rocket and recoiling gun accelerate because of the reaction forces created by the "bullets" they fire—air or no air. In fact, rockets work better above the atmosphere where there is no air drag.

Using Newton's third law, we can understand how a helicopter gets its lifting force. The whirling blades are shaped to force air particles downward (action), and the air forces the blades upward (reaction). This upward reaction force is called *lift*. When lift equals the weight of the craft, the helicopter hovers in midair. When lift is greater, the helicopter climbs upward.

Birds and airplanes also fly because of action and reaction forces. When a bird is soaring, the shape of its wings deflects air downward. The air in turn pushes the bird up. The slightly tilted wings of an airplane also deflect oncoming air downward and produce lift. Airplanes must continuously push air downward to maintain lift and remain airborne. This continuous supply of air is produced by the forward motion of the aircraft, which results from jets or propellers that push air backward. When the engines push air back, the air in turn pushes the engines and the plane forward. We will learn later how the curved surface of an airplane wing enhances the lifting force.

■ Question

Can you identify the action and reaction forces of an object falling in the vacuum of outer space?

6.5 Do Action and Reaction Forces Cancel?

Since action and reaction forces are equal and opposite, why don't they cancel to zero? To answer this interesting question, we first consider the system involved. Consider the force pair between the apple and the orange in Figure 6.10. Suppose we ignore the apple and everything else and concentrate only on the orange. We draw an imaginary circle around the orange and call it the *system*. The pull from the apple supplies a force on the system, and the system

■ Answer

To identify a pair of action-reaction forces in any situation, first identify the pair of interacting objects involved. In this case, the interaction is the gravitational attraction between the falling object and another object, possibly a distant planet. So the planet pulls down on the object *(action)*, while the object pulls up on the planet *(reaction)*.

 **Figure 6.10**
An apple pulls on an orange and the orange accelerates—period! The orange pulls back on the apple, but this affects the apple and not the orange.

accelerates. In this case, the interaction is between the system (the orange) and something external (the apple), so the action and reaction forces don't cancel. The fact that the orange simultaneously exerts a force on the apple, which is external to the system, affects the apple but not the orange.

However, if we consider the system to be both the orange and the apple, the force pair is internal to the system. In this case, the forces do cancel each other. Similarly, the many force pairs between the molecules in a baseball may hold the ball together into a cohesive solid, but they play no role at all in accelerating the ball. A force external to the ball is needed to accelerate it. Similarly, a force external to both the apple and orange is needed to produce acceleration of the entire system (like friction of the floor on the apple's feet). If action and reaction forces are internal to a system, they cancel each other and produce no acceleration of the system. Action and reaction forces do not cancel each other when either is external to the system being considered. If this is confusing, it may be well to note that Newton had difficulties with the third law himself.

Now, if two people kick the same ball with equal and opposite force at the same time, as shown in Figure 6.12, there are two interactions to be considered. In this case, there are two forces acting on the ball and the net force on the ball is zero—but not for the single interaction with one foot.

Figure 6.11 ▲
A acts on B and B accelerates.

Figure 6.12 ▲
Both A and C act on B. They can cancel each other so B does not accelerate.

■ Question

Suppose a friend who hears about Newton's third law says that you can't move a football by kicking it because the reaction force by the kicked ball would be equal and opposite to your kicking force. The net force would be zero, so no matter how hard you kick, the ball won't move! What do you say to your friend?

■ Answer

Tell your friend that if you kick a football, it will accelerate. Does this acceleration contradict Newton's third law? No! Your kick acts on the ball. No other force has been applied to the ball. The net force on the ball is very real, and the ball accelerates. What about the reaction force? Aha! That force doesn't act on the ball; it acts on your foot. The reaction force decelerates your foot as it makes contact with the ball. Tell your friend that you can't cancel a force on the ball with a force on your foot.

Rhonda

6.6 The Horse-Cart Problem

A situation similar to the kicked football is shown in the comic strip "Horse Sense." Here we think of the horse as believing its pull on the cart will be canceled by the opposite and equal pull by the cart on the horse, thus making acceleration impossible. This is a classic problem that stumps many college students. By thinking carefully, *you* can understand it.

The horse-cart problem can be looked at from three different points of view. First, consider the point of view of the farmer, who is concerned with getting his cart (the cart system) to market. Then, there is the point of view of the horse (the horse system). Finally, there is the point of view of the horse and cart together (the horse-cart system).

From the farmer's point of view, the only concern is with the force that is exerted on the cart system. The net force on the cart, divided by the mass of the cart, will produce a very real acceleration. The farmer doesn't care about the reaction on the horse.

Now look at the horse system. It's true that the opposite reaction force by the cart on the horse restrains the horse. Without this force, the horse could freely gallop to the market. This force tends to hold the horse back. So how does the horse move forward? The horse moves forward by interacting with the ground. When the horse pushes backward on the ground, the ground simultaneously pushes forward on the horse. If the horse pushes the ground with a greater force than it pulls on the cart, there will be a net force on the horse and acceleration occurs. When the cart is up to speed, the horse need only push against the ground with enough force to offset the friction between the cart wheels and the ground.

Finally, look at the horse-cart system as a whole. From this viewpoint, the pull of the horse on the cart and the reaction of the cart on the horse are internal forces, or forces that act and react within the system. They contribute nothing to the acceleration of the horse-cart system. They cancel and can be neglected. To move across the ground, there must be an interaction between the horse-cart system and the ground. For example, if your car is stalled, you can't get it

Figure 6.13 ▲
All the pairs of forces that act on the horse and cart are shown: (1) the pull *P* of the horse and the cart on each other; (2) the push *F* of the horse and the ground on each other; and (3) the friction *f* between the cart wheels and the ground. Notice that there are two forces applied to the cart and to the horse. Can you see that the acceleration of the horse-cart system is due to the net force *F − f*?

■ Questions

1. What is the net force that acts on the cart in Figure 6.13? On the horse? On the ground?

2. Once the horse gets the cart up to the desired speed, must the horse continue to exert a force on the cart?

■ Answers

1. The net force on the cart is *P − f;* on the horse, *F − P;* on the ground, *F − f.*

2. Yes, but only enough to counteract wheel friction and air resistance.

moving by sitting inside and pushing on the dashboard. You must interact with the ground outside. You must get outside and make the ground push the car. The horse-cart system is similar. It is the outside reaction by the ground that pushes the system.

Figure 6.14 ▲
If you hit the wall, it will hit you equally hard.

6.7 Action Equals Reaction

This chapter began with a discussion of how a wall pushes back on you when you push against it. Suppose that for some reason, you punch the wall. Bam! Your hand is hurt. Your friends see your damaged hand and ask what happened. What can you say truthfully? You can say that the wall hit your hand. How hard did the wall hit your hand? It hit just as hard as you hit the wall. You cannot hit the wall any harder than the wall can hit back on you.

Hold a sheet of paper in midair and tell your friends that the heavyweight champion of the world could not strike the paper with a force of 200 N (45 pounds). You are correct, because a 200-N interaction between the champ's fist and the sheet of paper in midair isn't possible. The paper is not capable of exerting a reaction force of 200 N, and you cannot have an action force without a reaction force. Now, if you hold the paper against the wall, that's a different story. The wall will easily assist the paper in providing 200 N of reaction force, and more if needed!

For every interaction between things, there is always a pair of oppositely directed forces that are equal in strength. If you push hard on the world, for example, the world pushes hard on you. If you touch the world gently, the world will touch you gently in return. The way you touch others is the way others touch you.

Figure 6.15 ▶
You cannot touch without being touched—Newton's third law.

6 Chapter Review

Concept Summary

An interaction between two things produces a pair of forces.

- Interacting things exert a force on each other.

- The two interacting forces are called the action force and the reaction force.

- Action and reaction forces are equal in strength and opposite in direction.

Important Terms

action force (6.2)   Newton's third law (6.2)

interaction (6.1)   reaction force (6.2)

Review Questions

1. In the interaction between a hammer and the nail it hits, is a force exerted on the nail? On the hammer? How many forces occur in this interaction? (6.1)

2. When a hammer exerts a force on a nail, how does the amount of force compare with that of the nail on the hammer? (6.1)

3. When you walk along a floor, what pushes you along? (6.2)

4. When swimming, you push the water backward—call this *action*. What is the reaction force? (6.2)

5. If the action is a bowstring acting on an arrow, identify the reaction force. (6.3)

6. When you jump up, the world really does recoil downward. Why can't this motion of the world be noticed? (6.4)

7. When a rifle is fired, how does the size of the force of the rifle on the bullet compare with the force of the bullet on the rifle? How does the acceleration of the rifle compare with that of the bullet? Defend your answer. (6.4)

8. How can a rocket be propelled above the atmosphere where there is no air to "push against"? (6.4)

Questions 9–11 refer to the apple-orange system in Figure 6.10. Consider only horizontal forces.

9. In the interaction between an apple and an orange, how many forces are exerted on the apple? On the orange? Are these forces equal in strength? Are these forces opposite in direction? (6.5)

10. Consider the orange system. Do action and reaction forces cancel each other in the orange system? Does the orange system accelerate? (6.5)

11. Consider the orange-apple system. Do action and reaction forces cancel each other in this system? Do the orange and apple accelerate away from each other, or do they remain together? (6.5)

Questions 12–15 refer to the horse-cart system in Figure 6.13. Consider only horizontal forces.

12. a. In the horizontal direction, how many forces are exerted on the cart?

    b. What is the net horizontal force on the cart? (6.6)

13. a. How many horizontal forces are exerted on the horse?

    b. What is the net horizontal force on the horse?

    c. How many horizontal forces are exerted by the horse on other objects? (6.6)

14. a. How many horizontal forces are exerted on the horse-cart system?

    b. What is the net horizontal force on the horse-cart system? (6.6)

15. In order to increase its speed, why must the horse push harder against the ground than it pulls on the wagon? (6.6)

16. If you hit a wall with a force of 200 N, how much force is exerted on you? (6.7)

17. Why can't you hit a feather in midair with a force of 200 N? (6.7)

18. How does the saying "You get what you give" relate to Newton's third law? (6.7)

Think and Explain

1. Your weight is the result of the gravitational force of the earth on your body. What is the corresponding reaction force?

2. If you walk on a log that is floating in the water, the log moves backward. Why?

3. Why is it easier to walk on a carpeted floor than on a smooth, polished floor?

4. If you step off a ledge, you accelerate noticeably toward the earth because of the gravitational interaction between you and the earth. Does the earth accelerate toward you as well? Explain.

5. Suppose you're weighing yourself while standing next to the bathroom sink. Using the idea of action and reaction, explain why

the scale reading will be less when you push down on the top of the sink. Why will the scale reading be more if you pull up on the bottom of the sink?

6. When a high jumper leaves the ground, what is the source of the upward force that accelerates her? What force acts after her feet are no longer in contact with the ground?

7. What is the reaction force to an action force of 1000 N exerted by the earth on an orbiting communications satellite?

8. If action equals reaction, why isn't the earth pulled into orbit around a communications satellite?

9. If a bicycle and a massive truck have a head-on collision, upon which vehicle is the impact force greater? Which vehicle undergoes the greater change in its motion? Defend your answers.

10. A speeding bus makes contact with a bug that splatters onto the windshield. Because of the sudden force, the unfortunate bug undergoes a sudden deceleration. Is the corresponding force that the bug exerts against the windshield greater, less, or the same? Is the resulting deceleration of the bus greater than, less than, or the same as that of bug?

11. Some people used to think that a rocket could not travel to the moon because it would have no air to push against once it left the earth's atmosphere. We now know that idea was mistaken. What force propels a rocket when it is in a vacuum?

12. Since the force that acts on a bullet when a gun is fired is equal and opposite to the force that acts on the gun, does this imply a zero net force and therefore the impossibility of an accelerating bullet? Explain.

13. Suppose you exert 200 N on your refrigerator and push it across the kitchen floor at constant velocity. What friction force acts between the refrigerator and the floor? Is the friction force equal and opposite to your 200-N push? Does the friction force make up the reaction force to your push?

14. A pair of 50-N weights are attached to a spring scale as shown. Does the spring scale read 0, 50, or 100 N? *(Hint:* Would it read any differently if one of the strings were held by your hand instead of being attached to the 50-N weight?)

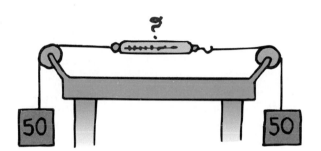

15. The strong man can withstand the tension force exerted by the two horses pulling in opposite directions. How would the tension compare if only one horse pulled and the left rope were tied to a tree? How would the tension compare if the two horses pulled in the same direction, with the left rope tied to the tree?

7 Momentum

Inertia *and* motion—that's momentum!

H ave you ever wondered how a karate expert can break a stack of cement bricks with the blow of a bare hand? Or why falling on a wooden floor hurts less than falling on a cement floor? Or why follow-through is important in golf, baseball, and boxing? To understand these things, you need to recall the concept of inertia introduced and developed when we discussed Newton's laws of motion. Inertia was discussed both in terms of objects at rest and objects in motion. In this chapter we are concerned only with the concept of inertia in motion—momentum.

7.1 Momentum

We know that it's harder to stop a large truck than a small car when both are moving at the same speed. We say the truck has more momentum than the car. By **momentum,** we mean *inertia in motion,* or more specifically, the mass of an object multiplied by its velocity.

$$momentum = mass \times velocity$$

or, in abbreviated notation,

$$momentum = mv$$

When direction is not an important factor, we can say

$$momentum = mass \times speed$$

which we still abbreviate mv.

We can see from the definition that a moving object can have a large momentum if it has a large mass, a high speed, or both. A moving truck has more momentum than a car moving at the same speed because the truck has more mass. But a fast car can have more momentum than a slow truck. And a truck at rest has no momentum at all.

Figure 7.1 ▲
A truck rolling down a hill has more momentum than a roller skate with the same speed, because the truck has more mass. But if the truck is at rest and the roller skate moves, then the skate has more momentum because only it has speed.

Can you think of a case where the roller skate and the truck shown in Figure 7.1 would have the same momentum?

7.2 Impulse Changes Momentum

If the momentum of an object changes, either the mass or the velocity or both change. If the mass remains unchanged, as is most often the case, then the velocity changes and acceleration occurs. What produces acceleration? We know the answer is *force*. The greater the force acting on an object, the greater its change in velocity, and hence, the greater its change in momentum.

How long the force acts is also important. Apply a brief force to a stalled automobile, and you produce a change in its momentum. Apply the same force over an extended period of time and you produce a greater change in the automobile's momentum. A force sustained for a long time produces more change in momentum than does the same force applied briefly. So both force and time are important in changing momentum.

The quantity *force* × *time interval* is called **impulse.** In short-hand notation

$$\text{impulse} = Ft$$

The greater the impulse exerted on something, the greater will be the change in momentum. The exact relationship is

$$\text{impulse} = \text{change in momentum}$$

or*

$$Ft = \Delta(mv)$$

The impulse-momentum relationship helps us to analyze a variety of situations where the momentum changes. Consider the familiar examples of impulse in the following cases of increasing and decreasing momentum.

■ **Answer**

The roller skate and truck can have the same momentum if the speed of the roller skate is much greater than the speed of the truck. How much greater? As many times greater as the truck's mass is greater than the roller skate's mass. Get it? For example, a 1000-kg truck backing out of a driveway at 0.01 m/s has the same momentum as a 1-kg skate going 10 m/s. Both have momentum = 10 kg m/s.

* This relationship is derived by rearranging Newton's second law to make the time factor more evident. If we equate the formula for acceleration, $a = F/m$, with what acceleration actually is, $a = \Delta v/\Delta t$, we get $F/m = \Delta v/\Delta t$. From this we derive $F\Delta t = \Delta(mv)$. Calling Δt simply t, the time interval, $Ft = \Delta(mv)$.

Case 1: Increasing Momentum

To increase the momentum of an object, it makes sense to apply the greatest force possible for as long as possible. A golfer teeing off and a baseball player trying for a home run do both of these things when they swing as hard as possible and follow through with their swing.

The forces involved in impulses usually vary from instant to instant. For example, a golf club that strikes a golf ball exerts zero force on the ball until it comes in contact with it; then the force increases rapidly as the ball becomes distorted (Figure 7.2). The force then diminishes as the ball comes up to speed and returns to its original shape. So when we speak of such impact forces in this chapter, we mean the *average* force of impact. (Be careful to distinguish between *impact* and *impulse*. Impact refers to a *force* and is measured in newtons; impulse is *impact force* × *time* and is measured in newton-seconds.)

Figure 7.2 ▲
The force of impact on a golf ball varies throughout the duration of impact.

Case 2: Decreasing Momentum

If you were in a car that was out of control and had to choose between hitting a concrete wall or a haystack, you wouldn't have to call on your knowledge of physics to make up your mind. Common sense tells you to choose the haystack. But knowing the physics helps you to understand *why* hitting a soft object is entirely different from hitting a hard one. In the case of hitting either the wall or the haystack and coming to a stop, your momentum is decreased by the same impulse. The same impulse does not mean the same

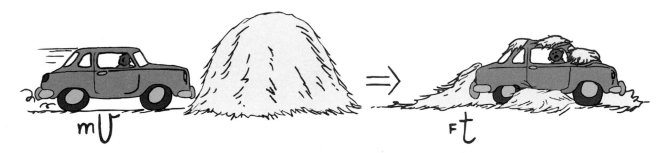

Figure 7.3 ▲
If the change in momentum occurs over a long time, the force of impact is small.

amount of force or the same amount of time; rather it means the same *product* of force and time. By hitting the haystack instead of the wall, you extend the impact time—*the time during which your momentum is brought to zero*. A longer impact time reduces the force of the impact and decreases the resulting deceleration. For example, if the time of impact is extended 100 times, the force of impact is reduced 100 times. Whenever we wish the force of impact to be small, we extend the time of impact.

We know that a padded dashboard in a car is safer than a rigid metal one and that airbags save lives. We also know that to catch a fast-moving ball safely with your bare hand, you extend your hand forward so there's plenty of room for it to move backward after making contact with the ball. When you extend the time of impact, you reduce the force of impact.

When jumping from an elevated position down to the ground, what would happen if you kept your legs straight and stiff? Ouch! Instead, you know to bend your knees when your feet make contact with the ground. By doing so you extend the time during which your momentum decreases by 10 to 20 times that of a stiff-legged, abrupt landing. The resulting force on your bones is reduced by 10 to 20 times. A wrestler thrown to the floor tries to extend his time of impact with the mat by relaxing his muscles and spreading the impact into a series of smaller ones as his foot, knee, hip, ribs, and shoulder successively hit the mat. Of course, falling on a mat is preferable to falling on a solid floor because the mat also increases the impact time.

We know a glass dish is more likely to survive if it is dropped on a carpet rather than a sidewalk because the carpet has more "give" than the sidewalk. Ask why a surface with more give makes for a safer fall and you will get a puzzled response from most people. They may simply say, "Because it gives more." However, your question is, *"Why* is a surface with more give safer for the dish?" In this case, a common explanation isn't really an explanation at all. A deeper explanation is needed.

To bring the dish or its fragments to rest, the carpet or the sidewalk must provide an impulse, which you know involves two variables—impact force and impact time. Since impact time is longer on the carpet than on the sidewalk, a smaller impact force results. The shorter impact time on the sidewalk results in a greater impact force.

Figure 7.5 ▲
In both cases the impulse provided by the boxer's jaw must counteract the momentum of the punch. (Left) When the boxer moves away from the punch, he increases the time of impact and reduces the force of impact. (Right) When the boxer unwisely moves toward the punch, the time of impact is reduced and the force of impact is increased. Ouch!

Bungee Jumping

The impulse-momentum relationship is put to a thrilling test during bungee jumping. Be glad the rubber cord stretches when the jumper's fall is brought to a halt, because the cord has to apply an impulse equal to the jumper's momentum in order to stop the jumper—hopefully above ground level.

Note how $Ft = \Delta(mv)$ applies here. The momentum, mv, we wish to change is the amount gained before the cord begins stretching. Ft is the impulse the cord supplies to reduce the momentum to zero.

Because the rubber cord stretches for a long time, a large time interval t ensures that a small average force F acts on the jumper. Elastic cords typically stretch to about twice their original length during the fall.

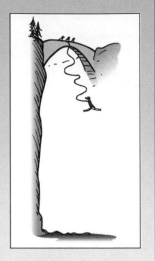

The safety net used by circus acrobats is a good example of how to achieve the impulse needed for a safe landing. The saftey net reduces the impact force on a fallen acrobat by substantially increasing the time interval of the impact.

Sometimes a difference in impact time is important even if you can't notice the give in a surface. For example, a wooden floor and a concrete floor may both seem rigid, but the wooden floor can have enough give to make quite a difference in the forces that these two surfaces exert.

■ Questions

1. When a dish falls, will the impulse be less if it lands on a carpet than if it lands on a hard floor?

2. If the boxer in Figure 7.5 is able to make the impact time five times longer by "riding" with the punch, how much will the force of impact be reduced?

■ Answers

1. No. The impulse would be the same for either surface because the same momentum change occurs for each. It is the *force* that is less for the impulse on the carpet because of the greater time of momentum change. If you answered this question incorrectly, you probably did not distinguish between impulse and impact. They sound the same, but they're not!

2. Since the time of impact increases five times, the force of impact will be reduced five times.

7.3 Bouncing

If a flower pot falls from a shelf onto your head, you may be in trouble. If it bounces from your head, you may be in more serious trouble. Why? Because impulses are greater when an object bounces. The impulse required to bring an object to a stop and then to "throw it back again" is greater than the impulse required merely to bring the object to a stop. Suppose, for example, that you catch the falling pot with your hands. You provide an impulse to reduce its momentum to zero. If you throw the pot upward again, you have to provide additional impulse. It takes a greater impulse to catch the pot *and* throw it back up than merely to catch it. This increased amount of impulse is supplied by your head if the pot bounces from it.

◀ **Figure 7.6**
Is the karate chop delivered in a short time or a long time? If the hand bounces upon impact, is the change in momentum greater? Is the impulse greater?

The fact that impulses are greater when bouncing takes place was used with great success during the California Gold Rush. The waterwheels used in gold mining operations were not very effective. A man named Lester A. Pelton saw that the problem had to do with

IMPULSE

◀ **Figure 7.7**
The Pelton Wheel. The curved blades cause water to bounce and make a U-turn, producing a large impulse that turns the wheel.

the flat paddles on the waterwheel. He designed a curve-shaped paddle that caused the incoming water to make a U-turn upon impact with the paddle. Because the water "bounced," the impulse exerted on the waterwheel was increased. Pelton patented his idea and probably made more money from his invention, the Pelton Wheel, than any of the gold miners earned. Physics can indeed make you rich!

7.4 Conservation of Momentum

From Newton's second law you know that to accelerate an object, a net force must be applied to it. This chapter says much the same thing, but in different language. If you wish to change the momentum of an object, exert an impulse on it.

In either case, the force or impulse must be exerted on the object by something outside the object. Internal forces won't work. For example, the molecular forces within a basketball have no effect on the momentum of the basketball, just as a push against the dashboard of a car you're sitting in does not affect the momentum of the car. Molecular forces within the basketball and a push on the dashboard are internal forces. They come in balanced pairs that cancel within the object. To change the momentum of the basketball or the car, an outside push or pull is required. If no outside force is present, no change in momentum is possible.

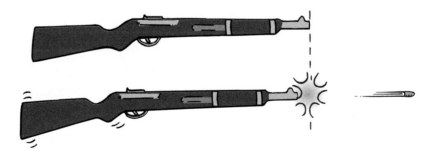

Figure 7.8 ▲
The momentum before firing is zero. After firing, the net momentum is still zero because the momentum of the rifle is equal and opposite to the momentum of the bullet.

Consider the rifle being fired in Figure 7.8. The force on the bullet inside the rifle barrel is equal and opposite to the force causing the rifle to recoil. Recall Newton's third law about action and reaction forces. These forces are internal to the system comprising the rifle and the bullet, so they don't change the momentum of the rifle-bullet system. Before the firing, the system is at rest and the momentum is zero. After the firing the net momentum, or total momentum, is *still* zero. Net momentum is neither gained nor lost. Let's consider the effects of internal and external forces carefully.

Momentum, like the quantities velocity and force, has both direction and magnitude. It is a *vector quantity.* Like velocity and force, momentum can be canceled. So, although the bullet in the preceding example gains momentum when fired and the recoiling rifle gains momentum in the opposite direction, the rifle-bullet *system* gains none. The momenta (plural form of momentum) of the bullet and the rifle are equal in magnitude and opposite in direction. Therefore, these momenta cancel each other out for the system as a whole. No external force acted on the system before or during firing. Since no net force acts on the system, there is no net impulse on the system and there is no net change in the momentum. You can see that *if no net force or net impulse acts on a system, the momentum of that system cannot change.*

In every case, the momentum of a system cannot change unless it is acted on by external forces. A system will have the same momentum before some internal interaction as it has after the inter-action occurs. When momentum, or any quantity in physics, does not change, we say it is **conserved.** The idea that momentum is conserved when no external force acts is elevated to a central law of mechanics, called the **law of conservation of momentum,** which states

> In the absence of an external force, the momentum of a system remains unchanged.

If a system undergoes changes wherein all forces are internal, as for example, in atomic nuclei undergoing radioactive decay, cars colliding, or stars exploding, the net momentum of the system before and after the event is the same.

■ Questions

1. Newton's second law states that if no net force is exerted on a system, no acceleration occurs. Does it follow that no change in momentum occurs?

2. Newton's third law states that the force a rifle exerts on a bullet is equal and opposite to the force the bullet exerts on the rifle. Does it follow that the *impulse* the rifle exerts on the bullet is equal and opposite to the *impulse* the bullet exerts on the rifle?

■ Answers

1. Yes, because no acceleration means that no change occurs in velocity or in momentum (mass × velocity). Another line of reasoning is simply that no net force means there is no net impulse and thus no change in momentum.

2. Yes, because the rifle acts on the bullet and the bullet reacts on the rifle during the same *time* interval. Since time is equal and force is equal and opposite for both, the impulse, *Ft,* is also equal and opposite for both. Impulse is a vector quantity and can be canceled.

Skateboards and Momentum

Stand at rest on a skateboard and throw a massive object forward or backward. Notice that you recoil in the opposite direction. The recoil is understandable because the momentum before the throw is zero and the net momentum just after the throw is also zero.

Your recoil momentum is equal and opposite to the momentum of the thrown object. Observe that momentum is conserved. Now repeat the throwing motion with the same object, but this time don't let go of it. Do you still recoil? Explain.

Activity

7.5 Collisions

The collision of objects clearly shows the conservation of momentum. Whenever objects collide in the absence of external forces, the net momentum of both objects before collision equals the net momentum of both objects after collision.

$$\text{net momentum}_{\text{before collision}} = \text{net momentum}_{\text{after collision}}$$

Elastic Collisions

When a moving billiard ball collides head-on with a ball at rest, the first ball comes to rest and the second ball moves away with a velocity equal to the initial velocity of the first ball. We see that

Figure 7.9 ▼

Elastic collisions. (a) A moving ball strikes a ball at rest. (b) A head-on collision between two moving balls. (c) A collision of two balls moving in the same direction. In all cases, momentum is simply transferred or redistributed without loss or gain.

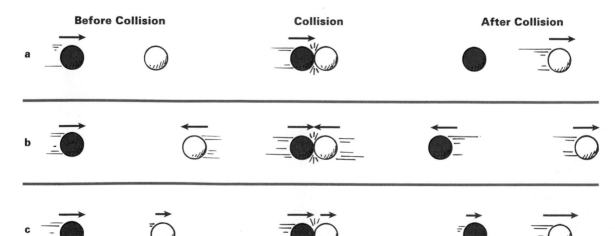

momentum is transferred from the first ball to the second ball. When objects collide without being permanently deformed and without generating heat, the collision is said to be an **elastic collision.** Colliding objects bounce perfectly in perfect elastic collisions, as shown in Figure 7.9. Note that the sum of the momentum vectors are the same before and after each collision.

Inelastic Collisions

Momentum conservation holds true even when the colliding objects become distorted and generate heat during the collision. Whenever colliding objects become tangled or couple together, an **inelastic collision** occurs. The freight train cars in Figure 7.10 provide an example. Suppose the freight cars are of equal mass m, and that one car moves at 4 m/s toward the other car that is at rest. Can you predict the velocity of the coupled cars after impact? From the conservation of momentum,

net momentum $_{\text{before collision}}$ = net momentum $_{\text{after collision}}$

Or in equation form,

$$(\text{net } mv)_{\text{before}} = (\text{net } mv)_{\text{after}}$$

$$(m)(4 \text{ m/s}) + (m)(0 \text{ m/s}) = (2m)(v_{\text{after}})$$

Since twice as much mass is moving after the collision, can you see that the velocity, v_{after}, must be one half of 4 m/s? Solving for the velocity after the collision, we find $v_{\text{after}} = 2$ m/s in the same direction as the velocity before the collision, v_{before}. The initial momentum is shared by both cars without loss or gain. Momentum is conserved.

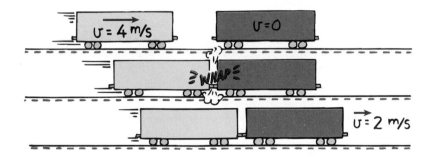

Figure 7.10 ▲
Inelastic collision. The momentum of the freight car on the left is shared with the freight car on the right.

Most collisions usually involve some external force. Billiard balls do not continue indefinitely with the momentum imparted to them. The moving balls encounter friction with the table and the air. These external forces are usually negligible during the collision, so the net momentum does not change during collision. The net momentum of two colliding trucks is the same before and just after the collision. As the combined wreck slides along the pavement, friction provides an

Figure 7.11 ▶
Conservation of momentum is nicely demonstrated with the use of an air track. Many small air jets provide a nearly frictionless cushion of air for the gliders to slide on.

impulse to decrease its momentum. Similarly, a pair of space vehicles docking in orbit have the same net momentum just before and just after contact. Since there is no air resistance in space, the combined momentum of the space vehicles after docking is then changed only by gravity.

■ Questions

Refer to the gliders on the air track in Figure 7.11 to answer the questions.

1. Suppose both gliders have the same mass. They move toward each other at the same speed and experience an elastic collision. Describe the motion after the collision.

2. Suppose both gliders have the same mass and stick together when they collide. The gliders move toward each other at equal speed. Describe their motion after the collision.

3. Suppose one glider is at rest and is loaded so that it has three times the mass of the moving glider. Again, the gliders stick together when they collide. Describe their motion after the collision.

■ Answers

1. Since the collision is elastic, the gliders reverse directions after colliding and move away from each other at a speed equal to their initial speed.

2. Before the collision, the gliders have equal and opposite momenta because their equal masses are moving in opposite directions at the same speed. The net momentum of the two gliders as a system is zero. Since momentum is conserved, their net momentum after sticking together must also be zero. They slam to a dead halt.

3. Before collision, the net momentum equals the momentum of the unloaded, moving glider. After the collision, the net momentum is the same as before, but now the gliders are stuck together and moving as a single unit. The mass of the stuck-together gliders is four times that of the unloaded glider. Thus, the postcollision velocity of the stuck-together gliders is one-fourth of the unloaded glider's velocity before collision. This velocity is in the same direction as before, since the direction as well as the amount of momentum is conserved.

Consider a 6-kg fish that swims toward and swallows a 2-kg fish that is at rest. If the larger fish swims at 1 m/s, what is its velocity immediately after lunch? Momentum is conserved from the instant before lunch until the instant after (in so brief an interval, water resistance does not have time to change the momentum), so we can write

$$\text{net momentum}_{\text{before lunch}} = \text{net momentum}_{\text{after lunch}}$$

$$(\text{net } mv)_{\text{before}} = (\text{net } mv)_{\text{after}}$$

$$(6 \text{ kg})(1 \text{ m/s}) + (2 \text{ kg})(0 \text{ m/s}) = (6 \text{ kg} + 2 \text{ kg})(v_{\text{after}})$$

$$6 \text{ kg·m/s} = (8 \text{ kg})(v_{\text{after}})$$

$$v_{\text{after}} = \frac{6 \text{ kg·m/s}}{8 \text{ kg}}$$

$$v_{\text{after}} = \frac{3}{4} \text{ m/s}$$

We see that the small fish has no momentum before lunch because its velocity is zero. Using simple algebra we see that after lunch the combined mass of the two-fish system is 8 kg and its speed is $\frac{3}{4}$ m/s in the same direction as the large fish's direction before lunch.

Suppose the small fish is not at rest but is swimming toward the large fish at 2 m/s. Now we have opposing directions. If we consider the direction of the large fish as positive, then the velocity of the small fish is –2 m/s. We pay attention to the negative sign and see that

$$(\text{net } mv)_{\text{before}} = (\text{net } mv)_{\text{after}}$$

$$(6 \text{ kg})(1 \text{ m/s}) + (2 \text{ kg})(-2 \text{ m/s}) = (6 \text{ kg} + 2 \text{ kg})(v_{\text{after}})$$

$$(6 \text{ kg·m/s}) + (-4 \text{ kg·m/s}) = (8 \text{ kg})(v_{\text{after}})$$

$$\frac{2 \text{ kg·m/s}}{8 \text{ kg}} = v_{\text{after}}$$

$$v_{\text{after}} = \frac{1}{4} \text{ m/s}$$

The negative momentum of the small fish is very effective in slowing the large fish. If the small fish were swimming at –3 m/s, then both fish would have equal and opposite momenta. Zero momentum before lunch would equal zero momentum after lunch, and both fish would come to a halt.

More interestingly, suppose the small fish swims at –4 m/s.

$$(\text{net } mv)_{\text{before}} = (\text{net } mv)_{\text{after}}$$

$$(6 \text{ kg})(1 \text{ m/s}) + (2 \text{ kg})(-4 \text{ m/s}) = (6 \text{ kg} + 2 \text{ kg})(v_{\text{after}})$$

$$(6 \text{ kg·m/s}) + (-8 \text{ kg·m/s}) = (8 \text{ kg})(v_{\text{after}})$$

$$\frac{-2 \text{ kg·m/s}}{8 \text{ kg}} = v_{\text{after}}$$

$$v_{\text{after}} = -\frac{1}{4} \text{ m/s}$$

The minus sign tells us that after lunch the two-fish system moves in a direction opposite to the large fish's direction before lunch.

Perfectly elastic collisions are not common in the everyday world. We find in practice that some heat is generated during collisions. Drop a ball and after it bounces from the floor, both the ball and the floor are a bit warmer. Even a dropped superball will not bounce to its initial height. At the microscopic level, however, perfectly elastic collisions are commonplace. For example, electrically charged particles bounce off one another without generating heat; they don't even touch in the classic sense of the word. Later chapters will show that the concept of touching needs to be considered differently at the atomic level.

7.6 Momentum Vectors

Momentum is conserved even when interacting objects don't move along the same straight line. To analyze momentum in any direction, we use the vector techniques we've previously learned. We'll look at momentum conservation involving angles by briefly considering the three following examples.

Notice in Figure 7.12 that the momentum of car A is directed due east and that of car B is directed due north. If their momenta are

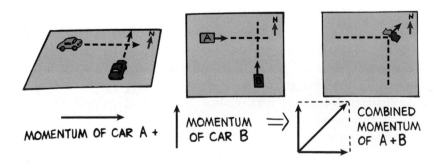

Figure 7.12 ▲
Momentum is a vector quantity. The momentum of the wreck is equal to the vector sum of the momenta of car A and car B before the collision.

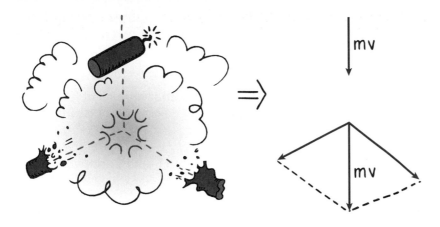

Figure 7.13 ▲
When the firecracker bursts, the vector sum of the momenta of its fragments add up to the firecracker's momentum just before bursting.

equal in magnitude, after colliding their combined momentum will be in a northeast direction with a magnitude $\sqrt{2}$ times the momentum either vehicle had before the collision (just as the diagonal of a square is $\sqrt{2}$ times the length of a side).

Figure 7.13 shows a falling firecracker that explodes into two pieces. The momenta of the fragments combine by vector rules to equal the original momentum of the falling firecracker.

Figure 7.14 shows tracks made by subatomic particles in a spark chamber. The mass of these particles can be computed by applying both the conservation of momentum and conservation of energy laws—the conservation of energy law will be discussed in the next chapter. The conservation laws are extremely useful to experimenters in the atomic and subatomic realms. A very important feature of their usefulness is that forces do not show up in the equations. Forces in collisions, however complicated, are not a concern.

Conservation of momentum and, as the next chapter will discuss, conservation of energy, are the two most powerful tools of mechanics. Their application yields detailed information that ranges from understanding the interactions of subatomic particles to entire galaxies.

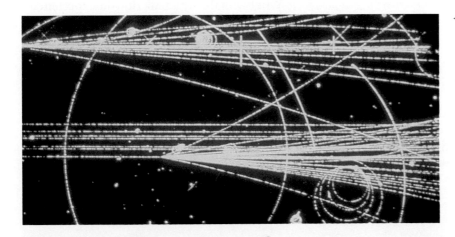

◀ **Figure 7.14**
Momentum is conserved for the high-speed elementary particles, as shown by the tracks they leave in a bubble chamber. The relative mass of the particles is determined by, among other things, the paths they take after collision.

7 Chapter Review

Concept Summary

The momentum of an object is the product of its mass and its velocity.

- The change in momentum depends on the force that acts and the length of time it acts.

- Impulse is average force multiplied by the time during which it acts.

- The impulse exerted on something is equal to the change in momentum it produces.

The law of conservation of momentum states that momentum is conserved when there is no net external force.

- When objects collide in the absence of external forces, momentum is conserved no matter whether the collision is elastic or inelastic.

Momentum is a vector quantity.

- Momenta combine by vector rules.

Important Terms

conserved (7.4)
elastic collision (7.5)
impulse (7.2)
inelastic collision (7.5)
law of conservation of momentum (7.4)
momentum (7.1)

Review Questions

1. Distinguish between *mass* and *momentum*. Which is inertia and which is inertia in motion? (7.1)

2. **a.** Which has the greater mass, a heavy truck at rest or a rolling skateboard?

   b. Which has greater momentum? (7.1)

3. Distinguish between *impact* and *impulse*. Which designates a force and which is force multiplied by time? (7.2)

4. When the force of impact on an object is extended in time, does the impulse increase or decrease? (7.2)

5. Distinguish between *impulse* and *momentum*. Which is force × time and which is inertia in motion? (7.2)

6. Does impulse equal momentum, or a *change* in momentum? (7.2)

7. For a constant force, suppose the duration of impact on an object is doubled.

   a. How much is the impulse increased?

   b. How much is the resulting change in momentum increased? (7.2)

8. In a car crash, why is it advantageous for an occupant to extend the time during which the collision takes place? (7.2)

9. If the time of impact in a collision is extended by four times, how much does the force of impact change? (7.2)

10. Why is it advantageous for a boxer to ride with the punch? Why should he avoid moving into an oncoming punch? (7.2)

11. Visualize yourself on a skateboard.

    a. When you throw a ball, do you experience an impulse?

    b. Do you experience an impulse when you catch a ball of the same speed?

    c. Do you experience an impulse when you catch it and then throw it out again?

    d. Which impulse is greatest? (7.3)

12. Why is more impulse delivered during a collision when bouncing occurs than during one when it doesn't? (7.3)

13. Why is the Pelton Wheel an improvement over paddle wheels with flat blades? (7.3)

14. In terms of momentum conservation, why does a gun kick when fired? (7.4)

15. What does it mean to say that momentum is conserved? (7.4)

16. Distinguish between an elastic and an inelastic collision. (7.5)

17. Imagine that you are hovering next to the space shuttle in an earth-orbit and your buddy of equal mass who is moving at 4 km/h with respect to the ship bumps into you. If he holds onto you, how fast do you both move with respect to the ship? (7.5)

18. Is momentum conserved for colliding objects that are moving at angles to one another? Explain. (7.6)

Plug and Chug

1. **a.** What is the momentum of an 8-kg bowling ball rolling at 2 m/s?

   b. If the bowling ball rolls into a pillow and stops in 0.5 s, calculate the average force it exerts on the pillow.

   c. What average force does the pillow exert on the ball?

2. **a.** What is the momentum of a 50-kg carton that slides at 4 m/s across an icy surface?

   b. The sliding carton skids onto a rough surface and stops in 3 s. Calculate the force of friction it encounters.

3. **a.** What impulse occurs when an average force of 10 N is exerted on a cart for 2.5 s?

   b. What change in momentum does the cart undergo?

   c. If the mass of the cart is 2 kg and the cart is initially at rest, calculate its final speed.

4. A 2-kg blob of putty moving at 3 m/s slams into a 2-kg blob of putty at rest.

   a. Calculate the speed of the two stuck-together blobs of putty immediately after colliding.

   b. Calculate the speed of the two blobs if the one at rest was 4 kg.

Think and Explain

1. When you ride a bicycle at full speed and the bike stops suddenly, why do you have to push hard on the handlebars to keep from flying forward?

2. In terms of impulse and momentum, why are air bags in automobiles a good idea?

3. Why is it difficult for a firefighter to hold a hose that ejects large amounts of water at high speed?

4. You can't throw a raw egg against a wall without breaking it, but you can throw it at the same speed into a sagging sheet without breaking it. Explain.

5. Suppose you roll a bowling ball into a pillow and the ball stops. Now suppose you roll it against a spring and it bounces back with an equal and opposite momentum.

   a. Which object exerts a greater impulse, the pillow or the spring?

   b. If the time it takes the pillow to stop the ball is the same as the time of contact of the ball with the spring, how do the average forces each exerts on the ball compare?

6. If you topple from your treehouse, you'll continuously gain momentum as you fall to the ground below. Doesn't this violate the conservation of momentum? Defend your answer.

7. A bug and the windshield of a fast-moving car collide. Indicate whether each of the following statements is true or false.

   a. The forces of impact on the bug and on the car are the same size.

   b. The impulses on the bug and on the car are the same size.

   c. The change in speed of the bug and the car is the same.

   d. The changes in momentum of the bug and of the car are the same size.

8. What difference in kick would you expect in firing a bullet versus firing a blank cartridge from the same gun? Explain.

9. A group of playful astronauts, each with a bag full of balls, form a circle as they free-fall in space. Describe what happens when they begin tossing balls simultaneously to one another.

10. A proton from an accelerator strikes an atom. An electron is observed flying forward in the same direction the proton was moving and at a speed much greater than the speed of the proton. What conclusion can you draw about the relative mass of a proton and an electron?

Think and Solve

1. A 1000-kg car moving at 20 m/s slams into a building and comes to a halt. Consider questions **a** and **b** below. Which question can be answered using the given information and which one cannot be answered? Explain.

   a. What impulse acts on the car?

   b. What is the force of impact on the car?

2. A car with a mass of 1000 kg moves at 20 m/s. What braking force is needed to bring the car to a halt in 10 s?

3. Assume an 8-kg bowling ball moving at 2 m/s bounces off a spring at the same speed that it had before bouncing.

   a. What is its momentum of recoil?

   b. What is its change in momentum? (*Hint:* What is the change in temperature when something goes from 1° to –1°?)

   c. If the interaction with the spring occurs in 0.5 s, calculate the average force the spring exerts on it.

   d. Compare this force with the force on the pillow in Plug and Chug 1.

4. A railroad diesel engine weighs four times as much as a freight car. If the diesel engine coasts at 5 km/h into a freight car that is at rest, how fast do the two coast after they couple?

5. A 5-kg fish swimming 1 m/s swallows an absent-minded 1-kg fish at rest. What is the speed of the large fish immediately after lunch? What would its speed be if the small fish were swimming toward it at 4 m/s?

6. Comic-strip hero Superman meets an asteroid in outer space and hurls it at 100 m/s. The asteroid is a thousand times more massive than Superman is. In the strip, Superman is seen at rest after the throw. Taking physics into account, what would be his recoil speed? What is this in miles per hour?

8 Energy

The mechanical energy of the wind can be harnessed to produce electrical power.

Energy is the most central concept underlying all of science. Surprisingly, the idea of energy was unknown to Isaac Newton, and its existence was still being debated in the 1850s. Even though the concept of energy is relatively new, today we find it ingrained not only in all branches of science, but in nearly every aspect of human society. We are all quite familiar with energy. Energy comes to us from the sun in the form of sunlight, it is in the food we eat, and it sustains life. Energy may be the most familiar concept in science, yet it is one of the most difficult to define. Persons, places, and things have energy, but we observe only the effects of energy when something is happening—only when energy is being transferred from one place to another or transformed from one form to another. We begin our study of energy by observing a related concept, work.

8.1 Work

The previous chapter showed that the change in an object's motion is related to both force and how long the force acts. "How long" meant time. Remember, the quantity *force × time* is called *impulse.* But "how long" need not always mean time. It can mean distance also. When we consider the quantity *force × distance,* we are talking about an entirely different concept. This concept is called **work.**

We do work when we lift a load against the earth's gravity. The heavier the load or the higher we lift it, the more work we do. Two things enter into every case where work is done: (1) the *application of a force,* and (2) the *movement of something* by that force.

Let's look at the simplest case, in which the force is constant and the motion takes place in a straight line in the direction of the force.

103

Figure 8.1 ▲
Work is done in lifting the barbell. If the barbell could be lifted twice as high, the weight lifter would have to do twice as much work.

Then the work done on an object by an applied force is the product of the force and the distance through which the object is moved.*

$$\text{work} = \text{force} \times \text{distance}$$

In equation form,

$$W = Fd$$

If we lift two loads up one story, we do twice as much work as we would in lifting one load the same distance, because the *force* needed to lift twice the weight is twice as great. Similarly, if we lift one load two stories instead of one story, we do twice as much work because the *distance* is twice as great.

Notice that the definition of work involves both a force *and* a distance. A weight lifter holding a barbell weighing 1000 N over his head does no work on the barbell. He may get really tired holding it, but if the barbell is not moved by the force he exerts, he does no work on the barbell. Work may be done on the muscles by stretching and squeezing them, which is force times distance on a biological scale, but this work is not done on the barbell. Lifting the barbell, however, is a different story. When the weight lifter raises the barbell from the floor, he is doing work on it.

Work generally falls into two categories. One of these is the work done against another force. When an archer stretches her bowstring, she is doing work against the elastic forces of the bow. Similarly, when the ram of a pile driver is raised, work is required to raise the ram against the force of gravity. When you do push-ups, you do work against your own weight. You do work on something when you force it to move against the influence of an opposing force—often friction.

The other category of work is work done to change the speed of an object. This kind of work is done in bringing an automobile up to speed or in slowing it down.

The unit of measurement for work combines a unit of force, N, with a unit of distance, m. The resulting unit of work is the newton-meter (N·m), also called the **joule** (rhymes with cool) in honor of James Joule. One joule (J) of work is done when a force of 1 N is exerted over a distance of 1 m, as in lifting an apple over your head. For larger values we speak of kilojoules (kJ)—thousands of joules—or megajoules (MJ)—millions of joules. The weight lifter in Figure 8.1 does work on the order of kilojoules. To stop a loaded truck going at 100 km/h takes megajoules of work.

8.2 Power

The definition of work says nothing about how long it takes to do the work. When carrying a load up some stairs, you do the same amount of work whether you walk or run up the stairs. So why are you more

---

* For the more general case, work is the product of the *component* of force acting in the direction of motion and the distance moved.

tired after running upstairs in a few seconds than after walking upstairs in a few minutes? To understand this difference, we need to talk about how fast the work is done, or **power.** Power is the rate at which work is done. It equals the amount of work done divided by the time interval during which the work is done.

$$power = \frac{work\ done}{time\ interval}$$

A high-power engine does work rapidly. An automobile engine that delivers twice the power of another automobile engine does not necessarily produce twice as much work or go twice as fast as the less powerful engine. Twice the power means the engine can do twice the work in the same amount of time or the same amount of work in half the time. A powerful engine can get an automobile up to a given speed in less time than a less powerful engine can.

The unit of power is the joule per second, also known as the **watt**, in honor of James Watt, the eighteenth-century developer of the steam engine. One watt (W) of power is expended when one joule of work is done in one second. One kilowatt (kW) equals 1000 watts. One megawatt (MW) equals one million watts. In the United States, we customarily rate engines in units of horsepower and electricity in kilowatts, but either may be used. In the metric system of units, automobiles are rated in kilowatts. One horsepower, hp, is the same as 0.75 kW, so an engine rated at 134 hp is a 100-kW engine.

Figure 8.2 ▲
The three main engines of the space shuttle can develop 33 000 MW of power when fuel is burned at the enormous rate of 3400 kg/s. This is like emptying an average-size swimming pool in 20 seconds!

■ **Question**

If a forklift is replaced with a new forklift that has twice the power, how much greater a load can it lift in the same amount of time? If it lifts the same load, how much faster can it operate?

8.3 Mechanical Energy

When work is done by an archer in drawing a bowstring, the bent bow acquires the ability to do work on the arrow. When work is done to raise the heavy ram of a pile driver, the ram acquires the ability to do work on the object it hits when it falls. When work is done to wind a spring mechanism, the spring acquires the ability to do work on various gears to run a clock, ring a bell, or sound an alarm.

In each case, something has been acquired that enables the object to do work. It may be in the form of a compression of atoms in the material of an object; a physical separation of attracting bodies; or a rearrangement of electric charges in the molecules of a substance.

■ **Answer**

The forklift that delivers twice the power will lift twice the load in the same time, or the same load in half the time. Either way, the owner of the new forklift is happy.

This "something" that enables an object to do work is **energy**.* Like work, energy is measured in joules. It appears in many forms that will be discussed in the following chapters. For now we will focus on the two most common forms of **mechanical energy**—the energy due to the position of something, or the movement of something. Mechanical energy can be in the form of either potential energy or kinetic energy.

8.4 Potential Energy

An object may store energy by virtue of its position. The energy that is stored and held in readiness is called **potential energy** (PE) because in the stored state it has the potential for doing work. A stretched or compressed spring, for example, has a potential for doing work. When a bow is drawn, energy is stored in the bow. The bow can do work on the arrow. A stretched rubber band has potential energy because of its position. If the rubber band is part of a slingshot, it is also capable of doing work.

The chemical energy in fuels is also potential energy. It is actually energy of position at the submicroscopic level. This energy is available when the positions of electric charges within and between molecules are altered, that is, when a chemical change takes place. Any substance that can do work through chemical action possesses potential energy. Potential energy is found in fossil fuels, electric batteries, and the food we eat.

---

* Strictly speaking, that which enables an object to do work is called its *available energy*, because not all the energy of an object can be transformed into work.

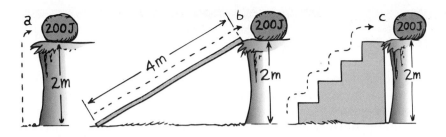

◀ **Figure 8.3**
The potential energy of the 100-N boulder with respect to the ground below is 200 J in each case because the work done in elevating it 2 m is the same whether the boulder is (a) lifted with 100 N of force, (b) pushed up the 4-m incline with 50 N of force, or (c) lifted with 100 N of force up each 0.5-m stair. No work is done in moving it horizontally, neglecting friction.

Work is required to elevate objects against the earth's gravity. The potential energy due to elevated positions is called *gravitational potential energy*. Water in an elevated reservoir and the raised ram of a pile driver have gravitational potential energy.

The amount of gravitational potential energy possessed by an elevated object is equal to the work done against gravity in lifting it. The work done equals the force required to move it upward times the vertical distance it is moved (remember $W = Fd$). The upward force required while moving at constant velocity is equal to the weight, mg, of the object, so the work done in lifting it through a height h is the product mgh.

$$\text{gravitational potential energy} = \text{weight} \times \text{height}$$

$$PE = mgh$$

Note that the height is the distance above some chosen reference level, such as the ground or the floor of a building. The gravitational potential energy, mgh, is relative to that level and depends only on mg and h. We can see in Figure 8.3 that the potential energy of the boulder at the top of the ledge does not depend on the path taken to get it there.

■ Questions

1. How much work is done on a 100-N boulder that you carry horizontally across a 10-m room? How much PE does the boulder gain?

2. a. How much work is done on a 100-N boulder when you lift it 1 m?

    b. What power is expended if you lift the boulder a distance of 1 m in a time of 1 s?

    c. What is the gravitational potential energy of the boulder in the lifted position?

■ Answers

1. You do no work on the boulder moved horizontally, because you apply no force (except for the tiny bit to start and stop it) in its direction of motion. It has no more PE across the room than it had initially.

2. a. You do 100 J of work when you lift it 1 m, since $Fd = 100$ N·m = 100 J

    b. Power = $\dfrac{100 \text{ J}}{1 \text{ s}}$ = 100 W

    c. It depends. Relative to its starting position, the boulder's PE is 100 J. Relative to some other reference level, its PE would be some other value.

Figure 8.4 ▲
The potential energy of the drawn bow equals the work (average force × distance) done by drawing the arrow into position. When released, potential energy will become the kinetic energy of the arrow.

8.5 Kinetic Energy

Push on an object and you can set it in motion. If an object is moving, then it is capable of doing work. It has energy of motion, or **kinetic energy** (KE). The kinetic energy of an object depends on the mass of the object as well as its speed. It is equal to half the mass multiplied by the square of the speed.

$$\text{kinetic energy} = \frac{1}{2}\,\text{mass} \times \text{speed}^2$$

$$\text{KE} = \frac{1}{2}mv^2$$

When you throw a ball, you do work on it to give it speed as it leaves your hand. The moving ball can then hit something and push it, doing work on what it hits. The kinetic energy of a moving object is equal to the work required to bring it to that speed from rest, or the work the object can do while being brought to rest.

$$\text{net force} \times \text{distance} = \text{kinetic energy}$$

or in equation notation,*

$$Fd = \frac{1}{2}mv^2$$

Note that the speed is squared, so if the speed of an object is doubled, its kinetic energy is quadrupled ($2^2 = 4$). Consequently, it takes four times the work to double the speed. Also, an object moving twice as fast takes four times as much work to stop. Accident investigators are aware that a car going 100 km/h has four times the kinetic energy it would have at 50 km/h. So a car going 100 km/h will skid four times as far when its brakes are locked than it would at 50 km/h, because speed is squared for kinetic energy.

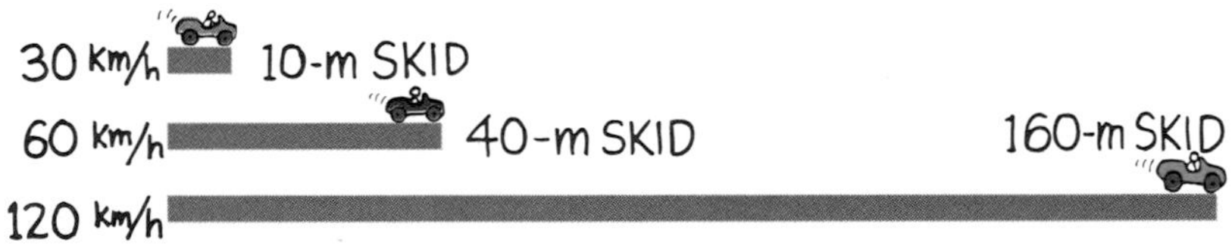

Figure 8.5 ▶
Typical stopping distance for cars traveling at various speeds. The work done to stop the car is friction force × distance of slide. Notice how work depends on the square of the speed. The distances would be even greater if the driver's reaction time were taken into account.

30 km/h ▭ 10-m SKID

60 km/h ▭ 40-m SKID

120 km/h ▭ 160-m SKID

Kinetic energy exists in various other forms of energy, such as thermal energy (random molecular motion), acoustical energy (molecules vibrating in rhythmic patterns), and radiant energy (originating from the motions of electrons within atoms). There is much in common among the various forms of energy.

---

* This formula is derived algebraically as follows. Multiply both sides of $F = ma$, which is Newton's second law, by d to get $Fd = mad$. Recall that for motion in a straight line at constant acceleration, $d = \frac{1}{2}at^2$. So $Fd = ma(\frac{1}{2}at^2) = \frac{1}{2}maat^2 = \frac{1}{2}m(at)^2$. Substituting $v = at$, we get $Fd = \frac{1}{2}mv^2$.

The Sweet Spot

The sweet spot of a tennis racquet or a baseball bat is the place where the ball's impact produces minimum vibrations in the racquet or bat. Strike a ball at the sweet spot and it goes faster and farther. Strike a ball in another part of the racquet or bat, and vibrations can occur that sting your hand! From an energy point of view, there is energy in the vibrations of the racquet or bat. There is energy in the ball after being struck. Energy that is not in vibrations is energy available to the ball. Do you see why a ball will go faster and farther when struck at the sweet spot?

■ **Question**

When the brakes of a motorcycle traveling at 60 km/h become locked, how much farther will the motorcycle skid than if it were traveling at 20 km/h?

8.6 Conservation of Energy

More important than knowing *what energy is*, is understanding how it behaves—*how it transforms.* We can better understand nearly every process or change that occurs in nature if we analyze it in terms of a transformation of energy from one form to another.

As you draw back the stone in a slingshot, you do work stretching the rubber band. The rubber band then has potential energy. When released, the stone has kinetic energy equal to this potential energy. It delivers this energy to its target, perhaps a wooden fence post. The slight distance the post is moved multiplied by the average force of impact doesn't quite match the kinetic energy of the stone. The energy score doesn't balance. But if you investigate further, you'll find that both the stone and fence post are a bit warmer. By how much? By the energy difference. Energy changes from one form to another. It transforms without net loss or net gain.

■ **Answer**

Nine times farther; the motorcycle has nine times as much energy when it travels three times as fast. KE $= \frac{1}{2}m(3v)^2 = \frac{1}{2}m9v^2 = 9(\frac{1}{2}mv^2)$. Ordinarily, the friction force will be the same in either case. Therefore, to do nine times the work requires nine times more sliding distance.

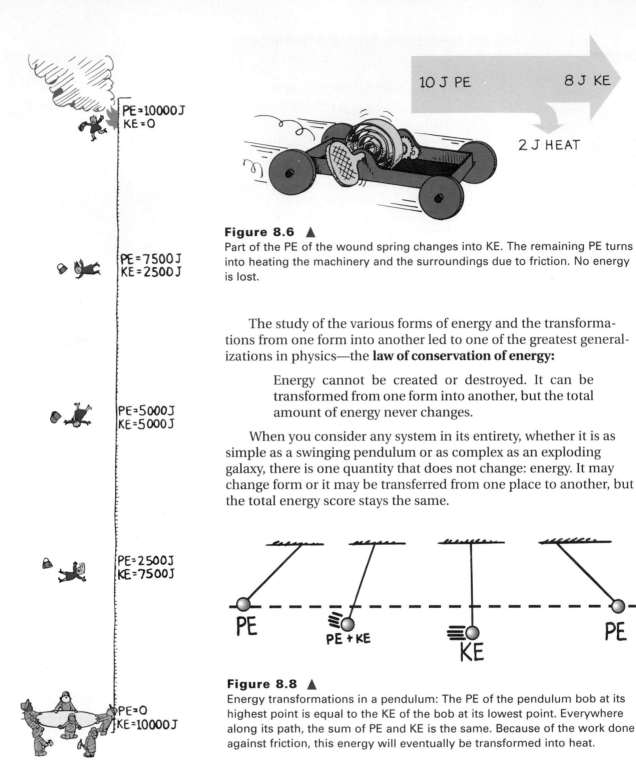

10 J PE   **8 J KE**

2 J HEAT

Figure 8.6 ▲
Part of the PE of the wound spring changes into KE. The remaining PE turns into heating the machinery and the surroundings due to friction. No energy is lost.

The study of the various forms of energy and the transformations from one form into another led to one of the greatest generalizations in physics—the **law of conservation of energy:**

> Energy cannot be created or destroyed. It can be transformed from one form into another, but the total amount of energy never changes.

When you consider any system in its entirety, whether it is as simple as a swinging pendulum or as complex as an exploding galaxy, there is one quantity that does not change: energy. It may change form or it may be transferred from one place to another, but the total energy score stays the same.

PE   PE + KE   KE   PE

Figure 8.8 ▲
Energy transformations in a pendulum: The PE of the pendulum bob at its highest point is equal to the KE of the bob at its lowest point. Everywhere along its path, the sum of PE and KE is the same. Because of the work done against friction, this energy will eventually be transformed into heat.

This energy score takes into account the fact that each atom that makes up matter is a concentrated bundle of energy. When the nuclei (cores) of atoms rearrange themselves, enormous amounts of energy can be released. The sun shines because some of its nuclear energy is transformed into radiant energy. In nuclear reactors, nuclear energy is transformed into heat.

PE = 10000 J
KE = 0

PE = 7500 J
KE = 2500 J

PE = 5000 J
KE = 5000 J

PE = 2500 J
KE = 7500 J

PE = 0
KE = 10000 J

Figure 8.7 ▲
When the lady in distress leaps from the burning building, note that the sum of her PE and KE remains constant at each successive position— $\frac{1}{4}$, $\frac{1}{2}$, $\frac{3}{4}$, and all the way down to the ground.

Enormous compression due to gravity in the deep hot interior of the sun causes hydrogen nuclei to fuse and become helium nuclei. This high-temperature welding of atomic nuclei is called *thermonuclear fusion* and will be covered later, in Chapter 40. This process releases radiant energy, some of which reaches the earth. Part of this energy falls on plants, and some of the plants later become coal. Another part supports life in the food chain that begins with microscopic marine animals and plants, and later gets stored in oil. Part of the sun's energy is used to evaporate water from the ocean. Some water returns to the earth as rain that is trapped behind a dam. By virtue of its elevated position, the water behind the dam has potential energy that is used to power a generating plant below the dam. The generating plant transforms the energy of falling water into electric energy. Electric energy travels through wires to homes where it is used for lighting, heating, cooking, and operating electric toothbrushes. How nice that energy is transformed from one form to another!

■ Question

Suppose a car with a miracle engine is able to convert into work 100% of the energy released when gasoline burns (40 million joules per liter). If the air drag and overall frictional forces on the car traveling at highway speed is 500 N, what is the upper limit in distance per liter of gasoline the car could cover at highway speed?

8.7 Machines

A **machine** is a device used to multiply forces or simply to change the direction of forces. The concept that underlies every machine is the conservation of energy. Consider one of the simplest machines, the lever, shown in Figure 8.9. At the same time we do work on one end of the **lever,** the other end does work on the load. We see that the direction of force is changed. If we push *down*, the load is lifted *up*. If the heat from friction is small enough to neglect, the work input will be equal to the work output.

■ Answer

Work is defined as force × distance. By rearranging the formula, we get distance = work ÷ force. If all 40 million joules of energy in one liter of gas were used to do the work of overcoming the air drag and frictional forces, the distance would be

$$\text{distance} = \frac{\text{work}}{\text{force}} = \frac{40\ 000\ 000\ \text{J}}{500\ \text{N}} = 80\ 000\ \text{m} = 80\ \text{km}$$

The important point here is that even with a perfect engine, fuel economy has an upper limit that is dictated by the conservation of energy.

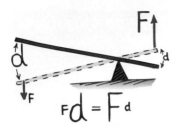

Figure 8.9 ▲
The lever. The work (force × distance) done at one end is equal to the work done on the load at the other end.

work input = work output

Since work equals force times distance, we can say

$$(\text{force} \times \text{distance})_{\text{input}} = (\text{force} \times \text{distance})_{\text{output}}$$

A little thought will show that the pivot point, or **fulcrum**, of the lever can be relatively close to the load. Then a small input force exerted through a large distance will produce a large output force over a correspondingly short distance. In this way, a lever can multiply forces. However, no machine can multiply work or energy. That's a conservation of energy no-no!

Figure 8.10 ▲
The output force (80 N) is eight times the input force (10 N), while the output distance (1/8 m) is one-eighth of the input distance (1 m).

Consider the ideal weightless lever in Figure 8.10. The child pushes down 10 N and lifts an 80-N load. The ratio of output force to input force for a machine is called the **mechanical advantage**. Here the mechanical advantage is (80 N)/(10 N), or 8. Notice that the load moves only one-eighth of the distance the input force moves. Neglecting friction, the mechanical advantage can also be determined by the ratio of input distance to output distance.

Figure 8.11 ▶
The three basic types of levers. Notice that the direction of the force is changed in type 1.

TYPE 1 TYPE 2 TYPE 3

Three common ways to set up a lever are shown in Figure 8.11. A type 1 lever has the fulcrum between the force and the load, or between input and output. This kind of lever is commonly seen in a playground seesaw with children sitting on each end of it. Push down on one end and you lift a load at the other. You can increase force at the expense of distance. Note that the directions of input and output are opposite.

For a type 2 lever, the load is between the fulcrum and the input force. To lift a load, you *lift* the end of the lever. One example is placing one end of a long steel bar under an automobile frame and lifting on the free end to raise the automobile. Again, force on the load is increased at the expense of distance. Since the input and output forces are on the same side of the fulcrum, the forces have the same direction.

In the type 3 lever, the fulcrum is at one end and the load is at the other. The input force is applied between them. Your biceps muscles are connected to the bones in your forearm in this way. The fulcrum is your elbow and the load is in your hand. The type 3 lever increases distance at the expense of force. When you move your biceps muscles a short distance, your hand moves a much greater distance. Notice that the input and output forces are on the same side of the fulcrum and therefore they have the same direction.

A **pulley** is basically a kind of lever that can be used to change the direction of a force. Properly used, a pulley or system of pulleys can multiply forces.

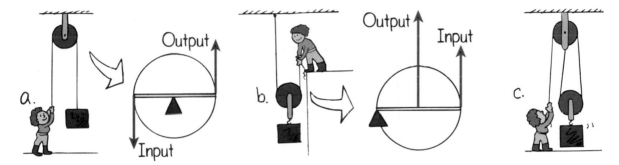

The single pulley in Figure 8.12a behaves like a type 1 lever. The axis of the pulley acts as the fulcrum, and both lever distances (the radius of the pulley) are equal so the pulley does not multiply force. It simply changes the direction of the applied force. In this case, the mechanical advantage equals 1. Notice that the input distance equals the output distance the load moves.

In Figure 8.12b, the single pulley acts as a type 2 lever. Careful thought will show that the fulcrum is at the left end of the "lever" where the supporting rope makes contact with the pulley. The load is suspended halfway between the fulcrum and the input end of the lever, which is on the right end of the "lever." Each newton of input will support two newtons of load, so the mechanical advantage is 2. This number checks with the distances moved. To raise the load 1 m, the woman will have to pull the rope up 2 m. We can say the mechanical advantage is 2 for another reason: the load is now supported by two strands of rope. This means each strand supports half the load. The force the woman applies to support the load is therefore only half of the weight of the load.

The mechanical advantage for simple pulley systems is the same as the number of strands of rope that actually support the load. In Figure 8.12a the load is supported by one strand and the mechanical advantage is 1. In Figure 8.12b the load is supported by two strands

Figure 8.12 ▲
A pulley can (a) change the direction of a force as input is exerted downward and load moves upward, (b) multiply force as input is now half the load, and (c) when combined with another pulley, both change the direction and multiply force.

Figure 8.13 ▲
In an idealized pulley system, applied force × input distance = output force × output distance.

and the mechanical advantage is 2. Can you use this rule to state the mechanical advantage of the pulley system in Figure 8.12c?*

The mechanical advantage of the simple system in Figure 8.12c is 2. Notice that although three strands of rope are shown, only two strands actually support the load. The upper pulley serves only to change the direction of the force. Actually experimenting with a variety of pulley systems is much more beneficial than reading about them in a textbook, so try to get your hands on some pulleys, in or out of class. They're fun.

The pulley system shown in Figure 8.13 is a bit more complex, but the principles of energy conservation are the same. When the rope is pulled 5 m with a force of 100 N, a 500-N load is lifted 1 m. The mechanical advantage is (500 N)/(100 N), or 5. Force is multiplied at the expense of distance. The mechanical advantage can also be found from the ratio of distances: (input distance)/(output distance) = 5.

No machine can put out more energy than is put into it. No machine can create energy. A machine can only transfer energy from one place to another or transform it from one form to another.

8.8 Efficiency

The previous examples of machines were considered to be *ideal*. All the work input was transferred to work output. An ideal machine would have 100% efficiency. In practice, 100% efficiency never happens, and we can never expect it to happen. In any machine, some energy is transformed into atomic or molecular kinetic energy—making the machine warmer. We say this wasted energy is dissipated as heat.**

When a simple lever rocks about its fulcrum, or a pulley turns about its axis, a small fraction of input energy is converted into thermal energy. We may put in 100 J of work on a lever and get out 98 J of work. The lever is then 98% efficient and we lose only 2 J of work input as heat. In a pulley system, a larger fraction of input energy is lost as heat. For example, if we do 100 J of work, the friction on the pulleys as they turn and rub on their axle can dissipate 40 J of heat energy. So the work output is only 60 J and the pulley system has an efficiency of 60%. The lower the efficiency of a machine, the greater is the amount of energy wasted as heat.

---

* The number-of-strands rule applies only to simple pulleys, where same-size pulleys are on the same shaft. The chain hoist popular in auto repair shops, for example, gets its mechanical advantage from different-size pulleys on the same shaft. We won't treat the chain hoist here.

** Energy of atomic or molecular motion is actually *thermal energy*, not *heat*. We'll see in Chapter 21 that heat is energy being transferred from one place to another by atomic and molecular motion. Heat is analogous to work; both involve energy transfer by motion.

Efficiency can be expressed as the ratio of useful work output to total work input.

$$\text{efficiency} = \frac{\text{useful work output}}{\text{total work input}}$$

An inclined plane is a machine. Sliding a load up an incline requires less force than lifting it vertically. Figure 8.14 shows a 5-m inclined plane with its high end elevated by 1 m. Using the plane to elevate a heavy load, we push the load five times farther than we lift it vertically. If friction is negligible, we need apply only one-fifth of the force required to lift the load vertically. The inclined plane has a *theoretical* mechanical advantage of 5.

◄ **Figure 8.14**
Figure 8.14
Pushing the block of ice 5 times farther up the incline than the vertical distance it's lifted requires a force of only one-fifth its weight. Whether pushed up the plane or simply lifted, the ice gains the same amount of PE.

A block of ice sliding along an icy plank, or a horse-pulled cart with well-lubricated wheels, might have an efficiency of almost 100%. However, when the load is a wooden crate sliding on a wooden plank, both the actual mechanical advantage and the efficiency will be considerably less. Efficiency can also be expressed as the ratio of actual mechanical advantage to theoretical mechanical advantage.

$$\text{efficiency} = \frac{\text{actual mechanical advantage}}{\text{theoretical mechanical advantage}}$$

Efficiency will always be a fraction less than 1. To convert efficiency to percent, we simply express it as a decimal and multiply by 100%. For example, an efficiency of 0.25 expressed in percent is 0.25 × 100%, or 25%.

The auto jack shown in Figure 8.15 is actually an inclined plane wrapped around a cylinder. You can see that a single turn of the handle raises the load a relatively small distance. If the circular distance the handle is moved is 500 times greater than the pitch, which is the distance between ridges, then the theoretical mechanical advantage of the jack is 500.* No wonder a child can raise a loaded moving van with one of these devices! In practice there is a great deal of friction

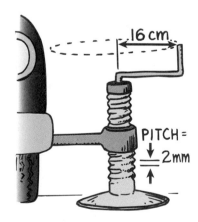

Figure 8.15 ▲
The auto jack is like an inclined plane wrapped around a cylinder. Every time the handle is turned one revolution, the load is raised a distance of one pitch.

---

* To raise a load by 2 mm the handle has to be turned once, through a distance equal to the circumference of the circular path of the 16-cm radius. This distance is 100 cm (since the circumference is $2\pi r = 2 \times 3.14 \times 16$ cm = 100 cm). A simple calculation will show that the 100-cm work-input distance is 500 times greater than the work-output distance of 2 mm. If the jack were 100% efficient, then the input force would be multiplied by 500 times. The theoretical mechanical advantage of the jack is 500.

in this type of jack, so the efficiency might be about 20%. Thus the jack actually multiplies force by about 100 times, so the actual mechanical advantage approximates an impressive 100. Imagine the value of one of these devices if it had been available when the great pyramids were being built!

■ Question

A child on a sled (total weight 500 N) is pulled up a 10-m slope that elevates her a vertical distance of 1 m.

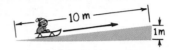

a. What is the theoretical mechanical advantage of the slope?

b. If the slope is without friction, and she is pulled up the slope at constant speed, what will be the tension in the rope?

c. Considering the practical case where friction *is* present, suppose the tension in the rope were actually 100 N. What is the actual mechanical advantage of the slope? What would the efficiency be?

An automobile engine is a machine that transforms chemical energy stored in fuel into mechanical energy. The molecules of the gasoline break up as the fuel burns. Burning is a chemical reaction in which atoms combine with the oxygen in the air. Carbon atoms from the gasoline combine with oxygen atoms to form carbon dioxide, hydrogen atoms combine with oxygen, and energy is released. The converted energy is used to run the engine.

It would be nice if all this energy were converted to mechanical energy, but as physicists learned in the nineteenth century, transforming 100% of thermal energy into mechanical energy is not possible. Some heat must flow from the engine. Friction adds more to the energy loss. Even the best-designed engines are unlikely to be more

■ Answer

a. The ideal, or theoretical, mechanical advantage is

$$\frac{\text{input distance}}{\text{output distance}} = \frac{10 \text{ m}}{1 \text{ m}} = 10$$

b. 50 N. With no friction, ideal mechanical advantage and actual mechanical advantage would be the same, 10. So the rope tension, or the input force, will be 1/10 the weight being raised, 500 N. (Note that this input force is the same as the component of the child's weight along the slope.)

c. The actual mechanical advantage is (weight being raised)/(input force) = (500 N)/(100 N) = 5. The efficiency would then be 0.5 or 50%, since (actual mechanical advantage)/(theoretical mechanical advantage) = 5/10 = 0.5. The efficiency can also be obtained from the ratio (useful work output)/(work input).

FUEL ENERGY IN = COOLING WATER LOSSES + ENGINE OUTPUT + EXHAUST HEAT
100 %   35 %   30 %   35 %

Figure 8.16
Only 30% of the energy produced by burning gasoline in a typical automobile engine becomes useful mechanical energy.

than 35% efficient. Some of the heat energy goes into the cooling system and is released through the radiator to the air. Some of it goes out the tailpipe with the exhaust gases, and almost half goes into heating engine parts as a result of friction. On top of these contributors to inefficiency, the fuel does not burn completely. A certain amount of it goes unused. We can look at inefficiency in this way: In any transformation there is a dilution of the amount of *useful energy*. Useful energy ultimately becomes thermal energy. Energy is not destroyed, it is simply degraded. Through heat transfer, thermal energy is the graveyard of useful energy.

8.9 Energy for Life

Every living cell in every organism is a machine. Like any machine, living cells need an energy supply. Most living organisms on this planet feed on various hydrocarbon compounds that release energy when they react with oxygen. Just as there is more energy stored in gasoline than in the products of its combustion, there is more energy stored in the molecules in food than there is in the reaction products after the food is digested. The energy difference is what sustains life.

The same principle of combustion occurs in the digestion of food in the body and the burning of fossil fuels in mechanical engines. The main difference is the rate at which the reactions take place. During digestion, the reaction rate is much slower and energy is released as it is needed by the body. Like the burning of fossil fuels, the reaction is self-sustaining once it starts. In digestion, carbon combines with oxygen to form carbon dioxide.

The reverse process is more difficult. Only green plants and certain one-celled organisms can make carbon dioxide combine with water to produce hydrocarbon compounds such as sugar. This process is *photosynthesis* and requires an energy input, which normally comes from sunlight. Sugar is the simplest food. All other foods, such as carbohydrates, proteins, and fats, are also synthesized compounds containing carbon, hydrogen, oxygen, and other elements. Aren't we very fortunate that green plants are able to use the energy of sunlight to make food that gives us and all other organisms energy? Because of this, there is life.

Energy Conservation

Most energy consumed in America comes from fossil fuels. Oil, natural gas, and coal supply the energy for almost all our industry and technology. Nearly 70% of electrical power in the United States comes from fossil fuels, with only 21% from nuclear power. Worldwide, fossil fuels also account for most energy consumption. We have grown to depend on fossil fuels because they have been plentiful and inexpensive. Until recently, our consumption was small enough that we could ignore their environmental impact.

But things have changed. Fossil fuels are being consumed at a rate that threatens to deplete the entire world supply. Locally and globally, our fossil fuel consumption is measurably polluting the air we breathe and the water we drink. Yet, despite these problems, many people consider fossil fuels to be as inexhaustible as the sun's glow and as acceptable as Mom's apple pie, because these fuels have lasted and nurtured us through the twentieth century. Financially, fossil fuels are still a bargain, but this is destined to change. Environmentally, the costs are already dramatic. Some other fuel must take the place of fossil fuels if we are to maintain the industry and technology to which we are accustomed. The French have chosen nuclear, with 74% of their electricity coming from nuclear power plants. What energy source would you choose as an alternative?

In the meantime, we shouldn't waste energy. As individuals, we should limit the consumption of useful energy by such measures as turning off unused electric appliances, using less hot water, going easy on heating and air conditioning, and driving energy-efficient automobiles. By doing these things, we are conserving *useful* energy. In how many reasonable ways can we limit energy consumption?

8 Chapter Review

Concept Summary

When a constant force moves an object in the direction of the force, the work done equals the product of the force and the distance the object is moved.

■ Power is the rate at which work is done.

The energy of an object enables it to do work.

■ Mechanical energy is due to the position of something (potential energy) or the movement of something (kinetic energy).

The law of conservation of energy states that energy cannot be created or destroyed.

■ Energy can be transformed from one form into another.

A machine is a device for multiplying force or changing the direction of force.

■ The lever, pulley, and inclined plane are simple machines.

■ The useful work output of a machine is less than the total work input.

Important Terms

efficiency (8.8)
energy (8.3)
fulcrum (8.7)
joule (8.1)
kinetic energy (8.5)
law of conservation of energy (8.6)
lever (8.7)

machine (8.7)
mechanical advantage (8.7)
mechanical energy (8.3)
potential energy (8.4)
power (8.2)
pulley (8.7)
watt (8.2)
work (8.1)

Review Questions

1. A force sets an object in motion. When the force is multiplied by the time of its application, we call the quantity *impulse*, which changes the *momentum* of that object. What do we call the quantity *force × distance*, and what quantity can this change? (8.1)

2. Work is required to lift a barbell. How many times more work is required to lift the barbell three times as high? (8.1)

3. Which requires more work, lifting a 10-kg load a vertical distance of 2 m or lifting a 5-kg load a vertical distance of 4 m? (8.1)

4. How many joules of work are done on an object when a force of 10 N pushes it a distance of 10 m? (8.1)

5. How much power is required to do 100 J of work on an object in a time of 0.5 s? How much power is required if the same work is done in 1 s? (8.2)

6. What are the two main forms of mechanical energy? (8.3)

7. **a.** If you do 100 J of work to elevate a bucket of water, what is its gravitational potential energy relative to its starting position?

   b. What would the gravitational potential energy be if the bucket were raised twice as high? (8.4)

8. A boulder is raised above the ground so that its potential energy relative to the ground is 200 J. Then it is dropped. What is its kinetic energy just before it hits the ground? (8.5)

9. Suppose an automobile has 2000 J of kinetic energy. When it moves at twice the speed, what will be its kinetic energy? What's its kinetic energy at three times the speed? (8.5)

10. What will be the *kinetic* energy of an arrow having a *potential* energy of 50 J after it is shot from a bow? (8.6)

11. What does it mean to say that in any system the total energy score stays the same? (8.6)

12. In what sense is energy from coal actually solar energy? (8.6)

13. How does the amount of work done on an automobile by its engine relate to the energy content of the gasoline? (8.6)

14. In what two ways can a machine alter an input force? (8.7)

15. In what way is a machine subject to the law of energy conservation? Is it possible for a machine to multiply energy or work input? (8.7)

16. What does it mean to say that a machine has a certain mechanical advantage? (8.7)

17. In which type of lever is the output force smaller than the input force? (8.7)

18. What is the efficiency of a machine that requires 100 J of input energy to do 35 J of useful work? (8.8)

19. Distinguish between theoretical mechanical advantage and actual mechanical advantage. How would these compare if a machine were 100% efficient? (8.8)

20. What is the efficiency of her body when a cyclist expends 1000 W of power to deliver mechanical energy to the bicycle at the rate of 100 W? (8.8)

Plug and Chug

1. Calculate the work done when a 20-N force pushes a cart 3.5 m.

2. Calculate the work done in lifting a 500-N barbell 2.2 m above the floor. What is the potential energy of the barbell when it is lifted to this height?

3. Calculate the power expended when the barbell above is lifted 2.2 m in 2 s.

4. **a.** Calculate the work needed to lift a 90-N block of ice a vertical distance of 3 m. What PE does it have?

b. When the same block of ice is raised the same vertical distance by pushing it up a 5-m

long inclined plane, only 54 N of force are required. Calculate the work done to push the block up the plane. What PE does it have?

5. Calculate the change in potential energy of 8 million kg of water dropping 50 m over Niagara Falls.

6. If 8 million kg of water flows over Niagara Falls each second, calculate the power available at the bottom of the falls.

7. **a.** Calculate the kinetic energy of a 3-kg toy cart that moves at 4 m/s.

b. Calculate the kinetic energy of the same cart at twice the speed.

8. A lever is used to lift a heavy load. When a 50-N force pushes one end of the lever down 1.2 m, the load rises 0.2 m. Calculate the weight of the load.

9. **a.** In raising a 5000-N piano with a pulley system, the workers note that for every 2 m of rope pulled down, the piano rises 0.4 m. Ideally, how much force is required to lift the piano?

b. If the workers actually pull with 2500 N of force to lift the piano, what is the efficiency of the pulley system?

10. Your European friends tell you they are pleased because their new car gives them 15 kilometers per liter of fuel. Translate this into miles per gallon so you can decide whether to be impressed. One mile is 1.6 km, and one gallon is 3.8 liters.

Think and Explain

1. State two reasons why a rock projected with a slingshot will go faster if the rubber is stretched an extra distance.

2. Does an object with momentum always have energy? Does an object with energy always have momentum? Explain.

3. If a mouse and an elephant both run with the same kinetic energy, can you say which is running faster? Explain in terms of the equation for KE.

4. An astronaut in full space gear climbs a vertical ladder on the earth. Later, the astronaut makes the same climb on the moon. In which location does the gravitational potential energy of the astronaut change more? Explain.

5. Most earth satellites follow an oval-shaped (elliptical) path rather than a circular path around the earth. The PE increases when the satellite moves farther from the earth. According to the law of energy conservation, does a satellite have its greatest speed when it is closest to or farthest from the earth?

6. Why does a small, lightweight car generally have better fuel economy than a big, heavy car? How does a streamlined design improve fuel economy?

7. Does using an automobile's air conditioner while driving increase fuel consumption? What about driving with the lights on? What about playing the car radio when parked with the engine off? Explain in terms of the conservation of energy.

8. What is the theoretical mechanical advantage for each of the three lever systems shown?

9. You tell your friend that no machine can possibly put out more energy than is put into it, and your friend states that a nuclear reactor puts out more energy than is put into it. What do you say?

10. The energy we require to live comes from the chemically stored potential energy in food, which is transformed into other energy forms during the digestion process. What happens to a person whose combined work and heat output is less than the energy consumed? What happens when the person's work and heat output is greater than the energy consumed? Can an undernourished person perform extra work without extra food? Defend your answers.

Think and Solve

1. A hammer falls off a rooftop and strikes the ground with a certain KE. If it fell from a roof that was four times higher, how would its KE of impact compare? Its speed of impact? (Neglect air resistance.)

2. A certain car can go from 0 to 100 km/h in 10 s. If the engine delivered twice the power to the wheels, how many seconds would it take?

3. If a car traveling at 60 km/h will skid 20 m when its brakes lock, how far will it skid if it is traveling at 120 km/h when its brakes lock? (This question is typical on some driver's license exams.)

4. How many kilometers per liter will a car go if its engine is 25% efficient and it encounters an average retarding force of 1000 N at highway speed? Assume the energy content of gasoline is 40 MJ per liter.

5. The pulley shown on the left has a mechanical advantage of 1. What is the mechanical advantage when it is used as shown on the right? Defend your answer (1) by considering the pulley to be a lever and (2) by tension in the rope.

9 Circular Motion

The linear speed of the hamster is related to the rotational speed of the track.

Which moves faster on a merry-go-round, a horse near the outside rail or one near the inside rail? Why don't the riders fall out of the rotating carnival ride in Figure 9.1 when the rotating platform is raised? If you swing a tin can at the end of a string in a circle over your head and the string breaks, does the can fly directly outward or does it continue its motion without changing its direction? Why do astronauts orbiting in a space shuttle float in a weightless condition, whereas astronauts orbiting in some future rotating space station will experience normal earth gravity? These questions indicate the flavor of this chapter. We begin by discussing the difference between rotation and revolution.

Figure 9.1 ▶
Why do the occupants of this carnival ride not fall out when it is tipped almost vertically?

9.1 Rotation and Revolution

Both the rotating platform of the carnival ride shown in Figure 9.1 and an ice skater doing a pirouette turn around an **axis.** An axis is the straight line around which rotation takes place. When an object turns about an internal axis—that is, an axis located within the body of the object—the motion is called **rotation,** or *spin*. Both the carnival ride and the skater rotate.

When an object turns about an *external* axis, the motion is called **revolution.** Although the platform of the carnival ride rotates, the riders along the platform's outer edge *revolve* about its axis.

The earth undergoes both types of rotational motion. It revolves around the sun once every $365\frac{1}{4}$ days,* and it rotates around an axis passing through its geographical poles once every 24 hours.**

Figure 9.2 ▲
The turntable *rotates* around its axis while a ladybug sitting at its edge *revolves* around the same axis.

9.2 Rotational Speed

We began this chapter by asking which moves faster on a merry-go-round, a horse near the outside rail or one near the inside rail. Similarly, which part of a turntable moves faster? On the pre-CD record player shown in Figure 9.2, which part of the record moves faster under the stylus—the groove at the outer part of the record or the groove near the center? If you ask people these questions you'll probably get more than one answer, because some people will think about linear speed while others will think about rotational speed.

Linear speed, which we simply called speed in Chapter 2, is the distance moved per unit of time. A point on the outer edge of a merry-go-round or turntable moves a greater distance in one complete rotation than a point near the center. The linear speed is greater on the outer edge of a rotating object than it is closer to the axis. The speed of something moving along a circular path can be called **tangential speed** because the direction of motion is always tangent to the circle. For circular motion, we can use the terms linear speed and tangential speed interchangeably.

Rotational speed (sometimes called angular speed) is the number of rotations per unit of time. All parts of the rigid merry-go-round and the turntable rotate about their axis *in the same amount of time*. Thus, all parts have the same rate of rotation,

---

* Can you see the reason for leap years? Since the earth takes $\frac{1}{4}$ day more than 365 days to circle the sun, an extra day is added to the calendar every fourth year.

** With respect to the sun, the earth rotates once each 24-hour period. A 24-hour day is the time required for a point on the earth that is located directly under the sun to rotate and reach that point again. But with respect to the stars, a complete rotation of the earth takes 23 hours and 56 minutes. Why? Because while the earth rotates, it revolves about a degree around the sun in its orbit.

or the same *number of rotations per unit of time.* It is common to express rotational speed in revolutions per minute (RPM).* For example, phonograph records that were common a few years ago rotate at $33\frac{1}{3}$ RPM. A ladybug sitting anywhere on the surface of the record revolves at $33\frac{1}{3}$ RPM.

Tangential speed and rotational speed are related. Have you ever ridden on a giant rotating platform in an amusement park? The faster it turns, the faster your tangential speed is. Tangential speed is directly proportional to the rotational speed and the radial distance from the axis of rotation. So we state**

Tangential speed ~ radial distance × rotational speed

At the center of the rotating platform, right at its axis, you have no tangential speed at all, but you do have rotational speed. You merely rotate. As you move away from the center, you move faster and faster—your tangential speed increases while your rotational speed stays the same. Move out twice as far from the center, and you have twice the tangential speed. Move out three times as far, and you have three times as much tangential speed. When a row of skaters locked arm in arm at the skating rink makes a turn, the motion of tail-end Charlie is evidence of this greater speed.

Figure 9.3 ▶
All parts of the turntable rotate at the same rotational speed, but ladybugs at different distances from the center travel at different linear speeds. A ladybug sitting twice as far from the center moves twice as fast.

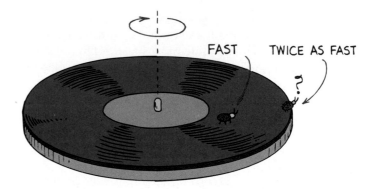

To summarize: In any rigidly rotating system, all parts have the same rotational speed. However, the linear or tangential speed varies. Tangential speed depends on rotational speed and the distance from the axis of rotation.

---

* Rotation rate is described as turns per unit of time, or as the angle turned (degrees or radians) per unit of time. The symbol for rotational speed is ω (the Greek letter omega). Examples of units for rotational speed are RPM, degrees per second, or radians per second. One degree is 1/360 of a full turn. One radian is about 1/6 of a full turn—precisely, it is $1/2\pi$ times 360 degrees, or 57.3 degrees. You may learn more about radians in a follow-up course.

** If you take a follow-up physics course you'll learn that when the proper units are used for tangential speed *v*, rotational speed ω, and radial distance *r*, the direct proportion of *v* to *r* and ω becomes the exact equation $v = r\omega$. This relationship applies only to a rotating system wherein all parts simultaneously have the same ω, like a rigid disk or rigid rod. It does not apply to a system of planets, for example, where each planet has a different rotational speed ω. (We will learn later that the innermost planets in a planetary system have both the greatest rotational speed *and* the greatest linear speed.)

DOING PHYSICS

Comparing Rolls

Roll a cylindrical can across a table. Note that the distance it rolls in each complete rotation equals the circumference of the can. Note also the can rolls in a straight-line path. Now roll an ordinary tapered drinking cup on the table (a paper or foam cup works fine). The wide end has a greater radius than the narrow end. Does the cup roll straight or does it curve? Does the wide end of the cup cover more distance as it rotates? Is the linear speed of the wide end greater? Doesn't this nicely show that linear speed depends on radius?

Activity

■ Questions

1. Which part of the earth's surface has the greatest rotational speed about the earth's axis? Which part has the greatest linear speed relative to the earth's axis?

2. On a particular merry-go-round, the horses along the outer rail are located three times farther from the axis of rotation than the horses along the inner rail. If a boy sitting on a horse near the inner rail has a rotational speed of 4 RPM and a tangential speed of 2 m/s, what will be the rotational speed and tangential speed of his sister who is sitting on a horse along the outer rail?

3. Trains ride on a pair of tracks. For straight-line motion, both tracks are the same length. But which track is longer for a curve, the one on the outside or the one on the inside of the curve?

■ Answers

1. Like a rotating turntable, all parts have the same rotational speed. The equatorial region has the greatest linear speed because it is farthest from the axis.

2. The rotational speed of the sister is also 4 RPM. Her tangential speed is 6 m/s. Since the merry-go-round is rigid, all the horses have the same rotational speed. But the horse at the outer rail is three times the distance from the center and thus has three times the tangential speed.

3. The outside track is longer, just as a circle with a greater radius has a longer circumference.

Figure 9.4 ▲
The only force that is exerted on the whirling can (neglecting gravity) is directed *toward* the center of circular motion. It is called a centripetal force. No outward force is exerted on the can.

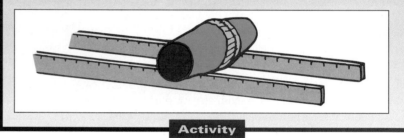
9.3 Centripetal Force

If you whirl a tin can on the end of a string, you'll notice that you must keep continuously pulling on the string as shown in Figure 9.4. You pull inward on the string to keep the can revolving over your head in a circular path. A force of some kind is required to maintain circular motion. Any force that causes an object to follow a circular path is called a **centripetal force.** *Centripetal* means "center-seeking," or "toward the center." The force that holds the occupants safely in the rotating carnival ride (Figure 9.1) is a center-directed force. Without it, the motion of the occupants would be along a straight line—they would not revolve.

Centripetal force is not a new kind of force. It is simply the name given to *any* force that is directed at a right angle to the path of a moving object and that tends to produce circular motion. Gravitational and electrical forces act across empty space as centripetal forces. Gravitational force directed toward the center of the earth holds the moon in an almost circular orbit around the earth. Electrons revolving around the nucleus of the atom are held in their orbits by an electrical force that is directed inward toward the nucleus.

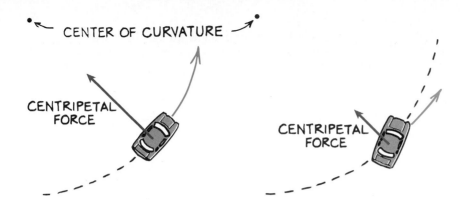

CENTER OF CURVATURE

CENTRIPETAL FORCE

CENTRIPETAL FORCE

◀ **Figure 9.5**
(Left) For the car to go around a curve, there must be sufficient friction to provide the required centripetal force. (Right) If the force of friction is not great enough, skidding occurs.

When an automobile rounds a corner, the sideways-acting friction between the tires and the road provides the centripetal force that holds the car on a curved path (Figure 9.5). If the force of friction is not great enough, the car fails to make the curve and the tires slide sideways. The car skids.

Centripetal force plays the main role in the operation of a centrifuge. A familiar example is the spinning tub in an automatic washing machine. The tub rotates at high speed during its spin cycle. The tub's inner wall exerts a centripetal force on the wet clothes, which are forced into a circular path. The tub exerts great force on the clothes, but the holes in the tub prevent the tub from exerting the same force on the water in the clothes. The water escapes through the holes. It is important to note that a force acts on the clothes, not the water. It is not a force that causes the water to escape. The water escapes because it tends to move by inertia in a straight-line path (Newton's first law) *unless* acted on by a centripetal force or any other force. So interestingly enough, the clothes are forced away from the water, and not the other way around.

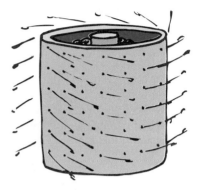

Figure 9.6 ▲
The clothes are forced into a circular path, but the water is not.

9.4 Centripetal and Centrifugal Forces

In the preceding examples, circular motion is described as being caused by a center-directed force. Sometimes an outward force is also attributed to circular motion. This outward force is called **centrifugal force.*** *Centrifugal* means "center-fleeing," or "away from the center." In the case of the whirling can, it is a common misconception to state that a centrifugal force pulls outward on the can. If the string holding the whirling can breaks (Figure 9.7), it is often wrongly stated that centrifugal force pulls the can from its circular path. But in fact, when the string breaks the can goes off in a tangential straight-line path because *no* force acts on it. We illustrate this further with another example.

Figure 9.7 ▲
When the string breaks, the whirling can moves in a straight line, tangent to—not outward from the center of—its circular path.

---

* Both centrifugal force and centripetal force depend on the mass m, tangential speed v, and radius of curvature r of the circularly moving object. If you take a follow-up physics course you'll learn the exact relationship $F = mv^2/r$.

Suppose you are the passenger in a car that suddenly stops short. If you're not wearing a seat belt you pitch forward toward the dashboard. When this happens, you don't say that something forced you forward. You know that you pitched forward because of the *absence* of a force, which a seat belt provides. Similarly, if you are in a car that rounds a sharp corner to the left, you tend to pitch outward against the right door. Why? Not because of some outward or centrifugal force, but rather because there is no centripetal force holding you in circular motion. The idea that a centrifugal force bangs you against the car door is a misconception.

So when you swing a tin can in a circular path, there is *no* force pulling the can outward. Only the force from the string acts on the can to pull the can inward. The outward force is *on the string,* not on the can.

Figure 9.8 ▶
The only force that is exerted *on* the whirling can (neglecting gravity) is directed *toward* the center of circular motion. This is a *centripetal* force. No outward force acts on the can.

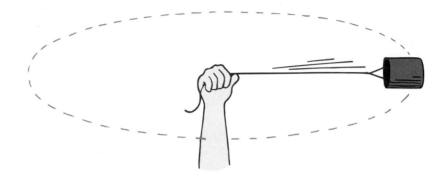

Now suppose there is a ladybug inside the whirling can (Figure 9.9). The can presses against the bug's feet and provides the centripetal force that holds it in a circular path. The ladybug in turn presses against the floor of the can. Neglecting gravity, the *only* force exerted *on the ladybug* is the force of the can on its feet. From our outside stationary frame of reference, we see there is no centrifugal force exerted on the ladybug, just as there was no centrifugal force exerted on the person who pitches against the car door. The "centrifugal-force effect" is attributed not to any real force but to inertia—the tendency of the moving body to follow a straight-line path. But try telling that to the ladybug!

CENTRIPETAL FORCE

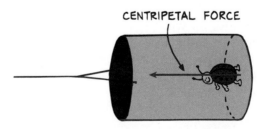

Figure 9.9 ▲
The can provides the centripetal force necessary to hold the ladybug in a circular path.

9.5 Centrifugal Force in a Rotating Reference

Our view of nature depends on the frame of reference from which we view it. For instance, when sitting on a fast-moving train, we have no speed at all relative to the train, but we have an appreciable speed relative to the reference frame of the stationary ground outside. From one frame of reference we have speed; from another we have none—likewise with force. Recall the ladybug inside the whirling can. From a stationary frame of reference outside the whirling can, we see there is *no* centrifugal force acting on the ladybug inside the whirling can. However, we do see centripetal force acting on the can, producing circular motion. Thus, from an outside stationary frame of reference the *only* force acting on the ladybug is the centripetal force exerted by the bottom of the can on the ladybug's feet.

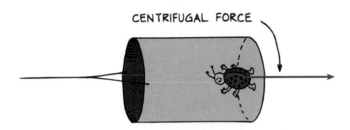

CENTRIFUGAL FORCE

◄ **Figure 9.10**
From the reference frame of the ladybug inside the whirling can, it is being held to the bottom of the can by a force that is directed away from the center of circular motion. The ladybug calls this outward force a *centrifugal* force, which is as real to it as gravity.

> **DOING PHYSICS**

Water-Bucket Swing

Half fill a bucket of water and swing it in a vertical circle. If you swing it fast enough, the water won't fall out at the top. Interestingly enough, although it won't fall *out,* it still falls. The trick is to swing the bucket fast enough so that the bucket falls as fast as the water inside falls. Can you see that because the bucket is revolving, the water moves tangentially as it falls—and stays in the bucket? Later we'll learn that an orbiting space shuttle similarly falls while in orbit. The trick is to give the shuttle sufficient tangential velocity so that it falls around the curved earth rather than into it.

Activity

But nature seen from the reference frame of the rotating system is different. In the rotating frame of reference of the whirling can, both centripetal force (supplied by the can) *and* centrifugal force act *on the ladybug.* The centrifugal force appears as a force in its own right, as real as the pull of gravity. However, there is a fundamental difference between the gravity-like centrifugal force and actual gravitational force. Gravitational force is always an interaction between one mass and another. The gravity we feel is due to the interaction between our mass and the mass of the earth. However, in a rotating reference frame the centrifugal force has no agent such as mass— there is no interaction counterpart. Centrifugal force is an effect of rotation. It is not part of an interaction and therefore it cannot be a true force. For this reason, physicists refer to centrifugal force as a

■ Questions

1. A heavy iron ball is attached by a spring to a rotating platform, as shown in the sketch. Two observers, one in the rotating frame and one on the ground at rest, observe its motion. Which observer sees the ball being pulled outward, stretching the spring? Which observer sees the spring pulling the ball into circular motion?

2. The spring in the sketch stretches 10 cm when the ball is located midway between the axis and the outer edge of the rotating circular platform. The spring support is then moved so the ball is located directly over the platform's outer edge, effectively doubling the ball's distance from the axis. Will the spring stretch more than, less than, or the same (10 cm) as it did before the support was moved?

■ Answers

1. The observer in the reference frame of the rotating platform states that centrifugal force pulls radially outward on the ball, which stretches the spring. The observer in the rest frame states that centripetal force supplied by the stretched spring pulls the ball into circular motion. (Only the observer in the rest frame can identify an action-reaction pair of forces; where action is spring-on-ball, reaction is ball-on-spring. The rotating observer can't identify a reaction counterpart to the centrifugal force because there isn't any.)

2. The ball has twice the linear speed when it's located at twice the distance from the rotational axis. With this greater speed there is a greater centripetal/centrifugal force. The spring therefore stretches more when the ball is near the edge. (We will see in Chapter 18 that the stretch of a spring is directly proportional to the applied force. This means that in this example, when the force is doubled, the stretch will double, from 10 cm to 20 cm.)

fictitious force, *un*like gravitational, electromagnetic, and nuclear forces. Nevertheless, to observers who are in a rotating system, centrifugal force is very real. Just as gravity is ever present at the earth's surface, centrifugal force is ever present within a rotating system.

9.6 Simulated Gravity

Consider a colony of ladybugs living inside a bicycle tire—the balloon kind used on mountain bikes that has plenty of room inside. If we toss the wheel through the air or drop it from an airplane high in the sky, the ladybugs will be in a weightless condition. They will seem to float freely while the wheel is in free fall. Now spin the wheel. The ladybugs will feel themselves pressed to the outer part of the tire's inner surface. If the wheel is spun at just the right speed, the ladybugs will experience *simulated gravity* that feels like the gravity they are accustomed to. Gravity is simulated by centrifugal force. The "down" direction to the ladybugs is what we call "radially outward," away from the center of the wheel.

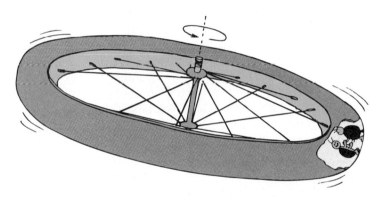

◀ **Figure 9.11**
If the spinning wheel freely falls, the ladybugs inside will experience a centrifugal force that feels like gravity when the wheel spins at the appropriate rate. To the occupants, the direction "up" is toward the center of the wheel and "down" is radially outward.

Today we live on the outer surface of our spherical planet, held here by gravity. Earth has been the cradle of humankind. But we will not stay in the cradle forever. We are on our way to becoming a spacefaring people. In the years ahead many people will likely live in huge lazily rotating space stations where centrifugal force simulates gravity. The simulated gravity will be provided so the people can function normally.

Occupants in today's space shuttles feel weightless because they lack a support force. They're not pressed against a supporting floor by gravity, nor do they experience a centrifugal force due to spinning. But future space travelers need not be subject to weightlessness. Their space habitats will probably spin, like the ladybug's spinning bicycle wheel, effectively supplying a support force and nicely simulating gravity.

The comfortable 1 *g* we experience at the earth's surface is due to gravity. Inside a rotating spaceship the acceleration experienced is the centripetal/centrifugal acceleration due to rotation. The magnitude of this acceleration is directly proportional to the radial distance and the square of the rotational speed. For a given RPM, the

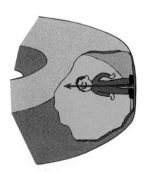

Figure 9.12 ▲
The interaction between the man and the floor as seen at rest outside the rotating system. The floor presses against the man (action) and the man presses back on the floor (reaction). The only force exerted on the man is by the floor. It is directed toward the center and is a centripetal force.

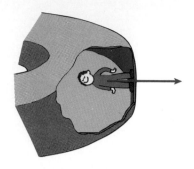

Figure 9.13 ▲

As seen from inside the rotating system, in addition to the man-floor interaction there is a centrifugal force exerted on the man at his center of mass. It seems as real as gravity. Yet, unlike gravity, it has no reaction counterpart—there is nothing out there that he can pull back on. Centrifugal force is not part of an interaction, but results from rotation. It is therefore called a fictitious force.

Figure 9.14 ▶

A NASA depiction of a rotational space colony.

acceleration, like the linear speed, increases with increasing radial distance. Doubling the distance from the axis of rotation doubles the centripetal/centrifugal acceleration. Tripling the distance triples the acceleration, and so forth. At the axis where radial distance is zero, there is no acceleration due to rotation.

Small-diameter structures would have to rotate at high speeds to provide a simulated gravitational acceleration of 1 g. Sensitive and delicate organs in our inner ears sense rotation. Although there appears to be no difficulty at a single revolution per minute (1 RPM) or so, many people have difficulty adjusting to rotational rates greater than 2 or 3 RPM (although some people easily adapt to 10 or so RPM). To simulate normal earth gravity at 1 RPM requires a large structure—one almost 2 km in diameter. This is an immense structure compared with the size of today's space shuttle vehicles. Economics will probably dictate that the size of the first inhabited structures be small. If these structures also do not rotate, the inhabitants will have to adjust to living in a seemingly weightless environment. Larger rotating habitats with simulated gravity will likely follow later.

If the structure rotates so that inhabitants on the inside of the outer edge experience 1 g, then halfway between the axis and the outer edge they would experience only 0.5 g. At the axis itself they would experience weightlessness at 0 g. The possible variations of g within the rotating space habitat holds promise for a most different and as yet unexperienced environment. We could perform ballet at 0.5 g; acrobatics at 0.2 g and lower g states; three-dimensional soccer and sports not yet conceived in very low g states. People will explore possibilities never before available to them. This time of transition from our earthly cradle to new vistas is an exciting time in which to live—especially for those who will be prepared to play a role in these new adventures.*

---

* At the risk of stating the obvious, this preparation can well begin by taking your study of physics very seriously.

9 Chapter Review

Concept Summary

An object rotates when it turns around an internal axis; it revolves when it turns around an external axis.

- Rotational speed is the number of rotations or revolutions made per unit of time.

A centripetal force pulls objects toward a center.

- An object moving in a circle is acted on by a centripetal force.

- When an object moves in a circle, there is no force pushing the object outward from the circle.

- From within a rotating frame of reference, there seems to be an outwardly directed centrifugal force, which can simulate gravity.

Important Terms

axis (9.1)
centrifugal force (9.4)
centripetal force (9.3)
linear speed (9.2)
revolution (9.1)
rotation (9.1)
rotational speed (9.2)
tangential speed (9.2)

Review Questions

1. Distinguish between a rotation and a revolution. (9.1)

2. Does a child on a merry-go-round revolve or rotate around the merry-go-round's axis? (9.1)

3. Distinguish between linear speed and rotational speed. (9.2)

4. What is linear speed called when something moves in a circle? (9.2)

5. At a given distance from the axis, how does linear (or tangential) speed change as rotational speed changes? (9.2)

6. At a given rotational speed, how does linear (or tangential) speed change as the distance from the axis changes? (9.2)

7. When you roll a cylinder across a surface it follows a straight-line path. A tapered cup rolled on the same surface follows a circular path. Why? (9.2)

8. When you whirl a can at the end of a string in a circular path, what is the direction of the force that acts on the can? (9.3)

9. Does the force that holds the riders on the carnival ride in Figure 9.1 act toward or away from the center? (9.3)

10. Does an inward force or an outward force act on the clothes during the spin cycle of an automatic washer? (9.3)

11. When a car makes a turn, do seat belts provide you with a centripetal force or centrifugal force? (9.4)

12. If the string that holds a whirling can in its circular path breaks, what causes the can to move in a straight-line path—centripetal force, centrifugal force, or a lack of force? What law of physics supports your answer? (9.4)

13. Identify the action and reaction forces in the interaction between the ladybug and the whirling can in Figure 9.10. (9.5)

14. A ladybug in the bottom of a whirling tin can feels a centrifugal force pushing it against the bottom of the can. Is there an outside source of this force? Can you identify this as the action force of an action-reaction pair? If so, what is the reaction force? (9.5)

15. Why is the centrifugal force the ladybug feels in the rotating frame called a fictitious force? (9.5)

16. For a rotating space habitat of a given size, what is the relationship between the magnitude of simulated gravity and the habitat's rate of spin? (9.6)

17. For space habitats spinning at the same rate, what is the relationship between simulated gravity and the radius of the habitat? (9.6)

18. Why will orbiting space stations that simulate gravity likely be large structures? (9.6)

Think and Explain

1. If you lose your grip on a rapidly spinning merry-go-round and fall off, in which direction will you fly?

2. A ladybug sits halfway between the axis and the edge of a rotating turntable. What will happen to the ladybug's linear speed if:

   a. the RPM rate is doubled?

   b. the ladybug sits at the edge?

   c. both **a** and **b** occur?

3. Which state in the United States has the greatest tangential speed as the earth rotates around its axis?

4. The speedometer in a car is driven by a cable connected to the shaft that turns the car's wheels. Will speedometer readings be more or less than actual speed when the car's wheels are replaced with smaller ones?

5. Keeping in mind the concept from the previous question, a taxi driver wishes to increase his fares by adjusting the air pressure in his tires. Should he inflate the tires at low pressure or high pressure?

6. Consider the pair of cups taped together as shown. Will this design self-correct its motion and keep the pair of wheels on track? Predict before you try it and see!

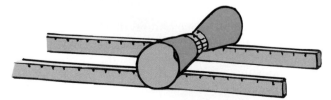

7. A motorcyclist is able to ride on the vertical wall of a bowl-shaped track, as shown. Does centripetal force or centrifugal force act on the motorcycle? Defend your answer.

8. When a soaring eagle turns during its flight, what is the source of the centripetal force acting on it?

9. A car resting on a level road has two forces acting on it: its weight (down) and the normal force (up). Recall from Chapter 4 that the normal force is the support force, which is always perpendicular to the supporting surface. If the car makes a turn on a level road, the normal force is still straight up. Friction between the wheels and road is the only centripetal force providing curved motion. Suppose the road is banked so the normal force has a component that provides centripetal force as shown in the sketch. Do you think the road could be banked so that for a given speed and a given radius of curvature, a vehicle could make the turn without friction? Explain.

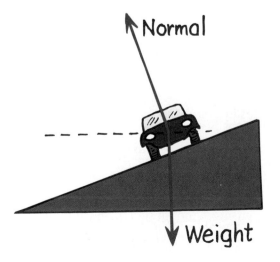

10. Does centripetal force do work on a rotating object? (Hint: Consider the direction of centripetal force and the direction of motion.) Defend your answer.

11. Explain why the faster the earth spins, the less a person weighs, whereas the faster a space station spins, the more a person weighs.

12. Occupants in a single space shuttle in orbit feel weightless. Describe a scheme whereby occupants in a pair of shuttles (or even one shuttle and another massive object in orbit) would be able to use a long cable to continuously experience a comfortably normal earth-like gravity.

Think and Solve

1. Mars is about twice the distance from the sun as is Venus. A Martian year, which is the time it takes Mars to go around the sun, is about three times as long as a Venusian year.

a. Which of these two planets has the greater rotational speed in its orbit?

b. Which planet has the greater linear speed?

2. A turntable that turns 10 revolutions each second is located on top of a mountain. Mounted on the turntable is a laser that emits a bright beam of light. As the turntable and laser rotate, the beam also rotates and sweeps across the sky. On a dark night the beam reaches some clouds 10 km away.

a. How fast does the spot of laser light sweep across the clouds?

b. How fast does the spot of laser light sweep across clouds that are 20 km away?

c. At what distance will the laser beam sweep across the sky at the speed of light (the speed of light is 300 000 km/s)? (In Chapter 15 we will learn that no material can travel at the speed of light; here we have a nonmaterial spot. Not only can it travel at the speed of

light, at greater radial distances the nonmaterial spot can travel *faster* than the speed of light!)

3. Consider a too-small space station that consists of a 4-m-radius rotating sphere. A man standing inside is 2 m tall and his feet are at 1 *g*. What is the acceleration of his head? Explain why rotating space structures need to be large.

4. Standing inside a rotating space station, your feet have a greater linear speed and a greater centripetal acceleration than your head. We say there is a difference in *g*s from your head to your feet, and this difference can be quite uncomfortable. Studies show, however, that a difference of $\frac{1}{100}$ *g* across the human body produces no discomfort. What, then, should be the radius of the space station compared to your height so that the difference in acceleration between your head and feet is only $\frac{1}{100}$ *g*?

10 Center of Gravity

Stability depends on the location of the center of gravity.

Why doesn't the famous Leaning Tower of Pisa topple over? How far can it lean before it does topple over? Why is it impossible for us to stand with our back and our heels against a wall and bend over and touch our toes without toppling forward? To answer these questions, we first need to know about center of gravity. Then we need to know how this concept can be applied to balancing and stability. Let's start with center of gravity.

10.1 Center of Gravity

Throw a baseball into the air, and it follows a smooth parabolic path. Throw a baseball bat into the air and its path is not smooth. The bat seems to wobble all over the place. But it wobbles about a special point. This point stays on a parabolic path, even though the rest of the bat does not; see Figure 10.1. The motion of the bat is the sum of two motions: (1) a spin around this point and (2) a movement through the air as if all the weight were concentrated at this point. This point is the **center of gravity** of the bat.

Figure 10.1 ▶
The centers of gravity of the baseball and of the spinning baseball bat each follow parabolic paths.

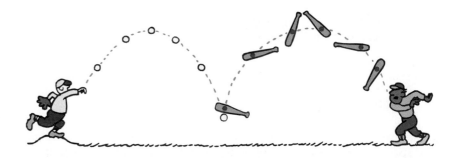

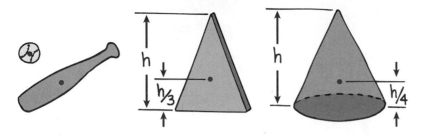

The center of gravity of an object is the point located at the object's average position of weight. For a symmetrical object, such as a baseball, this point is at the geometric center of the object. But an irregularly shaped object, such as a baseball bat, has more weight at one end than the other end, so the center of gravity is toward the heavier end. The center of gravity of a piece of tile cut into the shape of a triangle is located on the line passing through the center and the apex, one-third of the way up from the base. A solid cone's center of gravity is one-fourth of the way up from its base.

Objects not made of the same material throughout (that is, objects of varying density) may have the center of gravity quite far from the geometric center. Consider a hollow ball half-filled with lead. The center of gravity would not be at the geometric center; rather, it would be located somewhere within the lead part. The ball will always roll to a stop with its center of gravity as low as possible. Make the ball the body of a lightweight toy clown, and whenever it is pushed over, it will come back right-side up (Figure 10.3).

The multiple-flash photograph in Figure 10.4 shows the top view of a wrench sliding across a smooth horizontal surface. Notice that its center of gravity, which is marked by the white dot, follows a straight-line path. Other parts of the wrench rotate about this point as the wrench moves across the surface. Notice also that the center of gravity moves equal distances in the equal time intervals of the flashes (because no net force acts in the direction of motion). The motion of the wrench is a combination of straight-line motion of its center of gravity and rotation around its center of gravity.

Figure 10.3 ▲
The center of gravity of the toy is below its geometric center.

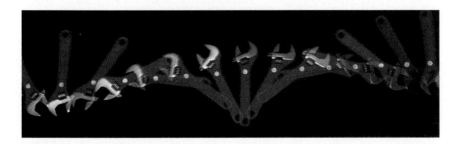

Figure 10.4 ▲
The center of gravity of the rotating wrench follows a straight-line path.

If the wrench were instead tossed into the air, no matter how it rotates its center of gravity would follow a smooth parabola. The same is true even for an exploding projectile, such as the fireworks rocket

Figure 10.5 ▶
The center of gravity of the fire-works rocket and its fragments move along the same path before and after the explosion.

shown in Figure 10.5. The internal forces during the explosion do not change the projectile's center of gravity. Interestingly enough, if air resistance is negligible, the center of gravity of the dispersed fragments as they fly through the air will be at any time where the center of gravity would have been if the explosion had never occurred.

10.2 Center of Mass

Center of gravity is often called **center of mass,** which is the average position of all the particles of *mass* that make up an object. For almost all objects on and near the earth, these terms are interchangeable. There can be a small difference between center of gravity and center of mass when an object is large enough for gravity to vary from one part to another. For example, the center of gravity of the World Trade Center is about 1 millimeter below its center of mass. This is due to the lower stories being pulled a little more strongly by the earth's gravity than the upper stories. For everyday objects (including tall buildings!) we can use the terms *center of gravity* and *center of mass* interchangeably.

If you threw a wrench so that it rotated as it moved through the air, you'd see it wobble about its center of gravity. The center of gravity itself would follow a parabolic path. Now suppose you threw a lopsided ball—one with its center of gravity off-center. You'd see it wobble also. The sun itself wobbles for a similar reason. As shown in Figure 10.6, the center of mass of the solar system can lie outside the massive sun, not at the sun's geometric center. Why? Because the masses of the planets contribute to the overall mass of the solar system. As the planets orbit at their respective distances, the sun actually wobbles off-center. Astronomers look for similar wobbles in nearby stars—the wobble is an indication of a star with a planetary system.

Figure 10.6 ▼
The center of mass of the solar system is not at the geometric center of the sun. If all the planets were lined up on one side of the sun, the center of mass would be about 2 solar radii from the sun's center.

10.3 Locating the Center of Gravity

The center of gravity (called the CG from here on) of a uniform object (such as a meterstick) is at the midpoint, its geometric center. The CG is the balance point. Supporting that single point supports the whole object. In Figure 10.7 the many small vectors represent the force of gravity along the meterstick. All of these can be combined into a resultant force that acts at the CG. The effect is as if the weight of the meterstick were concentrated at this point. That's why you can balance the meterstick by a single upward force directed at this point.

If you suspend any object (a pendulum, for example) at a single point, the CG of the object will hang directly below (or at) the point of suspension. To locate the object's CG, construct a vertical line beneath the point of suspension. The CG lies somewhere along that line. Figure 10.8 shows how a plumb line and bob can be used to construct a line that is exactly vertical. You can locate the CG by suspending the object from some other point and constructing a second vertical line. The CG is where the two lines intersect.

The CG of an object may be located where no actual material exists. The CG of a ring lies at the geometric center where no matter exists. The same holds true for a hollow sphere such as a basketball. The CG of even half a ring or half a hollow ball is still outside the physical structure. There is no material at the CG of an empty cup, bowl, or boomerang.

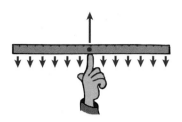

Figure 10.7 ▲
The weight of the entire stick behaves as if it were concentrated at its center.

Figure 10.8 ▲
Finding the CG for an irregularly shaped object with a plumb bob.

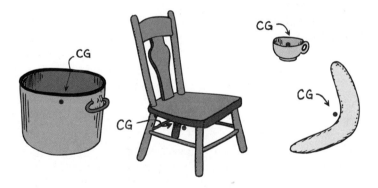

◄ **Figure 10.9**
There is no material at the CG of these objects.

■ Questions

1. Where is the CG of a donut?
2. Can an object have more than one CG?

■ Answers

1. In the center of the hole!

2. A rigid object has one CG. If it is nonrigid, such as a piece of clay or putty, and is distorted into different shapes, then its CG may change as its shape is changed. Even then, it has one CG for any given shape.

10.4 Toppling

Pin a plumb line to the center of a heavy wooden block and tilt the block until it topples over as shown in Figure 10.10. You can see that the block will begin to topple when the plumb line extends beyond the supporting base of the block.

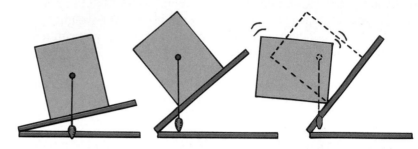

Figure 10.10 ▶
The block topples when the CG extends beyond its support base.

The rule for toppling is this: If the CG of an object is above the area of support, the object will remain upright. If the CG extends outside the area of support, the object will topple. This principle is dramatically employed in Figure 10.11.

Figure 10.11 ▶
A "Londoner" double-decker bus undergoing a tilt test. The bus must not topple when the chassis is tilted 28° with the top deck fully loaded and only the driver and conductor on the lower deck. Because so much of the weight of the vehicle is in the lower part, the load of the passengers on the upper deck raises the CG only a little, so the bus can be tilted well beyond this 28° limit without toppling.

Figure 10.12 ▲
The Leaning Tower of Pisa does not topple over because its CG lies above its base.

The Leaning Tower of Pisa does not topple because its CG does not extend beyond its base. As shown in Figure 10.12, a vertical line below the CG falls inside the base, and so the Leaning Tower has stood for centuries. If the tower leaned far enough so that the CG extended beyond the base, the tower would topple.

Males Versus Females

Try the following with a group of males and females. Stand exactly two footlengths away from a wall and place a chair between yourself and the wall. Bend over with a straight back and let your head lean against the wall, as shown. Then lift the chair off of the floor while your head is still leaning against the wall. Now attempt to straighten up. Give two reasons why females can generally do this while males cannot.

2 FOOTLENGTHS

The support base of an object does not have to be solid. The four legs of a chair bound a rectangular area that is the support base for the chair, as shown in Figure 10.13. Practically speaking, supporting props could be erected to hold the Leaning Tower up if it leaned too far. Such props would create a new support base. An object will remain upright if the CG is above its base of support.

■ Questions

1. When you carry a heavy load—such as a pail of water—with one arm, why do you tend to hold your free arm out horizontally?

2. To resist being toppled, why does a wrestler stand with (a) feet wide apart and (b) knees bent?

3. How will the support base of the chair in Figure 10.13 change if one of the front legs is removed? Will the chair topple?

■ Answers

1. You tend to hold your free arm outstretched to shift the CG of your body away from the load so your combined CG will more easily be above the base of support. To really help matters, divide the load in two if possible, and carry half in each hand. Or, carry the load on your head!

2. (a) Wide-apart feet increase the support base. (b) Bent knees lower the CG.

3. The support base for four legs is a rectangle. With three legs it's a triangle of half the area. The CG will be toward the rear of the chair because of the weight of the back and loss of weight of the front leg, and be within and above the triangular support base. So it will not topple—until somebody sits on it!

Tails

You can bend over only so far when trying to extend your horizontal reach. How far you can extend depends on keeping your CG within your support base. A monkey can reach proportionally much farther than you can without toppling. How? By extending its tail, thus keeping its CG above its feet. A tail gives an animal the ability to shift its CG and increase stability. The massive tails of dinosaurs tell us that they were able to extend their heads considerably beyond the support base of their feet.

Figure 10.13 ▲
The shaded area bounded by the bottom of the chair legs defines the support base of the chair.

Try balancing the pole end of a broom or mop on the palm of your hand. The support base is very small and relatively far beneath the CG, so it is difficult. But after some practice you can do it if you learn to make slight movements of your hand to exactly respond to variations in balance. You learn to avoid underresponding or over-responding to the slightest variations in balance. Similarly, high-speed computers help massive rockets remain upright when they are launched. Variations in balance are quickly sensed. The computers regulate the firings at multiple nozzles to make corrective adjustments, quite similar to the way your brain coordinates your adjustive action when balancing a long pole on the palm of your hand. Both feats are truly amazing.

10.5 Stability

It is nearly impossible to balance a pen upright on its point, while it is rather easy to stand it upright on its flat end, because the base of support is inadequate for the point and adequate for the flat end. But there is a second reason. Consider a solid wooden cone on a level table. As you can see in Figure 10.14, you cannot stand it on its tip. Even if you position it so that its center of gravity is exactly above its tip, the slightest vibration or air current will cause the cone to topple. When it does topple, will the center of gravity be raised, be lowered, or not change at all? The answer to this question provides the second reason for stability. A little thought will show that the CG is lowered by *any* movement. We say that an object balanced so that any displacement lowers its center of gravity is in **unstable equilibrium.**

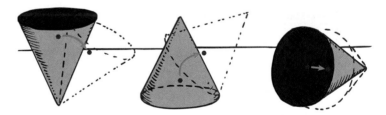

Figure 10.14 ▲
Equilibrium is (a) *unstable* when the CG is lowered with displacement, (b) *stable* when work must be done to raise the CG, and (c) *neutral* when displacement neither raises nor lowers the CG.

A cone balances easily on its base. To make it topple, its CG must be raised. This means the cone's potential energy must be increased, which requires work. We say an object that is balanced so that any displacement raises its center of gravity is in **stable equilibrium.**

Place the cone on its side and its CG is neither raised nor lowered with displacement. An object in this configuration is in **neutral equilibrium.**

Like the cone, the pen is in unstable equilibrium when it is on its point. When the pen is on its flat end, as in Figure 10.15, it is in stable equilibrium because the CG must be raised slightly to topple it over.

Figure 10.15 ▲
For the pen to topple when it is on its flat end, it must rotate over one edge. During the rotation, the CG rises slightly and then falls.

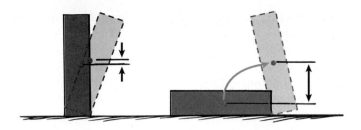

◄ **Figure 10.16**
Toppling the upright book requires only a slight raising of its CG; toppling the flat book requires a relatively large raising of its CG. Which requires more work?

Consider the upright book and the book lying flat in Figure 10.16. Both are in stable equilibrium. But you know the flat book is more stable. Why? Because it would take considerably more work to raise its CG to the point of toppling than to do the same for the upright book. An object with a low CG is usually more stable than an object with a relatively high CG.

The horizontally balanced pencil on the left in Figure 10.17 is in unstable equilibrium. Its CG is lowered when it tilts. But suspend a potato from each end and the pencil becomes stable, as shown on the right in Figure 10.17. Why? Because the CG is below the point of support, and is raised when the pencil is tilted.

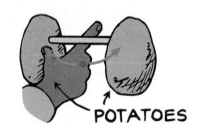

POTATOES

◄ **Figure 10.17**
(Left) A pencil balanced on the edge of a hand is in unstable equilibrium. (Right) When its ends are stuck into long potatoes that hang below, it is stable because its new CG rises when it is tipped.

■ Question

Explain why the toy in Figure 10.3 cannot remain on its side.

Some well-known balancing toys depend on this principle. Their secret is that they have been weighted so that the CG lies vertically underneath the point of support while most of the remainder of the toy is above it. See the example in Figure 10.18. A toy that hangs with its CG below its point of support is in stable equilibrium because the CG rises when the toy tilts.

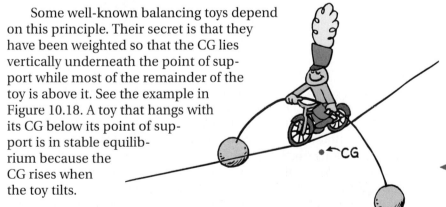

◄ **Figure 10.18**
The toy is in stable equilibrium because the CG rises when the toy tilts.

■ Answer

The CG is lowest when the toy is upright. Tipping the toy raises its CG and increases its potential energy. If it is pushed over, it will easily roll back with the aid of gravity until the CG is the lowest and its potential energy the least it can be.

Figure 10.19 ▲
The Seattle Space Needle can no more fall over than can a floating iceberg. In both, the CG is below the surface.

The CG of a building is lowered if much of the structure is below ground level. This is important for tall, narrow structures. An extreme example is the state of Washington's tallest freestanding structure, the Space Needle in Seattle. This structure is so "deeply rooted" that its center of gravity is actually below ground level. It cannot fall over intact. Why? Because falling would not lower its CG at all. If the structure were to tilt intact onto the ground, its CG would be raised!

The tendency for the CG to take the lowest position available can be seen by placing a very light object, such as a table-tennis ball, at the bottom of a box of dried beans or small stones. Shake the box, and the beans or stones tend to go to the bottom and force the ball to the top. By this process the CG of the whole system takes a lower position.

Figure 10.20 ▲
(Left) A table-tennis ball is placed at the bottom of a container of dried beans. (Right) When the container is shaken from side to side, the ball is nudged to the top. The result is a lower center of gravity.

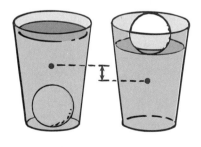

Figure 10.21 ▲
The CG of the glass of water is higher when the table-tennis ball is anchored to the bottom (left) and lower when the ball floats (right).

The same thing happens in water when an object rises to the surface and floats. If the object weighs less than an equal volume of water, the CG of the whole system will be lowered when the object is forced to the surface. This is because the heavier (more dense) water can then occupy the available lower space. If the object is heavier than an equal volume of water, it will be more dense than water and sink. In either case, the CG of the whole system is lowered. In the case where the object weighs the same as an equal volume of water (same density), the CG of the system is unchanged whether the object rises or sinks. The object can be at any level beneath the surface without affecting the CG. You can see that a fish must weigh the same as an equal volume of water (have the same density); otherwise it would be unable to remain at different levels in the water. We will return to these ideas in Chapter 19, where liquids are treated in more detail.

Shake a box of stones of different sizes and observe what happens. The shaking enables the small stones to slip down into the spaces between the larger stones and in effect lower the CG. The larger stones therefore tend to rise to the top. The same thing happens when a tray of berries is gently shaken—the larger berries tend to come to the top.

10.6 Center of Gravity of People

When you stand erect with your arms hanging at your sides, your CG is within your body. It is typically 2 to 3 cm below your navel, and midway between your front and back. The CG is slightly lower in

Figure 10.22 ▲
A high jumper executes a "Fosbury flop" to clear the bar while his CG passes beneath the bar.

Figure 10.23 ▲
When you stand, your CG is somewhere above the area bounded by your feet.

women than in men because women tend to be proportionally larger in the pelvis and smaller in the shoulders. In children, the CG is approximately 5% higher because of their proportionally larger heads and shorter legs.

Raise your arms vertically overhead. Your CG rises 5 to 8 cm. Bend your body into a U or C shape and your CG may be located outside your body altogether. This fact is nicely employed by the high jumper in Figure 10.22, who clears the bar while his CG nearly passes beneath the bar.

When you stand, your CG is somewhere above your support base, the area bounded by your feet. In unstable situations, as in standing in the aisle of a bumpy-riding bus, you place your feet farther apart to increase this area. Standing on one foot greatly decreases this area. In learning to walk, a baby must learn to coordinate and position the CG above a supporting foot. Many birds, pigeons for example, do this by jerking their heads back and forth with each step.

You can probably bend over and touch your toes without bending your knees. In doing so, you unconsciously extend the lower part of your body, as shown in Figure 10.24. In this way your CG, which is now outside your body, is nevertheless above your supporting feet. If you try it while standing with your back and heels to a wall, you may be in for a surprise. You cannot do it! This is because you are unable to adjust your body, and your CG protrudes beyond your feet. You are off balance and you topple over.

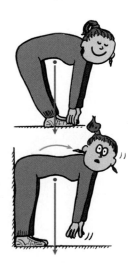

Figure 10.24 ▲
You can lean over and touch your toes without toppling only if your CG is above the area bounded by your feet.

Science and Pseudoscience

Science uses a powerful method of combining logic, observation, and experiment to find correlations, sometimes leading to a cause-and-effect relationship between things. It involves asking the kinds of questions science can handle, and searching for answers via careful, controlled experimentation. Only when repeated experiments produce consistent results and objective evidence is provided, is an idea scientifically valid. Such ideas reliably explain and predict many types of events.

A pseudoscience is a false science. It claims the power of science to explain and predict events, but it is not based on the careful methods of science. Often, "evidence" cited by a pseudoscientist to "prove"

PALM & CARD READINGS

his or her case is subjective. Also, in pseudoscience, cause-and-effect relationships may be claimed, but no detailed logical connections can be provided.

The danger of pseudoscience is that it can lead us to believe things that aren't true, or make us think we know things we don't. Thus, we may make unwise decisions. Nevertheless, pseudosciences appeal to many people. They can excite the imagination, simplify complex issues, and soothe anxiety about the unknown.

Are horoscopes that are seen frequently in newspapers and magazines an example of science or pseudoscience? Why or why not? How can you identify pseudoscience?

All the examples of toppling illustrate the rule: If the CG extends beyond the support base, toppling occurs. Knowing this rule can sharpen your perception of many familiar things—like the upright positions of toys, buildings, and tall trees, and how far they can lean and still remain stable. But knowing this rule is only a beginning, for this chapter has not taught you why things topple when they aren't above a support base. The reason has to do with the concept of torque, which begins the next chapter.

You don't need to take a course in physics to know where to balance a baseball bat, how to stand a pencil upright on its flat end, or that you can't lean over and touch your toes if your heels are against a wall. With or without physics, everybody knows that it is easier to hang by your hands below a supporting rope than it is to stand on your hands above a supporting floor. And you don't need a formal study of physics to balance like a gymnast. But maybe it's nice to know that physics is at the root of many things you already know about.

Knowing about things is not always the same as understanding things. Understanding begins with knowledge. So we begin by knowing about things, and then progress deeper to an understanding of things. That's where a knowledge of physics is very helpful.

10 Chapter Review

Concept Summary

The center of gravity (CG) of an object is the point at the center of its weight distribution.

- When an object is thrown through the air, its CG follows a smooth parabolic path, even if the object spins or wobbles.

- For everyday objects, the center of gravity is the same as the center of mass.

An object will remain upright if its CG is above the area of support.

- An object is in stable equilibrium when any displacement raises its CG.

Important Terms

center of gravity (10.1)
center of mass (10.2)
neutral equilibrium (10.5)
stable equilibrium (10.5)
unstable equilibrium (10.5)

Review Questions

1. Why is the CG of a baseball bat not at its midpoint? (10.1)

2. What part of an object follows a smooth path when the object is made to spin through the air or across a flat smooth surface? (10.1)

3. Describe the motion of the CG of a projectile, before and after it explodes in midair. (10.1)

4. When are the CG and center of mass of an object the same? When can the CG and center of mass be different? (10.2)

5. What is suggested by a star that wobbles off center? (10.2)

6. How can the CG of an irregularly shaped object be determined? (10.3)

7. Cite an example of an object that has a CG where no physical material exists. (10.3)

8. Why does the Leaning Tower of Pisa not topple? (10.4)

9. How far can an object be tipped before it topples over? (10.4)

10. How is balancing the pole end of a broom in an upright position on the palm of your hand similar to the launching of a rocket? (10.4)

11. Distinguish between unstable, stable, and neutral equilibrium. (10.5)

12. Is the gravitational potential energy more, less, or unchanged when the CG of an object is raised? (10.5)

13. Why is it easier to hang by your arms below a supporting cable than to do handstands on a supporting floor? (10.5)

14. What is the "secret" of balancing toys that exhibit stable equilibrium while appearing to be unstable? (10.5)

15. What accounts for the stability of the Space Needle in Seattle? (10.5)

16. If a container of dried beans with a table-tennis ball at the bottom is shaken, what happens to the CG of the container? (10.5)

17. What happens to the CG of a glass of water when a table-tennis ball is poked beneath the surface? (10.5)

18. Why do some high jumpers arch their bodies into a U shape when passing over the high bar? (10.6)

19. Why do you spread your feet farther apart when standing in a bumpy-riding bus? (10.6)

20. Why can you not successfully bend over and touch your toes when you stand with your back and heels against a wall? (10.6)

Activities

1. Suspend a belt from a piece of stiff wire that is bent as shown. Why does the belt balance as it does?

2. Rest a meterstick on two extended forefingers as shown. Slowly bring your fingers together. Note what occurs as your fingers meet at the stick's CG. Can you explain why this always happens, no matter where you start your fingers? Do you see the role of friction here? Do you see that when one finger gets ahead of the other, weight on it increases, enough to keep it from moving while the other finger, with less friction, slides and catches up? If you repeat with a small weight at one end of the stick, will your fingers meet at the system's CG?

3. Hang a hammer on a loose ruler as shown. Then explain why it doesn't fall.

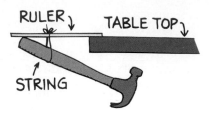

Think and Explain

1. To balance automobile wheels, particularly when tires have worn unevenly, lead weights are fastened to their edges. Where should the CG of the balanced wheel be located?

2. Why does a washing machine vibrate violently if the clothes are not evenly distributed in the tub?

3. A bottle rack that seems to defy common sense is shown in the figure. Where is the CG of the rack and bottle?

4. Which glass in the figure is unstable and will topple?

5. Which balancing act in the figure is in stable equilibrium? In unstable equilibrium? Nearly at neutral equilibrium?

6. How can the three bricks in the figure be stacked so that the top brick has maximum horizontal overhang above the bottom brick? For example, stacking them as the dotted lines suggest would be unstable and the bricks would topple. (*Hint:* Start with the top brick and think your way down. At every interface the CG of the bricks above must not extend beyond the end of the supporting brick.)

7. Why is it dangerous to roll open the top drawers of a fully loaded file cabinet that is not secured to the floor?

8. The CGs of the three trucks parked on a hill are shown by the Xs in the figure. Which truck(s) will topple over?

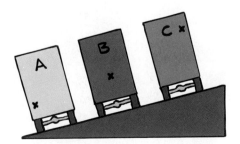

9. Why does a pregnant woman during the late stages of pregnancy or a man with a large paunch tend to lean backward when walking?

10. A long track balanced like a seesaw supports a golf ball and a more massive billiard ball with a compressed spring between the two as shown in the figure. The CG of the two-ball system is therefore directly above the point of support (the triangular fulcrum). When the spring is released, the balls move away from each other. As the balls roll outward, will the track remain in balance, or will it tip? What principles do you use for your explanation?

11 Rotational Mechanics

Rotational motion can produce linear motion.

Push on an object that is free to move, and you set it in motion. Some objects will move without rotating, some will rotate without moving, and others will do both. For example, a kicked football often tumbles end over end. What determines whether an object will rotate when a force acts on it? This chapter is about the factors that affect rotation. We will see that these factors explain most of the techniques used by gymnasts, ice skaters, and divers.

11.1 Torque

Every time you open a door, turn on a water faucet, or tighten a nut with a wrench, you exert a turning force. This turning force produces a *torque* (rhymes with fork). Torque is different from force. If you want to make an object move, apply a force. Forces tend to make things accelerate. If you want to make an object turn or rotate, apply a torque. Torques produce rotation.

Figure 11.1 ▶
A torque produces rotation.

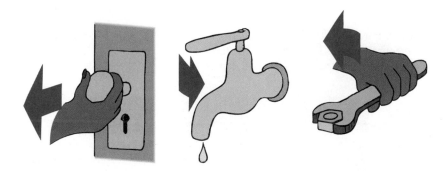

A torque is produced when a force is applied with "leverage." You use leverage when you use a claw hammer to pull a nail from a piece of wood. The longer the handle of the hammer, the greater the leverage and the easier the task. The longer handle of a crowbar provides even more leverage. You use leverage when you use a screwdriver or a table knife to open the lid of a paint can.

A torque is used when opening a door. A doorknob is placed far away from the turning axis at its hinges to provide more leverage when you push or pull on the doorknob. The direction of your applied force is important. In opening a door, you'd never push or pull the doorknob sideways to make the door turn. You push *perpendicular* to the plane of the door. Experience has taught you that a perpendicular push or pull gives more rotation for less effort.

If you have used both short- and long-handled wrenches, you also know that less effort and more leverage result with a long handle. When the force is perpendicular, the distance from the turning axis to the point of contact is called the **lever arm.** If the force is not at a right angle to the lever arm, then only the perpendicular component of the force, $F_\perp$, will contribute to the torque. **Torque** is defined as*

$$\text{torque} = \text{force}_\perp \times \text{lever arm}$$

So the same torque can be produced by a large force with a short lever arm, or a small force with a long lever arm. Greater torques are produced when both the force and lever arm are large.

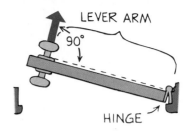

Figure 11.2 ▲
When a perpendicular force is applied, the lever arm is the distance between the doorknob and the edge with the hinges.

Figure 11.3 ▲
The doorknob on this door is placed at the center. What do physics types say about the practicality of this?

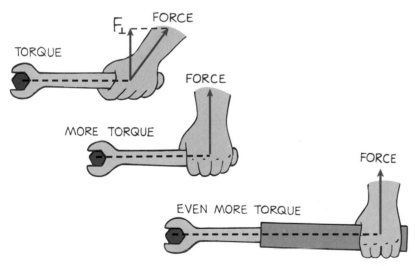

Figure 11.4 ▲
Although the magnitudes of the applied forces are the same in each case, the torques are different. Only the component of forces perpendicular to the lever arm contributes to torque.

---

* The unit of a torque is a newton-meter. Work is also measured in newton-meters (the same as joules), but work and torque are very different. What contributes to work is the force along the direction of motion; what contributes to torque is the force perpendicular to the lever arm.

1. If a doorknob were placed in the center of a door rather than at the edge, how much more force would be needed to produce the same torque for opening the door?

2. If you cannot exert enough torque to turn a stubborn bolt, would more torque be produced if you fastened a length of rope to the wrench handle as shown?

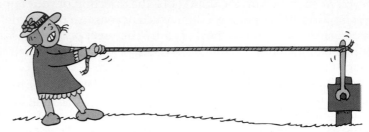

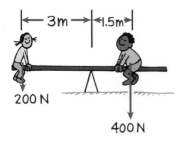

Figure 11.5 ▲
A pair of torques can balance each other.

11.2 Balanced Torques

Torques are intuitively familiar to youngsters playing on a seesaw. Children can balance a seesaw even when their weights are not equal. Weight alone does not produce rotation—torque does. Children soon learn that the distance they sit from the pivot point is as important as their weight (Figure 11.5). The heavier boy sits a shorter distance from the fulcrum (turning axis) while the lighter girl sits farther away. Balance is achieved if the torque that tends to produce clockwise rotation by the boy equals the torque that tends to produce counterclockwise rotation by the girl.

Scale balances that work with sliding weights are based on balanced torques, not balanced masses. The sliding weights are adjusted until the counterclockwise torque just balances the clockwise torque. Then the arm remains horizontal.

Figure 11.6 ▲
This scale relies on balanced torques.

11.3 Torque and Center of Gravity

If you attempt to lean over and touch your toes while standing with your back and heels to the wall, you will soon find yourself rotating. Recall from Chapter 10 that if there is no base of support beneath the

■ **Answers**

1. Twice as much, because the lever arm in the center of the door would be half as long. Mathematically, $2F \times d/2 = F \times d$, where F is the applied force, d is the distance from the hinges to the edge of the door, and $d/2$ is the distance to the center.

2. No, because the lever arm is the same. To increase the lever arm, a better idea would be to use a pipe that extends upward.

Suppose that a meterstick is supported at the center, and a 20-N block is hung at the 80-cm mark. Another block of unknown weight just balances the system when it is hung at the 10-cm mark. What is the weight of the second block?

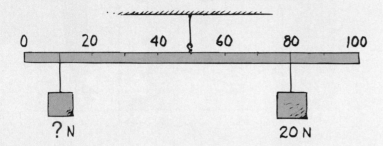

You can compute the unknown weight by applying the principle of balanced torques. The block of unknown weight tends to rotate the system of blocks and stick *counterclockwise* (ccw), and the 20-N block tends to rotate the system *clockwise* (cw). The system is in balance when the two torques are equal:

counterclockwise torque = clockwise torque

$$(F_\perp d)_{ccw} = (F_\perp d)_{cw}$$

The equation is rearranged to solve for the unknown weight:

$$F_{\perp ccw} = \frac{(F_\perp)_{cw} \times (d)_{cw}}{(d)_{ccw}}$$

It is important to note that the lever arm for the unknown weight is 40 cm, because the distance between the 10-cm mark and the pivot point at the 50-cm mark is 40 cm. The lever arm for the 20-N block is 30 cm, because its distance from the pivot point is 30 cm. Substituting these values into the equation, we determine the unknown weight:

$$F_{\perp ccw} = \frac{(20 \text{ N}) \times (30 \text{ cm})}{(40 \text{ cm})} = 15 \text{ N}$$

The unknown weight is thus 15 N. This makes sense. You can tell that the weight is less than 20 N because its lever arm is greater than that of the block of known weight. In fact, the unknown weight's lever arm is (40 cm) ÷ (30 cm) or 4/3 that of the first block, so its weight is 3/4 as much. Anytime you use physics to compute something, consider whether or not your answer makes sense. Computation without comprehension is not conceptual physics!

center of gravity (CG), an object will topple. When the area bounded by your feet is not beneath your CG, there is a torque. Now you can see that the cause of toppling is this torque.

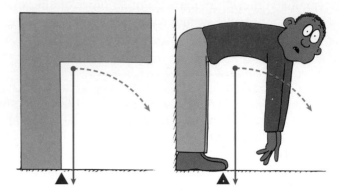

Figure 11.7 ▶
The L-shaped bracket will topple because of a torque. Similarly, when you stand with your back and heels to the wall and try to touch your toes, a torque is produced when your CG extends beyond your feet.

We began this chapter by asking what determines whether or not a football will tumble end over end when kicked. The answer involves CG, force, and torque. You know that a force is required to launch any projectile, whether it be a football or a Frisbee® flying disk. If the direction of the force is through the CG of the projectile, all the force can do is move the object as a whole. There will be no torque to turn the projectile around its own CG. However, if the force is directed "off center," then in addition to motion of the CG, the projectile will rotate about its CG.

For example, when you throw a ball and impart spin to it, or when you launch a Frisbee, a force must be applied off axis—to the edge of the object. This produces a torque that adds rotation to the projectile. If you wish to kick a football so that it sails through the air without tumbling, kick it in the middle (Figure 11.8 top). If you want it to tumble end over end in its trajectory, kick it below the middle (Figure 11.8 bottom) to impart torque as well as force to the ball.

Figure 11.8 ▲
If the football is kicked in line with its CG, it will move without rotating. If it is kicked above or below its CG, it will also rotate.

11.4 Rotational Inertia

In Chapter 3 you learned about inertia: An object at rest tends to stay at rest, and an object in motion tends to remain moving in a straight line. There is a similar law for rotation:

> An object rotating about an axis tends to keep rotating about that axis.

The resistance of an object to changes in its rotational motion is called **rotational inertia**.* Rotating objects tend to keep rotating, while nonrotating objects tend to stay nonrotating.

---

* Rotational inertia is sometimes called *moment of inertia.*

Just as it takes a force to change the linear state of motion of an object, a torque is required to change the rotational state of motion of an object. In the absence of a net torque, a rotating top keeps rotating, while a nonrotating top stays nonrotating.

Like inertia in the linear sense, rotational inertia depends on mass. But unlike inertia, rotational inertia depends on the distribution of the mass. The greater the distance between the bulk of the mass of an object and the axis about which rotation takes place, the greater the rotational inertia.

A long baseball bat held near its end has more rotational inertia than a short bat. Once moving, it has a greater tendency to keep moving, but it is harder to bring it up to speed. A short bat has less rotational inertia than a long bat, and is easier to swing. Baseball players sometimes "choke up" on a bat by grasping it closer to the more massive end than normal. Choking up on the bat reduces its rotational inertia and makes it easier to bring up to speed. A bat held at its end, or a long bat, doesn't "want" to swing as readily. Likewise with the legs of people and animals. Long-legged animals such as giraffes, horses, and ostriches normally run with a slower gait than hippos, dachshunds, and mice.

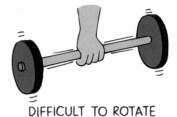
EASY TO ROTATE

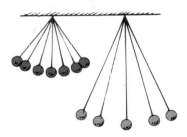
DIFFICULT TO ROTATE

Figure 11.9 ▲
Rotational inertia depends on the distance of mass from the axis.

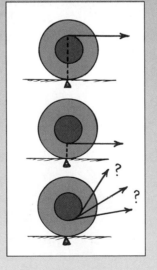

Figure 11.10 ▲
The short pendulum will swing back and forth more frequently than the long pendulum.

Figure 11.11 ▲
By holding a long pole, the tightrope walker increases his rotational inertia. This allows him to resist rotation and gives him plenty of time to readjust his CG.

It is important to note that the rotational inertia of an object is not necessarily a fixed quantity. It is greater when the mass within the object is extended from the axis of rotation. You can try this with your outstretched legs. Swing your outstretched leg back and forth from the hip. Now do the same with your leg bent. In the bent position it swings back and forth more easily. To reduce the rotational inertia of your legs, simply bend them. That's an important reason for running with your legs bent—bent legs are easier to swing back and forth.

Figure 11.12 ▲
For similar mass distributions, short legs have less rotational inertia than long legs. Animals with short legs run with quicker strides more easily than animals with long legs.

Figure 11.13 ▲
You bend your legs when you run to reduce their rotational inertia.

Flip Your Pencil

Flip your pencil back and forth between your fingers. Compare the ease of rotation when you flip it about its midpoint versus flipping it about one of its ends. For a third comparison, rotate the pencil between your thumb and forefinger about the pencil's long axis (so the lead is the axis). Do you see the three cases represented in Figure 11.14? Can you relate the relative ease of rotating the pencil with the formulas in the figure? In which case is rotation easiest? In this case is the small rotational inertia consistent with the small r?

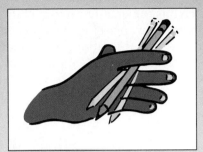

> **Activity**

Formulas for Rotational Inertia

When all the mass m of an object is concentrated at the same distance r from a rotational axis (as in a simple pendulum bob swinging on a string about its pivot point, or a thin wheel turning about its center), then the rotational inertia $I = mr^2$. When the mass is more spread out, as in your leg, the rotational inertia is less and the formula is different. Figure 11.14 compares rotational inertias for various shapes and axes. (It is not important for you to learn these values, but you can see how they vary with the shape and axis.)

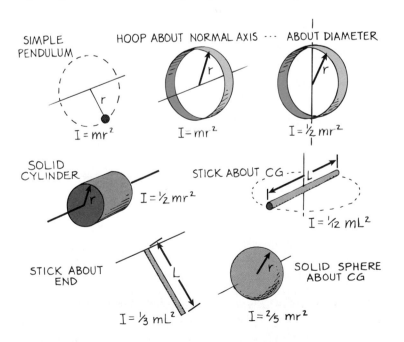

◀ **Figure 11.14**
Rotational inertias of various objects, each of mass m, about indicated axes.

Rolling

Which will roll down an incline with greater acceleration, a hollow cylinder or a solid cylinder of the same mass and radius? The answer is, the cylinder with the smaller rotational inertia. Why? Because the cylinder with the greater rotational inertia requires more time to get rolling. Remember that inertia of any kind is a measure of "laziness." Which has the greater rotational inertia—the hollow or the solid cylinder? The answer is, the one with its mass concentrated farthest from the axis of rotation—the hollow cylinder. So a hollow cylinder has a greater rotational inertia than a solid cylinder of the same radius and mass and will be more "lazy" in gaining speed. The solid cylinder will roll with greater acceleration.

Figure 11.15 ▲
A solid cylinder rolls down an incline faster than a hollow one, whether or not they have the same mass or diameter.

Interestingly enough, any solid cylinder will roll down an incline with more acceleration than any hollow cylinder, regardless of mass or radius. A hollow cylinder has more "laziness per mass" than a solid cylinder. Objects of the same shape but different sizes accelerate equally when rolled down an incline. You should experiment and see this for yourself. If started together, the smaller shape, whether it be a ball, disk, or hoop, rotates more times than the larger shape, but both reach the bottom of the incline in the same time. Why? Because all objects of the same shape have the same "laziness per mass" ratio. Similarly, recall from Chapter 5 how the same "weight per mass" ratio of all freely falling objects accounted for their equal acceleration: $a = F/m$.

■ Questions

1. When swinging your leg from your hip, why is the rotational inertia of the leg less when it is bent?

2. A heavy iron cylinder and a light wooden cylinder, similar in shape, roll down an incline. Which will have more acceleration?

■ Answers

1. The rotational inertia of any object is less when its mass is concentrated closer to the axis of rotation. Can you see that a bent leg satisfies this requirement?

2. The cylinders have different masses, but the *same rotational inertia per mass,* so both will accelerate equally down the incline. Their different masses make no difference, just as the acceleration of free fall is not affected by different masses. All objects of the same shape have the same "laziness per mass" ratio.

11.5 Rotational Inertia and Gymnastics

Consider the human body. You can rotate freely about three principal axes of rotation (Figure 11.16). These axes are each at right angles to the others (mutually perpendicular). Each axis coincides with a line of symmetry of the body and passes through the center of gravity. The rotational inertia of the body differs about each axis.

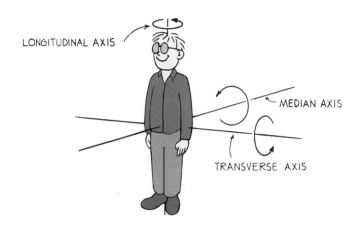

LONGITUDINAL AXIS

MEDIAN AXIS

TRANSVERSE AXIS

◀ **Figure 11.16**
The human body's principal axes of rotation.

Rotational inertia is least about the longitudinal axis, which is the vertical head-to-toe axis, because most of the mass is concentrated along this axis. Thus, a rotation of your body about your longitudinal axis is the easiest rotation to perform. An ice skater executes this type rotation when going into a spin. Rotational inertia is increased by simply extending a leg or the arms (Figure 11.17). The rotational inertia when both arms are extended is about three times more than when the arms are tucked in, so if you go into a spin with outstretched arms, you will triple your spin rate when you draw your arms in. With your leg extended as well, you can vary your spin rate by as much as six times.

You rotate about your transverse axis when you perform a somersault or a flip. Figure 11.18 shows the rotational inertia of different positions, from the least (when your arms and legs are drawn inward in the tuck position) to the greatest (when your arms and legs are fully extended in a line). The relative magnitudes of rotational inertia stated in the caption are with respect to the body's center of gravity.

Figure 11.17 ▼
Rotations about the longitudinal axis. Rotational inertia in position (d) is about 5 or 6 times as great as in position (a), so spinning in position (d) and then changing to position (a) will increase the spin rate about 5 or 6 times.

a

b

c

d

Figure 11.18 ▲
Rotations about the transverse axis. Rotational inertia is least in (a), the tuck position. It is about 1.5 times as great in (b), 3 times as great in (c), and 5 times as great in (d).

Rotational inertia is greater when the axis is through the hands, such as when doing a somersault on the floor or swinging from a horizontal bar with your body fully extended. The rotational inertia of a gymnast is up to 20 times greater when she is swinging in a fully extended position from a horizontal bar than after dismount when she somersaults in the tuck position. Rotation transfers from one axis to another, from the bar to a line through her center of gravity, and she automatically increases her rate of rotation by up to 20 times. This is how she is able to complete two or three somersaults before contact with the ground.

Figure 11.19 ▶
The rotational inertia of a body is with respect to the rotational axis. When the gymnast pivots about the bar (a), she has a greater rotational inertia than when she spins freely about her center of gravity (b).

The third axis of rotation for the human body is the front-to-back axis, or medial axis. This is a less common axis of rotation and is used in executing a cartwheel. Like rotations about the other axes, rotational inertia can be varied with different body configurations.

Figure 11.20 ▲
When the gymnast pivots around the horizontal bar with her body fully extended, her rotational inertia is greatest.

■ Question

Why is the rotational inertia greater for a gymnast when she swings by her hands from a horizontal bar with her body fully extended than when she executes a somersault with her body fully extended?

■ Answer

When the gymnast pivots about the bar, her mass is concentrated far from the axis of rotation. Her rotational inertia is therefore greater. When she somersaults, her mass is concentrated close to the axis because she is rotating about her center of gravity.

11.6 Angular Momentum

Anything that rotates, whether it be a colony in space, a cylinder rolling down an incline, or an acrobat doing a somersault, keeps on rotating until something stops it. A rotating object has an "inertia of rotation." Recall from Chapter 7 that all moving objects have "inertia of motion," or momentum, which is the product of mass and velocity. To be clear, let us call this kind of momentum **linear momentum.** Similarly, the "inertia of rotation" of rotating objects is called **angular momentum.**

Like linear momentum, angular momentum is a vector quantity and has direction as well as magnitude. When a direction is assigned to rotational speed, we call it **rotational velocity.** Rotational velocity is a vector whose magnitude is the rotational speed. (By convention, the rotational velocity vector, as well as the angular momentum vector, have the same direction and lie along the axis of rotation.)

Angular momentum is defined as the product of rotational inertia and rotational velocity.

angular momentum = rotational inertia × rotational velocity

or in equation form,

$$\text{angular momentum} = I \times \omega$$

It is the counterpart of linear momentum:

$$\text{linear momentum} = \text{mass} \times \text{velocity}$$

In this book, we won't treat the vector nature of angular momentum (or even of torque, which also is a vector) except to acknowledge the remarkable action of the gyroscope. The rotating bicycle wheel in Figure 11.23 shows what happens when a torque by Earth's gravity acts to change the direction of the bicycle wheel's angular momentum (which is along the wheel's axle). The pull of gravity that acts to topple the wheel over and change its rotational axis causes it instead to *precess* in a circular path about a vertical axis. You must do this yourself while standing on a turntable to fully *believe* it. Full *understanding* will likely not come until a later time.

Figure 11.21 ▲
A turntable has more angular momentum when it is turning at 45 RPM than at $33\frac{1}{3}$ RPM. It has even more angular momentum if a load is placed on it so its rotational inertia is greater.

Figure 11.22 ▲
A gyroscope. Low-friction swivels can be turned in any direction without exerting a torque on the whirling gyroscope. As a result, it stays pointed in the same direction.

◀ **Figure 11.23**
Angular momentum keeps the wheel axle almost horizontal when a torque supplied by earth's gravity acts on it. Instead of toppling, the torque causes the wheel to slowly precess about a vertical axis.

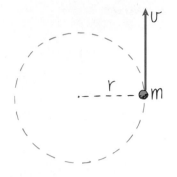

Figure 11.24 ▲
An object of concentrated mass *m* whirling in a circular path of radius *r* with a speed *v* has angular momentum *mvr*.

For the case of an object that is small compared with the radial distance to its axis of rotation, such as a tin can swinging from a long string or a planet orbiting in a circle around the sun, the angular momentum is simply equal to the magnitude of its linear momentum, *mv*, multiplied by the radial distance, *r*. In equation form,

$$\text{angular momentum} = mvr$$

Just as an external net force is required to change the linear momentum of an object, an external net torque is required to change the angular momentum of an object. We restate Newton's first law of inertia for rotating systems in terms of angular momentum:

> An object or system of objects will maintain its angular momentum unless acted upon by an unbalanced external torque.

Figure 11.25 ▲
Bicycles are easier to balance when their wheels have angular momentum.

We know it is easier to balance on a moving bicycle than on one at rest. The spinning wheels have angular momentum. When our center of gravity is not above a point of support, a slight torque is produced. When the wheels are at rest, we fall over. But when the bicycle is moving, the wheels have angular momentum, and a greater torque is required to change the direction of the angular momentum. The moving bicycle is easier to balance.

11.7 Conservation of Angular Momentum

Just as the linear momentum of any system is conserved if no net forces are acting on the system, angular momentum is conserved for systems in rotation. The **law of conservation of angular momentum** states

> If no unbalanced external torque acts on a rotating system, the angular momentum of that system is constant.

This means that with no external torque, the product of rotational inertia and rotational velocity at one time will be the same as at any other time.

An interesting example of angular momentum conservation is shown in Figure 11.26. The man stands on a low-friction turntable with weights extended. Because of the extended weights his overall rotational inertia is relatively large in this position. As he slowly turns, his angular momentum is the product of his rotational inertia and rotational velocity. When he pulls the weights inward, his overall rotational inertia is considerably decreased. What is the result? His rotational speed increases! This is best appreciated by the turning person who feels changes in rotational speed that seem to be mysterious. But it's straight physics! This procedure is used by a figure skater who starts to whirl with her arms and perhaps a leg extended, and then draws her arms and leg in to obtain a greater rotational speed. Whenever a rotating body contracts, its rotational speed increases.

Figure 11.26 ▲
Conservation of angular momentum. When the man pulls his arms and the whirling weights inward, he decreases his rotational inertia, and his rotational speed correspondingly increases.

163

Figure 11.27 ▲
Rotational speed is controlled by variations in the body's rotational inertia as angular momentum is conserved during a forward somersault.

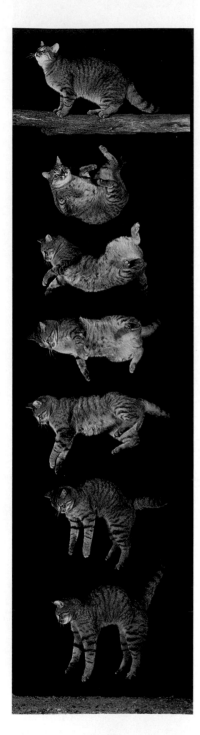

Figure 11.28 ▲
Time-lapse photo of a falling cat.

Similarly, when a gymnast is spinning freely with no unbalanced torques on his or her body, angular momentum does not change. However, as we have already seen, rotational speed can be changed by simply making variations in rotational inertia. This is done by moving some part of the body toward or away from the axis of rotation.

If a cat is held upside down and dropped, it is able to execute a twist and land upright even if it has no initial angular momentum. Zero-angular-momentum twists and turns are performed by turning one part of the body against the other. While falling, the cat rearranges its limbs and tail to change its rotational inertia. Repeated reorientations of the body configuration result in the head and tail rotating one way and the feet the other, so that the feet are downward when the cat reaches the ground.

During this maneuver the total angular momentum remains zero. When it is over, the cat is not turning. This maneuver rotates the body through an angle, but it does not create continuing rotation. To do so would violate angular momentum conservation.

Humans can perform similar twists without difficulty, though not as fast as a cat can. Astronauts have learned to make zero-angular-momentum rotations about any principal axis. They do these to orient their bodies in any preferred direction when floating freely in space.

The law of momentum conservation is seen in the planetary motions and in the shape of the galaxies. It will be a fact of everyday life to inhabitants of rotating space habitats who will head for these distant places.

11 Chapter Review

Concept Summary

Torque for an object being turned is the product of the lever arm and the component of the force perpendicular to the lever arm.

- When balanced torques act on an object, there is no change in rotation.

- When the center of gravity is not over the base of support, the gravitational force produces a torque that causes toppling.

The resistance of an object to changes in its rotational state of motion is called rotational inertia.

- The greater the rotational inertia, the harder it is to change the rotational speed of an object.

Angular momentum, the "inertia of rotation," is the product of rotational inertia and rotational velocity.

- Angular momentum is conserved when no external torque acts on an object.

Important Terms

angular momentum (11.6)
law of conservation of angular momentum (11.7)
lever arm (11.1)
linear momentum (11.6)
rotational inertia (11.4)
rotational velocity (11.6)
torque (11.1)

Review Questions

1. Compare the effects of a force and a torque exerted on an object. (11.1)

2. What is the lever arm of a force? (11.1)

3. In what direction should a force be applied to produce maximum torque? (11.1)

4. How do clockwise and counterclockwise torques compare in a balanced system? (11.2)

5. In terms of center of gravity, support base, and torque, why can't you stand with your heels and back to a wall and bend over to touch your toes without falling over? (11.3)

6. Where must a football be kicked so that it won't topple end over end as it sails through the air? (11.3)

7. What is the law of inertia for rotation? (11.4)

8. On what two quantities does rotational inertia depend? (11.4)

9. What is the effect of adding a weight to the end of a baseball bat used for practice swings? (11.4)

10. Why is it easier to swing your legs back and forth when they are bent? (11.4)

11. Which will have the greater acceleration rolling down an incline—a large ball or a small ball? (11.4)

12. Which will have the greater acceleration rolling down an incline—a hoop or a solid disk? (11.4)

13. How can a person vary his or her rotational inertia? (11.5)

14. Distinguish between rotational velocity and rotational speed. (11.6)

15. Distinguish between linear momentum and angular momentum. (11.6)

16. What motion does the torque produced by Earth's gravity impart to a vertically spinning bicycle wheel supported at only one end of its axle? (11.6)

17. What does it mean to say that angular momentum is conserved? (11.7)

18. When is angular momentum conserved? (11.7)

19. If a skater who is spinning pulls her arms in so as to reduce her rotational inertia to half, by how much will her rate of spin increase? (11.7)

20. What happens to a gymnast's angular momentum when he changes his body configuration during a somersault? What happens to his rotational speed? (11.7)

Activities

1. Roll various canned foods down an incline after predicting which will roll faster. Compare liquids (which slide or slosh rather than roll inside the can) and solids.

2. Place the ends of a uniform horizontal board on a pair of weighing scales. You'll note that each scale reads half the weight of the board. If a toy truck with a heavy brick in it is placed on the center of the board, the reading on each scale will increase by half the weight of the loaded truck. Why? How would you go about figuring how much weight is supported by each scale when the load is not in the middle? Make predictions for various places before you check by experiment. (Try to do this before you measure torques in the lab part of your course.)

Plug and Chug

1. a. Calculate the torque produced by a 50-N perpendicular force at the end of a 0.2-m-long wrench.

b. Calculate the torque produced by the same 50-N force when a pipe extends the length of the wrench to 0.5 m.

2. a. Calculate the individual torques produced by the weights of the girl and boy on the see-saw in the figure. What is the net torque?

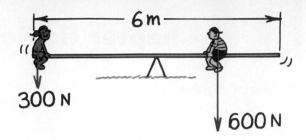

300 N

600 N

b. Calculate the distance a 600-N boy should sit from the fulcrum.

c. Calculate the distance a 300-N girl should sit when the boy weighs 400 N.

Think and Explain

1. Which is better for prying open a stuck cover from a can of paint—a screwdriver with a thick handle or one with a long handle? Which is better for turning stubborn screws? Provide an explanation.

2. When the spool is pulled to the right, in which direction does it accelerate? If the string is instead pulled straight upward, in which direction will the spool accelerate? Is there an angle where a gentle pull produces no acceleration? Show with a sketch.

3. If you know your own weight and have a see-saw and a meterstick available, how can you determine the approximate weight of a friend?

4. You cannot stand with heels and back to the wall and successfully lean over and touch your toes without toppling. Would either stronger legs or longer feet help you to do this? Defend your answer.

5. If you walked along the top of a fence, why would holding your arms out help you to balance?

6. Which will have the greater acceleration rolling down an incline—a bowling ball or a volleyball? Defend your answer.

7. The most popular gyroscope around is a Frisbee flying disk. What is one function, besides being a place for gripping and catching, of its somewhat thicker curved rim?

8. Consider two rotating bicycle wheels, one filled with air and the other filled with water. Which would be more difficult to stop rotating? Explain.

9. Why and how do you throw a football so that it spins about its long axis when traveling through the air?

10. You sit in the middle of a large, freely rotating turntable at an amusement park. If you crawled toward the outer rim, would the rotation speed increase, decrease, or remain unchanged? What law of physics supports your answer?

Think and Solve

1. What is the mass of the rock shown in the figure?

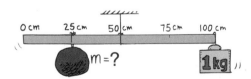

2. What is the mass of the meterstick shown in the figure?

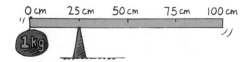

3. If a trapeze artist rotates twice each second while sailing through the air, and contracts to reduce her rotational inertia to one-third, how many rotations per second will result?

4. A pair of identical 1000-kg space pods in outer space are connected to each other by a 900-m-long cable. They rotate about a common point like a spinning dumbbell as shown in the figure. Calculate the rotational inertia of each pod about the axis of rotation. What is the rotational inertia of the two-pod system about its midpoint? Express your answers in $kg \cdot m^2$.

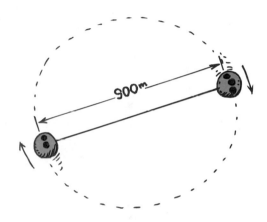

5. The two-pod system in the question above rotates 1.2 RPM to provide artificial gravity for its occupants. If one of the pods pulls in 100 m of cable (bringing the pods closer together), what will be the system's new rotation rate?

12 Universal Gravitation

Gravitation is universal.

Things such as leaves, rain, and satellites fall because of gravity. Gravity is what holds tea in a cup and what makes bubbles rise. It made the earth round, and it builds up the pressures that kindle every star that shines. These are things that gravity does. But what is gravity? In the early part of the twentieth century, Albert Einstein made the amazing discovery that gravity arises from the "warping" of space and time. Neither space nor time is perfectly smooth. Both are just a tiny bit lumpy (except near a black hole, which is a great big lump). Explaining what that means is beyond the scope of this book, but we can say this: Gravity is the way in which masses communicate with each other. Every mass in the universe reaches out to attract every other one, and every mass feels an attraction from every other one. So projectiles, satellites, planets, galaxies, and clusters of galaxies are influenced by gravity. This chapter covers the basic behavior of gravity. In the next chapter we shall investigate more of its consequences.

12.1 The Falling Apple

The idea that gravity extends throughout the universe is credited to Isaac Newton. According to popular legend, this idea occurred to Newton while he was sitting underneath an apple tree on his mother's farm pondering the forces of nature. Newton understood the concept of inertia developed earlier by Galileo; he knew that without an outside force, moving objects continue to move at constant speed in a straight line. He knew that if an object undergoes a change in speed or direction, then a force is responsible.

A falling apple triggered what was to become one of the most far-reaching generalizations of the human mind. Newton saw the apple fall, or maybe even felt it fall on his head—the story about this is not clear. Perhaps he looked up through the apple tree branches and noticed the moon. Newton was probably puzzled by the fact that the moon does not follow a straight-line path, but instead circles about the earth. He knew that circular motion is accelerated motion, which requires a force. But what was this force? Newton had the insight to see that the moon is falling toward the earth, just as the apple is. He reasoned that the moon is falling for the same reason the apple falls—they are both pulled by earth's gravity.

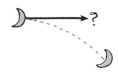

Figure 12.1 ▲
If the moon did not fall, it would follow the straight-line path. Because of its attraction to the earth, it falls along a curved path.

12.2 The Falling Moon

Newton developed this idea further. He compared the falling apple with the falling moon. Newton realized that if the moon did not fall, it would move off in a straight line and leave its orbit. His idea was that the moon must be falling *around* the earth. Thus the moon falls in the sense that it *falls beneath the straight line it would follow if no force acted on it.* He hypothesized that the moon was simply a projectile circling the earth under the attraction of gravity.

This concept is illustrated in an original drawing by Newton, shown in Figure 12.2. He compared motion of the moon to a cannonball fired from the top of a high mountain. He imagined that the mountaintop was above the earth's atmosphere, so that air resistance would not impede the motion of the cannonball. If a cannonball were fired with a small horizontal speed, it would follow a parabolic path and soon hit the earth below. If it were fired faster, its path would be less curved and it would hit the earth farther away. Newton reasoned that if the cannonball were fired fast enough, its path would become a circle and the cannonball would circle indefinitely. It would be in orbit.

Both the orbiting cannonball and the moon have a component of velocity parallel to the earth's surface. This sideways or *tangential velocity* is sufficient to ensure nearly circular motion *around* the earth rather than *into* it. If there is no resistance to reduce its speed, the moon will continue "falling" around and around the earth indefinitely.

Newton's idea seemed correct. But for the idea to advance from hypothesis to the status of a scientific theory, it would have to be tested. Newton's test was to see if the moon's "fall" beneath its otherwise straight-line path was in correct proportion to the fall of an apple or any object at the earth's surface. He reasoned that the mass of the moon should not affect how it falls, just as mass has no effect on the acceleration of freely falling objects on Earth. How far the moon falls, and how far an apple at the earth's surface falls, should relate only to their respective *distances* from the earth's center. If the distance of fall for the moon and the apple are in correct proportion, then the hypothesis that earth gravity reaches to the moon must be taken seriously.

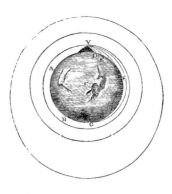

Figure 12.2 ▲
The original drawing by Isaac Newton showing how a projectile fired fast enough would fall around the earth and become an earth satellite. In the same way, the moon falls around the earth and is an earth satellite.

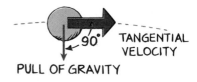

Figure 12.3 ▲
Tangential velocity is the "sideways" velocity—the component of velocity perpendicular to the pull of gravity.

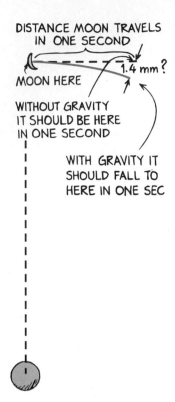

DISTANCE MOON TRAVELS
IN ONE SECOND

1.4 mm ?

MOON HERE

WITHOUT GRAVITY
IT SHOULD BE HERE
IN ONE SECOND

WITH GRAVITY IT
SHOULD FALL TO
HERE IN ONE SEC

Figure 12.4 ▲
If the force that pulls apples off trees also pulls the moon into orbit, the circle of the moon's orbit should fall 1.4 mm below a point along the straight line where the moon would other-wise be one second later.

The moon was already known to be 60 times farther from the center of the earth than an apple at the earth's surface. The apple will fall nearly 5 m in its first second of fall—or more precisely, 4.9 m. Newton reasoned that gravitational attraction to the earth must be "diluted" by distance. Does this mean the force of Earth's gravity would reduce to 1/60 at the moon's distance? No, it is much less than this. As we shall soon see, the influence of gravity should be diluted 1/60 of 1/60 or $1/(60)^2$. So in one second the moon should fall $1/(60)^2$ of 4.9 m, which is 1.4 millimeters.*

Using geometry, Newton calculated how far the circle of the moon's orbit lies below the straight-line distance the moon otherwise would travel in one second (Figure 12.4). His value turned out to be about the 1.4-mm distance accepted today. But he was unsure of the distance between the earth and moon, and whether or not the correct distance to use was the distance between their centers. At this time he hadn't proved mathematically that the gravity of the spherical earth (and moon) is the same as if all its mass were concentrated at its center. Because of this uncertainty, and also because of criticisms he had experienced in publishing earlier findings in optics, he placed his papers in a drawer, where they remained for nearly 20 years. During this period he laid the foundation and developed the field of geometrical optics for which he first became famous.

Newton finally returned to the moon problem at the prodding of his astronomer friend Edmund Halley (of Halley's comet fame). It wasn't until after Newton invented a new branch of mathematics, calculus, to prove his center-of-gravity hypothesis, that he published what is one of the greatest achievements of the human mind—the law of universal gravitation.** Newton generalized his moon finding to all objects, and stated that all objects in the universe attract each other.

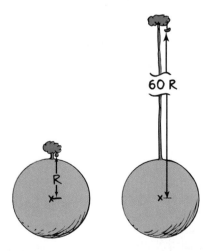

60 R

R

Figure 12.5 ▲
An apple falls 4.9 m during its first second of fall when it is near the earth's surface. Newton asked how far the moon would fall in the same time if it were 60 times farther from the center of the earth. His answer was (in today's units) 1.4 mm.

* Or working backward, (0.0014 m) × $(60)^2$ = 4.9 m.

12.3 The Falling Earth

Newton's theory of gravitation confirmed the Copernican theory of the solar system. No longer was the earth considered to be the center of the universe. The earth was not even the center of the solar system. The sun occupies the center, and it became clear that the earth and planets orbit the sun in the same way that the moon orbits the earth. The planets continually "fall" around the sun in closed paths. Why don't the planets crash into the sun? They don't because the planets have tangential velocities. What would happen if the tangential velocities of the planets were reduced to zero? The answer is simple enough: Their motion would be straight toward the sun and they would indeed crash into it. Any objects in the solar system with insufficient tangential velocities have long ago crashed into the sun; what remains is the harmony we observe.

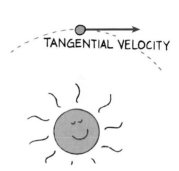

Figure 12.6 ▲
The tangential velocity of the earth about the sun allows it to fall around the sun rather than directly into it. If this tangential velocity were reduced to zero, what would be the fate of the earth?

■ Question

Since the moon is gravitationally attracted to the earth, why does it not simply crash into the earth?

■ Answer

The moon would crash into the earth if its tangential velocity were reduced to zero, but because of its tangential velocity, the moon falls around the earth rather than into it. We will return to this idea in more detail in Chapter 14.

** Compare Newton's painstaking effort to get everything right and nailed down mathematically with the lack of "doing one's homework," the hasty judgments, and the absence of cross-checking that so often characterize the pronouncement of less-than-scientific theories.

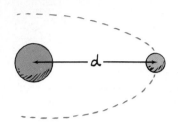

Figure 12.7 ▲
The force of gravity between objects depends on the distance between their centers of mass.

12.4 Newton's Law of Universal Gravitation

Newton did not discover gravity. What Newton discovered was that gravity is universal. Everything pulls on everything else in a beautifully simple way that involves only mass and distance. Newton's **law of universal gravitation** states that every object attracts every other object with a force that for any two objects is directly proportional to the mass of each object. The greater the masses, the greater the force of attraction between them. Newton deduced that the force decreases as the square of the distance between the centers of mass of the objects increases. The farther away the objects are from each other, the less the force of attraction between them.

The law can be expressed symbolically as

$$F \sim \frac{m_1\, m_2}{d^2}$$

where m_1 is the mass of one object, m_2 is the mass of the other, and d is the distance between their centers of gravity. The greater the masses m_1 and m_2, the greater the force of attraction between them.* The greater the distance d between the objects, the weaker the force of attraction.

The Universal Gravitational Constant, *G*

The proportionality form of the universal law of gravitation can be expressed as an exact equation when the constant of proportionality *G*, called the **universal gravitational constant,** is introduced. Then the equation is

$$F = G\frac{m_1\, m_2}{d^2}$$

In words, the force of gravity between two objects is found by multiplying their masses, dividing by the square of the distance between their centers, and then multiplying this result by the constant *G*. The magnitude of *G* is given by the magnitude of the force between two masses of 1 kilogram each, 1 meter apart: 0.000 000 000 0667 newton. This is an extremely weak force. The units of *G* are such as to make the force come out in newtons. In scientific notation,**

---

* In previous chapters we have treated mass as a measure of inertia, which is called inertial mass. Now we see mass as a measure of gravitational force, which in this context is called gravitational mass. Experiments show that the two are equal, and as a matter of principle, the equivalence of inertial and gravitational mass is the foundation of Einstein's general theory of relativity (which is beyond the scope of this book).

** The numerical value of *G* depends entirely on the units of measurement we choose for mass, distance, and time. Using SI units, the choice is: mass in kilograms; distance in meters; and time in seconds. Scientific notation is discussed in Appendix B at the end of this book.

$$G = 6.67 \times 10^{-11} \text{ N·m}^2/\text{kg}^2$$

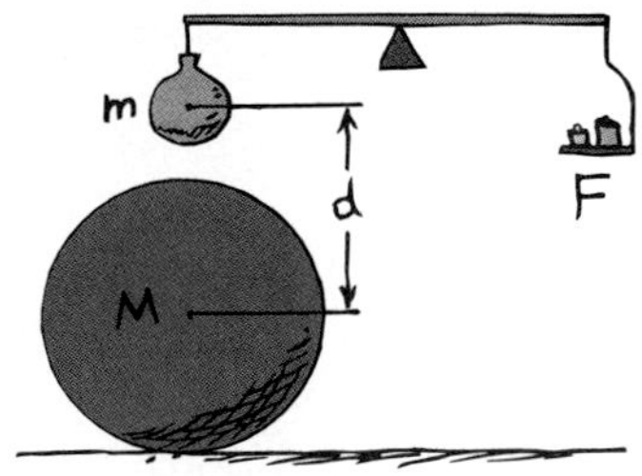

Figure 12.8 ▲
Von Jolly's method of measuring the attraction between two masses.

G was first measured 150 years after Newton's discovery of universal gravitation by an English physicist, Henry Cavendish. Cavendish accomplished this by measuring the tiny force between lead masses with an extremely sensitive torsion balance. A simpler method was later developed by Philipp von Jolly, who attached a spherical flask of mercury to one arm of a sensitive balance (Figure 12.8). After the balance was put in equilibrium, a 6-ton lead sphere was rolled beneath the mercury flask. The flask was pulled slightly downward. The gravitational force *F* between the lead masses was equal to the weight that had to be placed on the opposite end of the balance to restore equilibrium. Since the quantities *F*, m_1, m_2, and *d* were all known, the ratio *G* was calculated:

$$G = \frac{F}{m_1 m_2 / d^2} = 6.67 \times 10^{-11} \frac{\text{N}}{\text{kg}^2/\text{m}^2} = 6.67 \times 10^{-11} \text{ N·m}^2/\text{kg}^2$$

The value of *G* tells us that the force of gravity is a very weak force. It is the weakest of the presently known four fundamental forces. (The other three are the electromagnetic force and two kinds of nuclear forces.) We sense gravitation only when masses like that of the earth are involved. The force of attraction between you and a classmate is

Writing Very Large and Very Small Numbers

Large and small numbers are conveniently expressed in a mathematical abbreviation called *scientific notation*. An example of a large number is the equatorial radius of the earth: 6 370 000 m. This number can be obtained by multiplying 6.37 by 10, and again by 10, and so on until 10 has been used as a multiplier six times. So 6 370 000 can be written as 6.37×10^6. That's 6.37 million meters. A thousand million is a billion, 10^9. To better comprehend the size of a billion:

- a billion meters is slightly more than the earth-moon distance.
- a billion kilograms is the mass of the earth's oceans.
- a billion earths still wouldn't equal the mass of the sun.
- a billion seconds is 31.7 years.
- a billion minutes is 1903 years.
- a billion years ago there were no humans on Earth.
- a billion people live in China.
- a billion atoms make up the dot over this i.

Small numbers are expressed in scientific notation by dividing by 10 successive times. A millimeter (mm) is $\frac{1}{1000}$ m, or 1 m divided by 10 three times. In scientific notation, 1 mm = 10^{-3} m. The gravitational constant *G* is a very small number, 0.000 000 000 066 726 N·m²/kg². By dividing 6.6726 by 10 eleven times, and rounding off, it is 6.67×10^{-11} N·m²/kg².

too weak to notice (but it's there!). The force of attraction between you and the earth, however, is easy to notice. It is your weight.

In addition to your mass, your weight also depends on your distance from the center of the earth. At the top of a mountain your mass is the same as it is anywhere else, but your weight is slightly less than at ground level. Your weight is less because your distance from the center of the earth is greater.

Figure 12.9 ▶
Your weight is less at the top of a mountain because you are farther from the center of the earth.

Interestingly enough, Cavendish's first measure of G was called the "Weighing the Earth" experiment, because once the value of G was known, the mass of the earth was easily calculated. The force that the earth exerts on a mass of 1 kilogram at its surface is 9.8 newtons. The distance between the 1-kilogram mass and the center of mass of the earth is the earth's radius, 6.4×10^6 meters. Therefore, from $F = (Gm_1m_2/d^2)$, where m_1 is the mass of the earth,

$$9.8 \text{ N} = 6.67 \times 10^{-11} \text{ N·m}^2/\text{kg}^2 \times \frac{1 \text{ kg} \times m_1}{(6.4 \times 10^6 \text{ m})^2}$$

from which the mass of the earth $m_1 = 6 \times 10^{24}$ kilograms.

■ Question

If there is an attractive force between all objects, why do we not feel ourselves gravitating toward massive buildings in our vicinity?

■ Answer

We are gravitationally attracted to massive buildings and everything else in the universe. The 1933 Nobel Prize–winning physicist Paul A. M. Dirac put it this way: "Pick a flower on Earth and you move the farthest star!" How much you are influenced by buildings or how much interaction there is between flowers, is another story. The forces between you and buildings are relatively small because the masses are small compared with the mass of the earth. The forces due to the stars are small because of their great distances. These tiny forces escape our notice when they are overwhelmed by the powerful attraction to the earth.

12.5 Gravity and Distance: The Inverse-Square Law

We can understand how gravity is reduced with distance by considering an imaginary "butter gun" used in a busy restaurant for buttering toast (Figure 12.10). Imagine melted butter sprayed through a square opening in a wall. The opening is exactly the size of one piece of square toast. And imagine that a spurt from the gun deposits an even layer of butter 1 mm thick. Consider the consequences of holding the toast twice as far from the butter gun. You can see in Figure 12.10 that the butter would spread out for twice the distance and would cover twice as much toast vertically and twice as much toast horizontally. A little thought will show that the butter would now spread out to cover four pieces of toast. How thick will the butter be on each piece of toast? Since it has been diluted to cover four times as much area, its thickness will be one-quarter as much, or 0.25 mm.

Note what has happened. When the butter gets twice as far from the gun, it is only 1/4 as thick. More thought will show that if it gets 3 times as far, it will spread out to cover 3×3, or 9, pieces of toast. How thick will the butter be then? Can you see it will be 1/9 as thick? And can you see that 1/9 is the inverse *square* of 3? (The inverse of 3 is simply 1/3; the inverse square of 3 is $(1/3)^2$, or 1/9.) When a quantity varies as the inverse square of its distance from its source, it follows an **inverse-square law.** This law applies not only to the spreading of butter from a butter gun, and the weakening of gravity with distance, but to all cases where the effect from a localized source spreads evenly throughout the surrounding space. More examples are light, radiation, and sound.

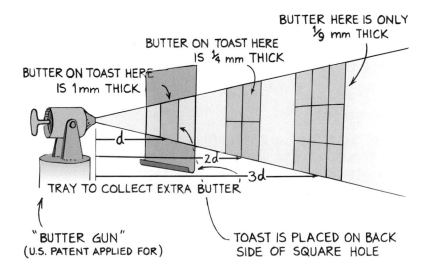

Figure 12.10 ▲
The inverse-square law. Butter spray travels outward from the nozzle of the butter gun in straight lines. Like gravity, the "strength" of the spray obeys an inverse-square law.

The greater the distance from the earth's center, the less an object will weigh (Figure 12.11). If your little sister weighs 300 N at sea level, she will weigh only 299 N atop Mt. Everest. But no matter how great the distance, the earth's gravity does not drop to zero. Even if you were transported to the far reaches of the universe, the gravitational influence of the earth would be with you. It may be overwhelmed by the gravitational influences of nearer and/or more massive objects, but it is there. The gravitational influence of every object, however small or far, is exerted through all space. That's impressive!

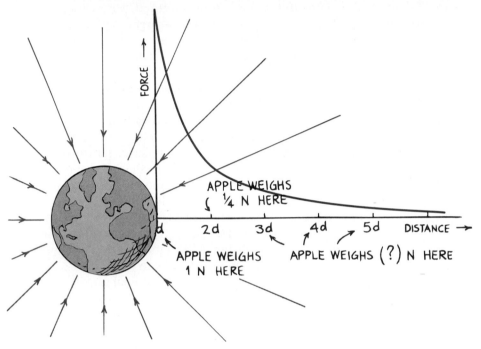

Figure 12.11 ▲
An apple that weighs 1 N at the earth's surface weighs only 0.25 N when located twice as far from the earth's center because the pull of gravity is only 1/4 as strong. When it is 3 times as far, it weighs only 1/9 as much, or 0.11 N. What would it weigh at 4 times the distance? Five times?

■ Question

Suppose that an apple at the top of a tree is pulled by earth gravity with a force of 1 N. If the tree were twice as tall, would the force of gravity on the apple be only 1/4 as strong? Explain your answer.

■ Answer

No, because the twice-as-tall apple tree is not twice as far from the earth's center. The taller tree would have to have a height equal to the radius of the earth (6370 km) before the weight of the apple would reduce to $\frac{1}{4}$ N. Before its weight decreases by 1%, an apple or any object must be raised 32 km—nearly four times the height of Mt. Everest, the tallest mountain in the world. So as a practical matter we disregard the effects of everyday changes in elevation.

Illusion Check

Hold your hands outstretched, one twice as far from your eyes as the other, and make a casual judgment about which hand looks

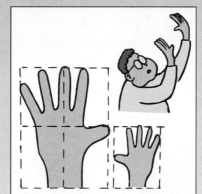

bigger. Most people see their hands to be about the same size, while many see the nearer hand as slightly bigger. Almost nobody upon casual inspection sees the nearer hand as four times as big. But by the inverse-square law, the nearer hand should appear twice as tall and twice as wide and therefore occupy four times as much of your visual field as the farther hand. Your belief that your hands are the same size is so strong that you likely overrule this information. Now, if you overlap your hands slightly and carefully view them with one eye closed, you'll see the nearer hand as clearly bigger. This raises an interesting question: What other illusions do you have that are not so easily checked?

Activity

12.6 Universal Gravitation

We all know that the earth is round. But *why* is the earth round? It is round because of gravitation. Since everything attracts everything else, the earth had attracted itself together before it became solid. Any "corners" of the earth have been pulled in so that the earth is a giant sphere. The sun, the moon, and the earth are all fairly spherical because they have to be (rotational effects make them somewhat wider at their equators).

If everything pulls on everything else, then the planets must pull on each other. The net force that controls Jupiter, for example, is not just from the sun, but from the planets also. Their effect is small compared with the pull of the more massive sun, but it still shows. When the planet Saturn is near Jupiter, for example, its pull disturbs the otherwise smooth path of Jupiter. Both planets deviate from their normal orbits. This deviation is called a **perturbation.**

Up until the middle of the last century astronomers were puzzled by unexplained perturbations of the planet Uranus. Even when the influences of the other planets were taken into account, Uranus was behaving strangely. Either the law of gravitation was failing at

Figure 12.12 ▲
Gravitation dictates the shape of the spiral arms in a galaxy.

this great distance from the sun, or some unknown influence such as another planet was perturbing Uranus.

The source of Uranus's perturbation was uncovered in 1845 and 1846 by two astronomers, John Adams in England and Urbain Leverrier in France. With only pencil and paper and the application of Newton's law of gravitation, both astronomers independently arrived at the same conclusion: A disturbing body beyond the orbit of Uranus was the culprit. They sent letters to their local observatories with instructions to search a certain part of the sky. The request by Adams was delayed by misunderstandings at Greenwich, England, but Leverrier's request to the director of the Berlin Observatory was heeded right away. The planet Neptune was discovered within a half hour.

Other perturbations of Uranus led to the prediction and discovery of the ninth planet, Pluto. It was discovered in 1930 at the Lowell Observatory in Arizona. Pluto takes 248 years to make a single revolution around the sun, so it will not be seen in its discovered position again until the year 2178.

The motion of double stars around one another and the shapes of distant galaxies are evidence that the law of gravitation extends beyond the solar system. Over still larger distances, gravitation dictates the fate of the entire universe.

Current scientific speculation is that the universe originated in the explosion of a primordial fireball some 10 to 15 billion years ago. This is the "Big Bang" theory of the origin of the universe. All the matter of the universe was hurled outward from this event and continues in an outward expansion. So it is that we find ourselves in an expanding universe.

This expansion may go on indefinitely, or it may be overcome by the combined gravitation of all the matter in the universe and come to a stop. Like a stone thrown upward, whose departure from the ground comes to an end when it reaches the top of its trajectory, and which then begins its descent to the place of its origin, the presently expanding universe may contract and fall back into a single unity. This would be the "Big Crunch." After that, the universe would presumably re-explode to produce a new universe. The same course of action might repeat itself in a series of explosion-implosion cycles. The process may well be cyclic. If this is true, we live in an oscillating universe.

We do not know whether the expansion of the universe is cyclic or indefinite, because we are uncertain about whether enough mass exists to halt the expansion. The period of oscillation is estimated to be somewhat less than 100 billion years. If the universe does oscillate, who can say how many times it has expanded and collapsed? We know of no way a civilization could leave a trace of ever having existed during a previous cycle, for all the matter in the universe would be reduced to bare subatomic particles during the Big Crunch. All the laws of nature, such as the law of gravitation, would then have to be rediscovered by the higher evolving life forms. And then students of these laws would read about them, as you are doing now. Think about that.

Few theories have affected science and civilization as much as Newton's theory of gravity. The successes of Newton's ideas ushered in the Age of Reason or Century of Enlightenment. Newton had demonstrated that by observation and reason, people could uncover the workings of the physical universe. How profound that all the moons and planets and stars and galaxies have such a beautifully simple rule to govern them, namely,

$$F = G\, \frac{m_1\, m_2}{d^2}$$

The formulation of this simple rule is one of the major reasons for the success in science that followed, for it provided hope that other phenomena of the world might also be described by equally simple and universal laws.

This hope nurtured the thinking of many scientists, artists, writers, and philosophers of the 1700s. One of these was the English philosopher John Locke, who argued that observation and reason, as demonstrated by Newton, should be our best judge and guide in all things. Locke urged that all of nature and even society should be searched to discover any "natural laws" that might exist. Using Newtonian physics as a model of reason, Locke and his followers modeled a system of government that found adherents in the 13 British colonies across the Atlantic. These ideas culminated in the Declaration of Independence and the Constitution of the United States of America.

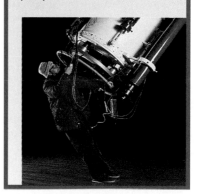

12 Chapter Review

Concept Summary

The moon and other objects in orbit around the earth are actually falling toward the earth but have great enough tangential velocity to avoid hitting the earth.

According to Newton's law of universal gravitation, everything pulls on everything else with a force that depends upon the masses of the objects and the distances between their centers of mass.

- The greater the masses, the greater is the force.
- The greater the distance, the smaller is the force.

Gravitation decreases according to the inverse-square law. The force of gravity weakens as the distance squared.

Important Terms

inverse-square law (12.5)
law of universal gravitation (12.4)
perturbation (12.6)
universal gravitational constant, G (12.4)

Review Questions

1. Why did Newton think that a force must act on the moon? (12.1)

2. What did Newton conclude about the force that pulls apples to the ground and the force that holds the moon in orbit? (12.1)

3. If the moon falls, why doesn't it get closer to the earth? (12.2)

4. What is meant by tangential velocity? (12.2)

5. How did Newton check his hypothesis that there is an attractive force between the earth and moon? (12.2)

6. What is required before a hypothesis (an educated guess) advances to the status of a scientific theory (organized knowledge)? (12.2)

7. Since the planets are pulled to the sun by gravitational attraction, why don't they simply crash into the sun? (12.3)

8. What did Newton discover about gravity? (12.4)

9. What does the very small value of the gravitational constant G (in standard units) tell us about the strength of gravitational forces? (12.4)

10. What are the two masses and the one distance that determine your weight? (12.4)

11. In what way is gravity reduced with distance from the earth? (12.5)

12. What would be the difference in your weight if you were five times farther from the center of the earth than you are now? Ten times? (12.5)

13. What makes the earth round? (12.6)

14. What causes planetary perturbations? (12.6)

Plug and Chug

1. Calculate the force of gravity on a 1-kg mass at the earth's surface. The mass of the earth is 6×10^{24} kg, and its radius is 6.4×10^6 m.

2. Calculate the force of gravity on the same 1-kg mass if it were 6.4×10^6 m above the earth's surface (that is, if it were 2 earth radii from the earth's center).

3. Calculate the force of gravity between the earth (mass = 6.0×10^{24} kg) and the moon (mass = 7.4×10^{22} kg). The average earth-moon distance is 3.8×10^8 m.

4. Calculate the force of gravity between the earth and the sun (sun's mass = 2.0×10^{30} kg; average earth-sun distance = 1.5×10^{11} m).

5. Calculate the force of gravity between a newborn baby (mass = 4 kg) and the planet Mars (mass = 6.4×10^{23} kg), when Mars is at its closest to the earth (distance = 8×10^{10} m).

6. Calculate the force of gravity between a newborn baby of mass 4 kg and the obstetrician of mass 75 kg, who is 0.3 m from the baby. Which exerts more gravitational force on the baby, Mars or the obstetrician? By how much?

Think and Explain

1. Comment on whether or not this label on a consumer product should be cause for concern. *CAUTION: The mass of this product affects every other mass in the universe, with an attracting force that is proportional to the product of the masses and inversely proportional to the square of the distance between them.*

2. The earth and the moon are gravitationally attracted to each other. Does the more massive earth attract the moon with a greater force, the same force, or less force than the moon attracts the earth?

3. What is the magnitude and direction of the gravitational force that acts on a woman who weighs 500 N at the surface of the earth?

4. If the gravitational forces of the sun on the planets suddenly disappeared, in what kind of paths would the planets move?

5. The moon "falls" 1.4 mm each second. Does this mean that it gets 1.4 mm closer to the earth each second? Would it get closer if its tangential velocity were reduced? Explain.

6. If the moon were twice as massive, would the attractive force of the earth on the moon be twice as large? Of the moon on the earth?

7. Which requires more fuel—a rocket going from the earth to the moon, or a rocket coming from the moon to the earth? Why?

8. Evidence indicates that the present expansion of the universe is slowing down. Is this consistent with, or contrary to, the law of gravity? Explain.

9. The planet Jupiter is about 300 times as massive as the earth, but an object on its surface would weigh only 2.5 times as much as it would on the earth. Can you come up with an explanation? (*Hint:* Let the terms in the equation for gravitational force guide your thinking.)

10. Some people dismiss the validity of scientific theories by saying they are "only" theories. The law of universal gravitation is a theory. Does this mean that scientists still doubt its validity? Explain.

Think and Solve

1. If the moon orbits twice as far from the earth, how far will it "fall" each second?

2. By what factor would your weight change if the earth's diameter were doubled and its mass were also doubled?

3. If you stood atop a ladder that was so tall that you were twice as far from the earth's center, how would your weight compare with its present value?

4. Estimate the size of Jupiter's diameter (compared with the earth's diameter). See Think and Explain 9.

5. To better comprehend the magnitude of the gravitational force between the earth and the moon, pretend gravity is turned off and the pull replaced by the tension in a steel cable joining them. How thick would such a cable need to be? You can estimate the diameter by knowing that the tensile strength of steel cable is about 5.0×10^8 N/m^2 (each square-meter cross section can support a force of 5.0×10^8 newtons).

13 Gravitational Interactions

All matter is under the influence of gravity.

Everyone knows that objects fall because of gravity. Even people before the time of Isaac Newton knew this. Contrary to popular belief, Newton did not discover gravity. What Newton discovered was that gravity is *universal*—that the same force that pulls an apple off a tree holds the moon in orbit, and that both the earth and moon are similarly held in orbit around the sun. And the sun revolves as part of a cluster of other stars around the center of our galaxy, the Milky Way. Newton discovered that all objects in the universe attract each other. This was discussed in the previous chapter. In this chapter we'll investigate the role of gravity at, below, and above the earth's surface. We will see how gravity affects the earth's oceans and its atmosphere, then we will look at gravity at its extreme—in stellar objects called black holes. We begin with the concept of the gravitational field.

Figure 13.1 ▲
We can say that the rocket is attracted to the earth, or that it is interacting with the gravitational field of the earth. Both are correct.

13.1 Gravitational Fields

If you have ever played with iron filings and a magnet, you're familiar with magnetic fields. A magnetic field is a **force field** that surrounds a magnet. A force field exerts a force on objects in its vicinity. A magnetic field exerts a magnetic force on magnetic substances. You can look ahead to Figure 36.4 and see how the iron filings around a magnet reveal the shape of its force field. The pattern of the filings shows the strength and direction of the magnetic field at different points in the space around the magnet. Where the filings are closest together, the field is strongest. Later we will learn about similar electric force fields that surround electric charges. But now we will investigate the kind of force field that surrounds massive objects—the **gravitational field.**

The earth's gravitational field can be represented by imaginary field lines (Figure 13.2). Like the iron filings around a magnet, the field lines are closer together where the gravitational field is stronger. The direction of the field at any point is along the line the point lies on. Arrows show the field direction. A particle, a spaceship, or any mass in the vicinity of the earth will be accelerated in the direction of the field line at that location.

The strength of the earth's gravitational field is the force per unit mass exerted by the earth on any object.* We define the gravitational field

$$\mathbf{g} = \frac{F}{m}$$

It has the same symbol as the acceleration due to gravity because it is equal to the free fall acceleration experienced by an object when only the force of gravity acts on it. Near the surface of the earth, the gravitational field strength is

$$\mathbf{g} = \frac{F}{m} = 9.8 \text{ N/kg} = 9.8 \text{ m/s}^2$$

Let's verify this value of **g.** At the earth's surface the gravitational force on an object is its weight—the gravitational attraction between the object's mass m and the mass of the earth, say M. The distance between centers of mass is the earth's radius, R. If we make these substitutions ($m_1 = m$, $m_2 = M$, $d = R$) in the law of gravity and divide by m, we get

$$\mathbf{g} = \frac{F}{m} = \frac{G\dfrac{mM}{R^2}}{m}$$

Canceling m, we get

$$\mathbf{g} = \frac{GM}{R^2}$$

If you have a calculator handy, evaluate **g,** using the following values for G, M, and R:

$G = 6.67 \times 10^{-11} \text{ N·m}^2/\text{kg}^2$

$M = 5.98 \times 10^{24} \text{ kg}$

$R = 6.37 \times 10^6 \text{ m}$

Multiply the first two numbers and divide twice by the third. Rounded off, your answer will be 9.8 N/kg. And, since 1 N equals 1 kg·m/s^2, the combination unit N/kg is the same as m/s^2.**

Thus, we can see that the numerical value of **g** at the earth's surface depends on the mass of the earth and its radius. If the earth had a different mass or radius, **g** at its surface would have a different

Figure 13.2 ▲
Field lines represent the gravitational field about the earth. Where the field lines are closer together, the field is stronger. Farther away, where the field lines are farther apart, the field is weaker.

Figure 13.3 ▲
The weight of the load is the gravitational force of attraction between the load and the earth.

---

* **g** is a vector quantity, for it has both magnitude (strength) and direction.

** The unit N/kg is equivalent to m/s^2. 1 N is the force that will cause 1 kg to accelerate 1 m/s^2. That is (using $F = ma$), 1 N = 1 kg·m/s^2. This means that the unit N/kg = (kg·m/s^2)/kg = m/s^2.

value. If you know the mass and radius of any planet, you can calculate the acceleration due to gravity at the surface of that planet.

The strength of the earth's gravitational field, like the strength of its force on objects, follows the inverse-square law outside the earth. So **g** weakens with increasing distance from the earth.

■ Questions

1. Why do all freely falling objects have the same acceleration?

2. The acceleration of objects on the surface of the moon is only 1/6 of 9.8 m/s². From this fact, is it correct to say that the mass of the moon is therefore 1/6 the mass of the earth?

3. How does **g** at the surface of Jupiter compare with **g** at the surface of the earth? Data: Jupiter's mass is about 300 times that of the earth, and its radius is about 10 times greater than the radius of the earth.

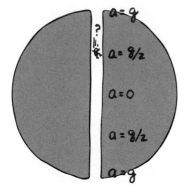

Figure 13.4 ▲
As you fall faster and faster into a hole bored through the earth, your acceleration diminishes because the pull of the mass above you partly cancels the pull below. At the earth's center the pulls cancel to zero and your acceleration is zero. Momentum carries you against a growing acceleration past the center to the opposite side where it is again **g**.

13.2 Gravitational Field Inside a Planet

The gravitational field of the earth exists inside the earth as well as outside. To investigate the gravitational field beneath the surface, imagine a hole drilled completely through the earth, say from the North Pole to the South Pole. Forget about impracticalities such as lava and high temperatures, and consider the kind of motion you would undergo if you fell into such a hole. If you started at the North Pole end, you'd fall and gain speed all the way down to the center, then overshoot and lose speed all the way up to the South Pole. You'd gain speed moving toward the center, and lose speed moving away

■ Answers

1. When we studied Newton's second law in Chapter 5, we learned that $a = F/m$. In free fall, the ratio F/m (weight/mass) is the same for all masses, so acceleration is the same. In this chapter we say the same thing from another point of view—the concept of the gravitational field **g,** which also equals F/m. From Newton's equation for the force of gravity, we see that F/m at the earth's surface equals 9.8 m/s². In the same gravitational field **g,** all freely falling objects have the same acceleration g.

2. No. We could conclude that the mass of the moon is 1/6 that of the earth's *only* if both the moon and the earth had the same radius. The radius of the moon (1.74×10^6 m) is in fact less than one third of the earth's radius, and its mass (7.36×10^{22} kg) is about 1/80 the mass of the earth.

3. For the earth, **g** = GM/R^2. The value of **g** on Jupiter's surface is $G(300M)/(10R)^2 = 300\ GM/(100R^2) = 3\ GM/R^2$, or 3 times the earth's **g**. (More precisely, Jupiter's **g** = 2.44 times the earth's **g** because its radius is nearly 11 times that of the earth.)

from the center. Without air drag, the trip would take nearly 45 minutes. If you failed to grab the edge, you'd fall back toward the center, overshoot, and return to the North Pole in the same amount of time.

Suppose you had some way to measure your acceleration during this trip. At the beginning of the fall, the gravitational field strength and your acceleration are **g,** but you'd find they steadily decrease as you continue toward the center of the earth.* Why? Because as you are being pulled "downward" toward the earth's center, you are also being pulled "upward" by the part of the earth that is "above" you. In fact, when you get to the center of the earth, the pull "down" is balanced by the pull "up." You are pulled in every direction equally, so the net force on you is zero. There is no acceleration as you whiz with maximum speed past the center of the earth. The gravitational field of the earth at its center is zero!

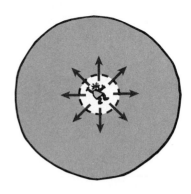

Figure 13.5 ▲
In a cavity at the center of the earth your weight would be zero, because you are pulled equally by gravity in all directions. The gravitational field at the earth's center is zero.

■ Questions

1. If you stepped into a hole bored completely through the earth and made no attempt to grab the edges at either end, what kind of motion would you experience?

2. Halfway to the center of the earth, would gravity pull on you more strongly or less strongly than at the surface of the earth?

■ Answers

1. You would oscillate back and forth, approximating *simple harmonic motion.* Each round trip would take nearly 90 minutes. Interestingly enough, we will see in the next chapter that an earth satellite in close orbit about the earth also takes the same 90 minutes to make a complete round trip.

2. Less strongly, because the part of the earth's mass that pulls you "down" is counteracted by mass above you that pulls you "up." If the earth were of uniform density, halfway to the center gravitational force would be exactly half that at the surface. But since the earth's core is so dense (about seven times the density of surface rock), the gravitational pull at this point would be somewhat more than half its strength at the surface. Exactly how much depends on how the earth's density varies with depth, information that is not known today.

* Interestingly enough, you'd gain acceleration during the first few kilometers beneath the earth's surface, because the density of surface material is much less than the density of the condensed center. This means you'd weigh slightly more for the first few kilometers beneath the earth's surface. Farther in, your weight would decrease and would diminish to zero at the earth's center.

More accurately, cancellation is of the entire surrounding "shell" of inner radius equal to your radial distance from the center. Halfway down, for example, all mass in the shell that surrounds you contributes zero gravitational force on you. You are pulled only by the mass within this shell—below you. It is this earth mass you use in Newton's equation for gravitation. At the earth's center, the whole earth is the shell and complete cancellation occurs. (Cancellation of a force field inside a shell is further developed for electrical forces in Section 33.3.)

13.3 Weight and Weightlessness

The force of gravity, like any force, causes acceleration. Objects under the influence of gravity are pulled toward each other and accelerate (as long as nothing prevents the acceleration). We are almost always in contact with the earth. For this reason, we think of gravity primarily as something that presses us against the earth rather than as something that accelerates us. The pressing against the earth is the sensation we interpret as weight.

Stand on a bathroom scale that is supported on a stationary floor. The gravitational force between you and the earth pulls you against the supporting floor and scale. By Newton's third law, the floor and scale in turn push upward on you. Located in between you and the supporting floor are springs inside the bathroom scale. The springs are compressed by this pair of forces. The weight reading on the scale is linked to the amount of compression of the springs.

If you repeated this weighing procedure in a moving elevator, you would find your weight reading would vary—not during steady motion, but during accelerated motion. If the elevator accelerated upward, the bathroom scale and floor would push harder against your feet, and the springs inside the scale would be compressed even more. The scale would show an increase in your weight.

Figure 13.6 ▶
The sensation of weight is equal to the force that you exert against the supporting floor. When the floor accelerates up or down, your weight seems to vary. You feel weightless when you lose your support in free fall.

NORMAL WEIGHT

GREATER WEIGHT

LESS WEIGHT

NO WEIGHT

If the elevator accelerated downward, the scale would show a decrease in your weight. The support force of the floor would now be less. If the elevator cable broke and the elevator fell freely, the scale reading would register zero. According to the scale, you would be weightless. And you would feel weightless, for your insides would no longer be supported by your legs and pelvic region. Your organs would respond as though gravity were absent. But gravity is not absent, so would you really be weightless? The answer to this question depends on your definition of weight.

Figure 13.7 ▲
The astronaut feels weightless all the time in orbit.

Rather than define your weight as the force of gravity that acts on you, it is more practical to define weight as the force you exert against a supporting floor (or weighing scales). According to this definition, you are as heavy as you feel. Thus, the condition of **weightlessness** is not the absence of gravity; rather, it is the absence of a support force. That queasy feeling you get when you are in a car that seems to leave the road momentarily when it goes over a hump, or worse, off a cliff, is not the absence of gravity. It is the absence of a support force. Astronauts in orbit are without a support force and are in a sustained state of weightlessness. Astronauts sometimes experience "space sickness" until they get used to a state of sustained weightlessness.

Figure 13.8 ▲
Both people experience weightlessness.

13.4 Ocean Tides

Seafaring people have always known there was a connection between the ocean tides and the moon, but no one could offer a satisfactory theory to explain why there are two high tides per day. Newton showed that the ocean tides are caused by *differences* in the gravitational pull of the moon on opposite sides of the earth. The moon's attraction is stronger on the earth's oceans closer to the moon, and weaker on the oceans farther from the moon. This is simply because the gravitational force is weaker with increased distance.

Since the moon's pull is stronger on the ocean nearer the moon and weaker on the opposite ocean farther away, why doesn't the ocean just pile up into a single bulge on the earth under the moon?

Figure 13.9 ▶
Twice a day, every point along
the ocean shore has a high tide.
In between the high tides is a
low tide.

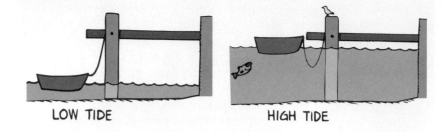

LOW TIDE   HIGH TIDE

Why isn't there one tide every day instead of two tides every day?
There would be only one tide *if* the earth were "nailed down" in one
place and held stationary except for its daily rotation. But the earth
is not nailed down; it is in orbit around the moon, just as the moon is
in orbit around the earth. Actually, both are circling about their combined center of mass, a point inside the earth about three-quarters
of the way from the earth's center to its surface (Figure 13.10).

What then happens is that the ocean nearest the moon is pulled
upward toward the moon, while the main body of the earth is pulled
toward the moon also—away from the ocean on the far side. This is
because the earth as a whole is closer to the moon than the far-side
ocean is. So the earth's waters get slightly elongated—at both ends. A
crude model of this elongation is shown in Figure 13.11. If a ball of
taffy is swung on the end of a string, it deforms, with "tidal bulges" on
the inner and outer sides. Although the actual earth-moon interaction differs from this simplified model, the result is similar. Both the
taffy and the earth are elongated. The earth's elongation is evident in
the pair of ocean bulges on opposite sides of the earth.*

Figure 13.10 ▲
Both the earth and the moon
orbit about a common point—
the center of mass of the earth-
moon system.

Figure 13.11 ▲
An initially spherical ball of gooey taffy will be elongated when it is spun in a
circular path.

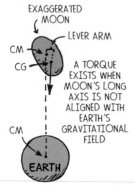

EXAGGERATED
MOON

LEVER ARM

CM

CG

A TORQUE
EXISTS WHEN
MOON'S LONG
AXIS IS NOT
ALIGNED WITH
EARTH'S
GRAVITATIONAL
FIELD

CM

EARTH

* The earth likewise causes tides on the moon, which means the solid moon is slightly
  football shaped. Its deviation from a sphere is enough so that its center of gravity is
  slightly displaced from its center of mass. Both lie along the moon's long axis. Whenever
  the moon's long axis is not lined up toward Earth (see sketch), the earth exerts a small
  torque on the moon. This tends to twist the moon toward aligning with the earth's gravitational field, like the torque that aligns a compass needle with a magnetic field. So we
  see there is a reason why the moon always shows us its same face! Interestingly
  enough, this "tidal lock" is also working on the earth. Our days are getting longer at the
  rate of 2 milliseconds per century. In a few billion years our day will be as long as a
  month and the earth will always show the same face to the moon. How about that?

The earth makes one complete turn per day beneath these ocean bulges. This produces two sets of ocean tides per day. Any part of the earth that passes beneath one of the bulges has a high tide. Worldwide, the average high tide is about 1 meter above the average surface level of the ocean. About six hours later, after the earth has made a quarter turn, the water level at the same part of the ocean is about 1 meter below the average sea level. This is low tide. The water that "isn't there" is under the bulges that make up the high tides elsewhere. So the earth turns beneath two tide bulges per rotation and we have two high tides and two low tides daily. While the earth spins, the moon moves in its orbit and appears at the same position in our sky every 24 hours and 50 minutes, so the two-high-tide cycle is actually at 24-hour-and-50-minute intervals. That's why tides do not occur at the same time every day.

Figure 13.12 ▲
Two tidal bulges remain relatively fixed with respect to the moon while the earth spins daily beneath them.

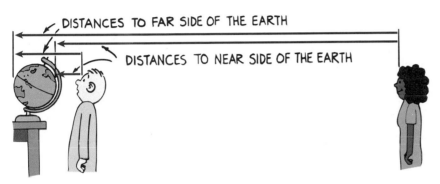

◀ **Figure 13.13**
When you stand close to the globe (as the moon is in relation to the earth), the closest part of the globe is noticeably nearer to you than the farthest part. When you stand far away (as the sun is in relation to the earth), the difference in distance between the closest and farthest parts of the globe is less significant.

The sun also contributes to ocean tides, but not as much as the moon. This may seem puzzling when you realize that the sun's pull on the earth is about 180 times stronger than the moon's pull on the earth. Why, then, doesn't the sun cause tides 180 times higher than lunar tides? The answer has to do with a key word: *difference*. Because of the sun's great distance from the earth, there is not much *difference* between the distance from the sun to the near part of the earth and the distance from the sun to the far part of the earth (Figure 13.13). This means there is not much difference in the gravitational pull between the sun and the part of the ocean nearer to it, and the sun and the part of the ocean farther from it. The small difference in solar pulls on opposite sides of the earth only slightly elongates the earth's shape and produces tidal bulges less than half those produced by the moon.*

---

* The relative difference in distance is only hinted at in Figure 13.13, which is way off scale. The actual distance between the earth and the sun is nearly 12 000 earth diameters. Using an ordinary globe 1/3 meter in diameter to judge the relative difference between closest and farthest parts of the globe from the moon, you'd have to stand 10 meters away. From the sun's position, you'd have to stand "across town," 4 kilometers away! (The percentage difference in the pull of the sun across the earth is only about 0.017%, compared with 6.7% across the earth by the moon. It is only because the sun's pull is 180 times stronger than the moon's pull that the sun's tides are almost half as high [180 × 0.017% = 3%, nearly half of 6.7%].)

Newton deduced that the difference in pulls decreases as the *cube* of the distance between the centers of the bodies—twice as far away produces 1/8 the tide; three times as far, only 1/27 the tide, and so on. Only relatively close distances result in appreciable tides, and so the nearby moon "out-tides" the enormously more massive but farther-away sun. The size of the tide also depends on the size of the body having tides. Although the moon produces a considerable tide in the earth's oceans, which are thousands of kilometers apart, it produces scarcely any in a lake. That's because no part of the lake is significantly closer to the moon than any other part of the lake, so there is no significant *difference* in the moon's pull on the lake. The same is true for the fluids in your body. Any tides in the fluids of your body caused by the moon are negligible. You're not tall enough compared with the moon's distance for tides. The microtides produced by a 1-kilogram book held 1 meter over your head are far greater than any microtides produced in your body by the moon.

■ Question

We know that both the moon and the sun produce our ocean tides. And we know the moon plays the greater role because it is closer. Does its closeness mean it pulls with more gravitational force than the sun on the earth's oceans?

Figure 13.14 ▶
When the sun, the moon, and the earth are aligned, spring tides occur.

When the sun, the earth, and the moon are all lined up, the tides due to the sun and the moon coincide. We then have higher-than-average high tides and lower-than-average low tides. These are called **spring tides** (Figure 13.14). (Spring tides have nothing to do with the spring season.)

■ Answer

No, the sun's pull is much stronger. Gravitational pull weakens with the square of the distance to the body that pulls. But the *difference* in pulls across the earth's oceans weakens with the distance cubed. When the distance to the sun is squared, gravitation from the sun is still stronger than gravitation from the closer moon—because of the sun's enormous mass. But when the distance to the sun is cubed, as is the case for tidal forces, the sun's influence is less than the moon's. Distance is the key to tidal forces. If the moon were closer to the earth, the tides on both the earth and the moon would increase with the inverse cube of this closer distance—which could be catastrophic, as the planetary rings of other planets suggest.

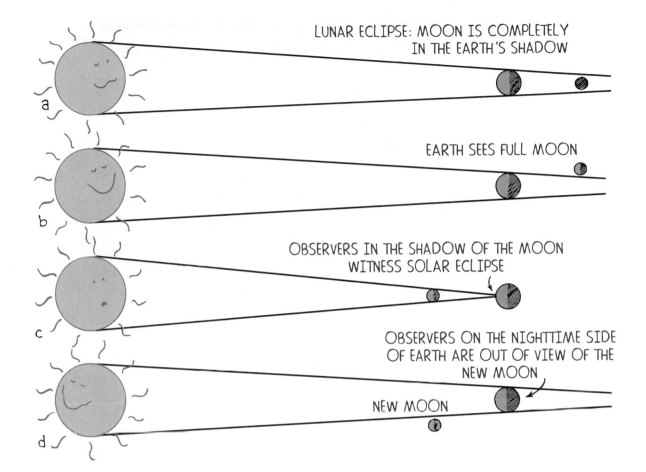

Figure 13.15 ▲
Detail of sun-earth-moon alignment. (a) Perfect alignment produces a lunar eclipse. (b) Nonperfect alignment produces a full moon. (c) Perfect alignment produces a solar eclipse. (d) Nonperfect alignment produces a new moon. Can you see that from the daytime side of the world, the new moon cannot be seen because the dark side faces the earth, and from the nighttime side of the world, the new moon is out of view altogether?

If the alignment is *perfect*, we have an eclipse. A **lunar eclipse** is produced when the earth is directly between the sun and moon (Figure 13.15a). A **solar eclipse** is produced when the moon is directly between the sun and the earth (Figure 13.15c). But the alignment is usually not perfect, because the plane of the moon's orbit is slightly tilted toward the plane of the earth's orbit about the sun. So each month when the earth is between the sun and the moon we have a full moon, and when the moon is between the sun and the earth we have a new (dark) moon. Consequently, spring tides occur at the times of a new moon and a full moon.

When we see a half-moon, that is, when the moon is halfway between a new moon and a full moon in either direction (Figure 13.16), the tides due to the sun and the moon partly cancel each other. Then, the high tides are lower-than-average high tides, and the low tides are higher-than-average low tides. These are called **neap tides.**

Figure 13.16 ▲
When the attractions of the sun and the moon are at right angles to one another (at the time of a half-moon), neap tides occur.

Power Production

Ocean tides have been producing electricity since 1968 in Brittany, France. A dam across an estuary gets its power from the rising and falling of the daily ocean tide. First the water is higher on one side of the dam, and is maintained at 3 meters higher than the lower side. Water then flows through a tube to the lower side, turning 24 huge turbines in the process. When the tide changes, the flow of water is in the reverse direction, again turning the turbines. The dam supports a highway that allows traffic to cross the estuary, and produces more than 200 MW of power for a city of 300 000 people.

Another factor affecting tides is the tilt of the earth's axis (Figure 13.17). Even though the opposite tidal bulges are equal, the earth's tilt causes the two daily high tides experienced in most parts of the ocean to be unequal most of the time.

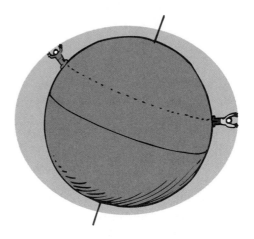

Figure 13.17 ▶
The earth's tilt causes the two daily high tides in the same place to be unequal.

Our treatment of tides is quite simplified here. We have ignored such complications as interfering land masses, tidal inertia, and friction with the ocean bottom, all of which result in a wide range of tides in different parts of the world. Although worldwide average tides are about 1 meter above and below the ocean's normal level, in some places the tides are much greater than this. In the Bay of Fundy in Nova Scotia and in some Alaskan fjords, for example, tides sometimes exceed 15 meters. This is largely due to the ocean floor, which funnels shoreward in a V-shape. The tide often comes in faster than a person can run. Don't dig clams near the water's edge at low tide in the Bay of Fundy!

13.5 Tides in the Earth and Atmosphere

The earth is not a rigid solid but, for the most part, is molten liquid covered by a thin solid and pliable crust. As a result, the moon-sun tidal forces produce earth tides as well as ocean tides. Twice each day the solid surface of the earth rises and falls by as much as 25 cm! We don't normally notice this, just as people on a ship in the middle of the ocean would not notice the gradual rising and falling of ocean tides. Satellite detectors measure both ocean and earth tidal movements. Interestingly enough, earthquakes and volcanic eruptions have a slightly higher probability of occurring when the earth is experiencing an earth spring tide—that is, near a full or new moon.

We live at the bottom of an ocean of air that also experiences tides. Atmospheric tides are relatively small. In the upper part of the atmosphere is the ionosphere, so named because it is made up of ions, electrically charged atoms that are the result of intense cosmic ray bombardment and ultraviolet radiation from the sun. Tidal effects in the ionosphere produce electric currents that alter the magnetic field that surrounds the earth. These are magnetic tides. They in turn regulate the degree to which cosmic rays penetrate into the lower atmosphere. The cosmic ray penetration affects the ionic composition of our atmosphere, which in turn is evident in subtle changes in the behaviors of living things. The highs and lows of magnetic tides are greatest when the atmosphere is having its spring tides—again, near the full and new moons.*

13.6 Black Holes

There are two main processes going on continously in stars like our sun. One process is gravitation, which tends to crunch all solar material toward the center. The other process is thermonuclear fusion consisting of reactions similar to those in a hydrogen bomb. These hydrogen bomb–like reactions tend to blow solar material outward. When the processes of gravitation and thermonuclear fusion balance each other, the result is the sun of a given size.

If the fusion rate increases, the sun will get hotter and bigger; if the fusion rate decreases, the sun will get cooler and smaller. What will happen when the sun runs out of fusion fuel (hydrogen)? The answer is, gravitation will dominate and the sun will start to collapse. For our sun, this collapse will ignite the nuclear ashes of fusion (helium) and fuse them into carbon. During this fusion process, the sun will expand to become the type of star known as a *red giant*. It will be so big that it will extend beyond the earth's orbit

LINK TO ASTRONOMY

Planetary Rings

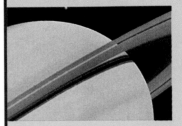

Four of the nine planets in the solar system have a system of planetary rings, Saturn's being predominant. Tidal forces may have caused the formation of these rings. A satellite experiences competing forces—the tidal forces that tend to tear it apart, and the self-gravitation that holds it together. Early in the life of the solar system, Saturn (and other outer planets) may have had one or more moons orbiting too close to the planet's surface. Powerful tidal forces could have stretched them and torn them apart. During billions of years fragments could have separated into billions of still smaller pieces spreading out to form the beautiful rings we see today. Our moon is sufficiently far away to resist this tidal disintegration. But if it were to come too close, within a few hundred kilometers of the earth, the increased tidal forces would tear the moon apart. Then the earth, like Saturn, Jupiter, Uranus, and Neptune, would have a system of planetary rings!

---

*   Could this be why some of your friends seem a bit weird at the time of a full moon?

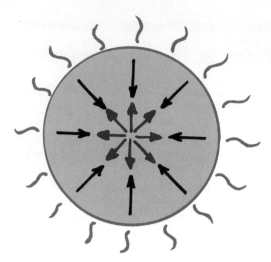

Figure 13.18 ▲
The size of the sun is the result of a "tug-of-war" between two opposing processes: nuclear fusion, which tends to blow it up (blue arrows), and gravitational contraction, which tends to crunch it together (black arrows).

and swallow the earth. Fortunately, this won't take place until some 5 billion years from now. When the helium is all "burned," the red giant will collapse and die out. It will no longer give off heat and light. It will then be the type of star called a *black dwarf*—a cool cinder among billions of others.

The story is a bit different for stars more massive than the sun. For a heavy star, one that is at least two to three times more massive than our sun,* once the flame of thermonuclear fusion is extinguished, gravitational collapse takes over—and it doesn't stop! The star not only caves in on itself, but the atoms that compose the stellar material also cave in on themselves until there are no empty spaces. According to theory, the collapse never stops and the density becomes literally infinite. Gravitation near these shrunken configurations is so enormous that nothing can get back out. Even light cannot escape. They have crushed themselves out of visible existence. They are called **black holes.**

Interestingly enough, a black hole is no more massive than the star from which it collapsed. The gravitational field near the black hole may be enormous, but the field beyond the original radius of the star is no different after collapse than before (Figure 13.19). The amount of mass has not changed, so there is no change in the field at any point beyond this distance. Black holes will be formidable only to future astronauts who venture too close.

The configuration of the gravitational field about a black hole represents the collapse of space itself. The field is usually represented as a warped two-dimensional surface, as shown in Figure 13.20. Astronauts could enter the fringes of this warp and, with a powerful spaceship, still escape. After a certain distance, however, they could

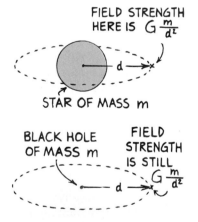

FIELD STRENGTH HERE IS $G\frac{m}{d^2}$

STAR OF MASS m

BLACK HOLE OF MASS m

FIELD STRENGTH IS STILL $G\frac{m}{d^2}$

Figure 13.19 ▲
The gravitational field strength near a giant star that collapses to become a black hole is the same before collapse (top) and after collapse (bottom).

---

* Astrophysicists don't yet have an exact number for the least massive star that will inevitably become a black hole. They believe it to be at least two solar masses.

Our sun is not massive enough to become a black hole. But if it
were, and if it collapsed from its present size to become a black
hole, would the earth be drawn into it?

not escape, and they would disappear from the observable universe.
Don't go too close to a black hole!

Although black holes can't be seen, their effects can be. Many
stars in the sky are binaries—pairs that orbit about each other. Some
binaries have invisible companions, and when these are closely situated black holes, matter drawn from the normal star emits X rays as
it accelerates into the neighboring black hole. Astrophysicists have
detected such. Fascinating!

Figure 13.20 ▲
A two-dimensional representation of the gravitational field around a black
hole. Anything that falls into the central warp disappears from the observable universe.

■ **Answer**

No, the gravitational force between the solar black hole and the earth would not change.
The earth and other planets would continue in their orbits. Observers outside the solar
system would see the planets orbiting about "nothing" and would likely deduce the presence of a black hole. Observations of astronomical bodies that orbit unseen partners indicate the presence of black holes.

13 Chapter Review

Concept Summary

The earth can be thought of as being surrounded by a gravitational field that interacts with objects and causes them to experience gravitational forces.

The gravitational field **g** is equal to the acceleration of a freely falling object.

Objects in orbit around the earth have a gravitational force acting on them even though they may appear to be weightless.

Ocean tides (and even tides within the solid earth and within the atmosphere) are caused by differences in the gravitational pull of the moon (and sun) on opposite sides of the earth.

When a star runs out of fuel for fusion, it collapses under gravitational forces. Sufficiently massive stars collapse to form a black hole.

Important Terms

black hole (13.6)
force field (13.1)
gravitational field (13.1)
lunar eclipse (13.4)
neap tide (13.4)
solar eclipse (13.4)
spring tide (13.4)
weightlessness (13.3)

Review Questions

1. We can think of a force field as a kind of extended aura that surrounds a body, spreading its influence to affect things. As later chapters will show, an electric field affects electric charges, and a magnetic field affects magnetic poles. What does a gravitational field affect? (13.1)

2. Which is correct—to say that a distant rocket interacts with the mass of the earth, or to say that it interacts with the gravitational field of the earth? Explain. (13.1)

3. How does the gravitational field strength at the surface of the earth compare with the acceleration of free fall? (13.1)

4. Upon what quantities does the acceleration of gravity on the surfaces of various planets depend? (13.1)

5. How does the gravitational field surrounding the earth vary with increasing distance? (13.1)

6. Where is your weight greatest—at the surface of the earth, deep below the surface, or above the surface? (13.2)

7. Why would your weight be less if you were deep beneath the earth's surface? (13.2)

8. What is the value of the gravitational field at the center of the earth? (13.2)

9. Does your apparent weight change when you ride an elevator moving at constant speed? An accelerating elevator? Explain. (13.3)

10. How does the support force of a floor you stand on relate to your weight? (13.3)

11. If the gravitational pull of the moon on the earth were the same over all parts of the earth, would there be any tides? Defend your answer. (13.4)

12. Why is *difference* a key word in explaining tides? (13.4)

13. Which force-pair is greater—that between the moon and earth, or that between the sun and earth? (13.4)

14. Which is more effective in raising ocean tides—the moon or the sun? Why does this answer not contradict your previous answer? (13.4)

15. Why are tides greater at the times of the full and new moons and at the times of lunar and solar eclipses? (13.4)

16. Distinguish between *spring* tides and *neap* tides. (13.4)

17. Do the sun and moon produce atmospheric tides on earth? (13.5)

18. What two major factors determine the size of a star? (13.6)

19. Distinguish between a stellar black *dwarf* and a black *hole*. (13.6)

20. Why would the earth not be sucked into the sun if it became a black hole? (13.6)

Plug and Chug

These Plug and Chug problems all relate to the following information.

A kilogram of water on the side of the earth nearest the moon is gravitationally attracted to the moon with a greater force than a kilogram of water on the side of the earth farthest from the moon. The difference in force per mass, the tidal force, is approximated by

$$T_F = \frac{4GMR}{d^3} ,$$

where G is the gravitational constant, M is the mass of the moon, R is the radius of the earth, and d is the distance between the centers of the moon and the earth.

1. Calculate T_F of the moon on the earth in units N/kg. ($M = 7.35 \times 10^{22}$ kg, $R = 6.4 \times 10^6$ m, and $d = 3.85 \times 10^8$ m)

2. If you stand on the earth directly under the moon, the moon's tidal pull will slightly stretch you, pulling harder on your head than on your feet. Your head and feet are not far apart like the earth's oceans are, so R in the tidal force equation is only about half your height rather than half the earth's diameter. Approximate the T_F of the moon on you.

3. The earth under your feet also stretches you ever so slightly by pulling harder on your feet than on your head. What is the earth's T_F on

you? (Now R is half your height and d is the distance from the center of the earth to you, 6.4×10^6 m. M will be the earth's mass, 6.0×10^{24} kg.) How does the earth's tidal force on you compare with the moon's?

4. Even a 1-kg melon held above your head exerts a tidal force on you. The above equation, however, is a good approximation only when d is much larger than R—certainly true for the previous cases, but not for the melon. The distance d between the center of the melon and your center is only about twice R. So the general tidal force equation (the source of the above equation) must be used, which is $T_F = (4GMdR)/(d^2 - R^2)^2$. Use this to calculate the melon's T_F on you. Here $M = 1$ kg, and if you're 2 m tall, $R = 1$ m, and $d = 2$ m. How does this compare with the tidal force of the moon on you?

Think and Explain

1. If the earth were the same size but twice as massive, how would the value of G change, if at all? How would the value of **g** change, if at all? (Why are your answers different? In this and the following question, let the equation $g = GM/R^2$ guide your thinking.)

2. If the radius of the earth somehow shrunk by half without any change in the earth's mass, what would be the value of **g** at the new surface? What would be the value of **g** above the new surface at a distance equal to the present radius?

3. The weight of an apple near the surface of the earth is 1 N. What is the weight of the earth in the gravitational field of the apple?

4. A friend proposes an idea for launching space probes that consists of boring a hole completely through the earth. Your friend reasons that a probe dropped into such a hole would accelerate all the way through and shoot like a projectile out the other side. Defend or oppose the reasoning of your friend.

5. If you stand on a shrinking planet, so that in effect you get closer to its center, your weight will increase. But if you instead burrow into the planet and get closer to its center, your weight will decrease. Explain.

6. If you were unfortunate enough to be in a freely falling elevator, you might notice the bag of groceries you were carrying is hovering in front of you, apparently weightless. Cite the frames of reference in which the groceries are falling, and in which they are not falling.

7. When would the moon be its fullest—just before a solar eclipse or just before a lunar eclipse? Defend your answer.

8. What would be the effect on the earth's tides if the diameter of the earth were larger than it is? If the earth were as it presently is, but the moon were larger—with the same mass?

9. The human body is more than 50% water. Is it likely that the moon's gravitational pull causes any significant biological tides—cyclic changes in water flow among the body's fluid compartments? (*Hint:* Is any part of your body appreciably closer to the moon than any other part? Is there a *difference* in lunar pulls?)

10. A black hole is no more massive than the star from which it collapsed. Why then, is gravitation so intense near a black hole?

Think and Solve

1. What would be the weight of a 1-kg mass on the surface of Mars? The mass of Mars is 0.11 that of the earth, and its radius 0.53 that of the earth.

2. Many people mistakenly believe that the astronauts that orbit the earth are "above gravity." Calculate **g** for space shuttle territory, 200 km above the earth's surface. Assume the earth's mass is 6.0×10^{24} kg, and its radius 6.38×10^6 m (6380 km). Your answer is what percentage of 9.8 m/s^2?

3. On the small and good planet Ballonius in a distant solar system, suppose we find the radius of the good planet to be 200 000 m. We also drop a rock and find that it travels 1.5 m in the first second of free fall, which means it accelerates at 3 m/s^2. Estimate the mass of Ballonius.

4. Choose any of the planets and calculate its tidal force on you. Compare your answer with your answer to Plug and Chug exercise 4. Then communicate your result to any of your friends who say the planets exert tidal forces on people and in so doing influence their lives!

14 Satellite Motion

If you drop a stone, it will fall in a straight-line path to the ground below. If you move your hand horizontally as you drop the stone, it will follow a curved path to the ground. If you move your hand faster, the stone will land farther away and the curvature of the path will be less pronounced. What would happen if the curvature of the path matched the curvature of the earth? The answer is simple enough: Without air resistance, you'd have an earth satellite!

All bodies in space are falling around other bodies.

Figure 14.1 ▲
The greater the stone's horizontal motion when released, the wider the arc of its curved path.

14.1 Earth Satellites

Simply put, an earth satellite is a projectile that falls *around* the earth rather than *into* it. Imagine yourself on a planet that is smaller than the earth (Figure 14.2). Because of the planet's small size and low mass, you would not have to throw the stone very fast to make its curved path match the surface curvature of the planet. If you threw the stone just right, it would follow a circular orbit.

Figure 14.2 ▲
If you toss the stone horizontally with the proper speed, its path will match the surface curvature of the asteroid.

How fast would the stone have to be thrown horizontally for it to orbit the earth? The answer depends on the rate at which the stone falls and the rate at which the earth curves. Recall from Chapter 2 that a stone dropped from rest accelerates 10 m/s² and falls a vertical distance of 5 meters during the first second (more precisely, these figures are 9.8 m/s² and 4.9 m respectively). Also recall from Chapter 3 that the same is true of any projectile as it starts to fall. Recall that in the first second a projectile will fall a vertical distance of 5 meters below the straight-line path it would have taken without gravity. (It may be helpful to refresh your memory and review Figure 3.10.)

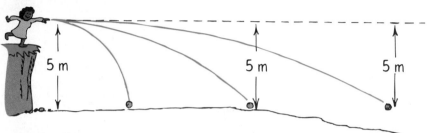

Figure 14.3 ▲
Throw a stone at any speed and one second later it will have fallen 5 m below where it would have been without gravity.

A geometric fact about the curvature of our earth is that its surface drops a vertical distance of nearly 5 meters for every 8000 meters tangent to its surface (Figure 14.4). This means that if you were swimming in a calm ocean, you would be able to see only the very top of a 5-m-tall mast on a ship that is 8 kilometers away.

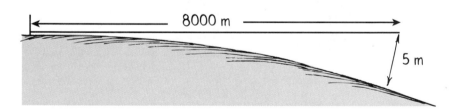

Figure 14.4 ▲
The earth's curvature (not to scale).

Can you see that a stone thrown fast enough to go a horizontal distance of 8 kilometers during the time (1 second) it takes to fall 5 meters, will follow the curvature of the earth? Isn't this speed simply 8 kilometers per second? So we see that the orbital speed for close orbit about the earth is 8 km/s. If this doesn't seem to be very fast, convert it to kilometers per hour; you'll see it is an impressive 29 000 km/h (or 18 000 mi/h). At that speed, atmospheric friction would burn an object to a crisp. That's why a satellite must stay about 150 kilometers or more above the earth's surface—to keep from burning up against the friction of the atmosphere like a "falling star."

14.2 Circular Orbits

Interestingly enough, in circular orbit the speed of a circling satellite is not changed by gravity. We can understand this by comparing a satellite in circular orbit to a bowling ball rolling along a bowling alley. Why doesn't the gravity that acts on the bowling ball change its speed? The answer is that gravity is pulling neither forward nor backward—it pulls straight downward, perpendicular to the ball's motion. The bowling ball has no component of gravitational force along the direction of the alley.

Figure 14.5 ▶
(Left) The force of gravity on the bowling ball does not affect its speed because there is no component of gravitational force horizontally. (Right) The same is true for the satellite in circular orbit. In both cases, the force of gravity is at right angles to the direction of motion.

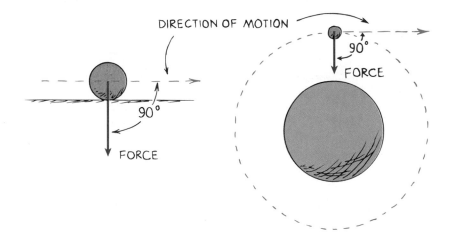

The same is true for a satellite in circular orbit. Here a satellite is always moving perpendicular to the force of gravity. It doesn't move in the direction of gravity, which would increase its speed, nor does it move in a direction against gravity, which would decrease its speed. Instead, the satellite exactly "criss-crosses" gravity, so that no change in speed occurs—only a change in direction. A satellite in circular orbit around the earth is always moving perpendicular to gravity and parallel to the earth's surface at constant speed.

For a satellite close to the earth, the time for a complete orbit around the earth, its **period,** is about 90 minutes. For higher altitudes, the orbital speed is less and the period is longer.* The moon is farther away, and has a 27.3-day period. Communications satellites are located in orbit 6.5 earth radii from the earth's center, so that their period is 24 hours. This period matches the earth's daily rotation. They are launched to orbit in the plane of the earth's equator, so they are always above the same place on the equator.

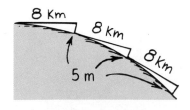

Figure 14.6 ▲
A satellite in circular orbit close to the earth moves tangentially at 8 km/s. During each second, it falls 5 m beneath each successive 8-km tangent.

---

* If you continue with your study of physics and take a follow-up course, you'll learn that the tangential speed v of a satellite in circular orbit is given by $v = \sqrt{GM/d}$, and the period T of satellite motion is given by $T = 2\pi\sqrt{d^3/GM}$, where G is the universal gravitational constant (see Chapter 12), M is the mass of the earth (or whatever body the satellite orbits), and d is the altitude of the satellite measured from the center of the earth or parent body.

■ Questions

1. There are usually alternate explanations for things. Is the following explanation valid? Satellites remain in orbit instead of falling to the earth because they are beyond the main pull of earth's gravity.

2. Satellites in close circular orbit fall about 5 m during each second of orbit. How can this be if the satellite does not get closer to the earth?

Recall from Chapter 12 that Isaac Newton understood satellite motion from his investigation of the moon's motion. He foresaw the launching of artificial satellites, for he reasoned that without air resistance, a cannonball could circle the earth and coast indefinitely if it had sufficient speed. He calculated this speed to be the same as 8 km/s. Since such speed was impossible then, he was not optimistic about people launching satellites. What Newton did not consider was multistage rockets—the idea of rockets carried piggyback style on other rockets to reach orbital speed by a succession of rocket firings.

14.3 Elliptical Orbits

A projectile just above the atmosphere at a horizontal speed somewhat more than 8 km/s will overshoot a circular path and trace an oval-shaped path—an **ellipse.**

An ellipse is a specific curve: the closed path taken by a point that moves in such a way that the sum of its distances from two fixed points (called **foci**) is constant. For a satellite orbiting a planet, the center of the planet is at one focus and the other focus could be inside or outside the planet. An ellipse can be easily constructed by using a pair of tacks, one at each focus, a loop of string, and a pencil, as shown in Figure 14.7. The closer the tacks, the closer the ellipse is to a circle. When both foci are together, the ellipse *is* a circle. A circle is a special case of an ellipse.

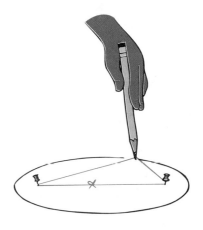

Figure 14.7 ▲
A simple method of constructing an ellipse.

■ Answers

1. No, no, a thousand times no! No mass can be beyond the pull of gravity. If any moving object were, it would move in a straight line and would not curve around the earth. Satellites remain in orbit because they *are* being pulled by gravity, not because they are beyond it. (Recall Think and Solve problem 2 in Chapter 13.)

2. In each second, the satellite falls about 5 m below the straight-line tangent it would have taken if there were no gravity. The earth's surface curves 5 m below an 8-km straight-line tangent. Since the satellite moves at 8 km/s, it "falls" at the same rate the earth "curves."

Figure 14.8 ▶
The shadows of the ball are all ellipses with one focus where the ball touches the table.

Satellite speed, which is constant in a circular orbit, *varies* in an elliptical orbit. When the initial speed is more than 8 km/s, the satellite overshoots a circular path and moves away from the earth, against the force of gravity. It therefore loses speed. Like a rock thrown into the air, the satellite slows to a point where it no longer

Figure 14.9 ▶
Elliptical orbit. When the satellite exceeds 8 km/s, it overshoots a circle (left) and travels away from the earth against gravity. At its maximum separation (center), it starts to come back toward the earth. The speed it lost in going away is gained in returning, and (right) the cycle repeats itself.

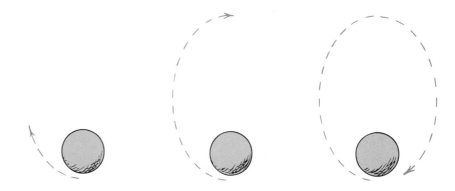

recedes, and begins falling back toward the earth. The speed lost in receding is regained as it falls back. The satellite then rejoins its path with the same speed it had initially (Figure 14.9). The procedure repeats over and over, and an ellipse is traced each cycle.

▶ **DOING PHYSICS**

Drawing Ellipses

Draw an ellipse with a loop of string, two tacks, and a pen or pencil, as shown in Figure 14.7. Try different tack spacings for a variety of ellipses. Or draw an ellipse by tracing the edge of the shadow cast by a circular disk on a flat surface. How can you move the disk to get different ellipses?

Activity

The orbit of a satellite is shown in the sketch. In which of the positions A through D does the satellite have the greatest speed? Least speed?

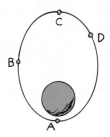

14.4 Energy Conservation and Satellite Motion

Recall from Chapter 8 that moving objects have kinetic energy (KE). An object above the earth's surface has potential energy (PE) due to its position. Everywhere in its orbit, a satellite has both KE and PE. The sum of the KE and PE everywhere is constant.

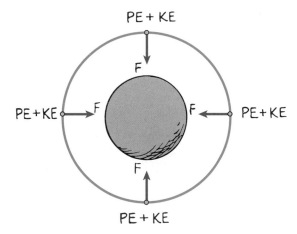

Figure 14.10 ▲
The force of gravity on the satellite is always toward the center of the body it orbits. For a satellite in circular orbit, no component of force acts along the direction of motion. The speed, and thus the KE, cannot change.

■ **Answer**

The satellite has its greatest speed as it whips around A. It has its least speed at C. Beyond C, it gains speed as it falls back to A to repeat its cycle.

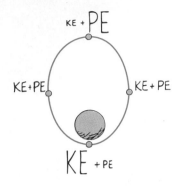

Figure 14.11 ▲
The sum of KE and PE for a satellite is a constant at all points along an elliptical orbit.

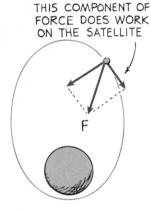

Figure 14.12 ▲
In elliptical orbit, a component of force exists along the direction of the satellite's motion. This component changes the speed and, thus, the KE. (The perpendicular component changes only the direction.)

In a circular orbit the distance between a planet's center and the satellite's center is constant. This means that the PE of the satellite is the same everywhere in orbit. So, by the law of conservation of energy, the KE is also constant. Thus, the speed is constant in any circular orbit.

In an elliptical orbit the situation is different. Both speed and distance vary. The PE is greatest when the satellite is farthest away (at the **apogee**) and least when the satellite is closest (at the **perigee**). Correspondingly, the KE will be least when the PE is most; and the KE will be most when the PE is least. At every point in the orbit, the sum of the KE and PE is constant.

At all points on the orbit—except at the apogee and perigee—there is a component of gravitational force parallel to the direction of satellite motion. This component changes the speed of the satellite. Or we can say: (this component of force) × (distance moved) = change in KE. Either way we look at it, when the satellite gains altitude and moves against this component, its speed and KE decrease. The decrease continues to the apogee. Once past the apogee, the satellite moves in the same direction as the component, and the speed and KE increase. The increase continues until the satellite whips past the perigee and repeats the cycle.

■ Questions

1. The orbital path of a satellite is shown in the sketch. In which of the positions *A* through *D* does the satellite have the most KE? Most PE? Most total energy?

2. Why does the force of gravity change the speed of a satellite when it is in an elliptical orbit, but not when it is in a circular orbit?

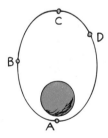

■ Answers

1. The KE is maximum at *A;* the PE is maximum at *C;* the total energy is the same anywhere in the orbit.

2. At any point on its path, the direction of motion of a satellite is always tangent to its path. If a component of force exists along this tangent, then there is a corresponding acceleration along this tangent—the satellite's tangential speed changes. In circular orbit the gravitational force is everywhere perpendicular to the satellite's direction of motion, just as every part of the circumference of a circle is perpendicular to the radius. So in circular orbit there is no component of gravitational force along the tangent, and only the direction of motion changes—not tangential speed. From a work-energy point of view, only a force parallel to motion can change KE, which occurs only for elliptical orbits.

14.5 Escape Speed

When a payload is put into earth-orbit by a rocket, the speed and direction of the rocket are very important. For example, what would happen if the rocket were launched vertically and quickly achieved a speed of 8 km/s? Everyone had better get out of the way, because it would soon come crashing back at 8 km/s. To achieve orbit, the payload must be launched *horizontally* at 8 km/s once above air resistance. Launched vertically, the old saying "What goes up must come down" becomes a sad fact of life.

But isn't there some vertical speed that is sufficient to ensure that what goes up will escape and not come down? The answer is yes. Neglecting air resistance, fire anything at any speed greater than 11.2 km/s, and it will leave the earth, going more and more slowly, but never stopping.* Let's look at this from an energy point of view.

How much work is required to move a payload against the force of earth gravity to a distance very, very far ("infinitely far") away? We might think that the PE would be infinite because the distance is infinite. But gravity diminishes rapidly with distance via the inverse-square law. The force of gravity is strong only close to the earth. Most of the work done in launching a rocket, for example, occurs near the earth. It turns out that the value of PE for a 1-kilogram mass infinitely far away is 62 million joules (MJ). So to put a payload infinitely far from the earth's surface requires at least 62 MJ of energy per kilogram of load. We won't go through the calculation here, but a KE per unit mass of 62 MJ/kg corresponds to a speed of 11.2 km/s. This is the value of the **escape speed** from the surface of the earth.**

If we give a payload any more energy than 62 MJ/kg at the surface of the earth or, equivalently, any greater speed than 11.2 km/s, then, neglecting air resistance, the payload will escape from the earth never to return. As it continues outward, its PE increases and its KE decreases. Its speed becomes less and less, though it is never reduced to zero. The payload outruns the gravity of the earth. It escapes.

The escape speeds of various bodies in the solar system are shown in Table 14.1. Note that the escape speed from the sun is 620 km/s at the surface of the sun. Even at a distance equaling that of the earth's orbit, the escape speed from the sun is 42.2 km/s. The escape speed values in the table ignore the forces exerted by other bodies. A projectile fired from the earth at 11.2 km/s, for example, escapes the earth but not necessarily the moon, and certainly not the sun. Rather than recede forever, it will take up an orbit around the sun.

Figure 14.13 ▲
The initial thrust of the rocket lifts it vertically. Another thrust tips it from its vertical course. When it is moving horizontally, it is boosted to the required speed for orbit.

---

* In a more advanced physics course you would learn how the value of escape speed v, from any planet or any body, is given by $v = \sqrt{2GM/d}$, where G is the universal gravitational constant, M is the mass of the attracting body, and d is the distance from its center. (At the surface of the body d would simply be the radius of the body.)

** Interestingly enough, this might well be called the *maximum falling speed*. Any object, however far from earth, released and allowed to fall to earth only under the influence of earth's gravity would not exceed 11.2 km/s. (With air friction, it would be less.)

The first probe to escape the solar system, *Pioneer 10,* was launched from Earth in 1972 with a speed of only 15 km/s. The escape was accomplished by directing the probe into the path of oncoming Jupiter. It was whipped about by Jupiter's great gravitational field, picking up speed in the process—just as the speed of a ball encountering an oncoming bat is increased when it departs from the bat. Its speed of departure from Jupiter was increased enough to exceed the sun's escape speed at the distance of Jupiter. *Pioneer 10* passed the orbit of Pluto in 1984. Unless it collides with another body, it will continue indefinitely through interstellar space. Like a note in a bottle cast into the sea, *Pioneer 10* contains information about the earth that might be of interest to extraterrestrials, in hopes that it will one day wash up and be found on some distant "seashore."

It is important to point out that the escape speeds for different bodies refer to the initial speed given by a brief thrust, after which

Table 14.1 Escape Speeds at the Surface of Bodies in the Solar System			
Astronomical Body	Mass (earth masses)	Radius (earth radii)	Escape Speed (km/s)
Sun	333 000	109	620
Sun (at a distance of the earth's orbit)	23	500	42.2
Jupiter	318	11	60.2
Saturn	95.2	9.2	36.0
Neptune	17.3	3.47	24.9
Uranus	14.5	3.7	22.3
Earth	1.00	1.00	11.2
Venus	0.82	0.95	10.4
Mars	0.11	0.53	5.0
Mercury	0.055	0.38	4.3
Moon	0.0123	0.27	2.4

SCIENCE, TECHNOLOGY, AND SOCIETY

Communications Satellites

The electromagnetic signals that are broadcast into space to carry television programs or telephone conversations travel in straight lines. In times past these straight-line (often called line-of-sight) communications required tall receiving antenna towers and signal-boosting relay stations on high buildings or mountains. Today many television and telephone signals bounce to us from satellites. These communications satellites are in equatorial orbits with 24-hour periods. Because they revolve once each time the earth rotates once, they appear stationary when we look up at them.

Dish-shaped antennas almost anywhere on Earth are on a line of sight from one or more communications satellites.

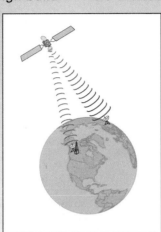

Because communications satellites are in equatorial orbit, dish antennas on the equator may tilt east or west, but they don't tilt north or south. An equatorial dish right under a communications satellite is looking straight up. If it held water, it would resemble a birdbath filled to the brim. Antennas north of the equator must be tilted to the south (and perhaps east or west, too). Those south of the equator must be tilted to the north, and likely east or west as well. Unless you live on the equator, all of the antennas that you see look like partly emptied bowls. In Antarctica or near the North Pole, a dish antenna is tipped so far over that it could hold no water at all.

there is no force to assist motion. But we could escape the earth at any *sustained* speed greater than zero, given enough time. Suppose a rocket is going to a destination such as the moon. If the rocket engines burn out while still close to the earth, the rocket will need a minimum speed of 11.2 km/s. But if the rocket engines can be sustained for long periods of time, the rocket could go to the moon without ever attaining 11.2 km/s.

It is interesting to note that the accuracy with which an unpiloted rocket reaches its destination is accomplished not by staying on a pre-planned path, or by getting back on that path if it strays off course. No attempt is made to return the rocket to its planned path. Instead, by communication with the control center, the rocket in effect asks, "Where am I now, and where do I want to go? What is the best way to get there from here, given my present situation?" With the aid of high-speed computers, the answers to these questions are used to find a *new* path. Corrective thrusters put the rocket on this new path. This process is repeated continuously along the way until the rocket reaches its destination.

Is there a lesson to be learned here? Suppose you find that you are "off course." You may, like the rocket, find it more fruitful to take a course that leads to your goal as best plotted from your present position and circumstances, rather than try to get back on the course you plotted from a previous position and in, perhaps, different circumstances.

209

Concept Summary

An earth satellite is a projectile that moves fast enough tangentially so that it falls around the earth rather than into it.

- The speed of a satellite in a circular orbit is not changed by gravity.

- The speed of a satellite in an elliptical orbit decreases as it recedes from the earth and increases as it approaches the earth.

- The sum of the kinetic and potential energies is constant in any orbit.

- If an object is launched from Earth at a speed exceeding 11.2 km/s, or equivalently, with more than 62 MJ of kinetic energy per kilogram of mass, it will escape the earth (overlooking the effect of air resistance).

Important Terms

apogee (14.4)   focus (pl. foci) (14.3)
ellipse (14.3)   perigee (14.4)
escape speed (14.5)   period (14.2)

Review Questions

1. If we drop a ball from rest, how far will it fall vertically in the first second? If we instead move our hand horizontally and drop it (throw it), how far will it fall vertically in the first second? (14.1)

2. What do the distances 8000 m and 5 m have to do with a line tangent to the earth's surface? (14.1)

3. How does the direction of motion of a satellite in circular orbit compare with the curve of the earth's surface? (14.2)

4. Why doesn't gravitational force change the speed of a satellite in circular orbit? (14.2)

5. Does the period of a satellite increase or decrease as its distance from the earth increases?

6. Describe an ellipse. (14.3)

7. Why does gravitational force change the speed of a satellite in elliptical orbit? (14.4)

8. a. Where in an elliptical orbit is the speed of a satellite maximum?

   b. Where is it minimum? (14.4)

9. The sum of PE and KE for a satellite in a circular orbit is constant. Is this sum also constant for a satellite in an elliptical orbit? (14.4)

10. Why does the force of gravity do no work on a satellite in circular orbit, but does do work on a satellite in an elliptical orbit? (14.4)

11. Neglecting air resistance, what will happen to a projectile that is fired vertically at 8 km/s? At 12 km/s? (14.5)

12. a. How fast would a particle have to be ejected from the sun to leave the solar system?

    b. What speed would be needed if an ejected particle started at a distance from the sun equal to the earth's distance from the sun? (14.5)

13. What is the escape speed on the moon? (14.5)

14. Although the escape speed from the surface of the earth is 11.2 km/s, couldn't a rocket with enough fuel escape at any speed? Why or why not? (14.5)

Think and Explain

1. A satellite can orbit at 5 km above the moon, but not at 5 km above the earth. Why?

2. Does the speed of a satellite around the earth depend on its mass? Its distance from the earth? The mass of the earth?

3. If a cannonball is fired from a tall mountain, gravity changes its speed all along its trajectory. But if it is fired fast enough to go into circular orbit, gravity does not change its speed at all. Why?

4. Does gravity do any *net* work on a satellite in an elliptical orbit during one full orbit? Explain your answer.

5. If you stopped an earth satellite dead in its tracks, it would simply crash into the earth. Why, then, don't the communications satellites that "hover motionless" above the same spot on the earth crash into the earth?

6. Would you expect the speed of a satellite in close circular orbit about the moon to be less than, equal to, or greater than 8 km/s? Why?

7. Why do you suppose that sites close to the equator are preferred for launching satellites? (*Hint:* Look at the spinning earth from above either pole and compare it to a spinning turntable.)

8. Why do you suppose that a space shuttle is sent into orbit by firing it in an easterly direction (the direction in which the earth spins)?

9. If an astronaut in an orbiting space shuttle wished to drop something to Earth, how could this be accomplished?

10. Why does most of the work done in launching a rocket take place when the rocket is still close to the earth's surface?

11. What is the maximum possible speed of impact upon the earth's surface for a far-away object initially at rest that falls to the earth due only to the earth's gravity?

12. If Pluto were somehow stopped short in its orbit, it would fall into the sun rather than around it. About how fast would it be moving when it hit the sun?

13. If the earth somehow acquired more mass, with no change in its radius, would escape speed be less than, equal to, or more than 11.2 km/s? Why?

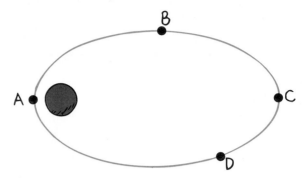

14. This question reviews several concepts of mechanics. A satellite travels the elliptical path shown. At which of the indicated positions *A* through *D* does the satellite experience the maximum (a) gravitational force? (b) speed? (c) velocity? (d) momentum? (e) kinetic energy? (f) gravitational potential energy? (g) total energy? (h) acceleration? (i) angular momentum?

Think and Solve

1. Calculate the speed at which the earth revolves around the sun in m/s. Note: The orbit is nearly circular.

2. Calculate the speed at which the moon revolves around the earth in m/s. Note: The orbit is nearly circular.

3. Escape speed at a distance *d* from the center of a body of mass *M* is

$$v_{escape} = \sqrt{\frac{2GM}{d}}$$

Calculate the escape speed from the moon's surface (moon radius = 1.74×10^6 m, moon mass = 7.35×10^{22} kg). Check your answer with Table 14.1.

15 Special Relativity— Space and Time

Space and time are related.

Everyone knows that we move in time, at the rate of 24 hours per day. And everyone knows that we can move through space, at rates ranging from a snail's pace to those of supersonic aircraft and space shuttles. But relatively few people know that motion through space is related to motion in time.

The first person to understand the relationship between space and time was Albert Einstein.* Einstein went beyond common sense when he stated in 1905 that in moving through space we also change our rate of proceeding into the future—time itself is altered. This view was introduced to the world in his **special theory of relativity.** This theory describes how time is affected by motion in space at constant velocity, and how mass and energy are related. Ten years later Einstein announced a similar theory, called the *general theory of relativity,* which encompasses accelerated motion as well. These theories have enormously changed the way scientists view the workings of the universe. This book discusses only the special theory and leaves the general theory for follow-up study later in your education.

This chapter will serve merely to acquaint you with the basic ideas of special relativity as they relate to space and time. Chapter 16 will continue with the relationship between mass and energy. These ideas, for the most part, are not common to your everyday experience. As a result, they don't agree with common sense. So please be patient with yourself if you find that you do not understand them. Perhaps your children or grandchildren will find them very much a part of their everyday experience. If so, they should find an understanding of relativity considerably less difficult.

---

* The concerns of Albert Einstein (1879–1955) were not limited to physics. As a German citizen in Nazi Germany he spoke out against Hitler's racial and political policies, which prompted his resignation from the University of Berlin. He fled Germany in 1933 and became an American citizen in 1940.

15.1 Space-Time

Newton and other investigators before Einstein thought of space as an infinite expanse in which all things exist. We are in space, and we move about in space. It was never clear whether the universe exists in space, or space exists within the universe. Is there space outside the universe? Or is space only within the universe? The same question could be raised for time. Does the universe exist in time, or does time exist only within the universe? Was there time before the universe came to be? Will there be time if and when the universe ceases to exist? Einstein's answer to these questions is that both space and time exist only within the universe. There is no time or space "outside."

◀ **Figure 15.1**
The universe does not exist in a certain part of infinite space, nor does it exist during a certain era in time. It is the other way around: space and time exist within the universe.

Einstein reasoned that space and time are two parts of one whole called **space-time.** To begin to understand this, consider your present knowledge that you are moving through time at the rate of 24 hours per day. This is only half the story. To get the other half, convert your thinking from "moving through time" to "moving through space-time." From the viewpoint of special relativity, you travel through a combination of space and time. You travel through space-time. When you stand still, then all your traveling is through time. When you move a bit, then some of your travel is through space and most of it is still through time. If you were somehow able to travel through space at the speed of light, what changes would

Figure 15.2 ▲
When you stand still, you are traveling at the maximum rate in time: 24 hours per day. If you traveled at the maximum rate through space (the speed of light), time would stand still.

Figure 15.3 ▲
The bag of groceries has an appreciable speed in the frame of reference of the building, but in the frame of reference of the freely falling elevator it has no speed at all.

you experience in time? The answer is that all your traveling would be through space, with no travel through time! You would be as ageless as light, for light travels through space only (not time) and is timeless. From the frame of reference of a photon traveling from one part of the universe to another, the journey takes no time at all!

Motion in space affects motion in time. Whenever we move through space, we to some degree alter our rate of moving into the future. This is **time dilation,** a stretching of time that occurs ever so slightly for everyday speeds, but significantly for speeds approaching the speed of light. If spacecraft of the future reach sufficient speed, people will be able to travel noticeably in time. They will be able to jump centuries ahead, just as today people can jump from the earth to the moon. To understand time dilation and how this can be, you first need to understand several ideas: the relativity of motion and the fundamental assumptions (postulates) of special relativity.

15.2 Motion Is Relative

Recall from Chapter 2 that whenever we discuss motion, we must specify the position from which the motion is being observed and measured. For example, you may walk along the aisle of a moving bus at a speed of 1 km/h relative to your seat, but at 100 km/h relative to the road outside. Speed is a relative quantity. Its value depends upon the place—the frame of reference—where it is observed and measured. An object may have different speeds relative to different frames of reference.

Suppose your friend always pitches a baseball at the same speed of 60 km/h. Neglecting air resistance and other small effects, the ball is moving at 60 km/h when you catch it. Now suppose your friend pitches the ball to you from the flatbed of a truck that moves toward you at 40 km/h. How fast does the ball meet you? You'll have to be sure to wear a catcher's mitt, because the speed of the ball will be 100 km/h (the 60 km/h relative to the truck plus the 40 km/h relative to the ground). Speed is relative.

Figure 15.4 ▲
Your speed is 1 km/h relative to your seat, and 100 km/h relative to the road.

TRUCK AT REST

TRUCK MOVES TOWARD YOU

TRUCK MOVES AWAY FROM YOU

◀ **Figure 15.5**
The ball is always pitched at 60 km/h relative to the truck. (a) When both you and the truck are at relative rest the ball is traveling at 60 km/h when you catch it. (b) When the truck moves toward you at 40 km/h, the ball is traveling at 100 km/h when you catch it. (c) When the truck moves away from you at the same speed, the ball is traveling at 20 km/h when you catch it.

Suppose the truck moves away from you at 40 km/h and your friend again pitches the ball to you. This time you will need no glove at all, for the ball will reach you at a speed of 20 km/h (since 60 km/h minus 40 km/h is 20 km/h). This is not surprising, for you expect that the ball will be traveling faster when the truck approaches you and that the ball will be traveling more slowly when the truck recedes.

The idea that speed is a relative quantity goes back to Galileo and was known long before the time of Einstein. As you will learn in this chapter, Einstein expanded the relativity of speed to include the relativity of things that seem unchangeable.

15.3 The Speed of Light Is Constant

Suppose you actually caught baseballs thrown off a moving truck out in a parking lot and found that no matter what the speed or direction of the truck, the ball always got to you at only one speed— 60 km/h. That is to say, if the truck zoomed toward you at 50 km/h and your friend pitched the ball at his speed of 60 km/h, you would catch the ball with the same speed as if the truck were not moving at all. Furthermore, if the truck moved away from you at whatever speed, the ball would still get to you at 60 km/h. This all seems quite impossible, for it is contrary to common experience. And if you *did* experience this, you would have to reevaluate your whole notion of reality. To put it mildly, you would be quite confused.

Baseballs do not behave this way. But it turns out that light does! Every measurement of the speed of light in empty space gives the

Figure 15.6 ▲
The speed of light is found to be the same in all frames of reference.

same value of 300 000 km/s, regardless of the speed of the source or the speed of the receiver.* We do not ordinarily notice this because light travels so incredibly fast.

The fact that light has only one speed in empty space was discovered at the end of the last century.** Light from an approaching source reaches an observer at the same speed as light from a receding source. And the speed of light is the same whether we move toward or away from a light source. How did the physics community regard this finding? They were as perplexed as you would be if you caught baseballs at only one speed no matter how they were thrown. Experiments were done and redone, and always the results were the same. Nothing could vary the speed of light. Various interpretations were proposed, but none were satisfactory. The foundations of physics were on shaky ground.

Albert Einstein looked at the speed of light in terms of the definition of speed. What is speed? It is the amount of *space* traveled compared to the *time* of travel. Einstein recognized that the classical ideas of space and time were suspect. He concluded that space and time were a part of a single entity—space-time. The constancy of the speed of light, Einstein reasoned, unifies space and time.

The special theory of relativity that Einstein developed rests on two fundamental assumptions, or **postulates.**

15.4 The First Postulate of Special Relativity

Einstein reasoned that there is no stationary hitching post in the universe relative to which motion should be measured. Instead, all motion is relative and all frames of reference are arbitrary. A spaceship cannot measure its speed relative to empty space, but only relative to other objects. If, for example, spaceship A drifts past spaceship B in empty space, spaceman A and spacewoman B will each observe only the relative motion. From this observation each will be unable to determine who is moving and who is at rest, if either.

---

* The presently accepted value for the speed of light is 299 792 km/s, which we round off to 300 000 km/s. This corresponds to 186 000 mi/s.

** In 1887 two American physicists, A. A. Michelson and E. W. Morley, performed an experiment to determine differences in the speed of light in different directions. They thought that the motion of the earth in its orbit about the sun would cause shifts in the speed of light. They thought that the speed of light should have been faster when it was going in the direction the earth was moving and slower when it was going opposite to the direction the earth was moving. Using a device called an *interferometer,* they found that the speed seemed to be the same in all directions. For Michelson's many experiments on the speed of light, he was the first American honored with a Nobel Prize.

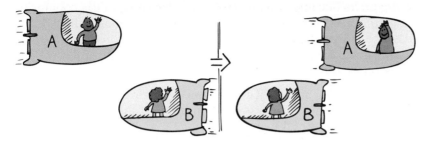

Figure 15.7 ▲

Spaceman A considers himself at rest and sees spacewoman B pass by. But spacewoman B considers herself at rest and sees spaceman A pass by. Who is moving and who is at rest?

This is a familiar experience to a passenger in a car at rest waiting for the traffic light to change. If you look out the window and see the car in the next lane begin moving backward, you may be surprised to find that the car you're observing is really at rest—your car is moving forward. If you could not see out the windows, there would be no way to determine whether your car was moving with constant velocity or was at rest.

In a high-speed jetliner we flip a coin and catch it just as we would if the plane were at rest. Coffee pours from the flight attendant's coffee pot as it does when the plane is standing on the runway. If we swing a pendulum, it will move no differently when the plane is moving uniformly (constant velocity) than when not moving at all. There is no physical experiment we can perform to determine our state of uniform motion. Of course, we can look outside and see the earth whizzing by, or send a radar signal out. However, no experiment confined within the cabin itself can determine whether or not there is uniform motion. The laws of physics within the uniformly moving cabin are the same as those in a stationary laboratory.

These examples illustrate one of the two building blocks of special relativity. It is Einstein's **first postulate of special relativity:**

> All the laws of nature are the same in all uniformly moving frames of reference.

Any number of experiments can be devised to detect *accelerated* motion, but none can be devised, according to Einstein, to detect the state of uniform motion.

Figure 15.8 ▲

A person playing pool on a smooth and fast-moving ocean liner does not have to make adjustments to compensate for the speed of the ship. The laws of physics are the same for the ship whether it is moving uniformly or is at rest.

15.5 The Second Postulate of Special Relativity

One of the questions that Einstein as a youth asked his schoolteacher was, "What would a light beam look like *if* you traveled along beside it?" According to classical physics, the beam would be at rest to such an observer. The more Einstein thought about this, the more convinced he became of its impossibility. He came to the conclusion that *if* an observer could travel *close* to the speed of light, he would measure the light as moving away from him at 300 000 km/s.

This is the idea that makes up Einstein's **second postulate of special relativity:**

> The speed of light in empty space will always have the same value regardless of the motion of the source or the motion of the observer.

The speed of light in all reference frames is always the same. Consider, for example, a spaceship departing from the space station shown in Figure 15.9. A flash of light is emitted from the station at 300 000 km/s—a speed we'll simply call *c*. No matter what the speed of the spaceship relative to the space station is, an observer on the spaceship will measure the speed of the flash of light passing her as *c*. If she sends a flash of her own to the space station, observers on the station will measure the speed of these flashes as *c*. The speed of the flashes will be no different if the spaceship stops or turns around and approaches. All observers who measure the speed of light will find it has the same value, *c*.

Figure 15.9 ▶
The speed of a light flash emitted by either the spaceship or the space station is measured as *c* by observers on the ship or the space station. Everyone who measures the speed of light will get the same value *c*.

The constancy of the speed of light is what unifies space and time. And for any observation of motion through space, there is a corresponding passage of time. The ratio of space to time for light is the same for all who measure it. The speed of light is a constant.

$$\frac{\text{SPACE}}{\text{TIME}} = \frac{\text{SPACE}}{\text{TIME}} = c$$

Figure 15.10 ▲
All space and time measurements of light are unified by *c*.

15.6 Time Dilation

Pretend you are in a spaceship at rest in a part of your "hometown" where a large public clock is displayed. Suppose the clock reads "12 noon." To say it reads "12 noon" is to say that light reflects from the clock and carries the information "12 noon" toward you in the direction of your line of sight. If you suddenly move your head to the side, instead of meeting your eye, the light carrying the information will continue past, presumably out into space. Out there an observer who *later* receives the light could say, "Oh, it's 12 noon on Earth now." But from your point of view it isn't. You and the distant observer will see 12 noon at different times. Now suppose your spaceship is moving as fast as the speed of light (just pretending). Then you'd keep up with the clock's information that says "12 noon." Traveling at the speed of light, then, tells you its always 12 noon back home. Time at home is frozen! So if your spaceship is not moving, you will see the hometown clock move into the future at the rate of 60 seconds per minute; if you could move at the speed of light, you'd see seconds on the clock taking infinite time. These are the two extremes. What's in between? How would the time readings appear to you moving at less than the speed of light? A little thought will show that the clock will be seen to run somewhere between the rate 60 seconds per minute and 60 seconds per an infinity of time if your speed is between zero and the speed of light. From your high-speed (but less than *c*) moving frame of reference, the clock and all events in the reference frame of the clock will be seen in slow motion. Time will be stretched. How much depends on speed. This is *time dilation*.

Special relativity turns around some of our conceptions about the world. We agree that speed is relative, that it depends on the speeds of the source and the observer. Yet, one speed, the speed of light, is absolute—independent of the speeds of the source or observer. Time, on the other hand, is usually thought of as absolute. It seems to pass at the same rate regardless of what is happening. Yet, our imaginary spaceship experience shows this is not true. Einstein proposed that time depends on the motion between the observer and the events being observed.

We measure time with a clock. A clock can be any device that measures periodic intervals, such as the swings of a pendulum, the oscillations of a balance wheel, or the vibrations of a quartz crystal. We are going to consider a "light clock," a rather impractical device, but one that will help to describe time dilation.

Imagine an empty tube with a mirror at each end (Figure 15.11). A flash of light bounces back and forth between the parallel mirrors. The mirrors are perfect reflectors, so the flash bounces indefinitely. If the tube is 300 000 km in length, each bounce will take 1 s in the frame of reference of the light clock. If the tube is 3 km long, each bounce will take 0.00001 s.

Suppose we view the light clock as it whizzes past us in a high-speed spaceship (Figure 15.12). We see the light flash bouncing up and down along a longer diagonal path.

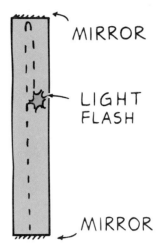

Figure 15.11 ▲
A stationary light clock. Light bounces between parallel mirrors and "ticks off" equal intervals of time.

Figure 15.12 ▲
(a) An observer moving with the spaceship observes the light flash moving vertically between the mirrors of the light clock. (b) An observer who is passed by the moving ship observes the flash moving along a diagonal path.

But remember the second postulate of relativity: The speed will be measured by *any* observer as *c*. Since the speed of light will not increase, we must measure more time between bounces! For us, looking in from the outside, one tick of the light clock takes longer than it takes for occupants of the spaceship. The spaceship's clock, according to our observations, has slowed down—although, for occupants of the spaceship, it has not slowed at all!

Figure 15.13 ▶
The longer distance taken by the light flash in following the diagonal path must be divided by a correspondingly longer time interval to yield an unvarying value for the speed of light.

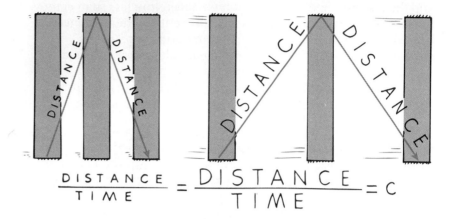

$$\frac{DISTANCE}{TIME} = \frac{DISTANCE}{TIME} = c$$

The slowing of time is not peculiar to the light clock. It is time itself in the moving frame of reference, as viewed from our frame of reference, that slows. The heartbeats of the spaceship occupants will have a slower rhythm. All events on the moving ship will be observed by us as slower. We say that time is "dilated."

How do the occupants on the spaceship view their own time? Do they perceive themselves moving in slow motion? Do they experience longer lives as a result of time dilation? As it turns out, they notice none of these things. Time for them is the same as when they do not appear to us to be moving at all. Recall Einstein's first postulate: All laws of nature are the same in all uniformly moving frames of reference. There is no way the spaceship occupants can tell uniform motion from rest. They have no clues that events on board are seen to be dilated when viewed from other frames of reference.

How do occupants on the spaceship view *our* time? Do they see our time as speeded up? The answer is no—motion is relative, and

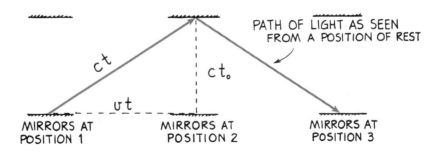

Figure 15.14 ▲
Mathematical detail of Figure 15.13.

The Time Dilation Equation*

Figure 15.14 shows three successive positions of the light clock as it moves to the right at constant speed v. The diagonal lines represent the path of the light flash as it starts from the lower mirror at position 1, moves to the upper mirror at position 2, and then back to the lower mirror at position 3.

The symbol t_o represents the time it takes for the flash to move between the mirrors as measured from a frame of reference fixed to the light clock. This is the time for straight up or down motion. Since the speed of light is always c, the light flash is observed to move a vertical distance ct_o in the frame of reference of the light clock. This is the distance between mirrors and is at right angles to the horizontal motion of the light clock. This vertical distance is the same in both reference frames.

The symbol t represents the time it takes the flash to move from one mirror to the other as measured from a frame of reference in which the light clock moves to the right with speed v. Since the speed of the flash is c and the time to go from position 1 to position 2 is t, the diagonal distance traveled is ct. During this time t, the clock (which travels horizontally at speed v) moves a horizontal distance vt from position 1 to position 2.

These three distances make up a right triangle in the figure, in which ct is the hypotenuse, and ct_o and vt are legs. A well-known theorem of geometry (the Pythagorean theorem) states that the square of the hypotenuse is equal to the sum of the squares of the other two sides. If we apply this to the figure, we obtain:

$$(ct)^2 = (ct_o)^2 + (vt)^2$$

$$(ct)^2 - (vt)^2 = (ct_o)^2$$

$$t^2[1 - (v^2/c^2)] = t_o^2$$

$$t^2 = \frac{t_o^2}{1 - (v^2/c^2)}$$

$$t = \frac{t_o}{\sqrt{1 - (v^2/c^2)}}$$

---

* The mathematical derivation of this equation for time dilation is included here mainly to show that it involves only a bit of geometry and elementary algebra. It is not expected that you master it! (If you take a follow-up physics course, you can master it then.)

1. Does time dilation mean that time *really* passes more slowly in moving systems or that it only *seems* to pass more slowly?

2. If you were moving in a spaceship at a high speed relative to the earth, would you notice a difference in your pulse rate? In the pulse rate of the people back on earth?

3. Will observers A and B agree on measurements of time if A moves at half the speed of light relative to B? If both A and B move together at 0.5c relative to the earth?

from *their* frame of reference it appears that *we* are the ones who are moving. They see our time running slow, just as we see their time running slow. Is there a contradiction here? Not at all. It is physically impossible for observers in different frames of reference to refer to one and the same realm of space-time. The measurements in one frame of reference need not agree with the measurements made in another reference frame. There is only one measurement they will always agree on: the speed of light.

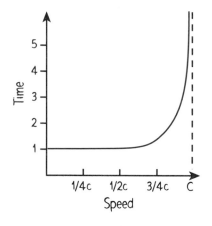

Figure 15.15 ▲
The graph shows how 1 second on a stationary clock is stretched out, as measured on a moving clock. Note that the stretching becomes significant only at speeds near the speed of light.

15.7 The Twin Trip

A dramatic illustration of time dilation is afforded by identical twins, one an astronaut who takes a high-speed round-trip journey while the other stays home on Earth. When the traveling twin returns, he is younger than the stay-at-home twin. How much younger depends on the relative speeds involved. If the traveling twin maintains a speed of 50% the speed of light for one year (according to clocks aboard the spaceship), 1.15 years will have elapsed on Earth. If the traveling twin maintains a speed of 87% the speed of light for a year, then 2 years will have elapsed on Earth. At 99.5% the speed of light,

■ **Answers**

1. The slowing of time in moving systems is not merely an illusion resulting from motion. Time really does pass more slowly in a moving system compared with one at relative rest, as we shall see in the next section. Read on!

2. There would be no relative speed between you and your own pulse, so no relativistic effects would be noticed. There would be a relativistic effect between you and people back on Earth. You would find their pulse rate slower than normal (and they would find your pulse rate slower than normal). Relativity effects are always attributed to "the other guy."

3. When A and B have different motions relative to each other, each will observe a slowing of time in the frame of reference of the other. So they will not agree on measurements of time. When they are moving in unison, they share the same frame of reference and will agree on measurements of time. They will see each other's time as passing normally, and they will each see events on Earth in the same slow motion.

◄ **Figure 15.16**
The traveling twin does not age
as fast as the stay-at-home twin.

10 earth years would pass in one spaceship year. At this speed the traveling twin would age a single year while the stay-at-home twin ages 10 years.

The question arises, since motion is relative, why isn't it just as well the other way around—why wouldn't the traveling twin return to find his stay-at-home twin younger than himself? We will show that from the frames of reference of both the earthbound twin and the traveling twin, it is the earthbound twin who ages more.

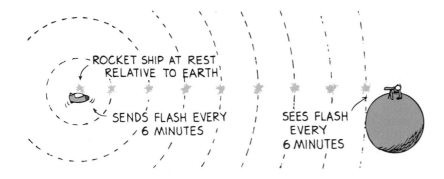

ROCKET SHIP AT REST
RELATIVE TO EARTH

SENDS FLASH EVERY
6 MINUTES

SEES FLASH
EVERY
6 MINUTES

Figure 15.17 ▲
When no motion is involved, the light flashes are received as frequently as the spaceship sends them.

First, consider a spaceship hovering at rest relative to a distant planet. Suppose the spaceship sends regularly spaced brief flashes of light to the planet (Figure 15.17). Some time will elapse before the flashes get to the planet, just as 8 minutes elapses before sunlight gets to the earth. The light flashes will encounter the receiver on the planet at speed *c*. Since there is no relative motion between the sender and receiver, successive flashes will be received as frequently as they are sent. For example, if a flash is sent from the ship every 6 minutes, then after some initial delay, the receiver will receive a flash every 6 minutes. With no motion involved, there is nothing unusual about this.

When motion is involved, the situation is quite different. It is important to note that the *speed* of the flashes will still be *c*, no matter how the ship or receiver may move. How *frequently* the flashes are seen, however, very much depends on the relative motion involved. When the ship travels *toward* the receiver, the receiver sees the flashes more often—that is, more frequently. This happens not only because time is altered due to motion, but mainly because each succeeding flash has less distance to travel as the ship gets closer to the receiver. If the spaceship emits a flash every 6 minutes, the flashes will be seen at intervals of less than 6 minutes. Suppose the ship is traveling fast enough for the flashes to be seen twice as frequently. Then they are seen at intervals of 3 minutes. Note in Figure 15.18 that the flashes for approach are closer together and equally spaced.

Figure 15.18 ▶
When the sender moves toward the receiver, the flashes are seen more frequently.

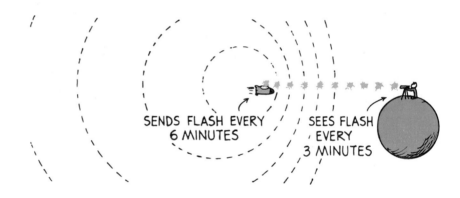

SENDS FLASH EVERY 6 MINUTES

SEES FLASH EVERY 3 MINUTES

If the ship *recedes* from the receiver at the same speed and still emits flashes at 6-min intervals, these flashes will be seen half as frequently by the receiver, that is, at 12-min intervals (Figure 15.19). This is mainly because each succeeding flash has a longer distance to travel as the ship gets farther away from the receiver.

The effect of moving away is just the opposite of moving closer to the receiver. So if the flashes are received twice as frequently when the spaceship is approaching (6-min flash intervals are seen every 3 min), they are received half as frequently when it is receding (6-min flash intervals are seen every 12 min).*

---

* The frequencies for approach and for recession are *reciprocals* of each other. That is, flashes that are seen 2 times as frequently for approach are seen $\frac{1}{2}$ as frequently for recession. If seen 3 times as frequently for approach, the flashes are seen $\frac{1}{3}$ as frequently for recession, and so on for higher speeds. This reciprocal relationship does not hold for waves that require a medium. In the case of sound waves, for example, a speed that results in a doubling of emitting frequency for approach produces the emitting frequency for recession.

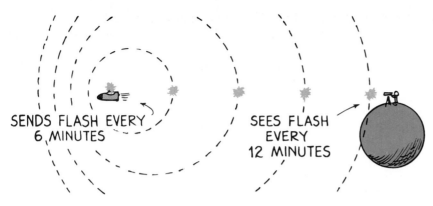

SENDS FLASH EVERY
6 MINUTES

SEES FLASH
EVERY
12 MINUTES

The light flashes make up a light clock, or timer. In the frame of reference of the receiver, something like taking a bath or cooking pancakes that takes 6 min in the spaceship, is seen to take 12 min when the spaceship recedes, and only 3 min when the ship is approaching.

■ Questions

1. Here's a simple arithmetic question: If a spaceship travels for one hour and emits a flash every 6 min, how many flashes will be emitted?

2. A spaceship sends equally spaced flashes while approaching a receiver at constant speed. Will the flashes be equally spaced when they encounter the receiver?

3. A spaceship emits flashes every 6 min for one hour. If the receiver sees these flashes at 3-min intervals, how much time will occur between the first and the last flash (in the frame of reference of the receiver)?

Let's apply this doubling and halving of flash intervals to the twins. Suppose the traveling twin recedes from the earthbound twin at the same high speed for one hour, then quickly turns around and returns in one hour. The traveling twin takes a round trip of two hours, according to all clocks aboard the spaceship. This trip will *not* be seen to take two hours from the earth frame of reference,

■ Answers

1. Ten flashes, since (60 min)/(6 min/flash) = 10 flashes.

2. Yes, as long as the ship moves at constant speed, the equally spaced flashes will be seen equally spaced but more frequently. (If the ship accelerated while sending flashes, then they would not be seen at equally spaced intervals.)

3. All 10 flashes will be seen in 30 min [(10 flashes) × (3 min/flash) = 30 min].

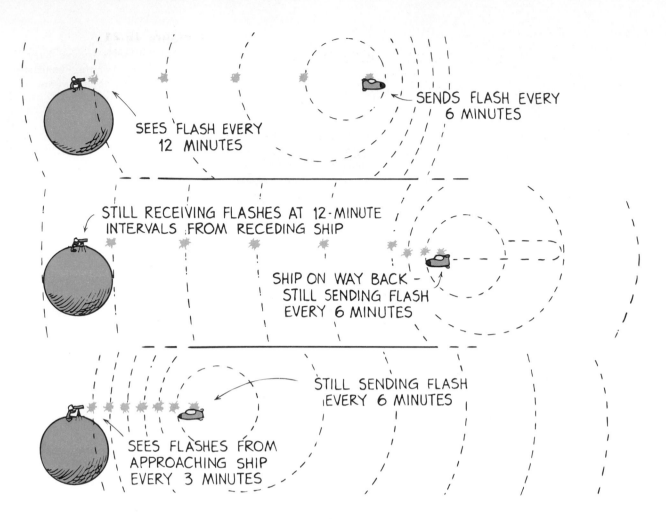

Figure 15.20 ▲
The spaceship emits flashes each 6 minutes during a two-hour trip. During the first hour, it recedes from the earth. During the second hour, it approaches the earth.

however. We can see this with the help of the flashes from the ship's light clock.

As the ship recedes from the earth, it emits a flash of light every 6 min. These flashes are received on Earth every 12 min. During the hour of going away from the earth, a total of 10 flashes are emitted. If the ship departs from the earth at noon, clocks aboard the ship will read 1 p.m. when the tenth flash is emitted. What time will it be on Earth when this tenth flash reaches the earth? The answer is 2 p.m. Why? Because the time it takes the earth to receive 10 flashes at 12-min intervals is (10 flashes) × (12 min/flash), or 120 min (= 2 h).

Suppose the spaceship is somehow able to suddenly turn around in a negligibly short time and return at the same high speed. During the hour of return it emits 10 more flashes at 6-min intervals. These flashes are received every 3 min on Earth, so all 10 come in 30 min. A clock on Earth will read 2:30 p.m. when the spaceship completes its two-hour trip. We see that the earthbound twin has aged a half hour more than the twin aboard the spaceship!

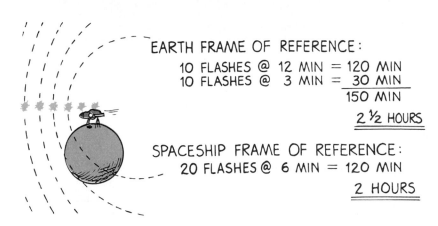

EARTH FRAME OF REFERENCE:
10 FLASHES @ 12 MIN = 120 MIN
10 FLASHES @ 3 MIN = 30 MIN
150 MIN
2 ½ HOURS

SPACESHIP FRAME OF REFERENCE:
20 FLASHES @ 6 MIN = 120 MIN
2 HOURS

The result is the same from either frame of reference. Consider the same trip again, only this time with flashes emitted *from the earth* at regularly spaced 6-min intervals in earth time. From the frame of reference of the receding spaceship, these flashes are received at 12-min intervals (Figure 15.22a). This means that 5 flashes are seen by the spaceship during the hour of receding from the earth. During the spaceship's hour of approaching, the light flashes are seen at 3-min intervals (Figure 15.22b), so 20 flashes will be seen.

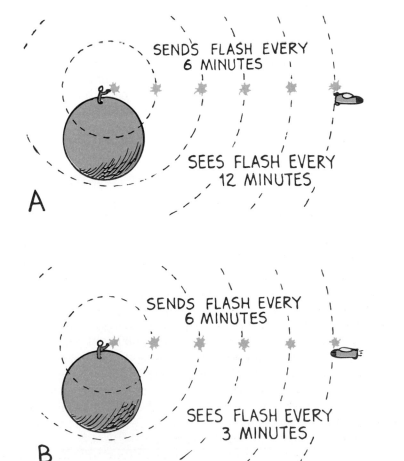

◀ **Figure 15.22**
Flashes sent from Earth at 6-min intervals are seen at 12-min intervals by the ship when it recedes, and at 3-min intervals when it approaches.

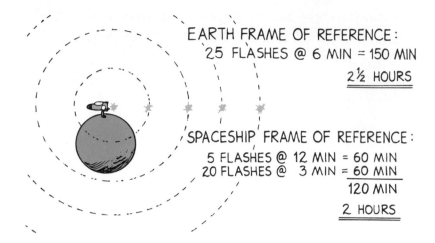

EARTH FRAME OF REFERENCE:
25 FLASHES @ 6 MIN = 150 MIN
<u>2½ HOURS</u>

SPACESHIP FRAME OF REFERENCE:
5 FLASHES @ 12 MIN = 60 MIN
20 FLASHES @ 3 MIN = <u>60 MIN</u>
120 MIN
<u>2 HOURS</u>

Figure 15.23 ▲
A time interval of 2.5 hours on Earth is seen to be 2 hours in the spaceship's frame of reference.

Hence, the spaceship receives a total of 25 flashes during its two-hour trip. According to clocks on the earth, however, the time it took to emit the 25 flashes at 6-min intervals was (25 flashes) × (6 min/flash), or 150 min (= 2.5 h). This is shown in Figure 15.23.

So both twins agree on the same results, with no dispute as to who ages more than the other. While the stay-at-home twin remains in a single reference frame, the traveling twin has experienced two different frames of reference, separated by the acceleration of the spaceship in turning around. The spaceship has in effect experienced two different realms of time, while the earth has experienced a still different but single realm of time. The twins can meet again at the same place in space only at the expense of time.

15.8 Space and Time Travel

Before the theory of special relativity was introduced, it was argued that humans would never be able to venture to the stars. It was thought that our life span is too short to cover such great distances—at least for the distant stars. Alpha Centauri is the nearest star to Earth, after the sun, and it is 4 light-years away.* It was therefore thought that a round-trip even at the speed of light would require 8 years. The center of our galaxy is some 30 000 light-years away, so it was reasoned that a person traveling even at the speed of light would have to survive for 30 000 years to make such a voyage! But these arguments fail to take into account time dilation. Time for a person on Earth and time for a person in a high-speed spaceship are not the same.

---

* A light-year is the distance that light travels in one year (9.46×10^{12} km).

A person's heart beats to the rhythm of the realm of time it is in. One realm of time seems the same as any other realm of time to the person, but not to an observer who is located outside the person's frame of reference—for she sees the difference. As an example, astronauts traveling at 99% the speed of light could go to the star Procyon (11.4 light-years distant) and back in 23.0 years in earth time. It would take light itself 22.8 years in earth time to make the same round-trip. Because of time dilation, it would seem that only 3 years had gone by for the astronauts. All their clocks would indicate this, and biologically they would be only 3 years older. It would be the space officials greeting them on their return who would be 23 years older.

At higher speeds the results are even more impressive. At a speed of 99.99% the speed of light, travelers could travel slightly more than 70 light-years in a single year of their own time. At 99.999% the speed of light, this distance would be pushed appreciably further than 200 years. A 5-year trip for them would take them farther than light travels in 1000 earth-time years.

Such journeys seem impossible to us today. The amounts of energy involved to propel spaceships to such relativistic speeds are billions of times the energy used to put the space shuttles into orbit. The problems of shielding radiation induced by these high speeds seems formidable. The practicalities of such space journeys are prohibitive, so far.

If and when these problems are overcome and space travel becomes routine, people will have the option of taking a trip and returning in future centuries of their choosing. For example, one might depart from Earth in a high-speed ship in the year 2150, travel for 5 years or so, and return in the year 2500. One might live among earthlings of that period for a while and depart again to try out the year 3000 for style. People could keep jumping into the future with some expense of their own time, but they could not trip into the past. They could never return to the same era on Earth that they bid farewell to.

Time, as we know it, travels only one way—forward. Here on Earth we constantly move into the future at the steady rate of 24 hours per day. An astronaut leaving on a deep-space voyage must live with the fact that, upon her return, much more time will have elapsed on Earth than she has experienced on her voyage. Star travelers will not bid "so long, see you later" to those they leave behind but, rather, a permanent "good-bye."

We can see into the past, but we cannot go into the past. When we look at stars in the night skies, the starlight we see left those stars dozens, hundreds, even millions of years ago. What we see is the stars as they were long ago. We are thus eyewitnesses to ancient history. We can only speculate about what may have happened to the stars since then.

When we think of time and the universe, we may wonder what went on before the universe began. We wonder what will happen if the universe ceases to exist in time. But the concept of time applies to events within the universe. Time is "in" the universe; the universe is not "in" time. Without the universe, there is no time; no before, no after. Likewise, space is "in" the universe; the universe is not "in" a region of space. There is no space "outside" the universe. We see that space-time exists within the universe. Think about that!

Figure 15.24 ▲
From the earth frame of reference, light takes 30 000 years to travel from the center of the Milky Way galaxy to our solar system. From the frame of reference of a high-speed spaceship, the trip takes less time. From the frame of reference of light itself, the trip takes no time. There is no time in a speed-of-light frame of reference.

15 Chapter Review

Concept Summary

According to Einstein's special theory of relativity, time is affected by motion in space at constant velocity.

■ Time appears to pass more slowly in a frame of reference that is moving relative to the observer.

All the laws of nature are the same in all uniformly moving frames of reference.

■ No experiment can be devised to detect whether the observer is moving at constant velocity.

The speed of light in empty space is the same in all frames of reference.

■ The speed of light has the same value regardless of the motion of the source or the motion of the observer.

Important Terms

first postulate of special relativity (15.4)
postulate (15.3)
second postulate of special relativity (15.5)
space-time (15.1)
special theory of relativity (15.0)
time dilation (15.1)

Review Questions

1. What is space-time? (15.1)

2. Can you travel while remaining in one place in space? Explain. (15.1)

3. Does light travel through space? Through time? Through both space and time? (15.1)

4. What is time dilation? (15.1)

5. What does it mean to say that motion is relative? (15.2)

6. The speed of a ball you catch that is thrown from a moving truck depends on the speed and direction of the truck. Does the speed of light caught from a moving source similarly depend on the speed and direction of the source? Explain. (15.3)

7. What does it mean to say that the speed of light is a constant? (15.3)

8. What is the first postulate of special relativity? (15.4)

9. What is the second postulate of special relativity? (15.5)

10. The ratio of velocity gain to time for a freely falling body is g. Similarly, what is the ratio of distance to time for light waves? (15.5)

11. The path of light in a vertical "light clock" in a high-speed spaceship is seen to be longer when viewed from a stationary frame of reference. Why, then, does the light not appear to be moving faster? (15.6)

12. If we view a passing spaceship and see that the inhabitants' time is running slow, how do they see our time running? (15.6)

13. When a flashing light source approaches you, does the speed of light, or the frequency of arrival of light flashes, or both, increase? (15.7)

14. **a.** How many frames of reference does the stay-at-home twin experience in the twin trip?

    b. How many frames of reference does the traveling twin experience? (15.7)

15. Is it possible for a person with a 70-year life span to travel farther than light travels in 70 years? Explain. (15.8)

Plug and Chug

1. If a spaceship moves away from you at half the speed of light and fires a rocket away from you at half the speed of light relative to the spaceship, common sense may tell you the rocket moves at the speed of light relative to you. But it doesn't! The relativistic addition of velocities (not covered in the chapter) is given by

$$V = \frac{v_1 + v_2}{1 + \dfrac{v_1 v_2}{c^2}}$$

Substitute $0.5c$ for both v_1 and v_2 and show that the velocity V of the rocket relative to you is $0.8c$.

2. If the spaceship of the previous question somehow travels at c relative to you, and it somehow fires its rocket at c relative to itself, use the equation to show that the speed of the rocket relative to you is still c!

3. Substitute small values of v_1 and v_2 in the preceding equation and show for everyday speeds that V is practically equal to $v_1 + v_2$.

Think and Explain

1. If you were in a smooth-riding train with no windows, could you sense the difference between uniform motion and rest? Between accelerated motion and rest? Explain how you could do this with a bowl filled with water.

2. Suppose you're playing catch with a friend in a moving train. When you toss the ball in the direction the train is moving, how does the speed of the ball appear to an observer standing at rest outside the train? (Does it increase or appear the same as if the observer were riding on the train?)

3. Suppose you're shining a light while riding on a train. When you shine the light in the direction the train is moving, how would the speed of light appear to an observer standing at rest outside the train? (Does it increase or appear the same as if the observer were riding with the train?)

4. Light travels a certain distance in, say, 10 000 years. Can an astronaut travel more slowly than the speed of light and yet travel the same distance in a 10-year trip? Explain.

5. Can you get younger by traveling at speeds near the speed of light?

6. Explain why it is that when we look out into the universe, we see into the past.

7. One of the fads of the future might be "century hopping," where occupants of high-speed spaceships would depart from the earth for a few years and return centuries later. What are the present-day obstacles to such a practice?

8. If you were in a high-speed spaceship traveling away from the earth at a speed close to that of light, would you measure your normal pulse to be slower, the same, or faster? How would your measurements of pulses of friends back on Earth be if you could monitor them from your ship? Explain.

9. Is it possible for a person to be biologically older than one's parents? Explain.

Think and Solve

1. Jose and Josie are twins. Suppose Josie takes a ride in a spaceship to a distant star some 4 light-years distant, traveling at a constant speed of $0.8c$. After arriving at the star Josie quickly turns around and immediately comes back at the same speed.

   a. How much older is Jose when Josie returns?

   b. How much older is Josie when she returns?

   c. Why isn't the reverse true, that Josie is older than Jose?

2. Joe Burpy is 30 years old and has a daughter who is 6 years old. Joe leaves on a Greyhound space bus and takes a 5-year (space-bus time) round-trip at $0.99c$. How old will he and his daughter be when he returns?

16 Special Relativity—Length, Momentum, and Energy

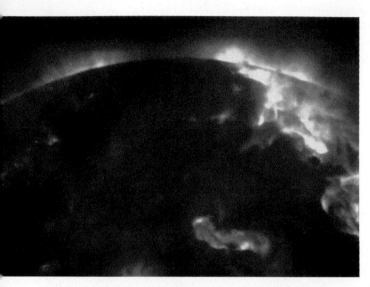

Mass converts to energy and vice versa.

The speed of light is the speed limit for all matter. Suppose that two spaceships are each traveling at nearly the speed of light and they are moving directly toward each other. The realms of space-time for each spaceship differ in such a way that the relative speed of approach is still less than the speed of light! For example, if both spaceships are traveling toward each other at 80% the speed of light with respect to the earth, an observer on each spaceship would measure her speed of approach as 98% the speed of light. There are *no* circumstances where the relative speeds of any material objects surpass the speed of light.

Why is the speed of light the universal speed limit? To understand this, we must know how motion through space affects the length, momentum, and energy of moving objects.

16.1 Length Contraction

For moving objects, space as well as time undergoes changes. When viewed by an outside observer, moving objects appear to contract along the direction of motion. The amount of contraction is related to the amount of time dilation. For everyday speeds, the amount of contraction is much too small to be measured. For relativistic speeds, the contraction would be noticeable. A meterstick aboard a spaceship whizzing past you at 87% the speed of light, for example, would appear to you to be only 0.5 meter long. If it whizzed past at 99.5% the speed of light, it would appear to you to be contracted to one tenth its original length. The width of the stick, perpendicular to the direction of travel (Figure 16.1), doesn't change. As relative speed gets closer and closer to the speed of light, the measured lengths of objects contract closer and closer to zero.

Figure 16.1 ▲
A meterstick traveling at 87% the speed of light relative to an observer would be measured as only half as long as normal.

Do people aboard the spaceship also see their metersticks—and everything else in their environment—contracted? The answer is no. People in the spaceship see nothing at all unusual about the lengths of things in their own reference frame. If they did, it would violate the first postulate of relativity. Recall that all the laws of physics are the same in all uniformly moving reference frames. Besides, there is no relative speed between the people on the spaceship and the things they observe in their own reference frame. However, there is a relative speed between themselves and *our* frame of reference, so they will see *our* metersticks contracted—and us as well. A rule of relativity is that changes due to alterations of space-time are always seen in the frame of reference of the "other guy."

Figure 16.2 ▲
In the frame of reference of the meterstick, its length is 1 meter. Observers from this frame see *our* metersticks contracted. The effects of relativity are always attributed to "the other guy."

The contraction of speeding objects is the contraction of space itself. Space contracts in only one direction, the direction of motion. Lengths along the direction perpendicular to this motion are the same in the two frames of reference. So if an object is moving horizontally, no contraction takes place vertically (Figure 16.3).

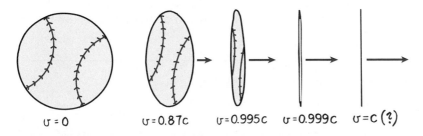

$v = 0$ $v = 0.87c$ $v = 0.995c$ $v = 0.999c$ $v = c\,(?)$

Figure 16.3 ▲
As relative speed increases, contraction in the direction of motion increases. Lengths in the perpendicular direction do not change.

Relativistic length contraction is stated mathematically:

$$L = L_0\sqrt{1 - (v^2/c^2)}$$

In this equation, v is the speed of the object relative to the observer, c is the speed of light, L is the length of the moving object as measured by the observer, and L_0 is the measured length of the object at rest.*

Suppose that an object is at rest, so that $v = 0$. When 0 is substituted for v in the equation, we find $L = L_0$, as we would expect. It was stated earlier that if an object were moving at 87% the speed of light, it would contract to half its length. When $0.87c$ is substituted for v in the equation, we find $L = 0.5L_0$. Or when $0.995c$ is substituted for v, we find $L = 0.1L_0$, as stated earlier. If the object could reach the speed c, its length would contract to zero. This is one of the reasons that the speed of light is the upper limit for the speed of any material object.

■ Question

A spacewoman travels by a spherical planet so fast that it appears to her to be an ellipsoid (egg shaped). If she sees the short diameter as half the long diameter, what is her speed relative to the planet?

16.2 Momentum and Inertia in Relativity

If we push an object that is free to move, it will accelerate. If we maintain a steady push, it will accelerate to higher and higher speeds. If we push with a greater and greater force, we expect the acceleration in turn to increase. It might seem that the speed should increase without limit, but there is a speed limit in the universe— the speed of light. In fact, we cannot accelerate any material object enough to reach the speed of light, let alone surpass it.

We can understand this from Newton's second law, which Newton originally expressed in terms of momentum: $F = \Delta mv/\Delta t$ (which reduces to the familiar $F = ma$, or $a = F/m$). The momentum form, interestingly, remains valid in relativity theory. Recall from Chapter 7, that the change of momentum of an object is equal to the

■ Answer

The spacewoman passes the spherical planet at 87% the speed of light.

* This equation (and those that follow) is simply stated as a "guide to thinking" about the ideas of special relativity. The equations are given here without any explanation as to how they are derived.

LINK TO BIOLOGY

Muons and Mutations

When cosmic rays bombard atoms at the top of the atmosphere, new particles are made. Some are *muons,* radioactive particles that streak downward toward the earth's surface. A muon's average lifetime is only two millionths of a second, seemingly too brief to reach the ground below before decaying. But because muons move at nearly the speed of light, length contraction dramatically shortens their distance to the earth. You are hit by hundreds of muons every second! Muon impact, like that of all high-speed elementary particles, causes biological mutations. So we see a link between the effects of relativity and the evolution of living creatures on earth.

impulse applied to it. Apply more impulse and the object acquires more momentum. Double the impulse and the momentum doubles. Apply ten times as much impulse and the object gains ten times as much momentum. Does this mean that momentum can increase without any limit, even though speed cannot? Yes, it does.

We learned that momentum equals mass times velocity. In equation form, $p = mv$ (we use p for momentum). To Newton, infinite momentum would mean infinite speed. Not so in relativity. Einstein showed that a new definition of momentum is required. It is

$$p = \frac{mv}{\sqrt{1 - (v^2/c^2)}}$$

where v is the speed of an object and c is the speed of light. Notice that the square root in the denominator looks just like the one in the formula for time dilation in the previous chapter. It tells us that the **relativistic momentum** of an object of mass m and speed v is larger than mv by a factor of $1/\sqrt{1 - (v^2/c^2)}$.

At relativistic speed, momentum increases dramatically. As v approaches c, the denominator of the equation approaches zero. This means that the momentum approaches infinity! An object pushed to the speed of light would have infinite momentum and would require an infinite impulse, which is clearly impossible. So nothing that has mass can be pushed to the speed of light, much less beyond it. Here is another reason that c is the speed limit in the universe.

What if v is much less than c? Then the denominator of the equation is nearly equal to 1 and p is nearly equal to mv. Newton's definition of momentum is valid at low speed.

We often say that a particle pushed close to the speed of light acts *as if* its mass were increasing, because its momentum—its "inertia in motion"—increases more than its speed increases. The quantity m in the equation above is called the **rest mass** of the object. It is a true constant, a property of the object no matter what speed it has.

Subatomic particles are routinely pushed to nearly the speed of light. The momenta of such particles may be thousands of times more than the Newton expression mv predicts. One way to look at the momentum of a high-speed particle is in terms of the "stiffness" of its trajectory. The more momentum it has, the harder it is to deflect it—the "stiffer" is its trajectory. If it has a lot of momentum, it more greatly resists changing course.

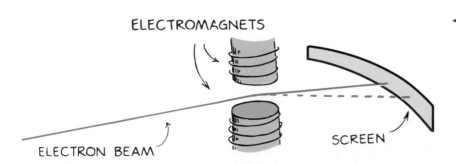

ELECTROMAGNETS

ELECTRON BEAM

SCREEN

◀ **Figure 16.4**
If the momentum of the electrons were equal to the Newtonian value *mv*, the beam would follow the dashed line. But because the relativistic momentum, or inertia in motion, is greater, the beam follows the "stiffer" trajectory shown by the solid line.

This can be seen when a beam of electrons is directed into a magnetic field. Charged particles moving in a magnetic field experience a force that deflects them from their normal paths. For a particle with a small momentum, the path curves sharply. For a particle with a large momentum, the path curves only a little—its trajectory is "stiffer" (Figure 16.4). Even though one particle may be moving only a little faster than another one—say 99.9% of the speed of light instead of 99% of the speed of light—its momentum will be considerably greater and it will follow a straighter path in the magnetic field. Through such experiments, physicists working with subatomic particles at atomic accelerators verify every day the correctness of the relativistic definition of momentum and the speed limit imposed by nature.

16.3 Equivalence of Mass and Energy

The most remarkable insight of Einstein's special theory of relativity is his conclusion that mass is simply a form of energy. A piece of matter, even if at rest and even if not interacting with anything else, has "energy of being." This is called its **rest energy.** Einstein concluded that it takes energy to make mass and that energy is released when mass disappears. Rest mass is, in effect, a kind of potential energy. Mass stores energy, just as a boulder rolled to the top of a hill stores energy. When the mass of something decreases, as it can do in nuclear reactions, energy is released, just as the boulder rolling to the bottom of the hill releases energy.

The amount of rest energy E_o is related to the mass m by the most celebrated equation of the twentieth century

$$E_o = mc^2$$

where c is again the speed of light. This equation gives the total energy content of a piece of stationary matter of mass m.

In ordinary units of measurement, the speed of light c is a large quantity and its square is even larger. This means that a small amount of mass stores a large amount of energy. The quantity c^2 is a "conversion factor." It converts the measurement of mass to the measurement of equivalent energy. Or it is the ratio of rest energy to mass: $E_o/m = c^2$. Its appearance in either form of this equation has nothing to do with light and nothing to do with motion. The magnitude of c^2 is 90 quadrillion (9×10^{16}) joules per kilogram. One kilogram of matter has an "energy of being" equal to 90 quadrillion joules. Even a speck of matter with a mass of only 1 milligram has a rest energy of 90 billion joules.

Rest energy, like any form of energy, can be converted to other forms. When we strike a match, for example, a chemical reaction occurs and heat is released. Phosphorus atoms in the match head rearrange themselves and combine with oxygen in the air to form new molecules. The resulting molecules have very slightly less mass than the separate phosphorus and oxygen molecules. From a mass standpoint, the whole is slightly less than the sum of its parts, but not by

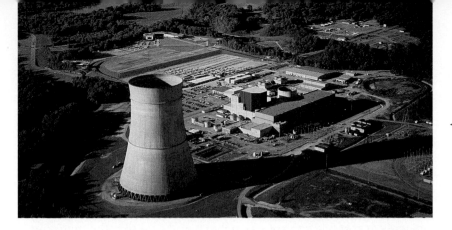

◀ **Figure 16.5**
Saying that a power plant delivers 90 million megajoules of energy to its consumers is equivalent to saying that it delivers 1 gram of energy to its consumers, because mass and energy are equivalent.

very much—by only about one part in a billion. For all chemical reactions that give off energy, there is a corresponding decrease in mass.

In nuclear reactions, the decrease in rest mass is considerably more than in chemical reactions—about one part in a thousand. This decrease of mass in the sun by the process of thermonuclear fusion bathes the solar system with radiant energy and nourishes life. The present stage of thermonuclear fusion in the sun has been going on for the past 5 billion years, and there is sufficient hydrogen fuel for fusion to last another 5 billion years. It is nice to have such a big sun!

The equation $E_o = mc^2$ is not restricted to chemical and nuclear reactions. A change in energy of any object at rest is accompanied by a change in its mass. The filament of a lightbulb has more mass when it is energized with electricity than when it is turned off. A hot cup of tea has more mass than the same cup of tea when cold. A wound-up spring clock has more mass than the same clock when unwound. But these examples involve incredibly small changes in mass—too small to be measured by conventional methods. No wonder the fundamental relationship between mass and energy was not discovered until this century.

The equation $E_o = mc^2$ is more than a formula for the conversion of rest mass into other kinds of energy, or vice versa. It states that energy and mass are the *same thing*. Mass is simply congealed energy. If you want to know how much energy is in a system, measure its mass. For an object at rest, its energy *is* its mass. Shake a massive object back and forth; it is energy itself that is hard to shake.

Figure 16.6 ▲
In one second, 4.5 million tons of rest mass are converted to radiant energy in the sun. The sun is so massive, however, that in a million years only one ten-millionth of the sun's rest mass will have been converted to radiant energy.

■ **Question**

Can we look at the equation $E_o = mc^2$ in another way and say that matter transforms into pure energy when it is traveling at the speed of light squared?

■ **Answer**

No, no, no! Matter cannot be made to move at the speed of light, let alone the speed of light squared (which is not a speed!). The equation $E_o = mc^2$ simply means that energy and mass are "two sides of the same coin."

16.4 Kinetic Energy in Relativity

Einstein dealt also with the energy of moving matter. His formula for the total energy of a moving piece of matter of mass m is

$$E = \frac{mc^2}{\sqrt{1-(v^2/c^2)}}$$

Notice the by-now familiar square root in the denominator. If the object is at rest, we can set v equal to 0 and find that the denominator is then equal to 1, leading to the famous rest-mass formula $E_0 = mc^2$. But if the object is moving, the denominator is less than 1 and the total energy E is greater than mc^2.

Consider what happens when a subatomic particle—or some possible future spacecraft—moves at a speed close to the speed of light. Then the denominator becomes quite small and the total energy E becomes much greater than the rest energy mc^2. If the speed v could be pushed to the speed of light, E would become infinite. Once again we see why no piece of matter can be made to travel at the speed of light. It would take infinite energy to do it. No scientist can imagine how the "warp speeds" of science fiction will ever become reality.

Since the revised formula for total energy applies to objects that are moving, it is natural to associate kinetic energy with the *difference* between total energy and rest energy. So kinetic energy is

$$KE = \frac{mc^2}{\sqrt{1-(v^2/c^2)}} - mc^2$$

This equation looks a bit complicated and certainly quite different from the equation $KE = \frac{1}{2}mv^2$. Yet it can be demonstrated mathematically that, for ordinary low speeds, this relativistic equation for kinetic energy reduces to the familiar $KE = \frac{1}{2}mv^2$. But at higher speeds, the actual kinetic energy is greater than $\frac{1}{2}mv^2$.

There are two important points in the discussion of the last few pages.

(1) Even at rest, an object has energy, which is locked in its mass.

(2) As the speed of an object approaches the speed of light, both its momentum and its energy approach infinity, so there is no way that the speed of light can be reached.*

---

* At least one thing reaches the speed of light—light itself! But light has no rest mass. The equations that apply to it are different. The theory of relativity tells us that light travels always at the same speed. A material particle can never be brought to the speed of light. Light can never be brought to rest.

16.5 The Correspondence Principle

If a new theory is to be valid, it must account for the verified results of the old theory. New theory and old must overlap and agree in the region where the results of the old theory have been fully verified. This requirement is known as the **correspondence principle.** It was advanced as a principle by Niels Bohr earlier in this century when Newtonian mechanics was being challenged by both quantum theory and relativity. If the equations of special relativity (or any other new theory) are to be valid, they must correspond to those of Newtonian mechanics—classical mechanics—when speeds much less than the speed of light are considered.

The relativity equations for time dilation, length contraction, and momentum are

$$t = \frac{t_o}{\sqrt{1 - (v^2/c^2)}}$$

$$L = L_o \sqrt{1 - (v^2/c^2)}$$

$$p = \frac{mv}{\sqrt{1 - (v^2/c^2)}}$$

We can see that these equations each reduce to Newtonian values for speeds that are very small compared with c. Then, the ratio $(v/c)^2$ is very small, and for everyday speeds may be taken to be zero. The relativity equations become

$$t = \frac{t_o}{\sqrt{1 - 0}} = t_o$$

$$L = L_o \sqrt{1 - 0} = L_o$$

$$p = \frac{mv}{\sqrt{1 - 0}} = mv$$

So for everyday speeds, the time scales and length scales of moving objects are essentially unchanged. Also, the Newtonian equation for momentum holds true (and so does the Newtonian equation for kinetic energy). But when the speed of light is approached, things change dramatically. Near the speed of light Newtonian mechanics change completely. The equations of special relativity hold for all speeds, although they are significant only for speeds near the speed of light.

So we see that advances in science take place not by discarding the current ideas and techniques, but by extending them to reveal new implications. Einstein never claimed that proven laws of physics were wrong, but instead showed that the laws of physics implied something that hadn't before been appreciated.

SCIENCE, TECHNOLOGY, AND SOCIETY

Scientists and Social Responsibility

The atomic bomb was an outcome of Einstein's physics. Einstein and other scientists were horrified by the bomb's destructive power, and they spoke out about it. Their work resulted in the bomb, so they felt partly responsible for its creation and use. Many scientists today feel responsible for the social consequences of their work.

But other scientists feel the scientist's role is to focus on science alone. They say scientists have no particular expertise in matters of public policy and that society is best served when scientific investigation remains unrestricted.

The power of science is vast.

Major General Leslie Groves and Dr. J. R. Oppenheimer (director of the Los Alamos Atomic Bomb Project) view the test site of the first atomic bomb explosion.

The power of science to change the world is vast. That power can be used wisely or foolishly, which often involves distinguishing between short-range and long-range benefits, policies, and goals. The scientist is the channel through which the power of science flows.

To what extent do you think scientists are obligated to consider social consequences of their work? Should scientists be considered more qualified than other citizens to guide public policy?

Einstein's theory of relativity has raised many philosophical questions. What, exactly, is time? Can we say it is nature's way of seeing to it that everything does not all happen at once? And why does time seem to move in one direction? Has it always moved forward? Are there other parts of the universe where time moves backward? Perhaps these unanswered questions will be answered by the physicists of tomorrow. How exciting!

16 Chapter Review

Concept Summary

When an object moves at very high speed relative to an observer, its measured length in the direction of motion is contracted.

When an object moves at very high speed relative to an observer, its momentum—its "inertia of motion"—is greater than the Newtonian value mv.

Mass and energy are equivalent—anything with mass also has energy. Rest energy is given by the equation $E_o = mc^2$.

Only when the release of energy is very great is the release of mass large enough to be detected.

During any reaction, the total energy remains constant if the energy equivalent of the change in rest mass is accounted for.

When the speed of an object approaches the speed of light, both the momentum and the total energy of the object approach infinity.

According to the correspondence principle, the equations of relativity must agree with the well-tested equations of Newtonian mechanics when the speed involved is small compared with the speed of light.

Important Terms

correspondence principle (16.5)
relativistic momentum (16.2)
rest energy (16.3)
rest mass (16.2)

Review Questions

1. If we witness events in a frame of reference moving past us, time appears to be stretched out (dilated). How do the lengths of objects in that frame appear? (16.1)

2. How long would a meterstick appear if it were thrown like a spear at 99.5% the speed of light? (16.1)

3. How long would a meterstick appear if it were traveling at 99.5% the speed of light, but with its length perpendicular to its direction of motion? (Why are your answers to this and the last question different?) (16.1)

4. If you were traveling in a high-speed spaceship, would metersticks on board appear contracted to you? Defend your answer. (16.1)

5. What would be the momentum of an object if it were pushed to the speed of light? (16.2)

6. What is meant by rest mass? (16.3)

7. What relativistic effect is evident when a beam of high-speed charged particles bends in a magnetic field? (16.3)

8. What is meant by the equivalence of mass and energy? That is, what does the equation $E_o = mc^2$ mean? (16.3)

9. What is the numerical quantity of the ratio rest energy/rest mass? (16.3)

10. Does the equation $E_o = mc^2$ apply only to reactions that involve the atomic nucleus? (16.3)

11. What evidence is there for the equivalence of mass and energy? (16.3)

12. What effect does solar energy have on the mass of the sun? (16.3)

13. An object at rest has an "energy of being," $E_o = mc^2$. When the same object is moving, is its total energy the same as, more than, or less than $E_o = mc^2$? (16.4)

14. Does the relativistic equation for kinetic energy mathematically reduce to the Newtonian equation for kinetic energy at low speeds? (16.4)

15. How much kinetic energy would a particle have if it could move at the speed of light? (16.4)

16. What is the correspondence principle? (16.5)

17. What results when low everyday speeds are used in the relativistic equations for time and length? (16.5)

18. Do the equations of Newton and Einstein overlap, or is there a sharp break between them? (16.5)

Think and Explain

1. Suppose your spaceship passes the earth at nearly the speed of light, and earth observers tell you that your ship appears to be contracted. Comment on the idea of checking their observation by measuring the spaceship yourself.

2. From the earth, the distance to the center of our galaxy is 24 000 light-years. From the frame of reference of a photon of light traveling from the earth to the center of our galaxy, what is this distance?

3. According to Newton's laws, if you apply an impulse on an object, acceleration occurs. What prevents an acceleration to and beyond the speed of light?

4. The two-mile-long linear accelerator at Stanford University in California is less than a meter long to the electrons that travel in it. Explain.

5. Pretend you can travel with the electrons in the Stanford accelerator after they are accelerated and are coasting toward their target at nearly the speed of light.

   a. What would the momentum of an electron be in your frame of reference? Its energy?

   b. What could you say about the length of the accelerator in your frame of reference? What could you say about the motion of the target?

6. The electrons that illuminate your TV screen travel at about $0.25c$. At this speed, relativity gives them an extra momentum that can be interpreted as an effective mass increase of about 3%. Does this relativistic effect tend to increase, decrease, or have no effect on your electric bill?

7. Since there is an upper limit on the speed of a particle, does it follow that there is therefore an upper limit on its kinetic energy or momentum? Defend your answer.

8. A friend says that the equation $E_0 = mc^2$ has relevance to nuclear power plants, but not to fossil fuel power plants. Another friend looks to see if you agree. What do you say?

9. Is this label on a consumer product cause for alarm? *CAUTION: The mass of this product contains the energy equivalent of 3 million tons of TNT per gram.*

10. Give three reasons why we say that there is a speed limit for particles in the universe.

Think and Solve

1. Josie takes a ride in a spaceship moving at $0.8c$ to a star 4 light-years away. From Josie's frame of reference, what distance in light-years does she travel to the distant star?

2. An electron, with a rest mass of 9.11×10^{-31} kg, shoots down a 1-km-long accelerator at an average speed of $0.95c$.

   a. From the electron's frame of reference, how long is the accelerator?

   b. From the electron's frame of reference, how long does it take to make the trip?

3. A 100-watt lightbulb consumes 100 joules of energy every second. How long could you burn that lightbulb from the energy in one penny, which has a mass of 0.003 kg? (Assume all the penny's mass is converted to energy.)

Atoms are so tiny that I inhale billions of trillions with each breath, nearly a trillion times more atoms than the total population of people since time zero! Every time I breathe in, I inhale atoms exhaled by every person **who ever lived,** except for newborn babies far away. Every time I breathe out, or sweat, I release atoms to the air that spread everywhere and become part of everybody else. New babies and all who follow will be made of the atoms that are now part of me. So we're all one!

17 The Atomic Nature of Matter

Atoms in the air you breathe become part of you. Atoms you exhale become part of everyone else.

Suppose you break apart a large boulder with a heavy sledgehammer. You break the boulder into rocks. Then you break the rocks into stones, and the stones into gravel. You keep going and break the gravel into sand, and the sand into a powder of fine crystals. Each fine crystal is composed of many billions of smaller particles called **atoms.** Atoms are the building blocks of most matter. Everything you see, hear, touch, taste, feel, or smell in the world around you is made of atoms. Shoes, ships, mice, lead, and people are all made of atoms.*

17.1 Elements

Just as dots of light of only three colors combine to form almost every conceivable color on a television screen, only about 100 distinct kinds of atoms combine to form all the materials we know about. Atoms are the building blocks of matter, yet atoms are not all alike. Atoms of the same kind make up what is called an **element.**

To date more than 110 elements are known. Of these, 90 occur in nature. The others are made in the laboratory with high-energy atomic accelerators and nuclear reactors. These laboratory-produced elements are too unstable (radioactive) to occur naturally in appreciable amounts.

From a pantry containing less than 100 elements, we have the atoms that make up almost every simple, complex, living, or nonliving substance in the known universe. More than 99% of the material

---

*   The word *atom* comes from a Greek word meaning "indivisible." But atoms are not truly indivisible. Under extreme conditions, such as in particle accelerators or in the center of the sun, atoms can be further broken down into smaller, "subatomic" particles.

on Earth is formed from only about a dozen of the elements. The majority of elements are relatively rare. Living things, for example, are composed primarily of five elements: oxygen (O), carbon (C), hydrogen (H), nitrogen (N), and calcium (Ca). The letters in parentheses represent the chemical symbols for these elements. Table 17.1 lists the 16 most common elements on Earth. Most of these elements, not just the five most common ones, are critical for life.

Table 17.1 The 16 Most Common Elements on Earth	
Aluminum (Al)	Nitrogen (N)
Calcium (Ca)	Oxygen (O)
Carbon (C)	Phosphorus (P)
Chlorine (Cl)	Potassium (K)
Fluorine (F)	Silicon (Si)
Hydrogen (H)	Sodium (Na)
Iron (Fe)	Sulfur (S)
Magnesium (Mg)	Titanium (Ti)

The lightest element of all is hydrogen. In the universe at large, it is the most abundant element—over 90% of the atoms in the known universe are hydrogen. Helium, the second-lightest element, makes up most of the remaining 10% of atoms in the universe, although it is rare on Earth. The heavier, naturally formed atoms that we find about us were manufactured by hydrogen fusion reactions in the hot, high-pressure cauldrons deep within stars. The heaviest elements are formed when huge stars implode then explode—supernovas. Nearly all the elements on earth are remnants of stars that exploded long before the solar system came into being.

◀ **Figure 17.1**
Both Leslie and you are made of stardust—in the sense that the carbon, oxygen, nitrogen, and other atoms that make up your body originated in the deep interior of ancient stars, which have long since exploded.

17.2 Atoms Are Recyclable

Atoms are much older than the materials they compose. Some atoms are nearly as old as the universe itself. Most atoms that make up our world are at least as old as the sun and the earth.

Atoms in your body have been around since long before the solar system came into existence. They cycle and recycle among innumerable forms, both living and nonliving. Every time you breathe, for example, only some of the atoms that you inhale are exhaled in your next breath. The remaining atoms are taken into your body to become part of you, and most leave your body sooner or later—to become part of everything else.

Strictly speaking, you don't "own" the atoms that make up your body—you're simply their present caretaker. There will be many others who later will care for the atoms that presently compose you. We all share from the same atom pool, as atoms migrate around, within, and throughout us. So some of the atoms in the ear you scratch today may have been part of your neighbor's breath yesterday!

Most people know we are all made of the same *kinds* of atoms. But what most people don't know is that we are made of the *same* atoms—atoms that cycle from person to person as we breathe and as our perspiration is vaporized.

■ Question

World population grows each year. Does this mean the mass of the earth increases each year?

17.3 Atoms Are Small

There are about 10^{23} atoms in a gram of water (a thimbleful). The number 10^{23} is an enormous number, more than the number of drops of water in all the lakes and rivers of the world. So there are more atoms in a thimbleful of water than there are drops of water in the world's lakes and rivers. Atoms are so small that there are about

■ Answer

The mass of the earth does increase by the incidence of roughly 40 000 tons of interplanetary dust each year. But the increasing number of people does not increase the mass of the earth. The atoms that make up our bodies are the same atoms that were here before we were born—we are but dust, and unto dust we shall return. Human cells are merely rearrangements of material previously present. The atoms that make up a forming baby in the mother's womb must be supplied by the food she eats. And those atoms were formed in stars that long since exploded (Figure 17.1).

DOING PHYSICS

Atom Migration

Into any small container of water put some drops of food coloring, ink, or any colored liquid. Notice that the colored drops dilute and spread to all parts of the water in a fairly short time. The atoms or molecules in the drop, like the H_2O molecules of the water, are perpetually moving. The faster they move, the greater the rate of migration. Repeat using hot water. What happens to the migration rate?

Activity

as many atoms in the air in your lungs at any moment as there are breathfuls of air in the atmosphere of the whole world.

Atoms are perpetually moving. They also migrate from one location to another. In solids the rate of migration is low, in liquids it is greater, and in gases migration is greatest. Drops of food coloring in a glass of water soon spread to color the entire glass of water. Likewise for a cupful of toxic material thrown into an ocean that spreads around and is later found in every part of the world's oceans. The same is true of materials released into the atmosphere.

It takes about six years for one of your exhaled breaths to become evenly mixed in the atmosphere. At that point, every person in the world inhales an average of one of your exhaled atoms in a single breath. And this occurs for *each* breath you exhale! When you take into account the many thousands of breaths that people exhale, at any time—like right now—you have hordes of atoms in your lungs that were once in the lungs of every person who ever lived. We are literally breathing one another's breaths.

Atoms are too small to be seen—at least with visible light. You could connect microscope to microscope and never "see" an atom. This is because light is made up of waves, and atoms are smaller than the wavelengths of visible light. The size of a particle visible under the highest magnification must be larger than the wavelength of light. (More about this in Chapter 31.)

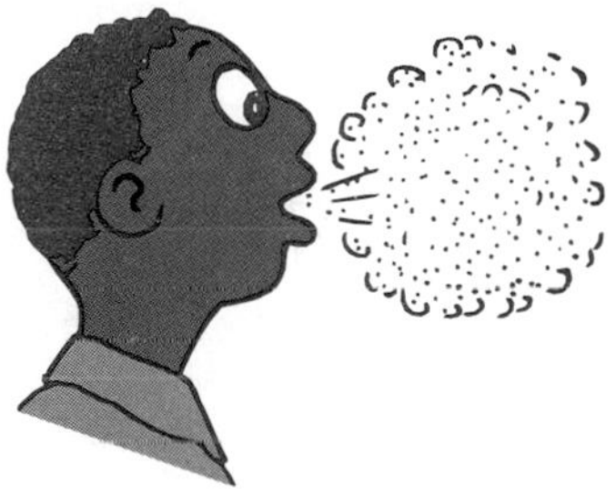

Figure 17.2 ▲
There are as many atoms in a normal breath of air as there are breathfuls of air in the atmosphere of the world.

■ Question

Is your brain made of atoms that were once part of Albert Einstein?

■ Answer

Yes, and of Charlie Chaplin too. However, these atoms are combined differently than they were before. The next time you have one of those days when you feel like you'll never amount to anything significant, take comfort in the thought that the atoms that now compose you will be part of the bodies of all the people on Earth who are yet to be!

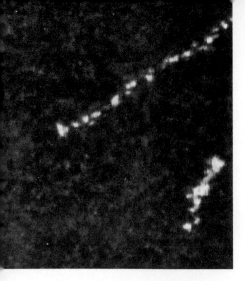

Figure 17.3 ▲
The strings of dots are chains of thorium atoms photographed with a scanning electron microscope by researchers at the University of Chicago's Enrico Fermi Institute.

Figure 17.4 ▶
Scanning tunneling microscope photograph of uranium atoms.

17.4 Evidence for Atoms

The first fairly direct evidence for the existence of atoms was unknowingly discovered in 1827. A Scottish botanist, Robert Brown, was studying pollen grains under a microscope. He noticed that the spores were in a constant state of agitation, always jiggling about. At first, Brown thought that the spores were some sort of moving life forms. Later, he found that inanimate dust particles and grains of soot also showed this kind of motion. The perpetual jiggling of particles that are just large enough to be seen is called **Brownian motion.** Brownian motion is now known to result from the motion of neighboring atoms and molecules.

More direct evidence for the existence of atoms is available today. A photograph of individual atoms is shown in Figure 17.3. The photograph was made not with visible light but with an electron beam. A familiar example of an electron beam is the one that sprays the picture on your television screen. Although an electron beam is a stream of tiny particles (electrons), it has wave properties. It so happens that a high-energy electron beam has a wavelength more than a thousand times smaller than the wavelength of visible light. With such a beam, atomic detail can be seen. The historic (1970) photograph in Figure 17.3 was taken with a very thin electron beam in a scanning electron microscope. It is the first photograph of clearly distinguishable atoms.

In the mid-1980s, researchers developed a different kind of microscope—the scanning tunneling microscope, small enough to be held in your hand. In Figure 17.4, you can see individual atoms. Even greater detail is possible with newer types of imaging devices that are presently revolutionizing microscopy.

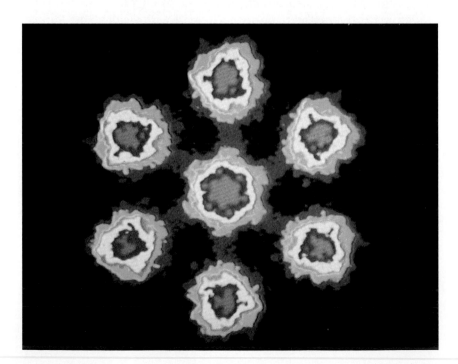

17.5 Molecules

Atoms can combine to form larger particles called **molecules.** For example, two atoms of hydrogen (H) combine with a single atom of oxygen (O) to form a water molecule (H_2O). The gases nitrogen and oxygen, which make up most of the atmosphere, are each made of simple two-atom molecules (N_2 and O_2). In contrast, the double helix of deoxyribonucleic acid (DNA), the blueprint of life, is composed of millions of atoms.

Matter that is a gas or liquid at room temperature is usually made of molecules. Matter made of molecules may contain all the same kind of molecule, or it may be a mixture of different kinds of molecules. Purified water contains almost entirely H_2O molecules, whereas clean air contains molecules belonging to several different substances.

 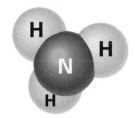 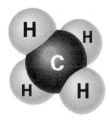

Figure 17.5 ▲
Models of the simple molecules O_2, NH_3, and CH_4. The atoms that compose a molecule are not just mixed together, but are bonded in a well-defined way.

But not all matter is made of molecules. Metals and crystalline minerals (including common table salt) are made of atoms that are not joined in molecules.

Like atoms, individual molecules are too small to be seen with optical microscopes.* More direct evidence of tiny molecules is seen in electron microscope photographs. The photograph in Figure 17.6 is of virus molecules, each composed of thousands of atoms. These giant molecules are visible with a short-wavelength electron beam, but are still too small to be seen with visible light.

We are able to detect molecules through our sense of smell. Noxious gases such as sulfur dioxide, ammonia, and ether are clearly sensed by the organs in our nose. The smell of perfume is the result of molecules that rapidly evaporate from the liquid and jostle around completely haphazardly in the air until some of them accidentally get close enough to our noses to be inhaled. The perfume molecules are certainly not attracted to our noses! They wander aimlessly in all directions from the liquid perfume to become a small fraction of the randomly jostling molecules in the air.

Figure 17.6 ▲
An electron microscope photograph of rubella virus molecules. The white dots are the virus errupting on the surface of an infected cell.

---

* An exception is DNA, a macromolecule that can be seen with an optical microscope. More dramatically, a diamond may be regarded as one huge carbon molecule, and this can be seen with the naked eye!

17.6 Compounds

A **compound** is a substance that is made of atoms of different elements combined in a fixed proportion. The **chemical formula** of the compound tells the proportions of each kind of atom. For example, in the gas carbon dioxide, the formula CO_2 indicates that for every carbon (C) atom there are two oxygen (O) atoms. Water, table salt, and carbon dioxide are all compounds. Air, wood, and salty water are not compounds, because the proportions of their atoms vary.

A compound may or may not be made of molecules. Water and carbon dioxide are made of molecules. On the other hand, table salt (NaCl) is made of different kinds of atoms arranged in a regular pattern. Every chlorine atom is surrounded by six sodium atoms (see Figure 17.7). In turn, every sodium atom is surrounded by six chlorine atoms. As a whole, there is one sodium atom for each chlorine atom, but there are no separate sodium-chlorine groups that can be labeled molecules.

Compounds have different properties from the elements of which they are made. At ordinary temperatures, water is a liquid, whereas hydrogen and oxygen are both gases. Salt is an edible solid, whereas chlorine is a poisonous gas.

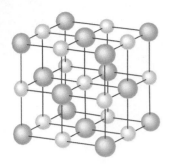

Figure 17.7 ▲
Table salt (NaCl) is a compound that is not made of molecules. The sodium and chlorine atoms are arranged in a repeating pattern. Each atom is surrounded by six atoms of the other kind. (We'll soon see that the atoms in NaCl are actually *ions*.)

17.7 The Atomic Nucleus

An atom is mostly empty space. Almost all its mass is packed into the central region called the **nucleus.** The New Zealander physicist Ernest Rutherford discovered this in 1911 in his now-famous gold foil experiment. Rutherford's group directed a beam of electrically charged particles (alpha particles) from a radioactive source through a very thin gold foil. They measured the angles at which the particles were deflected from their straight-line paths when they emerged. This was accomplished by noting spots of light on a zinc-sulfide screen that nearly surrounded the gold foil (Figure 17.8). Most

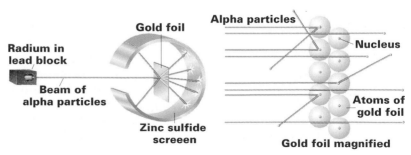

Figure 17.8 ▲
The occasional large-angle scattering of alpha particles from the gold atoms led Rutherford to the discovery of the small, very massive nucleus at their centers.

particles continued in a more or less straight-line path through the thin foil. But, surprisingly, some particles were widely deflected. Some were even scattered back almost along their incoming path. It was as surprising, Rutherford said, as firing a 15-inch shell at a piece of tissue paper and having it come back and hit you.

Rutherford reasoned that within the atom there had to be a positively charged object with two special properties. It had to be very small compared with the size of the atom, and it had to be massive enough to resist being shoved aside by heavy alpha particles. Rutherford had discovered the atomic nucleus.

Although the mass of an atom is primarily concentrated in the nucleus, the nucleus occupies less than a trillionth of the volume of an atom. Atomic nuclei (plural of *nucleus*) are extremely compact and extremely dense. If bare atomic nuclei could be packed against one another into a lump 1 cm in diameter (about the size of a small grape), the lump would weigh about a billion tons!

Huge electrical forces of repulsion prevent such close packing of atomic nuclei because each nucleus is electrically charged and repels the other nuclei. Only under special circumstances are the nuclei of two or more atoms squashed into contact. When this happens, the violent reaction known as nuclear fusion takes place. Fusion occurs in the core of the sun and in a hydrogen bomb.

The principal building blocks of the nucleus are **nucleons.*** Nucleons in an electrically neutral state are **neutrons.** Nucleons in an electrically charged state are **protons.** All neutrons are identical; they are copies of each other. Similarly, all protons are identical. Atoms of various elements differ from each other by their numbers of protons. Atoms with the same number of protons all belong to the same element.

For a given element, however, the number of neutrons will vary. Atoms of the same element having different numbers of neutrons are called **isotopes** of that element. The nucleus of the common hydrogen atom has a single proton. When this proton is accompanied by a neutron, we have *deuterium*, an isotope of hydrogen. When two neutrons are in a hydrogen nucleus, we have the isotope *tritium*. Every element has a variety of isotopes. Lighter elements usually have an equal number of protons and neutrons, and heavier elements usually have more neutrons than protons.

Atoms are classified by **atomic number.** The atomic number is the number of protons in the nucleus. The nucleus of a hydrogen atom has one proton, so its atomic number is 1. Helium has two protons, so its atomic number is 2. Lithium has three protons, so its atomic number is 3, and so on, in sequence up to the heaviest elements.

Electric charge comes in two kinds, positive and negative. Protons in the atom's nucleus are positive, and electrons orbiting the nucleus are negative. Positive and negative refer to a basic property of matter—electric *charge.* (Much more about electric charge in Unit V.) Like kinds of charge repel one another and unlike kinds attract one

---

* Nucleons are composed of still smaller particles called *quarks,* which are discussed in Chapter 39.

another. Protons repel protons but attract electrons. Electrons repel electrons but attract protons. Inside the nucleus, protons are held together by a *strong nuclear force*. This force is extremely intense but acts only across tiny distances. (More about the strong nuclear force in Chapter 38.)

17.8 Electrons in the Atom

Electrons that orbit the atomic nucleus are the same kind of electrons that flow in the wires of electric circuits. They are negatively charged subatomic particles. A proton or neutron has about 1800 times the mass of an electron, so electrons do not significantly contribute to the atom's mass.

In an electrically neutral atom, the number of negatively charged electrons always equals the number of positively charged protons in the nucleus. When the number of electrons is more, or less, than the number of protons in an atom, the atom is no longer neutral and has a net charge. It is an **ion.**

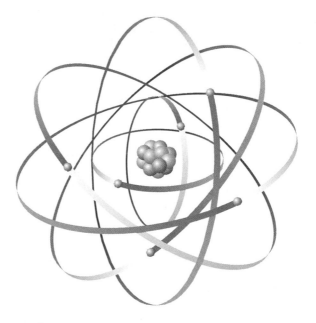

Figure 17.9 ▲
The classic model of the atom consists of a tiny nucleus surrounded by orbiting electrons.

Attraction between a proton and an electron can cause a *bond* between atoms to form a molecule. For example, two atoms can be held together by the sharing of electrons (covalent bond). Atoms can also be held together when electrons are transferred between atoms. In this case (ionic bond), ions of opposite charge are formed, and these ions are held together by simple electric forces.

**Hydrogen – One electron
in one shell**

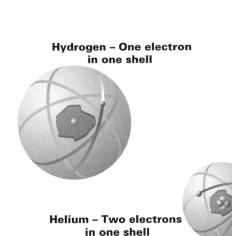

**Lithium – Three
electrons in
two shells**

**Helium – Two electrons
in one shell**

**Aluminum – Thirteen
electrons in
three shells**

Figure 17.10 ▲
The shell model of the atom pictures the electrons orbiting in concentric, spherical shells around the nucleus.

Just like our solar system, the atom is mostly empty space. The nucleus and surrounding electrons occupy only a tiny fraction of the atomic volume. Yet the electrons, because of their wave nature, form a kind of cloud as they cruise throughout the atom. Compressing this electron cloud takes great energy and means that when two atoms come close together, they repel each other. If it were not for this repulsive force between atoms, solid matter would be much more dense than it is. We and the solid floor are mostly empty space, because the atoms making up these and all materials are themselves mostly empty space. But we don't fall through the floor. The forces of repulsion keep atoms from caving in on one another under pressure.

To explain how atoms of different elements interact to form compounds, scientists have produced the **shell model of the atom.** Electrons are pictured as orbiting in spherical shells around the nucleus. There are seven different shells and each shell has its own capacity for electrons. For example, the first and innermost shell has a capacity for 2 electrons while the seventh and outermost shell has a capacity for 32 electrons. The arrangement of electrons in the shells about the atomic nucleus dictates such properties as melting and freezing temperatures, electric conductivity, and the taste, texture, appearance, and color of substances. The arrangement of electrons quite literally gives life and color to the world.

The **periodic table** is a chart that lists atoms by their atomic number and by their electron arrangements (see Figure 17.11). As you read across from left to right, each element has one more proton and electron than the preceding element. As you go down, each element has one more shell filled to its capacity than the element above.

Elements in the same column have similar chemical properties, reacting with other elements in similar ways to form new compounds and materials. Elements in the same column are said to belong to the same **group** of elements. Elements of the same group have similar chemical properties because their outermost electrons are arranged in a similar fashion.

Periodic Table of the Elements

1A	IIA	IIIB	IVB	VB	VIB	VIIB	VIII			IB	IIB	IIIA	IVA	VA	VIA	VIIA	0
1 **H** Hydrogen 1.008																	2 **He** Helium 4.003
3 **Li** Lithium 6.94	4 **Be** Beryllium 9.012											5 **B** Boron 10.81	6 **C** Carbon 12.011	7 **N** Nitrogen 14.007	8 **O** Oxygen 15.999	9 **F** Fluorine 18.998	10 **Ne** Neon 20.17
11 **Na** Sodium 22.990	12 **Mg** Magnesium 24.305											13 **Al** Aluminum 26.98	14 **Si** Silicon 28.09	15 **P** Phosphorus 30.974	16 **S** Sulfur 32.06	17 **Cl** Chlorine 35.453	18 **Ar** Argon 39.948
19 **K** Potassium 39.098	20 **Ca** Calcium 40.08	21 **Sc** Scandium 44.956	22 **Ti** Titanium 47.90	23 **Va** Vanadium 50.942	24 **Cr** Chromium 51.996	25 **Mn** Manganese 54.938	26 **Fe** Iron 55.847	27 **Co** Cobalt 58.933	28 **Ni** Nickel 58.71	29 **Cu** Copper 63.546	30 **Zn** Zinc 65.38	31 **Ga** Gallium 69.735	32 **Ge** Germanium 72.59	33 **As** Arsenic 74.992	34 **Se** Selenium 78.96	35 **Br** Bromine 79.904	36 **Kr** Krypton 83.80
37 **Rb** Rubidium 85.467	38 **Sr** Strontium 87.62	39 **Y** Yttrium 88.906	40 **Zr** Zirconium 91.22	41 **Nb** Niobium 92.906	42 **Mo** Molybdenum 95.94	43 **Tc** Technetium 98.906	44 **Ru** Ruthenium 101.07	45 **Rh** Rhodium 102.91	46 **Pd** Palladium 106.4	47 **Ag** Silver 107.868	48 **Cd** Cadmium 112.41	49 **In** Indium 114.82	50 **Sn** Tin 118.69	51 **Sb** Antimony 121.75	52 **Te** Tellurium 127.60	53 **I** Iodine 126.904	54 **Xe** Xenon 131.30
55 **Cs** Cesium 132.905	56 **Ba** Barium 137.33	57–71* Rare Earths	72 **Hf** Hafnium 178.49	73 **Ta** Tantalum 180.947	74 **W** Tungsten 183.85	75 **Re** Rhenium 186.207	76 **Os** Osmium 190.02	77 **Ir** Iridium 192.22	78 **Pt** Platinum 195.09	79 **Au** Gold 196.967	80 **Hg** Mercury 200.59	81 **Tl** Thallium 204.37	82 **Pb** Lead 207.2	83 **Bi** Bismuth 208.98	84 **Po** Polonium (209)	85 **At** Astatine (210)	86 **Rn** Radon (222)
87 **Fr** Francium (233)	88 **Ra** Radium 226.03	89–103† Actinides	104 **Unq** (261)	105 **Unp** (262)	106 **Unh** (263)	107 **Uns** (264)	108 **Uno** (265)	109 **Une** (266)	110 (267)								

***Rare Earths (Lanthanide series)**	57 **La** Lanthanum 139.91	58 **Ce** Cerium 140.12	59 **Pr** Praseodymium 140.91	60 **Nd** Neodymium 144.24	61 **Pm** Promethium (145)	62 **Sm** Samarium 150.36	63 **Eu** Europium 151.96	64 **Gd** Gadolinium 157.25	65 **Tb** Terbium 158.93	66 **Dy** Dysprosium 162.50	67 **Ho** Holmium 164.93	68 **Er** Erbium 167.26	69 **Tm** Thulium 168.93	70 **Yb** Ytterbium 173.04	71 **Lu** Lutetium 174.967
†Actinide series	89 **Ac** Actinium 227.028	90 **Th** Thorium 232.038	91 **Pa** Proctactinium 231.036	92 **U** Uranium 238.029	93 **Np** Neptunium 237.048	94 **Pu** Plutonium (244)	95 **Am** Americium (243)	96 **Cm** Curium (247)	97 **Bk** Berkelium (247)	98 **Cf** Californium (251)	99 **Es** Einsteinium (254)	100 **Fm** Fermium (257)	101 **Md** Mendelevium (258)	102 **No** Nobelium (259)	103 **Lr** Lawrencium (260)

Figure 17.11 ▲
The periodic table of the elements. The atomic number, above the chemical symbol, is equal to the number of protons in the nucleus (and equivalently, the number of electrons that surround the nucleus in a neutral atom). The number below is the atomic mass. Each row in the periodic table corresponds to a different number of electron shells in the atom.

Note that the uppermost row consists of only two elements, hydrogen and helium. The electrons of helium complete the innermost shell. Elements are arranged vertically on the basis of similarity in the arrangement of outer electrons, which dictates similarities in physical and chemical properties of the elements and their compounds.

17.9 The Phases of Matter

Matter exists in four phases. You are familiar with the solid, liquid, and gaseous phases. In the **plasma** phase, matter consists of positive *ions* (atoms that are missing some electrons) and free electrons. The plasma phase exists only at high temperatures. Although the plasma phase is less common to our everyday experience, it is the predominant phase of matter in the universe. The sun and other stars as well as much of the intergalactic matter are in the plasma phase. Closer to home, the glowing gas in a fluorescent lamp is a plasma.

In all phases of matter, the atoms are constantly in motion. In the solid phase, the atoms and molecules vibrate about fixed positions. If the rate of molecular vibration is increased enough, molecules will shake apart and wander throughout the material, jostling in nonfixed positions. The shape of the material is no longer fixed but takes the shape of its container. This is the liquid phase. If more energy is put into the material so that the molecules move about at even greater rates, they may break away from one another and assume the gaseous phase.

All substances can be transformed from one phase to another. We often observe this changing of phase in the compound H_2O. When solid, it is ice. If we heat the ice, the increased molecular motion jiggles the molecules out of their fixed positions, and we have water. If we heat the water, we can reach a stage where continued increase in molecular motion results in a separation between water molecules, and we have steam. Continued heating causes the molecules to separate into atoms. If we heat these to temperatures exceeding 2000°C, the atoms themselves will be shaken apart, making a gas of ions and free electrons. Then we have a plasma.

The following three chapters treat the solid, liquid, and gaseous phases in turn.

◄ **Figure 17.12**
The aurora borealis. High-altitude gases in the northern sky are transformed into glowing plasmas by the bombardment of charged particles. Less spectacular plasmas are found in glowing fluorescent lamps and advertising signs.

Concept Summary

Most matter is made from only about 100 different kinds of atoms.

- Each kind of atom belongs to a different element.

- Most atoms have been recycling through matter even before the solar system came into being.

- Atoms are too small to see with visible light but can be photographed with an electron microscope.

A compound is a substance made of different elements combined in a fixed proportion.

- Some compounds are made of molecules, which are particles made of atoms joined together.

- Other compounds are made of different kinds of atoms arranged in a regular pattern.

The atom is mostly empty space. Its mass is almost entirely in its nucleus.

The nucleus is made of protons and neutrons.

- The number of protons determines the element to which the atom belongs.

An electrically neutral atom has electrons outside the nucleus equal in number to the protons inside the nucleus.

- The shell model of the atom pictures electrons in spherical shells around the nucleus.

The periodic table is a chart of elements arranged according to similar atomic structure and similar properties.

Important Terms

atom (17.0)
atomic number (17.7)
Brownian motion (17.4)
chemical formula (17.6)
compound (17.6)

element (17.1)
group (17.8)
ion (17.8)
isotope (17.7)
molecule (17.5)

neutron (17.7)
nucleon (17.7)
nucleus (17.7)
periodic table (17.8)

plasma (17.9)
proton (17.7)
shell model of the atom (17.8)

Review Questions

1. Approximately how many elements are known today? (17.1)

2. Which element has the lightest atoms? (17.1)

3. How does the age of most atoms compare with the age of the solar system? (17.2)

4. What is meant by the statement that you don't "own" the atoms that make up your body? (17.2)

5. How does the approximate number of atoms in the air in your lungs compare with the number of breaths of air in the atmosphere of the whole world? (17.3)

6. How do the sizes of atoms compare with the wavelengths of visible light? (17.3)

7. What causes dust particles to move with Brownian motion? (17.4)

8. Individual atoms cannot be seen with visible light; yet there is a photograph of individual atoms in Figure 17.3. Explain. (17.4)

9. Distinguish between an atom and a molecule. (17.5)

10. a. How many elements compose pure water?

    b. How many individual atoms are there in a water molecule? (17.5)

11. a. Cite an example of a substance that is made of molecules.

b. Cite a substance that is not made of molecules. (17.5)

12. True or false: We smell things because certain molecules are attracted to our noses. (17.5)

13. a. What is a compound?

b. Cite the chemical formulas for at least three compounds. (17.6)

14. What did Rutherford discover when his group bombarded a thin foil of gold with subatomic particles? (17.7)

15. How does the mass of an atomic nucleus compare with the mass of the whole atom? (17.7)

16. How does the size of an atomic nucleus compare with the size of the whole atom? (17.7)

17. What are the two kinds of nucleons? (17.7)

18. a. What is an isotope?

b. Give two examples of isotopes. (17.7)

19. How does the atomic number of an element compare with the number of protons in its nucleus? With the number of electrons that normally surround the nucleus? (17.7)

20. How does the mass of an electron compare with the mass of a nucleon? (17.8)

21. a. What is an ion?

b. Give two examples. (17.8)

22. At the atomic level, a solid block of iron is mostly empty space. Explain. (17.8)

23. What is the periodic table of the elements? (17.8)

24. According to the shell model of the atom, how many electron shells are there in the hydrogen atom? The lithium atom? The aluminum atom? (17.8)

25. What are the four phases of matter? (17.9)

Think and Explain

1. Identify which of the following chemical formulas represent pure elements: H_2, H_2O, He, Na, NaCl, Au, U.

2. Which are older, the atoms in the body of an elderly person, or those in the body of a baby?

3. Suppose you smell the shaving lotion your brother is wearing almost immediately after he walks into the room. From an atomic point of view, exactly what is happening?

4. In what way does the number of protons in an atomic nucleus dictate the chemical properties of the element?

5. Atoms are mostly empty space, and structures such as a floor are composed of atoms and are therefore also empty space. So why don't you fall through the floor?

6. What element will result if a proton is added to the nucleus of carbon? (See periodic table.)

7. What element will result if two protons and two neutrons are ejected from a uranium nucleus?

8. You could swallow a capsule of the element germanium without harm. But if a proton were added to each of the germanium nuclei, you would not want to swallow the capsule. Why?

9. Assuming that all the atoms stay in the atmosphere, what are the chances that at least one of the atoms you exhaled in your very first breath as a baby will be inhaled in your next breath?

10. What does the addition or subtraction of heat have to do with whether or not a substance is a solid, liquid, gas, or plasma?

18 Solids

H$_2$O can be in the form of a solid, liquid, or gas.

Humans have been classifying and using solid materials for many thousands of years. The names Stone Age, Bronze Age, and Iron Age tell us the importance of solid materials in the development of civilization. Wood and clay were perhaps the first materials important to early peoples, and gems were put to use for art and adornment. While the number and uses of materials multiplied over the centuries, there was little progress in understanding their true nature. Not until recent times did the discovery of atoms and their interactions make it possible to understand the structure of materials. We have progressed from being finders and assemblers of materials to actual makers of materials. In today's laboratories chemists, metallurgists, and materials scientists routinely design and produce new materials to meet specific needs. Solid-state physicists explore solids such as semiconductors and tailor them to meet the needs of an information society.

18.1 Crystal Structure

When we look at samples of minerals such as quartz, mica, or galena, we see many smooth, flat surfaces at angles to each other within the mineral. The mineral samples are made of **crystals,** or regular geometric shapes. Each sample is made of many crystals, assembled in various directions. The samples themselves may have very irregular shapes, as if they were tiny cubes or other small units glued together to make a free-form solid sculpture.

Not all crystals are evident to the naked eye. Their existence in many solids was not discovered until X rays became a tool of research early in the twentieth century. The X-ray pattern caused by the crystal structure of common table salt (sodium chloride) is shown in Figure 18.1. Rays from the X-ray tube are blocked by a lead screen except for a narrow beam that hits the crystal of sodium chloride. The radiation that penetrates the crystal produces the pattern shown on the photographic film beyond the crystal. The white spot in the center is caused by the main unscattered beam of X rays. The size and arrangement of the other spots indicate the arrangement of sodium and chlorine atoms in the crystal. All crystals of sodium chloride produce this same design. Every crystalline structure has its own unique X-ray pattern.

The patterns made by X rays on photographic film show that the atoms in a crystal have an orderly arrangement. For example, in a sodium chloride crystal, the atoms are arranged like a three-dimensional chess board or a child's jungle gym (Figure 18.2).

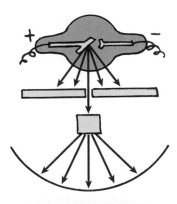

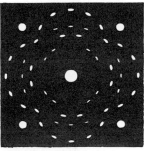

Figure 18.1 ▲
X-ray pattern caused by the crystal structure of common table salt (sodium chloride).

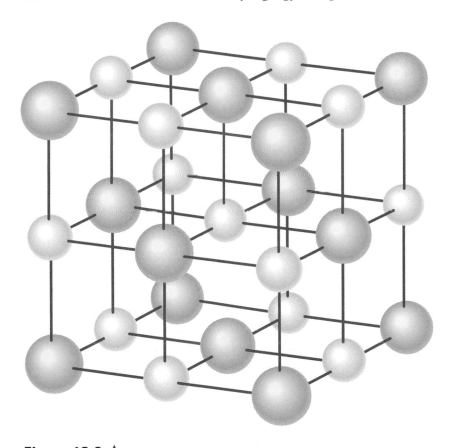

Figure 18.2 ▲
Model of a sodium chloride crystal. The large spheres represent chloride ions. The small ones represent sodium ions.

Metals such as iron, copper, and gold have relatively simple crystal structures. Tin and cobalt are only a little more complex. You can see metal crystals if you look carefully at a metal surface that has been cleaned (etched) with acid. You can also see them on the surface of galvanized iron that has been exposed to the weather, or on brass doorknobs that have been etched by the perspiration of hands.

Figure 18.3 ▲
Crystal structure is quite evident on galvanized (zinc-coated) metals.

18.2 Density

One of the properties of solids, as well as liquids and even gases, is the measure of how tightly the material is packed together: **density.** Density is a measure of how much matter is squeezed into a given space; it is the amount of mass per unit volume:

$$\text{density} = \frac{\text{mass}}{\text{volume}}$$

What happens to the density of a chocolate bar when you break it in two? The answer is, nothing. Each piece may have half the mass, but each piece also has half the volume. Density is not mass and it is not volume. Density is a ratio; it is the amount of mass per unit volume. A pure iron nail has the same density as a pure iron frying pan. The frying pan may have 100 times as many iron atoms and have 100 times as much mass, but its atoms will take up 100 times as much space. The mass per unit volume for the iron nail and the iron frying pan is the same.

Figure 18.4 ▲
When the loaf of bread is squeezed, its volume decreases and its density increases.

Both the masses of atoms and the spacing between atoms determine the density of materials. Osmium, a hard bluish-white metallic element, is the densest substance on Earth, even though the individual osmium atom is less massive than individual atoms of gold, mercury, lead, and uranium. The close spacing of osmium atoms in an osmium crystal gives it the greatest density. A cubic centimeter of osmium contains more atoms than a cubic centimeter of gold or uranium.

Table 18.1 Densities of a Few Substances

Material	Density (g/cm³)	Material	Density (g/cm³)
Solids		Liquids	
Osmium	22.6	Mercury	13.6
Platinum	21.4	Glycerin	1.26
Gold	19.3	Seawater	1.03
Uranium	19.0	Water at 4°C	1.00
Lead	11.3	Benzene	0.90
Silver	10.5	Ethyl alcohol	0.81
Copper	8.9		
Brass	8.6		
Iron	7.8		
Steel	7.8		
Tin	7.3		
Diamond	3.5		
Aluminum	2.7		
Graphite	2.25		
Ice	0.92		
Pine wood	0.50		
Balsa wood	0.12		

The densities of a few materials, in units of grams per cubic centimeter, are given in Table 18.1.* Density varies somewhat with temperature and pressure, so, except for water, densities are given at 0°C and atmospheric pressure. Note that water at 4°C has a density of 1.00 g/cm³. The gram was originally defined as the mass of a cubic centimeter of water at a temperature of 4°C. A gold brick, with a density of 19.3 g/cm³, is 19.3 times more massive than an equal volume of water.

A quantity known as **weight density** can be expressed by the amount of *weight* a body has per unit volume:

$$\text{weight density} = \frac{\text{weight}}{\text{volume}}$$

Weight density is commonly used when discussing liquid pressure (see next chapter).**

---

* Densities in kg/m³ are 1000 times greater. For instance, the density of water is 1000 kg/m³ and the density of osmium is 22 600 kg/m³.

** Weight density is common to British units, where one cubic foot of freshwater (almost 7.5 gallons) weighs 62.4 pounds. So freshwater has a weight density of 62.4 lb/ft³. Salt water is a bit denser, 64 lb/ft³.

A standard measure of density is **specific gravity**—the ratio of the mass (or weight) of a substance to the mass (or weight) of an equal volume of water. For example, if a substance weighs five times as much as an equal volume of water, its specific gravity is 5. Or put another way, specific gravity is a ratio of the density of a material to the density of water. So specific gravity has no units (density units divided by density units cancel). If you want to know the specific gravity of any material listed in Table 18.1, it's there. The magnitude of its density is its specific gravity.

■ Questions

1. Which has the greater density—1 kg of water or 10 kg of water?
2. Which has the greater density—5 kg of lead or 10 kg of aluminum?
3. Which has the greater density—1 g of uranium or the planet Earth?
4. The density of gold is 19.3 g/cm³. What is its specific gravity?

The Value of a Simple Computation

One of the reasons gold was used as money was that it is one of the densest of all substances and could therefore be easily identified. A merchant suspicious that gold was diluted with a less valuable substance had only to compute its density by measuring its mass and dividing by its volume. The merchant would then compare this value with the density of gold, 19.3 g/cm³.

Consider a gold nugget with a mass of 57.9 g and a volume of 3.00 cm³. Is the nugget pure gold? We compute its density as follows:

$$\text{density} = \frac{\text{mass}}{\text{volume}} = \frac{57.9 \text{ g}}{3.00 \text{ cm}^3} = 19.3 \text{ g/cm}^3$$

Its density matches that of gold, so the nugget can be presumed to be pure gold. (It is possible to get the same density by mixing gold with a platinum alloy, but this is unlikely since platinum has several times the value of gold.) In the next chapter, you'll learn an easy way to measure the volume of an irregularly shaped solid.

■ Answers

1. The density of *any* amount of water (at 4°C) is 1.00 g/cm³.
2. Any amount of lead always has a greater density than any amount of aluminum. The amount of material is irrelevant.
3. *Any* amount of uranium is more dense than the earth. The average density of the earth is actually 5.5 g/cm³, much less than the density of uranium (19.0 g/cm³).
4. Gold's specific gravity = density of gold/density of water = 19.3 g/cm³/1.0 g/cm³ = 19.3.

18.3 Elasticity

When we hang a weight on a spring, the spring stretches. When we add additional weights, the spring stretches still more. When we remove the weights, the spring returns to its original length. We say the spring is **elastic.**

When a batter hits a baseball, he temporarily changes the ball's shape. When an archer shoots an arrow, she first bends the bow, which springs back to its original form when the arrow is released. The spring, the baseball, and the bow are examples of elastic objects. **Elasticity** is that property of a body by which it experiences a change in shape when a deforming force acts on it, and by which it returns to its original shape when the deforming force is removed.

Not all materials return to their original shape when a deforming force is applied and then removed. Materials that do not resume their original shape after being distorted are said to be **inelastic.** Clay, putty, and dough are inelastic materials. Lead is also inelastic, since it is easy to distort it permanently.

By hanging a weight on a spring, we are applying a force to the spring. It is found that the stretch or compression is directly proportional to the applied force (Figure 18.6).

Figure 18.5 ▲
The bow is elastic. When the deforming force is removed, it returns to its original shape.

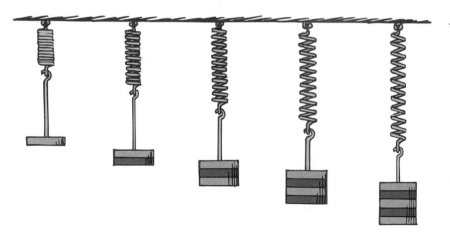

◀ **Figure 18.6**
The stretch of the spring is directly proportional to the applied force. When the weight is doubled, the spring stretches twice as much.

This relationship was noted by the British physicist Robert Hooke, a contemporary of Isaac Newton, in the midseventeenth century. It is called **Hooke's law**. The amount of stretch or compression, x, is directly proportional to the applied force F. Double the force and we double the stretch; triple the force and we get three times the stretch, and so on. In equation form,

$$F \sim x$$

If an elastic material is stretched or compressed more than a certain amount, it will not return to its original state. Instead, it will remain distorted. The distance at which permanent distortion occurs is called the **elastic limit.** Hooke's law holds only as long as the force does not stretch or compress the material beyond its elastic limit.

■ Questions

1. A certain tree branch is found to obey Hooke's law. When a 20-kg load is hung from the end of it, the branch sags a distance of 10 cm. If, instead, a 40-kg load is hung from the same place, by how much will the branch sag? How about if a 60-kg load were hung from the same place? (Assume that none of these loads makes the branch sag beyond its elastic limit.)

2. If a force of 10 N stretches a certain spring 4 cm, how much stretch will occur for an applied force of 15 N?

18.4 Compression and Tension

Steel is an excellent elastic material. It can be stretched and it can be compressed. Because of its strength and elastic properties, it is used to make not only springs but also construction girders. Vertical girders of steel used in the construction of tall buildings undergo only slight compression. A typical 25-meter-long vertical girder used in high-rise construction is compressed about a millimeter when it carries a 10-ton load. Most deformation occurs when girders are used horizontally, where the tendency is to sag under heavy loads.

A horizontal beam supported at one or both ends is under stress from the load it supports, including its own weight. It undergoes a stress of both compression and tension (stretching). Consider the beam supported at one end in Figure 18.7. It sags because of its own weight and because of the load it carries at its end.

Can you see that the top part of the beam is stretched? Atoms are tugged away from one another. The top part is slightly longer. And can you see that the bottom part of the beam is compressed? Atoms there are pushed toward one another, making the bottom part slightly shorter. So the top part of the beam is stretched, and the bottom part is compressed. A little thought will show that somewhere in

■ Answers

1. A 40-kg load has twice the weight of a 20-kg load. In accord with Hooke's law, $F \sim x$, two times the applied force will result in two times the stretch, so the branch should sag 20 cm. The weight of the 60-kg load will make the branch sag three times as much, or 30 cm. (When the elastic limit is exceeded, the amount of sag cannot be predicted with the information given.)

2. The spring will stretch 6 cm. By ratio and proportion,

$$\frac{10 \text{ N}}{4 \text{ cm}} = \frac{15 \text{ N}}{x}$$

which is read "10 N is to 4 cm as 15 N is to x." Solving for x gives $x =$ (15 N) × (4 cm)/(10 N) = 6 cm. In lab you will learn that the ratio of force to stretch is called the *spring constant* k (in this case $k = 2.5$ N/cm), and Hooke's law is expressed as the equation $F = kx$.

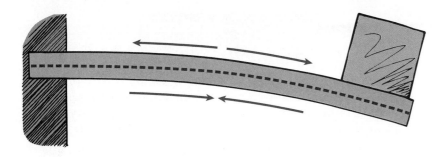

Figure 18.7 ▲
The top part of the beam is stretched and the bottom part is compressed. What happens in the middle portion, between top and bottom?

between the top and bottom, there will be a region where there is neither stretching nor compression. This is the *neutral layer.*

Consider the beam shown in Figure 18.8. It is supported at both ends, and carries a load in the middle. This time the top of the beam is in compression and the bottom is in tension. Again, there is a neutral layer along the middle portion of the length of the beam.

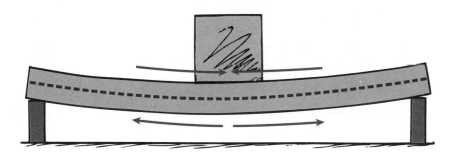

Figure 18.8 ▲
The top part of the beam is compressed and the bottom part is stretched. Where is the neutral layer (the part that is not under stress due to compression or stretching)?

Have you ever wondered why the cross section of many steel girders has the form of the letter I (Figure 18.9)? Most of the material in these I-beams is concentrated in the top and bottom parts, called the *flanges.* The piece joining the bars, called the *web,* is thinner. Why is this the shape?

The answer is that the stress is predominantly in the top and bottom flanges when the beam is used horizontally in construction. One flange tends to be stretched while the other tends to be compressed. The web between the top and bottom flanges is a region of low stress that acts principally to hold the top and bottom flanges apart. Heavier loads are supported by wider-apart flanges. For this purpose, comparatively little material is needed. An I-beam is nearly as strong as if it were a solid bar, and its weight is considerably less.

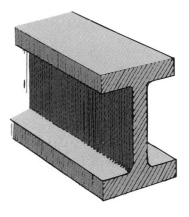

Figure 18.9 ▲
An I-beam is like a solid bar with some of the steel scooped from its middle where it is needed least. The beam is therefore lighter for nearly the same strength.

Constructing Triangles

The triangle is the strongest shape for building. You have probably noticed triangular shapes in steel bridges, sports domes, and the framed corners of homes. The triangle is the only shape that does not twist and collapse when under pressure. Let's verify the structural merits of the triangle. Nail or bolt three sticks together as shown. Now nail or bolt four sticks together. Compare how the two shapes resist collapse when you apply pressure to them.

 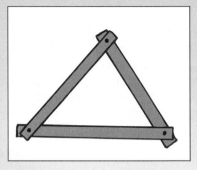

■ Question

If you had to make a hole horizontally through the tree branch shown, in a location that would weaken it the least, would you bore it through the top, the middle, or the bottom?

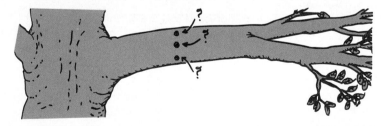

■ Answer

Drill the hole in the middle, through the neutral layer. Wood fibers in the top part of the branch are being stretched, and if you drilled the hole there, tension in that part may pull the branch apart. Fibers in the lower part are being compressed, and a hole there might crush under compression. In between, in the neutral layer, the hole will not affect the strength of the branch because fibers there are being neither stretched nor compressed.

18.5 Scaling

Did you ever notice how strong an ant is for its size? An ant can carry the weight of several ants on its back, whereas a strong elephant couldn't even carry one elephant on its back. How strong would an ant be if it were scaled up to the size of an elephant? Would this "super ant" be several times stronger than an elephant? Surprisingly, the answer is no. Such an ant would not be able to lift its own weight off the ground. Its legs would be too thin for its greater weight and would likely break.

Ants have thin legs and elephants have thick legs for a reason. The proportions of things in nature are in accord with their size. The study of how size affects the relationship between weight, strength, and surface area is known as **scaling.** As the size of a thing increases, it grows heavier much faster than it grows stronger. You can support a toothpick horizontally at its ends and you'll notice no sag. But support a tree of the same kind of wood horizontally at its ends and you'll see a noticeable sag. The tree is not as strong per unit mass as the toothpick is.

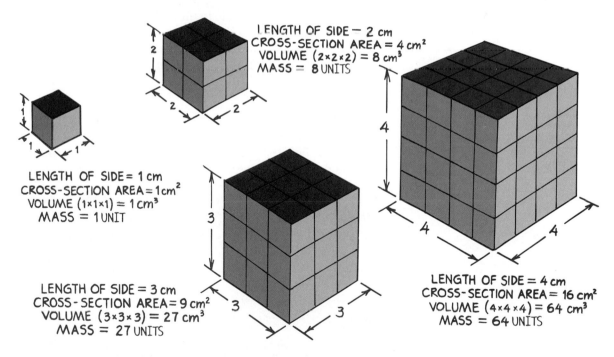

Figure 18.10 ▲
If the linear dimensions of an object are multiplied by some number, then the area will grow by the square of the number, and the volume (and mass and weight) will grow by the cube of the number. If the linear dimensions of the cube grow by 2, the area will grow by $2^2 = 4$, and the volume will grow by $2^3 = 8$. If the linear dimensions grow by 3, the area will grow by $3^2 = 9$, and the volume will grow by $3^3 = 27$.

Weight depends on volume, and strength comes from the area of the cross section of limbs—tree limbs or animal limbs. To understand this weight-strength relationship, let's consider the simple case of a solid cube of matter, 1 centimeter on a side.

A 1-cubic-centimeter cube has a cross section of 1 square centimeter. That is, if we sliced through the cube parallel to one of its faces, the sliced area would be 1 square centimeter. Compare this to a cube that has double the linear dimensions, a cube 2 centimeters on each side. Its cross-sectional area will be 2×2 (or 4) square centimeters and its volume will be $2 \times 2 \times 2$ (or 8) cubic centimeters. If it had the same density it would be eight times more massive. Careful investigation of Figure 18.10 shows that when linear dimensions are enlarged, the cross-sectional area (as well as the total area) grows as the square of the enlargement, whereas volume and weight grow as the cube of the enlargement.

The volume (and weight) increases much faster than the corresponding enlargement of cross-sectional area. Although the figure demonstrates the simple example of a cube, the principle applies to an object of any shape. Consider an athlete who can lift his weight with one arm. Suppose he could somehow be scaled up to twice his size—that is, twice as tall, twice as broad, his bones twice as thick and every linear dimension enlarged by a factor of 2. Would he be twice as strong? Would he be able to lift himself with twice the ease? The answer to both questions is no. Since his twice-as-thick arms would have four times the cross-sectional area, he would be four times as strong. At the same time, his volume would be eight times as great, so he would be eight times as heavy. So, for comparable

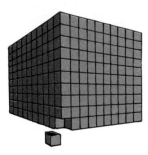

■ Questions

1. Suppose a cube 1 cm long on each side were scaled up to a cube 10 cm long on each edge. What would be the volume of the scaled-up cube? What would be its cross-sectional surface area? Its total surface area?

2. If an athlete were somehow scaled up proportionally to twice size, would he be stronger or weaker?

■ Answers

1. The volume of the scaled-up cube would be (length of side)^3 = (10 cm)^3, or 1000 cm^3. Its cross-sectional surface area would be (length of side)^2 = (10 cm)^2, or 100 cm^2. Its total surface area = 6 sides $\times$ area of side = 600 cm^2.

2. The scaled-up athlete would be four times as strong, because the cross-sectional area of his twice-as-thick bones and muscles increases by a factor of four. He could lift a maximum load four times as heavy as before. But his own weight is eight times as much as before, so he would be weaker in relation to his weight. Having four times the strength while carrying eight times the weight gives him a strength-to-weight ratio of only half its former value. This means that if he could just lift his own weight before, he could now lift only half his new weight. In sum, while his actual strength would increase, his strength-to-weight ratio would decrease.

effort, he could lift only half his weight. *In relation to his weight,* he would be weaker than before.

The fact that volume (and weight) grow as the cube of linear enlargement, while strength (and area) grow as the square of linear enlargement is evident in the disproportionately thick legs of large animals compared with those of small animals. Consider the different legs of an elephant and a deer, or a tarantula and a daddy longlegs.

So the great strengths attributed to King Kong and other fictional giants cannot be taken seriously. The fact that the consequences of scaling are conveniently omitted is one of the differences between science and science fiction.

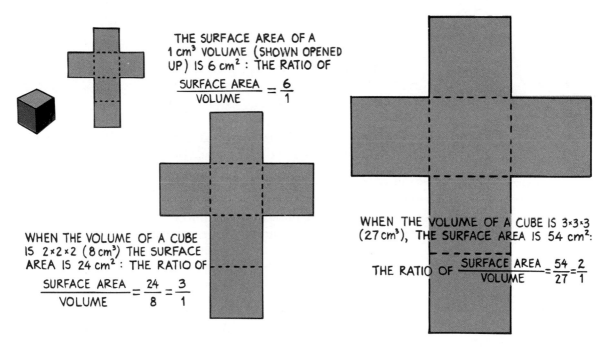

Figure 18.11 ▲
As an object grows proportionally in all directions, there is a greater increase in volume than in surface area. As a result, the ratio of surface area to volume decreases.

Important also is the comparison of total surface area with volume. A study of Figure 18.11 shows that as the linear size of an object increases, the volume grows faster than the total surface area. (Volume grows as the cube of the enlargement, and both cross-sectional area and total surface area grow as the square of the enlargement.) So as an object grows, its surface area and volume grow at different rates, with the result that the surface area to volume ratio *decreases.* In other words, both the surface area and the volume of a growing object increase, but the growth of surface area *relative to* the growth of volume decreases. Not many people really understand this idea. The following examples may be helpful.

An experienced cook knows that more skin results when peeling 5 kg of small potatoes than when peeling 5 kg of large potatoes.

Smaller objects have more surface area per kilogram. Since cooling occurs at the surfaces of objects, crushed ice will cool a drink much faster than a single ice cube of the same mass. This is because crushed ice presents more surface area to the beverage.

The rusting of iron is also a surface phenomenon. The greater the amount of surface exposed to the air, the faster rusting takes place. That's why small filings, or iron in the form of "steel wool," which have large surfaces compared with their volumes, are soon eaten away. The same mass of iron packed in a solid cube or sphere would undergo little rusting in comparison.

Chunks of coal burn, while coal dust explodes when ignited. Thin french fries cook faster in oil than fat fries. Flat hamburgers cook faster than meatballs of the same mass. Large raindrops fall faster than small raindrops, and large fish swim faster than small fish. These are all consequences of the fact that volume and area are not in direct proportion to each other.

The big ears of elephants are not for better hearing, but for cooling. They are nature's way of making up for the small ratio of surface area to volume for these large animals. The heat that an animal dissipates is proportional to its surface area. If an elephant did not have large ears, it would not have enough surface area to cool its huge mass. The large ears of the African elephant greatly increase overall surface area, and enable it to cool off in hot climates.

At the biological level, living cells must contend with the fact that the growth of volume is faster than the growth of surface area. A cell obtains nourishment by diffusion through its surface. As it grows, surface area enlarges, but not fast enough to keep up with volume. For example, when the surface area increases four times, the corresponding volume increases eight times. Eight times the mass must be sustained by only four times the nourishment. This puts a limit on the growth of a living cell. So cells divide, and there is life as we know it. That's nice.

Not so nice is the fate of large animals when they fall. The statement "the bigger they are, the harder they fall" holds true and is a consequence of the small ratio of surface area to weight. Air resistance to movement through the air depends on the surface area of the moving object. If you fell off a cliff, even with air resistance, your speed would increase at the rate of very nearly 1 *g*. You wouldn't have enough surface area relative to your weight—unless you wore a parachute. Small animals need no parachute. They have plenty of surface area relative to their small weights. An insect can fall from the top of a tree to the ground below without harm. The surface area per weight ratio is in the insect's favor—in a sense, the insect *is* its own parachute.

It is interesting to note that the rate of heartbeat in a mammal is related to the size of the mammal. The heart of a tiny shrew beats about twenty times as fast as the heart of an elephant. In general, small mammals live fast and die young; larger animals live at a leisurely pace and live longer. Don't feel bad about a pet hamster that doesn't live as long as a dog. All warm-blooded animals have about the same life span—not in terms of years, but in the average number of heartbeats (about 800 million). Humans are the exception: we live two to three times longer than other mammals of our size.

Figure 18.12 ▲
The African elephant has less surface area compared with its weight than other animals. It compensates for this with its large ears, which significantly increase its radiating surface area and promote cooling.

18 Chapter Review

Concept Summary

Many solids are made of crystals.

- The atoms in a crystal have an orderly arrangement.

One property of solids is density, the amount of mass per unit volume.

- Density is related both to the masses of the atoms and the spacing between atoms.
- Weight density is weight per unit volume.

Another property of solids is elasticity.

- Elastic solids return to their original shape when a deforming force is applied and removed, as long as they are not deformed beyond their elastic limit.
- By Hooke's law, stretch or compression is proportional to the applied force (within the elastic limit).
- Inelastic materials remain distorted after the force is removed.

Scaling is the study of how size affects the relationships among weight, strength, and surface.

Important Terms

crystal (18.1)
density (18.2)
elastic (18.3)
elastic limit (18.3)
elasticity (18.3)
Hooke's law (18.3)
inelastic (18.3)
scaling (18.5)
specific gravity (18.2)
weight density (18.2)

Review Questions

1. How does the arrangement of atoms differ in a crystalline and a noncrystalline substance? (18.1)

2. What evidence do we have for the microscopic crystal nature of some solids? (18.1)

3. What evidence do we have for the visible crystal nature of some solids? (18.1)

4. What happens to the density of a uniform piece of wood when we cut it in half? (18.2)

5. Uranium is the heaviest atom found in nature. Why isn't uranium metal the most dense material? (18.2)

6. Which has the greater density—a heavy bar of pure gold or a pure gold ring? (18.2)

7. Does the *mass* of a loaf of bread change when you squeeze it? Does its *volume* change? Does its *density* change? (18.2)

8. What is the difference between mass density and weight density? (18.2)

9. **a.** What is the evidence for the claim that steel is elastic?

   b. That putty is inelastic? (18.3)

10. What is Hooke's law? (18.3)

11. What is an elastic limit? (18.3)

12. A 2-kg mass stretches a spring 3 cm. How far does the spring stretch when it supports 6 kg? (Assume the spring has not reached its elastic limit.) (18.3)

13. Is a steel beam slightly shorter when it stands vertically? (18.4)

14. Where is the neutral layer in a horizontal beam that supports a load? (18.4)

15. Why is the cross section of a metal beam I-shaped and not rectangular? (18.4)

16. What is the weight-strength relationship in scaling? (18.5)

17. **a.** If the linear dimensions of an object are doubled, how much does the total area increase?

b. How much does the volume increase? (18.5)

18. True or false: As the volume of an object increases, its surface area also increases, but the *ratio* of surface area to volume decreases. Explain. (18.5)

19. Which will cool a drink faster—a 10-gram ice cube or 10 grams of crushed ice? (18.5)

20. **a.** Which has more skin—an elephant or a mouse?

b. Which has more skin *per body weight*—an elephant or a mouse? (18.5)

Think and Explain

1. Which has more volume—a kilogram of lead or a kilogram of aluminum?

2. Which has more weight—a liter of ice or a liter of water?

3. A certain spring stretches 1 cm for each kilogram it supports. If the elastic limit is not reached, how far will it stretch when it supports a load of 8 kg?

4. Suppose the spring in the preceding question is placed next to an identical spring so the two side-by-side springs equally share the 8-kg load. How far will each spring stretch?

5. Metal beams are not "solid" like wooden beams, but are "cut out" in the middle so that their cross section has an I-shape. What are the advantages of this shape?

6. Compression and tension stress occurs in a beam that supports a load (even when the load is its own weight). Show by means of a simple sketch an example where a horizontal load-carrying beam is in tension at the top and compression at the bottom. Then show a case where the opposite occurs: compression at the top and tension at the bottom.

7. Consider a model steel bridge that is 1/100 the exact scale of the real bridge that is to be built.

a. If the model bridge weighs 50 N, what will the real bridge weigh?

b. If the model bridge doesn't appear to sag under its own weight, is this evidence that the real bridge, if built exactly to scale, will not appear to sag either? Explain.

8. If you use a batch of cake batter for cupcakes instead of a cake and bake them for the time suggested for baking a cake, what will be the result?

9. Explain, in terms of scaling, why it is an advantage that natives of the hot African desert tend to be relatively tall and slender, and natives of the Arctic region tend to be short and stout. (*Hint:* A piece of wire will cool faster when stretched out than when rolled up into a ball.)

10. Animals lose heat through the surface areas of their skin. A small animal, such as a mouse, uses a much larger proportion of its energy to keep warm than does a large animal, like an elephant. Why is the rate of heat loss greater in a small animal than a large one?

19 Liquids

We live on the only planet in the solar system covered mostly by a liquid. The earth's oceans are made of H_2O in the liquid phase. If the earth were a little closer to the sun, the oceans would turn to vapor. If the earth were a little farther away, most of its surface, not just its polar regions, would be solid ice. It's nice that the earth is where it is.

In the liquid phase, molecules can flow freely from position to position by sliding over one another. The shape of a liquid takes the shape of its container.

H_2O in the form of liquid water.

19.1 Liquid Pressure

A liquid in a container exerts forces on the walls and bottom of the container. To investigate the interaction between the liquid and a surface, it is useful to discuss the concept of *pressure*. Recall from Chapter 5 that pressure is defined as the force per area on which the force acts.*

$$\text{pressure} = \frac{\text{force}}{\text{area}}$$

The pressure that a block exerts on a table is simply the weight of the block divided by its area of contact. Similarly, for a liquid in a cylindrical container like the one shown in Figure 19.1, the pressure the liquid exerts against the bottom of the container is the weight of

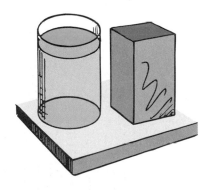

Figure 19.1 ▲
The liquid exerts a pressure against the bottom of its container, just as the block exerts a pressure against the table.

---

* Pressure may be measured in any unit of force divided by any unit of area. The standard international (SI) unit for pressure is newtons per square meter, called the *pascal* (Pa), named after the seventeenth-century theologian and scientist Blaise Pascal. A pressure of 1 Pa is very small, about that of a dollar bill resting flat on a table. Science types more often use kilopascals (1 kPa = 1000 Pa).

the liquid divided by the area of the container bottom. (We'll ignore for now the additional atmospheric pressure.)

How much a liquid weighs, and thus how much pressure it exerts, depends on its density. Consider two identical containers, one filled with mercury and the other filled to the same depth with water. For the same depth, the denser liquid exerts more pressure. Mercury is 13.6 times as dense as water. So for the same volume of liquid, the weight of mercury is 13.6 times the weight of water. Thus, the pressure of mercury on the bottom is 13.6 times the pressure of water.

For liquids of the same density, the pressure will be greater at the bottom of the deeper liquid. Consider the two containers in Figure 19.2. If the liquid in the first container is twice as deep as the liquid in the second container, then as with two blocks one atop the other, liquid pressure at the bottom of the first container will be twice that of the second container.

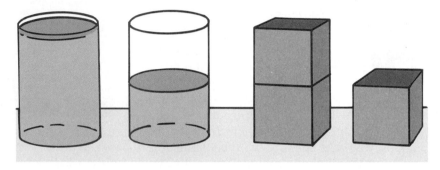

Figure 19.2 ▲
The two blocks exert twice as much pressure on the table as one block. Similarly, the liquid in the first container is twice as deep, so the pressure on the bottom is twice that exerted by the liquid in the second container.

It turns out that the pressure of a liquid at rest depends only on the density and the depth of the liquid, *not* on the shape of the container or the size of its bottom surface. Liquids are practically incompressible, so except for changes in temperature, the density of a liquid is normally the same at all depths. The pressure created by a liquid* is

$$\text{pressure due to liquid} = \text{weight density} \times \text{depth}$$

---

* This relationship is derived from the definitions of pressure and density. Consider an area at the bottom of a container of liquid. Pressure is produced by the weight of the column of liquid directly above this area. From the definition of weight density as weight divided by volume, this weight of liquid can be expressed as weight density times volume. The volume of the column is simply the area multiplied by the depth. Then we get

$$\text{pressure} = \frac{\text{force}}{\text{area}} = \frac{\text{weight}}{\text{area}} = \frac{\text{weight density} \times \text{volume}}{\text{area}}$$

$$= \frac{\text{weight density} \times \cancel{\text{area}} \times \text{depth}}{\cancel{\text{area}}} = \text{weight density} \times \text{depth}$$

Noting Blood Pressure

Note how your veins stand out on the back of your hands when you hold them as low as you can—perhaps when you bend over so your hands are the lowest part of your body, or when doing handstands. Then note the difference when you hold your hands above your head. This is "pressure depends on depth" in action. Why is your blood pressure measured in your upper arm, level with your heart?

Activity

At a given depth, a given liquid exerts the same pressure against *any* surface—the bottom or sides of its container, or even the surface of an object submerged in the liquid to that depth. The pressure a liquid exerts depends on its density and depth.

If you press your hand against a surface, and somebody else presses against your hand in the same direction, then the pressure against the surface is greater than if you pressed alone. Likewise with the atmospheric pressure that presses on the surface of a liquid. The total pressure of a liquid, then, is weight density × depth *plus* the pressure of the atmosphere. When this distinction is important we will use the term *total pressure*. Otherwise, our discussions of liquid pressure refer to pressure in addition to the normally ever-present atmospheric pressure. (More about atmospheric pressure in the next chapter.)

It may surprise you that the pressure of a liquid does not depend on the amount of liquid. Neither the volume nor even the total weight of liquid matters. For example, if you sampled water pressure at 1 meter beneath the surface of a large lake and 1 meter beneath the surface of a small pool, the pressures would be the same.*

---

* What would that pressure be? The density of freshwater is 1 gram per cubic centimeter, which equals 1000 kilograms per cubic meter in SI units. Since the weight (*mg*) of 1000 kilograms is (1000 kg) × (9.8 N/kg) = 9800 N, the weight density of water is 9800 newtons per cubic meter (9800 N/m^3). Water pressure in a pool or lake is simply equal to the product of the weight density of freshwater and the depth in meters. So at a depth of 1 meter, water pressure in a large lake or small pool is (9800 N/m^3) × (1 m) = 9800 N/m^2. In SI units, pressure is measured in pascals (1 Pa = 1 N/m^2), so our result would be 9800 Pa or, in the frequently used unit of kilopascals, 9.8 kPa. (This would be 10 kPa in seawater, which has a weight density of 10 000 N/m^3.) For the *total pressure* in these cases, add the pressure of the atmosphere, 101.3 kPa.

Figure 19.3 ▶

The water pressure is greater at the bottom of the deeper lake, not the lake with the most water. The dam holding back water twice as deep must withstand greater average water pressure, regardless of the total volume of water.

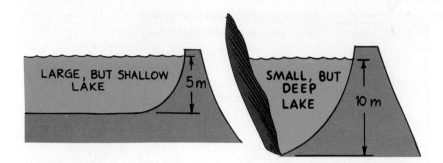

The dam that must withstand the greater pressure is the dam with the *deepest* water behind it, not the *most* water (Figure 19.3).

The fact that water pressure depends on depth and not on volume is nicely illustrated with "Pascal's vases" (Figure 19.4). Note that the water's surface in each of the connected vases is at the same level. This occurs because the pressures at equal depths beneath the surfaces are the same. At the bottom of each vase, for example, the pressures are equal. If they were not, liquid would flow until the pressures were equalized. This is why we say "water seeks its own level."

Figure 19.4 ▶

The pressure of the liquid is the same at any given depth below the surface, regardless of the shape of the container.

At any point within a liquid, the forces that produce pressure are exerted equally in all directions. For example, when you are swimming underwater, no matter which way you tilt your head, you feel the same amount of water pressure on your ears.

■ Questions

1. Is there more pressure at the bottom of a bathtub of water 30 cm deep or at the bottom of a pitcher of water 35 cm deep?

2. A brick mason wishes to mark the back of a building at the exact height of bricks already laid at the front of the building. How can he measure the same height using only a garden hose and water?

■ Answers

1. There is more pressure at the bottom of the pitcher, because the water in it is deeper. The fact that there is more water in the bathtub does not matter.

2. To measure the same height, the brick mason can extend a garden hose that is open at both ends from the front to the back of the house, and fill it with water until the water level reaches the height of bricks in the front. Since water seeks its own level, the level of water in the other end of the hose will be the same!

When the liquid is pressing against a surface, there is a net force directed perpendicular to the surface (Figure 19.5 top). If there is a hole in the surface, the liquid initially will move perpendicular to the surface (Figure 19.5 bottom). Gravity, of course, causes the path of the liquid to curve downward. At greater depths, the net force is greater and the horizontal velocity of the escaping liquid is greater.

19.2 Buoyancy

If you have ever lifted a submerged object out of water, you are familiar with **buoyancy,** the apparent loss of weight of objects when submerged in a liquid. It is a lot easier to lift a boulder submerged on the bottom of a riverbed than to lift it above the water's surface. The reason is that when the boulder is submerged, the water exerts an upward force that is opposite in direction to gravity. This upward force is called the **buoyant force.**

To understand where the buoyant force comes from, look at Figure 19.6. The arrows represent the forces by the liquid that produce pressure against the submerged boulder. The forces are greater at greater depth. The forces acting horizontally against the sides cancel each other, so the boulder is not pushed sideways. But the forces acting upward against the bottom are greater than those acting downward against the top because the bottom of the boulder is deeper. The difference in upward and downward forces is the buoyant force.

When the weight of a submerged object is greater than the buoyant force, the object will sink. When the weight is equal to the buoyant force, the submerged object will remain at any level, like a fish. When the weight is less than the buoyant force, the object will rise to the surface and float.

To further understand buoyancy, it helps to think more about what happens when an object is placed in water. If a stone is placed in a container of water, the water level will rise (Figure 19.7). Water is said to be **displaced,** or pushed aside, by the stone. A little thought will tell us that the volume—that is, the amount of space taken up or the number of cubic centimeters—of water displaced is equal to the

Figure 19.5 ▲
(Top) The forces in a liquid that produce pressure against a surface add up to a net force that is perpendicular to the surface. (Bottom) Liquid escaping through a hole initially moves perpendicular to the surface.

Figure 19.6 ▲
The upward forces against the bottom of a submerged object are greater than the downward forces against the top. There is a net upward force, the buoyant force.

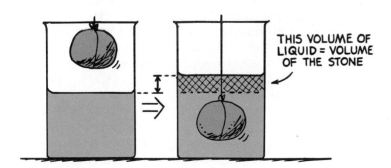

THIS VOLUME OF LIQUID = VOLUME OF THE STONE

Figure 19.7 ▲
When an object is submerged, it displaces a volume of water equal to the volume of the object itself.

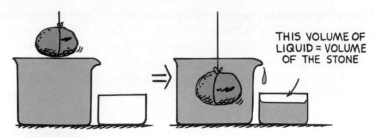

THIS VOLUME OF
LIQUID = VOLUME
OF THE STONE

Figure 19.8 ▲
When an object is submerged in a container that is initially brim full, the volume of water that overflows is equal to the volume of the object itself.

volume of the stone. *A completely submerged object always displaces a volume of liquid equal to its own volume.*

This gives us a good way to determine the volume of an irregularly shaped object. Simply submerge it in water in a measuring cup and note the increase in volume of the water. That increase in volume is equal to the volume of the submerged object. You'll find this technique handy whenever you want to determine the density of things like rocks that have irregular shapes.

19.3 Archimedes' Principle

The relationship between buoyancy and displaced liquid was discovered in ancient times by the Greek philosopher Archimedes (third century B.C.). It is stated as follows:

> An immersed object is buoyed up by a force equal to the weight of the fluid it displaces.

This relationship is called **Archimedes' principle.** It is true for liquids and gases, which are both fluids.

Immersed means "either completely or partially submerged." For example, if we immerse a sealed 1-liter container halfway into water, it will displace half a liter of water and be buoyed up by the weight of half a liter of water. If we immerse it all the way (submerge it), it will be buoyed up by the weight of a full liter of water (9.8 newtons). Unless the completely submerged container becomes compressed, the buoyant force will equal the weight of 1 liter of water at *any* depth.* Why? Because the container will displace the same volume of water, and hence the same weight of water, at any depth. The weight of this displaced water (not the weight of the submerged object!) is the buoyant force.

A 300-gram brick weighs about 3 N in air. Suppose the brick displaces 2 N of water when it is submerged (Figure 19.10). The buoyant force on the submerged brick will also equal 2 N. The brick will seem to weigh less under water than above water. In the water, its apparent

Figure 19.9 ▲
A liter of water occupies 1000 cubic centimeters, has a mass of 1 kilogram, and weighs 9.8 N. Any object with a volume of 1 liter will experience a buoyant force of 9.8 N when fully submerged in water.

---

*   Water is practically incompressible. A liter of water under great pressure far below the surface still weighs almost the same as a liter of water near the surface, approximately 9.8 N.

A brick weighs less in water than in air. The buoyant force on the submerged brick is equal to the weight of the water displaced. So the brick appears lighter under water by an amount equal to the weight of water (2 N) that has spilled into the smaller container. The apparent weight of the brick under water equals its weight in air minus the buoyant force (3 N – 2 N = 1 N).

weight will be 3 N minus the 2-N buoyant force, or 1 N. The apparent weight of a submerged object is its weight in air minus the buoyant force.

Perhaps your teacher will summarize Archimedes' principle by way of a numerical example to show that the upward force due to water pressure on the bottom of a submerged block, minus the downward force due to water pressure on the top equals the weight of liquid displaced. As long as the block is submerged, depth makes no difference. Why? Because although there is more pressure at greater depths, the *difference* in pressures on the bottom and top of the block is the same at any depth (Figure 19.11). Whatever the shape of a submerged object, the buoyant force equals the weight of liquid displaced.

■ Questions

1. A 1-liter (L) container filled with mercury has a mass of 13.6 kg and weighs 133 N. When it is submerged in water, what is the buoyant force on it?

2. A block is held suspended beneath the water in the three positions, A, B, and C, shown in Figure 19.11. In which position is the buoyant force on it greatest?

3. A stone is thrown into a deep lake. As it sinks deeper and deeper into the water, does the buoyant force on it increase, decrease, or remain unchanged?

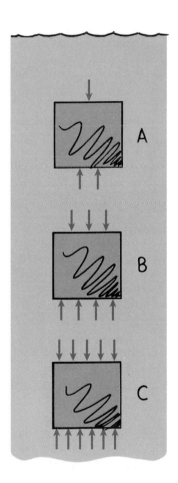

Figure 19.11 ▲
The difference in the upward and downward force acting on the submerged block is the same at any depth.

■ Answers

1. The buoyant force equals the weight of 1 L of water (about 10 N) because the *volume* of displaced water is 1 L. The mass or weight of mercury is irrelevant; 1 L of *anything* submerged in water displaces 1 L of water and is buoyed upward with a force of 10 N. Or look at it this way: when the container is put in the water, it pushes 1 L of water out of the way by its presence. In accordance with Newton's law of action and reaction, the water pushes back on the container with a force equal to the weight of 1 L of water. If the container had twice the volume, it would displace 2 L of water and be buoyed up with a force equal to the weight of 2 L of water. (The concepts involved with buoyant force are best dealt with if you *visualize* what is going on—*conceptual physics!*)

2. The buoyant force is the same at all three positions. Why? Because the amount of water displaced is the same in each case. Buoyant force equals the weight of displaced water, which is the same at positions A, B, and C.

3. Like the previous question, the volume of displaced water is the same at any depth. Water is practically incompressible, so its density is the same at any depth, and equal volumes of water weigh the same. The buoyant force on the stone remains unchanged as it sinks deeper and deeper.

Just as most of a floating iceberg is below the water's surface, most of a mountain is below ground level. Mountains "float" too! About 15% of a mountain is above the surrounding ground level. The rest extends deep into the earth, resting on the dense semiliquid mantle. If we could shave off the top of an iceberg, the iceberg would be lighter and float higher. Similarly, when mountains erode they float higher. That's why it takes so long for mountains to weather away. As the mountain wears away, it floats higher, pushed up from below. When a mile of mountain erodes away, 85% of it comes back.

19.4 Does It Sink, or Does It Float?

We have learned that the buoyant force on a submerged object depends on the object's volume. A smaller object displaces less water so a smaller buoyant force acts on it. A larger object displaces more water so a larger buoyant force acts on it. The submerged object's *volume*—not its *weight*—determines buoyant force. (A misunderstanding of this idea is at the root of a lot of confusion that you or your friends may have about buoyancy!)

So far we've focused on the weight of displaced fluid, not the weight of the submerged object. Now we consider its role.

Whether an object sinks or floats (or does neither) depends on both its buoyant force (up) and its weight (down)—how great the buoyant force is compared *with the object's weight*. Careful thought will show that when the buoyant force exactly equals the weight of a completely submerged object, then the object's weight must equal the weight of displaced water. Since the volumes of the object and of the displaced water are the same, the density of the object must equal the density of water.

This is true for a fish, whose density equals the density of water. The fish is "at one" with the water—it doesn't sink or float. If the fish were somehow bloated up, it would be less dense than water, and would float to the top. If the fish swallowed a stone and became more dense than water, it would sink to the bottom.

Place a beaker of water on a balance scale and adjust the balance to be level. Read the following and think before doing anything. What will happen to the balance if you lightly touch the surface of the water with your finger? If you push the water downward, even slightly, will the balance record this "push"? Is your push on the water communicated to the beaker and thus the balance? Doesn't displacing some water make the water slightly deeper? And doesn't slightly deeper water have a slightly greater pressure on the bottom? Is this transmitted to the balance? Think. Discuss. Then try it and see!

Activity

This can be summed up in three simple rules.

1. An object more dense than the fluid in which it is immersed sinks.

2. An object less dense than the fluid in which it is immersed floats.

3. An object with density equal to the density of the fluid in which it is immersed neither sinks nor floats.

From these rules, what do we say about people who, try as they may, cannot float?* They're simply too dense! To float more easily, you must reduce your density. Since weight density is weight divided by volume, you must either reduce your weight or increase your volume. Taking in a lung full of air can increase your volume (temporarily!). A life jacket does the job better. It increases volume while adding little to your weight.

The density of a submarine is controlled by the flow of water into and out of its ballast tanks. In this way the weight of the submarine can be varied to achieve the desired average density. A fish regulates its density by expanding or contracting an air sac that changes its volume. The fish can move upward by increasing its volume (which decreases density) and downward by contracting its volume (which increases density). A crocodile increases its density when it swallows stones. From 4 to 5 kg of stones have been found lodged in the front part of the stomach in large crocodiles. With its increased density, the crocodile swims lower in the water and exposes less of itself to its prey.

Figure 19.12 ▲
The wood floats because it is less dense than water. The rock sinks because it is more dense than water. The fish neither rises nor sinks because it has the same density as water.

◀ **Figure 19.13**
(Left) A crocodile coming toward you in the water. (Right) A crocodile with a belly full of stones coming toward you in the water.

■ Question

We know that if a fish makes itself more dense, it will sink; if it makes itself less dense, it will rise. In terms of buoyant force, why is this so?

■ Answer

When the fish increases its density by decreasing its volume, it displaces less water, so the buoyant force decreases. When the fish decreases its density by expanding, it displaces more water, and the buoyant force increases.

---

* Interestingly enough, the people who can't float are, 9 times out of 10, males. Most males are more muscular and slightly denser than females.

19.5 Flotation

Primitive peoples made their boats of wood. Could they have conceived of an iron ship? We don't know. The idea of floating iron might have seemed strange. Today it is easy for us to understand how a ship made of iron can float.

Consider a solid 1-ton block of iron. Iron is nearly eight times as dense as water, so when it is submerged, it will displace only 1/8 ton of water. The buoyant force will be far from enough to keep it from sinking. Suppose we reshape the same iron block into a bowl shape, as shown in Figure 19.14. The iron bowl still weighs 1 ton. If you lower the bowl into a body of water, it displaces a greater volume of water than before. The deeper the bowl is immersed, the more water is displaced and the greater is the buoyant force exerted on the bowl. When the weight of the displaced water equals the weight of the bowl, it will sink no farther. It will float because the buoyant force now equals the weight of the bowl.

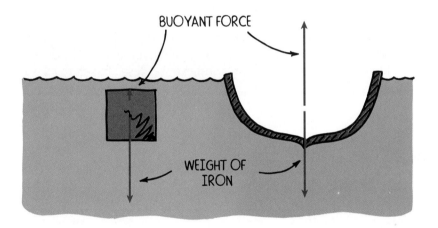

Figure 19.14 ▲
A solid iron block sinks, while the same block shaped to occupy at least eight times as much volume floats.

Figure 19.15 ▲
The weight of a floating object equals the weight of the water displaced by the submerged part.

This is an example of the **principle of flotation,** which states:*

> A floating object displaces a weight of fluid equal to its own weight.

Every ship must be designed to displace a weight of water equal to its own weight. Thus, a 10 000-ton ship must be built wide enough to displace 10 000 tons of water before it sinks too deep below the surface.

Think about a submarine beneath the surface. If it displaces a weight of water greater than its own weight, it will rise. If it displaces less, it will go down. If it displaces exactly its weight, it will remain at constant depth. Water has slightly different densities at different temperatures, so a submarine must make periodic adjustments as it moves through the ocean. As the next chapter shows, a hot-air balloon obeys the same rules.

Figure 19.16 ▲
A floating object displaces a weight of liquid equal to its own weight.

Figure 19.17 ▲
The same ship empty and loaded. How does the weight of its load compare with the weight of extra water displaced?

■ Questions

Complete the following statements by choosing the correct word for each.

1. The volume of a submerged object is equal to the __?__ of liquid displaced.

2. The weight of a floating object is equal to the __?__ of liquid displaced.

■ Answers

1. volume

2. weight

---

*   Note that this is a general statement for all fluids rather than just liquids. As the next chapter shows, the same principle applies to gases as well.

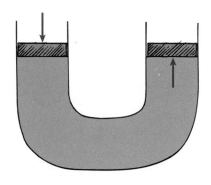

Figure 19.18 ▲
The force exerted on the left piston increases the pressure in the liquid and is transmitted to the right piston.

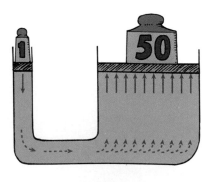

Figure 19.19 ▲
A 1-N load on the left piston will support 50 N on the right piston.

19.6 Pascal's Principle

Push a stick against a wall and we can exert pressure at a distance. Interestingly enough, we can do the same with a fluid. Whenever we change the pressure in one part of a fluid, this change is transmitted to other parts. For example, if the pressure of city water is increased at the pumping station by 10 units of pressure, the pressure everywhere in the pipes of the connected system will be increased by 10 units of pressure (when water is not moving). This rule is called **Pascal's principle:**

> Changes in pressure at any point in an enclosed fluid at rest are transmitted undiminished to all points in the fluid and act in all directions.

Pascal's principle was discovered in the seventeenth century by Blaise Pascal, for whom the SI unit of pressure is named. Pascal's principle is employed in a hydraulic press. If you fill a U-shaped tube with water and place pistons at each end, as shown in Figure 19.18, pressure exerted against the left piston will be transmitted throughout the liquid and against the bottom of the right piston. (The pistons are simply "plugs" that can freely slide snugly inside the tube.) The pressure the left piston exerts against the water will be exactly equal to the pressure the water exerts against the right piston if the levels are the same.

This is nothing to get excited about. But suppose you make the tube on the right side wider and use a piston of larger area; then the result is impressive. In Figure 19.19 the piston on the left has an area of 1 square centimeter, and the piston on the right has an area fifty times as great, 50 square centimeters. Suppose there is a 1-newton load on the left piston. Then an additional pressure of 1 newton per square centimeter (1 N/cm^2) is transmitted throughout the liquid and up against the larger piston. Here is where the difference between force and pressure comes in. The additional pressure of 1 N/cm^2 is exerted against *every* square centimeter of the larger piston. Since there are 50 square centimeters, the total extra force exerted on the larger piston is 50 newtons. Thus, the larger piston will support a 50-newton load. This is fifty times the load on the smaller piston!

This is quite remarkable, for we can multiply forces with such a device—1 newton input, 50 newtons output. By further increasing the area of the larger piston (or reducing the area of the smaller piston), we can multiply forces to any amount. Pascal's principle underlies the operation of the hydraulic press.

The hydraulic press does not violate energy conservation, for the increase in force is compensated for by a decrease in distance moved. When the small piston in the last example is moved downward 10 cm, the large piston will be raised only one-fiftieth of this, or 0.2 cm. Very much like a mechanical lever, the input force multiplied by the distance it moves is equal to the output force multiplied by the distance it moves. The hydraulic press is a "machine," much like those discussed in Section 8.7.

Pascal's principle applies to all fluids (gases and liquids). A typical application of Pascal's principle for gases and liquids is the automobile lift seen in many service stations (Figure 19.20). Compressed air exerts pressure on the oil in an underground reservoir. The oil in turn transmits the pressure to a cylinder, which lifts the automobile. The relatively low pressure that exerts the lifting force against the piston is about the same as the air pressure in the tires of the automobile, because a low pressure exerted over a relatively large area produces a considerable force.

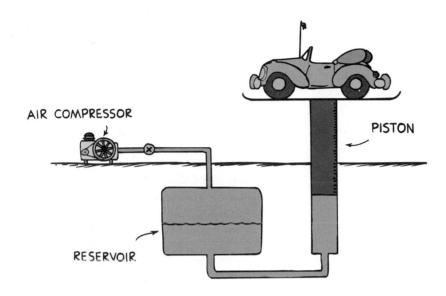

Figure 19.20 ▲
Pascal's principle in a service station.

■ Question

As the automobile in Figure 19.20 is being lifted, how does the change in oil level in the reservoir compare with the distance the automobile moves?

■ Answer

The car moves up a greater distance than the oil level drops, since the area of the piston is smaller than the surface area of the oil in the reservoir.

19 Chapter Review

Concept Summary

Pressure is the force per area on which the force acts.

- Liquids exert pressure equally in all directions at all points within the liquid.

- The pressure created by a liquid at any point is proportional to the density of the liquid times the depth of that point below the liquid surface.

- The total pressure in a liquid is the pressure created by the liquid plus the pressure of the air above it.

Buoyancy is the apparent loss of weight of an object immersed in a fluid.

- An immersed object displaces the fluid in which it is immersed.

- A completely submerged object always displaces a volume of fluid equal to its own volume.

- According to Archimedes' principle, an immersed object is buoyed up by a force equal to the weight of the fluid it displaces.

- When an object is denser than the fluid in which it is immersed, it sinks.

- When an object is less dense than the fluid in which it is immersed, it floats.

- When an object is as dense as the fluid in which it is immersed, it can remain suspended in the fluid, neither sinking nor floating.

- A floating object displaces a weight of fluid equal to its own weight.

According to Pascal's principle, changes in pressure at any point in an enclosed fluid at rest are transmitted undiminished to all points in the fluid and act in all directions.

- The hydraulic press, based on Pascal's principle, multiplies forces.

Important Terms

Archimedes' principle (19.3)
buoyancy (19.2)
buoyant force (19.2)
displacement (19.2)
Pascal's principle (19.6)
principle of flotation (19.5)

Review Questions

1. Distinguish between *pressure* and *force*. (19.1)

2. What is the relationship between liquid pressure and depth of a liquid? Between liquid pressure and density? (19.1)

3. **a.** By how much does the water pressure on a submarine change when the submarine dives to double its previous depth (neglect the very real effect of atmospheric pressure above)?

   b. If the submarine operated in freshwater, would the pressure it feels be greater or less than at the same depth in salt water? (19.1)

4. How does water pressure 1 meter below the surface of a small pond compare with water pressure 1 meter below the surface of a huge lake? (19.1)

5. If you immerse a tin can with a small hole in it in water so that water spurts through the hole, what will be the direction of water flow where the hole is? (19.1)

6. Why does the buoyant force act upward for an object submerged in water? (19.2)

7. How does the buoyant force that acts on a fish compare with the weight of the fish? (19.2)

8. Why does the buoyant force on submerged objects not act sideways? (19.2)

9. How does the volume of a completely submerged object compare with the volume of water displaced? (19.2)

10. When an object is said to be immersed in water, does this mean it is completely submerged? Does it mean it is partially submerged? Does the word *immersed* apply to either case? (19.3)

11. What is the mass of 1 liter of water in kilograms? What is its weight in newtons? (19.3)

12. **a.** Does the buoyant force on a submerged object depend on the weight of the object itself or on the weight of the fluid displaced by the object?

   b. Does it depend on the weight of the object itself or on its volume? Defend your answer. (19.3)

13. When the buoyant force on a submerged object is equal to the weight of the object, how do the densities of the object and water compare? (19.4)

14. When the buoyant force on a submerged object is more than the weight of the object, how do the densities of the object and water compare? (19.4)

15. When the buoyant force on a submerged object is less than the weight of the object, how do the densities of the object and water compare? (19.4)

16. **a.** How is the density of a submarine controlled?

   b. How is the density of a fish controlled? (19.4)

17. Does the buoyant force on a floating object depend on the weight of the object itself or on the weight of the fluid displaced by the object? Or are these both the same for the special case of floating? (19.5)

18. What is the buoyant force that acts on a 100-ton ship? (To make things simple, give your answer in tons.) (19.5)

19. According to Pascal's principle, what happens to the pressure in all parts of a confined fluid when you produce an increase in pressure in one part? (19.6)

20. When the pressure in a hydraulic press is increased by an additional 10 N/cm^2, how much extra load will the output piston support when its cross-sectional area is 50 square centimeters? (19.6)

Activities

1. Try to float an egg in water. Then dissolve salt in the water until the egg floats. How does the density of an egg compare with that of tap water? Salt water?

2. Make a Cartesian diver like the one shown. Completely fill a large, pliable plastic bottle with water. Partially fill a small pill bottle so that it just barely floats when capped, turned upside down, and placed in the large bottle. (You may have to experiment to get it just right.) Once the pill bottle is barely floating, secure the lid or cap on the large bottle so that it is airtight. When you press the sides of the large bottle, the pill bottle sinks; when you release it, the bottle returns to the top. Experiment by squeezing the bottle different ways to get different results. Can you explain the behavior you see?

3. Punch a couple of holes in the bottom of a water-filled container, and water will spurt out because of water pressure (left of figure). Now drop the container and note that as it falls freely the water no longer spurts out (right of figure)! If your friends don't understand this, can you figure it out and then explain it to them? (What happens to *g*, and hence weight, and hence weight density, and hence pressure in the reference frame of the falling container?)

Plug and Chug

1. Calculate the water pressure at the base of Hoover Dam. The depth of water behind the dam is 220 m. (Neglect the pressure due to the atmosphere.)

2. The top floor of a building is 30 m above the basement. Calculate how much greater the water pressure is in the basement compared with the pressure at the top floor.

3. An 8.6-kg piece of metal displaces 1 liter of water when submerged. Calculate its density.

4. A 4.7-kg piece of metal displaces 0.6 liter of water when submerged. Calculate its density.

Think and Explain

1. Next time you're near a farm, notice that a silo has metal bands around it to give it strength. Notice also that these bands are closer together near ground level and are spaced farther apart near the top of the silo. Why is this so?

2. There is a legend of a Dutch boy who bravely held back the whole Atlantic Ocean by plugging a hole in a dike with his finger. Is this possible and reasonable? Estimate the force if the hole were about 1 square centimeter in area and 1 meter below the water level.

3. Why is it inaccurate to say that heavy objects sink and light objects float?

4. A 1-kg block of iron and a 1-kg block of aluminum are submerged in water. Upon which does the greater buoyant force act? Why?

5. Compared with an empty ship, would a ship loaded with a cargo of foam insulation sink deeper into water or rise in water? Explain.

6. The density of a rock doesn't change when it is submerged in water, but *your* density does change when you are submerged. Why?

7. A fire truck carrying a load of firefighters and a large tank of water is about to cross a bridge that may not support the load. The chief suggests that some of the people aboard get into the tank and thereby lighten the load. Is this a good idea, or a poor idea? Explain.

8. A balloon is weighted so it is barely able to float in water. If it is pushed beneath the surface, will it come back to its starting point, stay at the depth to which it is pushed, or sink? Explain. (*Hint:* What change in density, if any, does the balloon undergo?)

9. If you were to float in the Dead Sea, you would float appreciably higher than in freshwater because of the water's greater density. Would the buoyant force on you be greater also? Why or why not?

10. a. A half-filled bucket of water is on a spring scale. If a live fish is placed in it, will the reading on the scale increase?

    b. Would your answer be different if the bucket were initially filled to the brim?

Think and Solve

1. Which do you suppose produces more pressure on the ground, an elephant or a lady standing on high heels? Make reasonable assumptions and approximate a rough calculation for each.

2. A 1-kg rock suspended above water weighs 9.8 N. When suspended beneath the surface of the water, its apparent weight is 7.8 N.

   a. What is the buoyant force on the rock?

   b. If the container of water on a bathroom-type scale weighs 9.8 N, what is the scale reading when the rock is suspended beneath the surface of the water?

   c. What is the scale reading when the rock is released and rests at the bottom of the container?

3. Calculate the approximate volume of a person of mass 100 kg who can just barely float in freshwater.

4. A gravel barge, rectangular in shape, is 4 m wide and 10 m long. When loaded, it sinks 2 m in the water. What is the weight of gravel in the barge?

5. Oak is 0.8 as dense as water and therefore floats in water.

   a. What weight of water will be displaced by a 50-kg floating oak beam?

   b. What additional force would be required to poke the oak beneath the surface so it is completely submerged?

6. The cross-sectional area of the output piston in a hydraulic device is ten times the input piston's area.

   a. By how much will the device multiply the input force?

   b. How far will the output piston move compared with the distance the input piston is moved? (Does this satisfy $F_1 d_1 = F_2 d_2$, the equation for conservation of energy?)

20 Gases

H$_2$O in its gaseous form is invisible; when it condenses to liquid it can be seen.

Gases are similar to liquids in that they flow; hence both are called *fluids.* The primary difference between gases and liquids is the distance between molecules. In a liquid, the molecules are close together, where they continually experience forces from the surrounding molecules. These forces strongly affect the motion of the molecules. In a gas, the molecules are far apart, allowing them to move freely between collisions. When two molecules in a gas collide, if one gains speed in the collision, the other loses speed, such that their total kinetic energy is unchanged.

A gas expands to fill all space available to it and takes the shape of its container. Only when the quantity of gas is very large, such as in the earth's atmosphere or in a star, does gravitation determine the shape of the gas.

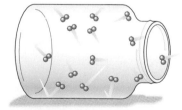

Figure 20.1 ▲
Molecules in the gaseous state are far apart, are in continuous motion, bounce off one another without net change in energy, and fill up and take the shape of their container.

20.1 The Atmosphere

We don't have to look far to find a sample of gas. We live in an ocean of gas, our atmosphere. Molecules in the air occupy space and extend many kilometers above the earth's surface. The molecules are energized by sunlight and kept in continual motion. Without the earth's gravity, they would fly off into outer space. And without the sun's energy, the molecules would just end up as matter on the ground. Fortunately, there is an energizing sun and there is gravity, so we have an atmosphere.

Unlike the ocean, which has a very definite surface, the earth's atmosphere has no definite surface. Unlike the uniform density of a liquid at any depth, the density of the atmosphere decreases with altitude. Air is more compressed at sea level than at higher altitudes.

The atmosphere is like a huge pile of feathers, where those at the bottom are more squashed than those nearer the top. The air gets thinner and thinner (less dense) the higher one goes; it eventually thins out into space.

Even in the vacuous regions of interplanetary space there is a gas density of about one molecule per cubic centimeter. This is primarily hydrogen, the most plentiful element in the universe.

Figure 20.2 shows how thin our atmosphere is. Note that 50% of the atmosphere is below 5.6 kilometers (18 000 ft). Not shown, 75% of the atmosphere is below 11 kilometers (56 000 ft). Note that 90% is below 17.7 kilometers, and 99% of the atmosphere is below an altitude of about 30 kilometers. Compared with the earth's radius, 30 kilometers is very small. To give you an idea of how small, the "thickness" of the atmosphere relative to the size of the world is like the thickness of condensed breath on a cold billiard ball. Our atmosphere is a delicate and finite life-sustaining thin shell of air; that's why we should care for it.

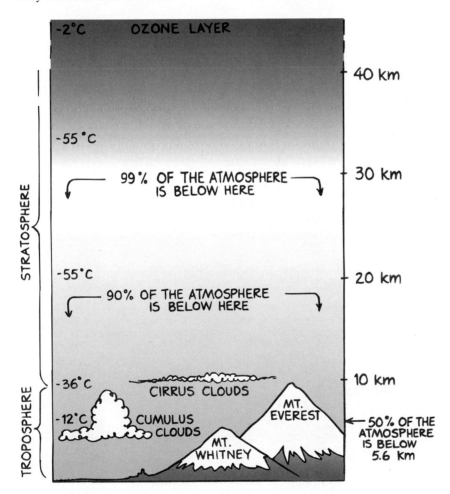

Figure 20.2 ▲
The atmosphere. Note how the temperature of the atmosphere drops as one goes higher (until it rises again at very high altitude).

Figure 20.3 ▲
You don't notice the weight of a bag of water while you're submerged in water. Similarly, you don't notice the weight of air.

20.2 Atmospheric Pressure

We live at the bottom of our ocean of air. The atmosphere, much like water in a lake, exerts a pressure. And just as water pressure is caused by the weight of water, atmospheric pressure is caused by the weight of air. We are so accustomed to the invisible air that we sometimes forget it has weight. Perhaps a fish "forgets" about the weight of water in the same way.

Table 20.1 Densities of Various Gases		
Gas		**Density (kg/m³)***
Dry air	0°C	1.29
	10°C	1.25
	20°C	1.21
	30°C	1.16
Helium		0.178
Hydrogen		0.090
Oxygen		1.43

*At sea-level atmospheric pressure and at 0°C (unless otherwise specified).

At sea level, 1 cubic meter of air at 20°C has a mass of about 1.2 kg. Calculate the number of cubic meters in your room, multiply by 1.2 kg/m³, and you'll have the mass of air in your room. Don't be surprised if it has more mass than your kid sister. Air is heavy if you have enough of it. If your kid sister doesn't believe air has weight,

Figure 20.4 ▲
It takes more than 1000 kg of additional air to fully pressurize a 747 jumbo jet.

maybe its because she's surrounded by air all the time. Hand her a plastic bag of water and she'll tell you it has weight. But hand her the same bag of water while she's submerged in a swimming pool, and she won't feel its weight because the bag is surrounded by water.

Consider a superlong bamboo pole that reaches up through the atmosphere for 30 kilometers. Suppose the inside cross-sectional area of the hollow pole is 1 square centimeter. If the density of air inside the pole matches the density of air outside, the enclosed mass of air would be about 1 kilogram. The weight of this much air is about 10 newtons. So air pressure at the bottom of the bamboo pole would be about 10 newtons per square centimeter (10 N/cm²). Of course, the same is true without the bamboo pole.

There are 10 000 square centimeters in 1 square meter, so a column of air 1 square meter in cross section that extends up through the atmosphere has a mass of about 10 000 kilograms. The weight of this air is about 100 000 newtons (10^5 N). This weight produces a pressure of 100 000 newtons per square meter, or equivalently, 100 000 pascals, or 100 kilopascals. More exactly, the average atmospheric pressure at sea level is 101.3 kilopascals (101.3 kPa).*

The pressure of the atmosphere is not uniform. Aside from variations with altitude, there are variations in atmospheric pressure at any one locality due to moving air currents and storms. Measurement of changing air pressure is important to meteorologists in predicting weather.

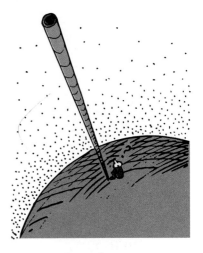

Figure 20.5 ▲
The mass of air that would occupy a bamboo pole that extends to the "top" of the atmosphere is about 1 kg. This air has a weight of 10 N.

■ Questions

1. About how many kilograms of air occupy a classroom that has a 200-square-meter floor area and a 4-meter-high ceiling?

2. Why doesn't the pressure of the atmosphere break windows?

■ Answers

1. 960 kg. The volume of air is (200 m²) × (4 m) = 800 m³. Each cubic meter of air has a mass of about 1.2 kg, so (800 m³) × (1.2 kg/m³) = 960 kg.

2. The atmospheric pressure doesn't normally break windows because it acts on *both* sides of a window. So no net force is exerted by the atmosphere on the windows.

* The average pressure at sea level used to be called one *atmosphere.* This term is still commonly used, but it is no longer acceptable with SI units. In British units, the average atmospheric pressure at sea level is 14.7 pounds/inch².

Figure 20.6 ▲
The weight of air that bears down on a 1-square-meter surface at sea level is about 100 000 newtons. So atmospheric pressure is about 100 000 newtons per square meter (10^5 N/m²), or about 100 kPa.

20.3 The Simple Barometer

An instrument used for measuring the pressure of the atmosphere is called a **barometer.** A simple mercury barometer is illustrated in Figure 20.7. A glass tube, longer than 76 cm and closed at one end, is filled with mercury and tipped upside down in a dish of mercury. The mercury in the tube runs out of the submerged open bottom until the level falls to about 76 cm. The empty space trapped above, except for some mercury vapor, is a vacuum. The vertical height of the mercury column remains constant even when the tube is tilted, unless the top of the tube is less than 76 cm above the level in the dish, in which case the mercury completely fills the tube.

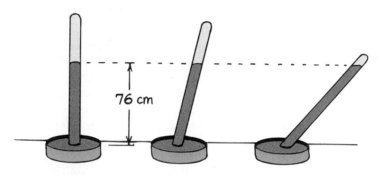

Figure 20.7 ▲
A simple mercury barometer. Variations above and below the average column height of 76 cm are caused by variations in atmospheric pressure.

Why does mercury behave this way? The explanation is similar to the reason a simple see-saw balances when the weights of people at its two ends are equal. The barometer "balances" when the weight of liquid in the tube exerts the same pressure as the atmosphere outside. Whatever the width of the tube, a 76-cm column of mercury weighs the same as the air that would fill a supertall 30-km tube of the same width. If the atmospheric pressure increases, then it will push the mercury column higher than 76 cm. The mercury is literally pushed up into the tube of a barometer by atmospheric pressure.

Could water be used to make a barometer? The answer is yes, but the glass tube would have to be much longer—13.6 times as long, to be exact. You may recognize this number as the density of mercury relative to that of water. A volume of water 13.6 times that of mercury is needed to provide the same weight as the mercury in the tube (or in the imaginary tube of air outside). So the height of the tube would have to be at least 13.6 times taller than the mercury column. A water barometer would have to be $13.6 \times (0.76 \text{ m})$, or 10.3 m high—too tall to be practical.

The operation of a barometer is similar to the process of drinking through a straw. By sucking, you reduce the air pressure in the straw that is placed in a drink. Atmospheric pressure on the liquid's surface pushes liquid up into the reduced-pressure region. Strictly speaking,

Liquid Transfer Technique

Lower a narrow glass tube or drinking straw in water, and place your finger over the top of the tube. Lift the tube from the water. Why does the water not fall out? Then lift your finger from the top of the tube. What happens? (You'll do this often if you enroll in a chemistry lab.)

Activity

the liquid is not *sucked* up; it is *pushed* up, by the pressure of the atmosphere. If the atmosphere is prevented from pushing on the surface of the drink, as in the party trick bottle with the straw through the airtight cork stopper, one can suck and suck and get no drink.

Figure 20.8 ▲
You cannot drink soda through the straw unless the atmosphere exerts a pressure on the surrounding liquid.

If you understand these ideas, you can understand why there is a 10.3-meter limit on the height water can be lifted with vacuum pumps. The old fashioned farm-type pump (Figure 20.9) operates by producing a partial vacuum in a pipe that extends down into the water below. The atmospheric pressure exerted on the surface of the water simply pushes the water up into the region of reduced pressure inside the pipe. Can you see that even with a perfect vacuum, the maximum height to which water can be lifted is 10.3 meters?

◀ **Figure 20.9**
The atmosphere pushes water from below up into a pipe that is evacuated of air by the pumping action.

20.4 The Aneroid Barometer

A popular classroom demonstration to illustrate atmospheric pressure is crushing a can with atmospheric pressure. A can containing a little water in it is heated until steam forms. Then the can is capped securely and removed from the source of heat. There is now less air inside the can than before it was heated. (Why? Because when the water boils and changes to steam, the steam pushes air out of the can.) When the sealed can cools, the pressure inside is reduced because steam inside the can liquefies when it cools. The greater pressure of the atmosphere outside the can then proceeds to crush the can (Figure 20.10). The pressure of the atmosphere is even more dramatically shown when a 50-gallon drum is crushed by the same procedure.

Figure 20.10 ▲
When the air pressure inside is reduced, the greater atmospheric pressure outside crushes the can.

A much more subtle application of atmospheric crushing is used in an **aneroid** ("without liquid") **barometer.** This small portable instrument (Figure 20.11) is more prevalent than the mercury barometer. It uses a small metal box that is partially exhausted of air and has a slightly flexible lid that bends in or out with atmospheric-pressure changes. The pressure difference between the inside and outside is less drastic than that of the crushed can of Figure 20.10. Motion of the lid is indicated on a scale by a mechanical spring-and-lever system. Since atmospheric pressure decreases with increasing altitude, a barometer can be used to determine the elevation. An aneroid barometer calibrated for altitude is called an *altimeter* ("altitude meter"). Some of these instruments are sensitive enough to indicate changes in elevation of less than a meter.

Figure 20.11 ▲
An aneroid barometer.

20.5 Boyle's Law

The air pressure inside the inflated tires of an automobile is considerably more than the atmospheric pressure outside. The density of air inside is also more than that of the air outside. To understand the relationship between pressure and density, think of the molecules of air* inside the tire.

Inside the tire, the molecules behave like tiny Ping-Pong® balls, perpetually moving helter-skelter and banging against the inner walls. Their impacts on the inner surface of the tire produce a jittery force that appears to our coarse senses as a steady push. This pushing force averaged over a unit of area provides the pressure of the enclosed air.

Suppose there are twice as many molecules in the same volume (Figure 20.12)—then the air density is doubled. If the molecules move at the same average speed—or, equivalently, if they have the same temperature—then to a close approximation, the number of collisions will double. This means the pressure is doubled. So pressure is proportional to density.

Figure 20.12 ▲
When the density of the air in the tire is increased, the pressure is increased.

The density of the air can also be doubled by simply compressing the air to half its volume. We increase the density of air in a balloon when we squeeze it, and likewise increase air density in the

Figure 20.13 ▲
When the volume of gas is decreased, the density—and therefore pressure—are increased.

---

* Air is composed of a mixture of gases—mainly nitrogen and oxygen, with some carbon dioxide. When we speak of *molecules of air,* we are referring to any of the different kinds of molecules found in air.

cylinder of a tire pump when we push the piston downward. Consider the cylinder with the movable piston in Figure 20.13. If the piston is pushed downward so that the volume is half the original volume, the density of molecules will be doubled, and the pressure will correspondingly be doubled. Decrease the volume to a third its original value, and the pressure will be increased by three, and so on.

Notice from these examples that the product of pressure and volume is the same. For example, a doubled pressure multiplied by a halved volume gives the same value as a tripled pressure multiplied by a one-third volume. In general, we can say that the product of pressure and volume for a given mass of gas is a constant as long as the temperature does not change. "Pressure × volume" for a quantity of gas at one time is equal to any "different pressure × different volume" at any other time. In equation form,

$$P_1 V_1 = P_2 V_2$$

where P_1 and V_1 represent the original pressure and volume, respectively, and P_2 and V_2 the second, or final, pressure and volume. This relationship is called **Boyle's law,** after Robert Boyle, the seventeenth-century physicist who is credited with its discovery.*

■ Questions

1. If you squeeze a balloon to one-third its volume, by how much will the pressure inside increase?

2. A piston in an airtight pump is withdrawn so that the volume of the air chamber is increased five times. What is the change in pressure?

3. A scuba diver 10.3 m deep breathes compressed air. If she holds her breath while returning to the surface, by how much does the volume of her lungs tend to increase?

■ Answers

1. The pressure in the balloon is increased three times. No wonder balloons break when you squeeze them!

2. The pressure in the piston chamber decreases to 1/5. This is the principle behind a mechanical vacuum pump.

3. Atmospheric pressure can support a column of water 10.3 m high, so the pressure in water due to the weight of the water alone equals atmospheric pressure at a depth of 10.3 m. Taking the pressure of the atmosphere at the water's surface into account, the total pressure at this depth is twice atmospheric pressure. Unfortunately for the scuba diver, her lungs will tend to inflate to twice their normal size if she holds her breath while rising to the surface. A first lesson in scuba diving is *not* to hold your breath when ascending. To do so can be fatal.

---

* A general law that takes temperature changes into account is $P_1 V_1/T_1 = P_2 V_2/T_2$, where T_1 and T_2 represent the initial and final *absolute temperatures,* measured in SI units called *kelvins* (Chapter 21).

20.6 Buoyancy of Air

In the last chapter you learned about buoyancy in liquids. All the rules for buoyancy were stated in terms of *fluids* rather than liquids. The reason is simple enough: the rules hold for gases as well as liquids. The physical laws that explain a dirigible aloft in the air are the same that explain a fish "aloft" in water. We can state Archimedes' principle for air:

> An object surrounded by air is buoyed up by a force equal to the weight of the air displaced.

Recall that a cubic meter of air at ordinary atmospheric pressure and room temperature has a mass of about 1.2 kg, so its weight is about 12 N. Therefore any 1-cubic-meter object in air is buoyed up with a force of 12 N. If the mass of the 1-cubic-meter object is greater than 1.2 kg (so that its weight is greater than 12 N), it will fall to the ground when released. If an object this size has a mass less than 1.2 kg, it will rise in the air. Any object that has a mass less than the mass of an equal volume of surrounding air will rise. Another way to say this is, any object less dense than the air around it will rise. Gas-filled balloons that rise in the air are less dense than the surrounding air.

When you next see a large dirigible airship aloft in the air, think of it as a giant fish. Both remain aloft as they swim through their fluids for the same reason; they both displace their own weights of fluid. When in motion, the dirigible may be raised or lowered by means of horizontal rudders or "elevators."

Figure 20.14 ▲
The dirigible and the fish both hover at a given level for the same reason.

Figure 20.15 ▲
Everything is buoyed up by a force equal to the weight of the air it displaces. Why, then, isn't everything held aloft like this balloon?

■ Questions

1. Is there a buoyant force acting on you? If there is, why are you not buoyed up by this force?

2. Two balloons are inflated to the same size, one with air and the other with helium. Which balloon experiences the greater buoyant force? Why does the air-filled balloon sink and the helium-filled balloon float?

■ Answers

1. There *is* a buoyant force acting on you, and you *are* buoyed upward by it. You don't notice it only because your weight is so much greater.

2. Both balloons are buoyed upward with the same buoyant force because they displace the same weight of air. The reason the air-filled balloon sinks in air is because it is heavier than the buoyant force that acts on it. The helium-filled balloon is lighter than the buoyant force that acts on it. Or put another way, the air-filled balloon is slightly more dense than the surrounding air (principally because it is filled with *compressed* air). Helium, even somewhat compressed, is appreciably less dense than air.

20.7 Bernoulli's Principle

The discussion of fluid pressure thus far has been confined to stationary fluids. Motion produces an additional influence.

Most people think that atmospheric pressure increases in a gale, tornado, or hurricane. Actually, the opposite is true. High-speed winds may blow the roof off your house, but the pressure within the winds is actually less than for still air of the same density. As strange as it may first seem, when the speed of a fluid increases, its pressure decreases. This is true for all fluids—liquids and gases alike.

Consider a continuous flow of water through a pipe. Because water doesn't "bunch up," the amount of water that flows past any given section of the pipe is the same as the amount that flows past any other section of the same pipe. This is true whether the pipe widens or narrows. As a consequence of continuous flow, the water in the wide parts will slow down, and in the narrow parts, it will speed up. You can observe this when you put your finger over the outlet of a water hose.

Daniel Bernoulli, a Swiss scientist of the eighteenth century, experimented with water flowing through pipes. He found that the greater the speed of flow, the less is the force of the water at right angles (sideways) to the direction of flow. The pressure at the walls of the pipes decreases when the speed of the water increases. Bernoulli found this to be a principle of both liquids and gases. **Bernoulli's principle** in its simplest form states:

> When the speed of a fluid increases, the pressure drops.

Bernoulli's principle is a consequence of the conservation of energy. For a steady flow of fluid there are three kinds of energy: kinetic energy due to motion, potential energy due to pressure, and gravitational potential energy due to elevation. In a steady fluid flow where no energy is added or taken away, the sum of these forms of energy remains constant.* If the elevation of the flowing fluid does not change, then an increase in speed means a decrease in pressure, and vice versa.

The decrease of fluid pressure with increasing speed may at first seem surprising, particularly if we fail to distinguish between the pressure in the fluid and the pressure by the fluid on something that interferes with its flow. The pressure within the fast-moving water in a fire hose is relatively low, whereas the pressure that the water can exert on anything in its path to slow it down may be huge.

Figure 20.16 ▲
Because the flow is continuous, water speeds up when it flows through the narrow and/or shallow part of the brook.

---

* In mathematical form: $\frac{1}{2}mv^2 + pV + mgy$ = constant, where m is the mass of some small volume V, v its speed, p its pressure, g the acceleration due to gravity, and y its elevation. If mass m is expressed in terms of density ρ, where $\rho = m/V$, and each term is divided by V, Bernoulli's equation takes the form $\frac{1}{2}\rho v^2 + p + \rho gy$ = constant. Then all three terms have units of pressure. If y does not change, then an increase in v means a decrease in p, and vice versa.

In steady flow, one small bit of fluid follows along the same path as a bit of fluid in front of it. The motion of a fluid in steady flow follows **streamlines,** which are represented by dashed lines in Figure 20.17 and later figures. Streamlines are the smooth paths, or trajectories, of the bits of fluid. The lines are closer together in the narrower regions, where the flow is faster and pressure is less.

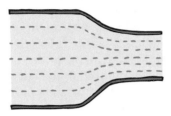

◀ **Figure 20.17**
A fluid speeds up when it flows into the narrow region. The constricted streamlines indicate increased speed and decreased internal pressure.

Bernoulli's principle holds only for steady flow. If the flow speed is too great, the flow may become turbulent and follow a changing, curling path known as an **eddy.** In that case, Bernoulli's principle does not hold.

20.8 Applications of Bernoulli's Principle

Bernoulli's principle accounts for the flight of birds and aircraft. The shape and orientation of the wings ensure that air passes somewhat faster over the top surface of the wing than beneath the lower surface. Pressure above the wing is less than pressure below the wing. The difference between these pressures produces a net upward force, appropriately called **lift.*** Even a small pressure difference multiplied by a large wing area can produce a considerable force. When lift equals weight, horizontal flight is possible. The lift is

Figure 20.18 ▲
The paper rises.

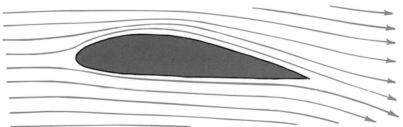

Figure 20.19 ▲
Air pressure is less above the wing than below it.

---

*   Pressure differences are only one way to understand wing lift. Another way uses Newton's third law. The wing forces air downward (action) and the air forces the wing upward (reaction). Air is deflected downward by the wing tilt, called the *angle of attack.* When riding in a car, place your hand out the window and pretend it's a wing. Tip it up slightly so air is forced downward. Up goes your hand! Air lift provides a nice example to remind us that often there is more than one way to understand the way nature behaves.

Figure 20.20 ▲
In high winds, air pressure above a roof can drastically decrease.

greater for higher speeds and larger wing areas. Hence, low-speed gliders have very large wings relative to the size of the fuselage. The wings of faster-moving aircraft are relatively small.

We began our discussion of Bernoulli's principle by stating that atmospheric pressure decreases in a strong wind. As Figure 20.20 shows, air pressure above a roof is less than air pressure inside the building when a wind is blowing. This produces a lift that may result in the roof being blown off. Roofs are usually constructed to withstand increased downward loads, the weight of snow for example, but not for increased upward forces. Unless the building is well vented, the stagnant air inside can push the roof off.

Bernoulli's principle is involved in the curved path of spinning balls. When a moving baseball, tennis ball, or any kind of ball spins, unequal air pressures are produced on opposite sides of the ball. Note in Figure 20.21 (right) that the streamlines are closer at B than at A for the direction of spin shown. Air pressure is greater at A, and the ball curves as indicated. Curving may be increased by threads or fuzz, which help to drag a thin layer of air with the ball and produce further crowding of streamlines on one side.

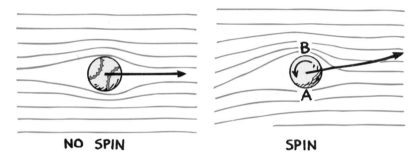

NO SPIN   **SPIN**

Figure 20.21 ▲
(Left) The streamlines are the same on either side of a nonspinning ball. (Right) A spinning ball produces a crowding of streamlines. It is pushed to one side, causing it to curve.

Figure 20.22 ▲
Pressure is greater in the stationary fluid (air) than in the moving fluid (water). The atmosphere pushes the ball into the region of reduced pressure.

You can demonstrate Bernoulli's principle quite interestingly in your kitchen sink (Figure 20.22). Tape a Ping-Pong ball to a string and allow the ball to swing into a stream of running water. You'll see that it will remain in the stream even when tugged slightly to the side, as shown. Pressure of stationary air on the ball is greater than pressure of moving water. The ball is pushed into the region of reduced pressure by the atmosphere.

A similar thing happens to a bathroom shower curtain when the shower water is turned on full blast. Air near the water stream flows into the lower-pressure stream and is swept downward with the falling water. Air pressure inside the curtain is thus reduced, and the atmospheric pressure outside pushes the curtain inward (providing an escape route for the downward-swept air). This effect is small compared with the convection produced by temperature differences, but nevertheless, the next time you're taking a shower and the curtain swings in against your legs, think of Daniel Bernoulli!

20 Chapter Review

Concept Summary

The earth's atmosphere is an ocean of air extending about 30 km above the earth's surface, with dilute air extending even higher.

- Air is more compressed at sea level than at higher altitudes.

- Air exerts pressure on everything; the pressure at sea level is about 100 kPa.

- Simple barometers measure atmospheric pressure in terms of how high a column of mercury in a closed tube can be supported by atmospheric pressure.

- Aneroid barometers work without liquids and measure the position of a movable lid against a box with low pressure inside.

Boyle's law states that at constant temperature, the pressure times the volume of an enclosed gas is constant; if one increases, the other decreases.

Bernoulli's principle states that the pressure of a fluid flowing horizontally decreases as the speed of the fluid increases.

- Bernoulli's principle holds only for steady flow, in which the flow follows streamlines.

- Bernoulli's principle explains lift forces.

Important Terms

aneroid barometer (20.4)
barometer (20.3)
Bernoulli's principle (20.7)
Boyle's law (20.5)
eddy (20.7)
lift (20.8)
streamline (20.7)

Review Questions

1. **a.** What is the energy source for the motion of gases in the atmosphere?

   b. What prevents atmospheric gases from flying off into space? (20.1)

2. How does the density of gases at different elevations in the atmosphere differ from the density of liquids at different depths? (20.1)

3. What causes atmospheric pressure? (20.2)

4. What is the mass of a cubic meter of air at 20°C at sea level? (20.2)

5. **a.** What is the mass of a column of air that has a cross-sectional area of 1 square centimeter and that extends from sea level to the top of the atmosphere?

   b. What is the weight of this air column?

   c. What is the pressure at the bottom of this column? (20.2)

6. Is the value for atmospheric pressure at the surface of the earth a constant? Explain. (20.2)

7. How does the pressure at the bottom of the 76-cm column of mercury in a barometer compare with the pressure due to the weight of the atmosphere? (20.3)

8. When you drink liquid through a straw, it is more accurate to say the liquid is *pushed* up the straw rather than *sucked* up the straw. What exactly does the pushing? Explain. (20.3)

9. Why will a vacuum pump not operate for a well that is deeper than 10.3 m? (20.3)

10. The atmosphere does not ordinarily crush cans. Yet it will crush a can after it has been heated, capped, and cooled. Why? (20.4)

11. Why can an aneroid barometer be used to measure altitude? (20.4)

12. When air is compressed, what happens to its density? (20.5)

13. a. How great is the buoyant force on a balloon that weighs 1 N when it is suspended by buoyancy in air?

b. What happens if the buoyant force decreases?

c. What happens if the buoyant force increases? (20.6)

14. When the speed of a fluid flowing in a horizontal pipe increases, what happens to the internal pressure in the fluid? (20.7)

15. a. What are streamlines?

b. Is the pressure greater or less in regions where streamlines are crowded? (20.7)

16. Does Bernoulli's principle provide a complete explanation for wing lift, or is there some other significant factor? (20.8)

17. Why does a spinning ball curve in flight? (20.8)

Activities

1. Try this in the bathtub or while washing dishes. Lower a glass, mouth downward, over a small floating object as shown. What happens? How deep would the glass have to be pushed to compress the enclosed air to half its volume? (*Hint:* You can't do this in your bathtub unless it's 10.3 m deep!)

2. Place a card over the open top of a glass filled to the brim with water, and invert it as shown on the left. Why does the card stay intact? Try turning the glass sideways as shown on the right.

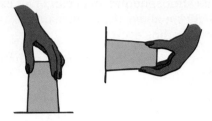

3. Fill a bottle with water and hold it partially under water so that its mouth is beneath the surface. Why does the water not run out? How tall would the bottle have to be before water ran out? (*Hint:* You can't do this indoors unless you have a ceiling 10.3 m high!)

4. Hold a spoon in a stream of water, as shown, and feel the effect of the differences in pressure.

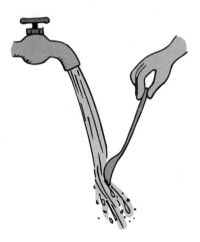

Plug and Chug

1. The "height" of the atmosphere is about 30 km. The radius of the earth is 6400 km. What percentage of the earth's radius is the height of the atmosphere?

2. The weight of the atmosphere above 1 square meter of the earth's surface is 100 000 newtons. If the density of the atmosphere were a *constant* 1.2 kg/m³, calculate where the top of the atmosphere would be.

3. Make a calculated estimate of the weight of air in your classroom.

Think and Explain

1. Which would weigh more—a bottle filled with helium gas, or the same bottle evacuated?

2. Relative to sea level, would it be slightly more difficult or somewhat easier to drink via a straw at the bottom of a deep mine? At the top of a high mountain? Explain.

3. If there were a liquid twice as dense as mercury, and if it were used to make a barometer, how tall would the column be?

4. Small bubbles of air are released by a scuba diver deep in the water. As the bubbles rise, do they become larger, smaller, or stay about the same size? Explain.

5. It is easy to breathe when snorkeling with only your face beneath the surface of the water, but quite difficult to breathe when you are submerged nearly a meter, and nearly impossible when you are more than a meter deep (even if your snorkel tube reaches to the surface). Figure out why, and explain carefully.

6. From Table 20.1, which filling would be more effective in making a balloon rise—helium or hydrogen? Why?

7. An inflated balloon sufficiently weighted with rocks will sink in water. What will happen to the size of the balloon as it sinks? Compared with its volume at the surface, what volume will it have when it is 10.3 m below the surface?

8. The buoyant force of air is considerably greater on an elephant than on a small helium-filled balloon. Why, then, does the elephant remain on the ground, while the balloon rises?

9. Estimate the buoyant force that the atmosphere exerts on you. (To do this, you can estimate your volume by knowing your weight and by assuming that your weight density is a bit less than that of water.)

10. Why is it that when cars pass each other at high speeds on the road, they tend to be "drawn" to each other?

11. Why does the fire in a fireplace burn more briskly on a windy day?

12. In a department store, an airstream from a hose connected to the exhaust of a vacuum cleaner blows upward at an angle and supports a beach ball in midair. Does the air blow under or over the ball to provide support?

13. The diameter of a fire hose varies with the flow rate of water inside. The hose may be relatively narrow, and at another time puffed up like a fat snake. In which case is water flowing fast, and when is water hardly flowing at all?

14. You overhear a conversation between two physics types. One says that birds couldn't fly before the time of Bernoulli. The other says, not so. That birds could fly before the time of Bernoulli, but couldn't fly before the time of Newton. Humor aside, what points are they making?

Think and Solve

1. Estimate the volume of a hydrogen-filled balloon that will carry a 300-kg load in air. Assume the density of hydrogen is 0.09 kg/m^3, and the density of the surrounding air 1.30 kg/m^3.

2. The density of liquid air at standard temperature and pressure is about 900 kg/m^3. What will be the volume of 1 m^3 of liquid air when it turns to its gaseous form?

3. How many newtons of lift are exerted on the wings of an airplane that have a total area of 100 m^2 when the difference in air pressure below and above the wings is 5% of atmospheric pressure?

Heat

On a cold morning, the pavement feels colder than the lawn to my bare feet. More heat energy goes from my foot to the pavement than to the lawn. At first thought, I tend to think the pavement has a lower temperature than the lawn—but does it? And on a hot afternoon the pavement feels much hotter than the lawn. Does it have a higher temperature than the lawn, or is there some other physics going on here? Perhaps differences in heat conductivities? I wonder about such things, and it's nice to find explanations in my physics book. It's even nicer to find I can understand them!

21 Temperature, Heat, and Expansion

The pie cools—the air warms.

All matter—solid, liquid, and gas—is composed of continually jiggling atoms or molecules. Because of this random motion, the atoms and molecules in matter have kinetic energy. The average kinetic energy of these individual particles causes an effect we can sense —warmth. Whenever something becomes warmer, the kinetic energy of its atoms or molecules has increased.

It's easy to increase the kinetic energy in matter. You can warm a penny by striking it with a hammer—the blow causes the molecules in the penny to jostle faster. If you put a flame to a liquid, the liquid also becomes warmer. Rapidly compress air in a tire pump and the air becomes warmer. When the atoms or molecules in matter move faster, the matter gets warmer. Its atoms or molecules have more kinetic energy. For brevity in this chapter, rather than saying *atoms and molecules,* we'll simply say molecules—by which we mean either.

So when you warm up by a fire on a cold winter night, you are increasing the molecular kinetic energy in your body.

21.1 Temperature

The quantity that tells how hot or cold something is compared with a standard is **temperature.** We express temperature by a number that corresponds to a degree mark on some chosen scale.

Nearly all matter expands when its temperature increases and contracts when its temperature decreases. A common thermometer measures temperature by showing the expansion and contraction of a liquid—usually mercury or colored alcohol—in a glass tube using a scale.

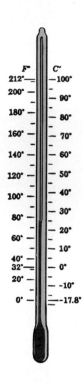

Figure 21.1 ▲
Fahrenheit and Celsius scales on a thermometer.

Figure 21.2 ▲
There is more molecular kinetic energy in the bucketful of warm water than in the small cupful of higher-temperature water.

On the most widely used temperature scale, the international scale, the number 0 is assigned to the temperature at which water freezes, and the number 100 to the temperature at which water boils (at standard atmospheric pressure). The gap between freezing and boiling is divided into 100 equal parts, called *degrees*. This temperature scale is the **Celsius scale.***

On the temperature scale used commonly in the United States, the number 32 designates the temperature at which water freezes, and the number 212 is assigned to the temperature at which water boils. This temperature scale is called the **Fahrenheit scale.** The Fahrenheit scale will become obsolete if and when the United States goes metric.

The scale used in scientific research is the SI scale—the **Kelvin scale.** Its degrees are the same size as the Celsius degree and are called "kelvins." On the Kelvin scale, the number 0 is assigned to the lowest possible temperature—**absolute zero.** At absolute zero a substance has no kinetic energy to give up. Zero on the Kelvin scale, or absolute zero, corresponds to –273°C on the Celsius scale. We will learn more about the Kelvin scale in Chapter 24.

Arithmetic formulas can be used for converting from one temperature scale to another and are often popular in classroom exams. Such arithmetic exercises are not really physics, so we will not be concerned with them here. Besides, a conversion from Celsius to Fahrenheit, or vice versa, can be very closely approximated by simply reading the corresponding temperature from the side-by-side scales in Figure 21.1.

Temperature and Kinetic Energy

Temperature is related to the random motions of the molecules in a substance. In the simplest case of an ideal gas, temperature is proportional to the *average* kinetic energy of molecular translational motion (that is, motion along a straight or curved path). In solids and liquids, where molecules are more constrained and have potential energy, temperature is more complicated. But it is still true that temperature is closely related to the average kinetic energy of translational motion of molecules. So the warmth you feel when you touch a hot surface is the kinetic energy transferred by molecules in the surface to molecules in your fingers.

Note that temperature is *not* a measure of the *total* kinetic energy of all the molecules in a substance. There is twice as much kinetic energy in 2 liters of boiling water as in 1 liter. But the temperatures of both liters of water are the same because the average kinetic energy of molecules in each is the same.

---

*   The Celsius scale is named in honor of the man who first suggested it, the Swedish astronomer Anders Celsius (1701–1744). It used to be called the centigrade scale, from *centi* ("hundredth") and *gradus* ("degree"). The Fahrenheit scale is named after the German physicist Gabriel Fahrenheit (1686–1736), and the Kelvin scale, after the British physicist Lord Kelvin (1824–1907).

21.2 Heat

If you touch a hot stove, energy will enter your hand from the stove because the stove is warmer than your hand. But if you touch ice, energy will pass out of your hand and into the colder ice. The direction of spontaneous energy transfer is always from a warmer substance to a cooler substance. The energy that transfers from one object to another because of a temperature difference between them is called **heat.**

It is common—but incorrect with physics types—to think that matter *contains* heat. Matter contains energy in several forms, but it does not contain heat. Heat is energy in transit from a body of higher temperature to one of lower temperature. Once transferred, the energy ceases to be heat.* In Chapter 8 we called the energy resulting from heat flow *thermal energy*, to make clear its link to heat and temperature. In this and following chapters, we will use the term that scientists prefer, *internal energy.*

When heat flows from one object or substance to another it is in contact with, the objects or substances are said to be in **thermal contact.** Given thermal contact, heat flows from the higher-temperature substance into the lower-temperature substance. However, heat will not necessarily flow from a substance with more total molecular kinetic energy to a substance with less total molecular kinetic energy. For example, there is more total molecular kinetic energy in a large bowl of warm water than there is in a red-hot thumbtack. Yet, if the tack is immersed in the water, heat does not flow from the water which has more total kinetic energy to the tack which has less. It flows from the hot tack to the cooler water. Heat flows according to temperature differences—that is, average molecular kinetic energy differences. Heat never flows on its own from a cooler substance into a hotter substance. We will return to this concept in Chapter 24 when we look at thermodynamics.

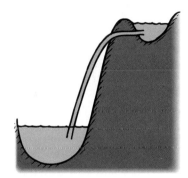

Figure 21.3 ▲
Just as water will not flow uphill by itself, regardless of the relative amounts of water in the reservoirs, heat will not flow from a cooler substance into a hotter substance by itself.

---

* Similarly, work is also energy in transit. A body does not *contain* work. It *does* work or has work done on it.

21.3 Thermal Equilibrium

After objects in thermal contact with each other reach the same temperature, no heat flows between them—we say the objects are in **thermal equilibrium.**

To read a thermometer we wait until it reaches thermal equilibrium with the substance being measured. When a thermometer is in contact with a substance, heat flows between them until they have the same temperature. We then know the temperature of the thermometer is also the temperature of the substance. So a thermometer, interestingly enough, shows only its own temperature.

A thermometer should be small enough that it does not appreciably alter the temperature of the substance being measured. If you are measuring the temperature of room air, then the heat absorbed by the thermometer will not lower the air temperature noticeably. But if you are trying to measure the temperature of a drop of water, the temperature of the drop after thermal contact may be quite different from its initial temperature.

Figure 21.4 ▲
Somewhat like water in the pipes seeking a common level (for which the pressures at equal elevations are the same), the thermometer and its immediate surroundings reach a common temperature (at which the average kinetic energy per particle is the same for both).

21.4 Internal Energy

In addition to the translational kinetic energy of jostling molecules in a substance, there is energy in other forms. There is rotational kinetic energy of molecules and kinetic energy due to internal movements of atoms within molecules. There is also potential energy due to the forces between molecules. The grand total of all energies inside a substance is called **internal energy.** A substance does not contain heat—it contains internal energy.

When a substance takes in or gives off heat, any of these energies may change. Thus, as a substance absorbs heat, this energy may or may not make the molecules jostle faster. In some cases, as when ice is melting, a substance absorbs heat without an increase in temperature. The substance changes phase, the subject of Chapter 23.

21.5 Measurement of Heat

So we see that heat is energy transferred from one substance to another by a temperature difference. The amount of heat transferred can be determined by measuring the temperature change of a known mass of water that absorbs the heat.

When a substance absorbs heat, the resulting temperature change depends on more than just the mass of the substance. The quantity of heat that brings a cupful of soup to a boil might raise the temperature of a pot of soup by only a few degrees. To quantify heat, we must specify the *mass* and *kind* of substance affected.

The unit of heat is defined as the heat necessary to produce some standard, agreed-on temperature change for a specified mass of material. The most commonly used unit for heat is the **calorie.** The calorie is defined as the amount of heat required to raise the temperature of 1 gram of water by 1°C.

The **kilocalorie** is 1000 calories (the heat required to raise the temperature of 1 kilogram of water by 1°C). The heat unit used in rating foods is actually a kilocalorie, although it's often referred to as the calorie. To distinguish it from the smaller calorie, the food unit is sometimes called a Calorie (written with a capital C).

It is important to remember that the calorie and Calorie are units of energy. These names are historical carryovers from the early idea that heat was an invisible fluid called *caloric.* We now know heat is a form of energy. The United States is in a period of transition to the International System of Units (SI), where quantity of heat is measured in joules, the SI unit for all forms of energy. The relationship between calories and joules is that 1 calorie equals 4.184 J. In this book we'll learn about heat with the conceptually simpler calorie—but in the lab you may use the joule equivalent, where an input of 4.184 joules raises the temperature of 1 gram of water by 1°C.*

HOT STOVE

◀ **Figure 21.5**
Although the same quantity of heat is added to both containers, the temperature of the container with the smaller amount of water increases more.

The energy value in food is determined by burning the food and measuring the energy that is released as heat. Food and other fuels are rated by how much energy a certain mass of the fuel gives off as heat when burned.

---

* Still another unit of heat is the British thermal unit (Btu). The Btu is defined as the quantity of heat required to change the temperature of 1 pound of water by 1°F. One Btu is equal to 1054 J.

Figure 21.6 ▲
To the weight watcher, the
peanut contains 10 Calories; to
the physicist, it releases 10 000
calories (or 41 840 joules) of
energy when burned or digested.

Computational Example: Dimensional Analysis

A woman with an average diet consumes and expends about 2000 Calories per day. The energy used by her body is eventually given off as heat. How many joules per second does her body give off? Or, in other words, what is her average thermal power output?

We find this by converting 2000 Calories per day to joules per second. We use the information that 1 Calorie = 4187 joules, 1 day = 24 hours, and 1 hour = 3600 seconds. The conversion is then set up as follows:

$$\frac{2000 \text{ Cal}}{1 \text{ d}} \times \frac{1 \text{ d}}{24 \text{ h}} \times \frac{1 \text{ h}}{3600 \text{ s}} \times \frac{4184 \text{ J}}{1 \text{ Cal}} = 96.8 \text{ J/s} = 96.8 \text{ W}$$

Notice that the original quantity (2000 Cal/d) is multiplied by a set of fractions in which the numerator equals the denominator. Since each fraction has the value 1, multiplying by it does not change the value of the original quantity. The rule for choosing which quantity to put in the numerator is that the units should cancel and reduce to those of the end result. (We call this technique "dimensional analysis.") So, on the average, the woman emits heat at the rate of 96.8 J/s, which is 96.8 watts. This is nearly the same as a glowing 100-W lamp! It's easy to see why a crowded room soon becomes warm! (Don't confuse the 96.8 watts given off by the woman with her internal temperature of 98.6°F. The closeness of the numerical values is a coincidence. A body's temperature and its rate of expending heat are entirely different from each other.)

■ Question

Suppose you use a flame to add a certain quantity of heat to 1 liter of water, and the water temperature rises by 2°C. If you add the same quantity of heat to 2 liters of water, by how much will its temperature rise?

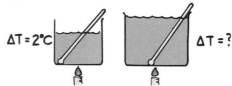

■ Answer

Its temperature will rise by 1°C, because there are twice as many molecules in 2 liters of water and each molecule receives only half as much energy on average. So average kinetic energy, and temperature, increases by half as much.

21.6 Specific Heat Capacity

Almost everyone has noticed that some foods remain hot much longer than others. Boiled onions and moist squash on a hot dish, for example, are often too hot to eat while mashed potatoes may be just right. The filling of hot apple pie can burn your tongue while the crust will not, even when the pie has just been taken out of the oven. The aluminum covering on a frozen dinner can be peeled off with your bare fingers as soon as it is removed from the oven. (But be careful of the food beneath it!)

Different substances have different capacities for storing internal energy. If we heat a pot of water on a stove, we may find that it requires 15 minutes to raise it from room temperature to its boiling temperature. But if we were to put an equal mass of iron on the same flame, we would find that it would rise through the same temperature range in only about 2 minutes. For silver, the time would be less than a minute. We find that specific materials require specific quantities of heat to raise the temperature of a given mass of the material by a specified number of degrees.

Absorbed energy can affect substances in different ways. Absorbed energy that increases the translational speed of molecules is responsible for increases in temperature. Absorbed energy may also increase the rotation of molecules, increase the internal vibrations within molecules, or stretch intermolecular bonds and be stored as potential energy. These kinds of energy, however, are not measures of temperature. Temperature is a measure only of the kinetic energy of translational motion. Generally, only part of the energy absorbed by a substance raises its temperature.

Whereas a gram of water requires 1 calorie of energy to raise the temperature 1°C, it takes only about one eighth as much energy to raise the temperature of a gram of iron by the same amount. Iron atoms in the iron lattice primarily shake back and forth in translational fashion, while water molecules soak up a lot of energy in rotations, internal vibrations, and bond stretching. So water absorbs more heat per gram than iron for the same change in temperature. We say water has a higher **specific heat capacity** (sometimes simply called *specific heat*).

Figure 21.7 ▲
You can touch the aluminum pan of the frozen dinner soon after it has been taken from the hot oven, but you'll burn your fingers if you touch the food it contains.

■ Question

Which has a higher specific heat capacity—water or sand?

■ Answer

Water has a greater heat capacity than sand. Water is much slower to warm in the hot sun and slower to cool in the cold night. Water has more thermal inertia. Sand's low heat capacity, as evidenced by how quickly the surface warms in the morning sun and how quickly it cools at night, affects local climates.

The specific heat capacity of any substance is defined as the quantity of heat required to raise the temperature of a unit mass of the substance by 1 degree.

We can think of specific heat capacity as thermal inertia. Recall that *inertia* is a term used in mechanics to signify the resistance of an object to change in its state of motion. Specific heat capacity is like a thermal inertia since it signifies the resistance of a substance to change in its temperature.

Computational Example: Heating Water

When we know the specific heat capacity c for a particular substance, the quantity of heat Q involved when the mass m of the substance undergoes a temperature change ΔT is $Q = mc\Delta T$. In words, heat transferred = mass × specific heat capacity × temperature change.

Suppose we wish to know the number of calories needed to raise the temperature of 1 liter of water by 15°C. The specific heat capacity for water, c, is 1 cal/g°C, and the mass of 1 liter of water is 1 kilogram, which is 1000 grams. Since c is expressed in calories per *gram* °C, we express the mass of water m in grams. Then,

$$Q = mc\Delta T$$

$$Q = (1000 \text{ g})(1 \text{ cal/g°C})(15°C) = 15\,000 \text{ calories}$$

Suppose we deliver this energy to the water with a 1000-watt immersion heater. How long will it take to heat the water? We know that 1000 watts delivers energy at the rate 1000 joules per second. Converting calories to joules,

$$15\,000 \text{ cal} \times 4.184 \text{ J/cal} = 62\,760 \text{ joules}$$

At the rate of 1000 joules per second, can you see that the time required for heating the water by 15°C is somewhat more than a minute?

21.7 The High Specific Heat Capacity of Water

Water has a much higher capacity for storing energy than most common materials. A relatively small amount of water absorbs a great deal of heat for a correspondingly small temperature rise. Because of this, water is a very useful cooling agent, and is used in cooling systems in automobiles and other engines. If a liquid of lower specific heat capacity were used in cooling systems, its temperature would rise higher for a comparable absorption of heat. (Of course, if the temperature of the liquid rises to the temperature of the engine, no further cooling will take place.) Water also takes longer to cool, a useful fact to your great-grandparents, who on cold winter nights likely used foot-warming hot-water bottles in their beds.

This property of water to resist changes in temperature improves the climate in many places. The next time you are looking at a world globe, notice the high latitude of Europe. If water did not have a high heat capacity, the countries of Europe would be as cold as the northeastern regions of Canada, for both Europe and Canada get about the same amount of the sun's energy per square kilometer. The Atlantic current known as the Gulf Stream brings warm water northeast from the Caribbean. It holds much of its internal energy long enough to reach the North Atlantic off the coast of Europe, where it then cools. The energy released (one calorie per degree for each gram of water that cools) is carried by the westerly winds over the European continent.

Similarly, the climates differ on the east and west coasts of North America. The winds in the latitudes of North America are westerly. On the west coast, air moves from the Pacific Ocean to the land. Because of water's high heat capacity, ocean temperature does not vary much from summer to winter. The water is warmer than the air in the winter, and cooler than the air in the summer. In winter, the water warms the air that moves over it and warms the western coastal regions of North America. In summer, the water cools the air and the western coastal regions are cooled. On the east coast, air moves from the land to the Atlantic Ocean. Land, with a lower specific heat capacity, gets hot in summer but cools rapidly in winter. As a result of water's high heat capacity and the wind directions, the west coast city of San Francisco is warmer in the winter and cooler in the summer than the east coast city of Washington, D.C., which is at about the same latitude.

The central interior of a large continent usually experiences extremes of temperature. For example, the high summer and low winter temperatures common in Manitoba and the Dakotas are largely due to the absence of large bodies of water. Europeans, islanders, and people living near ocean air currents should be glad that water has such a high specific heat capacity. San Franciscans are!

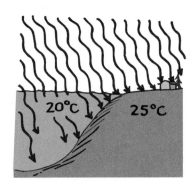

Figure 21.8 ▲
Water has a high specific heat and is transparent, so it takes more energy to heat up than land does. Why would its transparency be a factor?

21.8 Thermal Expansion

When the temperature of a substance is increased, its molecules jiggle faster and normally tend to move farther apart. This results in an *expansion* of the substance. With few exceptions, all forms of matter—solids, liquids, and gases—expand when they are heated and contract when they are cooled. For comparable pressures and comparable changes in temperature, gases generally expand or contract much more than liquids, and liquids expand or contract more than solids.*

---

* This rule is valid if the solid, liquid, and gas expand against constant pressure. A gas in a container can be prevented from expanding, but then its pressure is not constant.

If concrete sidewalks and highway paving were laid down in one continuous piece, cracks would appear due to the expansion and contraction brought about by the difference between summer and winter temperatures. To prevent this, the surface is laid in small sections, each one being separated from the next by a small gap that is filled in with a substance such as tar. On a hot summer day, expansion often squeezes this material out of the joints.

The *expansion* of materials must be allowed for in the construction of structures and devices of all kinds. A dentist uses filling material that has the same rate of expansion as teeth. The aluminum pistons of an automobile engine are smaller enough in diameter than the steel cylinders to allow for the much greater expansion rate of aluminum. A civil engineer uses steel of the same expansion rate as concrete for reinforcing concrete. Long steel bridges often have one end fixed while the other rests on rockers that allow for expansion. The roadway itself is segmented with tongue-and-groove-type gaps called expansion joints (Figure 21.9).

Figure 21.9 ▲
This gap is called an *expansion joint,* and allows the bridge to expand and contract.

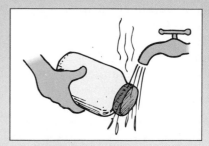

Different materials expand at different rates. In a **bimetallic strip,** two strips of different metals, say one of brass and the other of iron, are welded or riveted together (Figure 21.10). When the strip is heated, the difference in the amounts of expansion of brass and iron shows up easily. One side of the double strip becomes longer than the other, causing the strip to bend into a curve. On the other hand, when the strip is cooled, it bends in the opposite direction, because the metal that expands the most also contracts the most. The movement of the strip may be used to turn a pointer, regulate a valve, or operate a switch.

A **thermostat** is a practical application of a bimetallic strip (Figure 21.11). The back-and-forth bending of the bimetallic coil opens and closes an electric circuit. When the room becomes too cold, the coil bends toward the brass side, and in so doing it closes an electric switch that turns on the heat. When the room becomes too warm, the coil bends toward the iron side, which opens the switch and turns off the heating unit. Refrigerators are equipped with special thermostats to prevent them from becoming too hot or too cold. Bimetallic strips are used in oven thermometers, electric toasters, automatic chokes on carburetors, and other devices.

The amount of expansion of a substance depends on its change in temperature. If one part of a piece of glass is heated or cooled more rapidly than adjacent parts, the expansion or contraction that results may break the glass. This is especially true for thick glass. Heat-resistant glass is specially formulated to expand very little with increasing temperature.

Liquids expand appreciably with increases in temperature. When the gasoline tank of a car is filled at a gas station and the car is then parked for a while, the gasoline often overflows the tank. This occurs as the cold gasoline from the underground storage tanks warms up as it sits in the car's tank. As the gasoline warms, it expands and overflows the gas tank. Similarly, an automobile radiator filled to the brim with cold water overflows when heated.

In most cases, the expansion of liquids is greater than the expansion of solids. The gasoline overflowing a car's tank on a hot day is evidence for this. Similarly, a pot filled to the brim with water soon overflows when heated. Also, mercury rises in a thermometer when heated because the liquid mercury expands more than the glass.

■ Question

Why is it advisable to allow telephone lines to sag when stringing them between poles in summer?

■ Answer

Telephone lines are longer in summer, when they are warmer, and shorter in winter, when they are cooler. They therefore sag more on hot summer days than in winter. If the telephone lines are not strung with enough sag in summer, they might contract too much and snap during the winter.

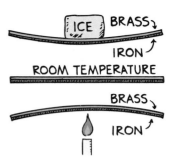

Figure 21.10 ▲
A bimetallic strip. Brass expands (or contracts) more when heated (or cooled) than does iron, so the strip bends as shown.

Figure 21.11 ▲
A thermostat. When the bimetallic coil expands, the mercury rolls away from the electrical contacts and breaks the circuit. When the coil contracts, the mercury rolls against the contacts and completes the electric circuit.

Figure 21.12 ▲
Place a dented Ping-Pong ball in boiling water, and you'll remove the dent. Why?

Computational Example: Ratio and Proportion

Steel changes in length about 1 part in 100 000 for each Celsius degree change in temperature. This is a *ratio*,

$$\frac{1}{100\ 000}$$

For different lengths of steel, expansion would follow the same proportion. For short lengths of steel, expansion may be negligible. But consider the expansion of a make-believe snugly fitting steel pipe that completely encircles the earth. How much longer would this 40-million-meter pipe be if its temperature increased by 1°C?

The ratio of its change in length X to its full size is the same as the ratio above, so for a 1°C temperature change we say

$$\frac{1}{100\ 000} = \frac{X\,\text{m}}{40\ 000\ 000\ \text{m}}$$

A little computation will show that the change in length X is 400 m. Here's the interesting part: If such a pipe were elongated by this 400 m, then there would be a gap between it and the earth's surface. Would the gap be big enough to put this book under? To crawl under? To drive a truck under? How big would this gap be?

We can find the gap by ratio and proportion. The ratio of circumference C to diameter D for any circle is equal to π (about 3.14). The ratio of the change in circumference ΔC to the change in diameter ΔD also has the same value. Inserting values, we have

$$\frac{\Delta C}{\Delta D} = \frac{400\ \text{m}}{\Delta D} = 3.14$$

Solving for ΔD gives

$$\Delta D = \frac{400\ \text{m}}{3.14} = 127.4\ \text{m}$$

This 127.4 m is the increase in *diameter* of the circular pipe. The size of the gap between the earth's surface and the expanded pipe is equal to the increase in radius, which is half the increase in diameter, or 63.7 m.

So if a steel pipe that fits snugly against the earth were increased in temperature by 1°C, perhaps by people all along its length breathing hard on it, the pipe would expand and stand an amazing 63.7 meters off the ground! Using ratio and proportion is a straightforward way to solve many problems. Another way to solve for the expansion of a material involves a formula ($L = \alpha L_o \Delta T$). You may encounter this formula in the lab part of your course, but we'll not treat it here in the text.

21.9 Expansion of Water

Almost all liquids will expand when they are heated. Ice-cold water, however, does just the opposite! Water at the temperature of melting ice, 0°C (or 32°F), *contracts* when the temperature is increased. This is most unusual. As the water is heated and its temperature rises, it continues to contract until it reaches a temperature of 4°C. With further increase in temperature, the water then begins to *expand;* the expansion continues all the way to the boiling point, 100°C. This odd behavior is shown graphically in Figure 21.13.

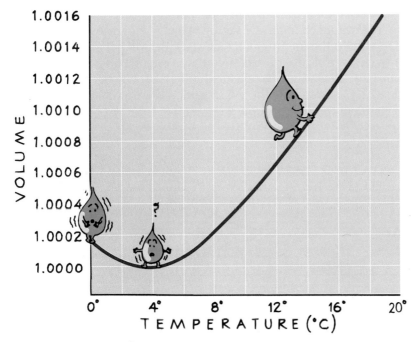

Figure 21.13 ▲
The change in volume of water with increasing temperature.

A given amount of water has its smallest volume—and thus its greatest density—at 4°C. The same amount of water has its largest volume—and smallest density—in its solid form, ice. (Remember, ice floats in water, so it must be less dense than water.) The volume of ice at 0°C is not shown in Figure 21.13. (If it were plotted to the same exaggerated scale, the graph would extend far beyond the top of the page.) After water has turned to ice, further cooling causes it to contract.

The explanation for this behavior of water has to do with the odd crystal structure of ice. The crystals of most solids are structured so that the solid state occupies a smaller volume than the liquid state. Ice, however, has open-structured crystals (Figure 21.14). These crystals result from the angular shape of the water molecules, plus the fact that the forces binding water molecules together are strongest at certain angles. Water molecules in this open structure occupy a greater volume than they do in the liquid state. Consequently, ice is less dense than water.

Figure 21.14 ▶
Water molecules in their crystal
form have an open-structured,
six-sided arrangement. As a
result, water expands upon freezing, and ice is less dense than
water.

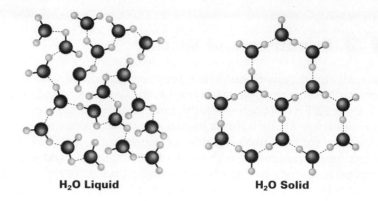

H₂O Liquid   **H₂O Solid**

The reason for the dip in the curve of Figure 21.13 is that two
types of volume changes are taking place. A decrease in volume
occurs due to the melting of ice crystals. Between 0°C and 10°C,
water—a "microscopic slush"—contains microscopic ice crystals. At
about 10°C all the ice crystals have collapsed. The left-hand graph in
Figure 21.15 shows how the volume of cold water changes due to the
collapsing of the microscopic ice crystals.

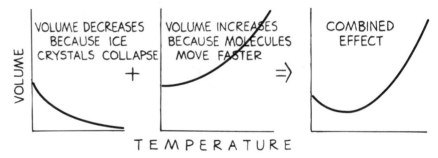

Figure 21.15 ▲
The collapsing of ice crystals (left) plus increased molecular motion with
increasing temperature (center) combine to make water most dense at
4°C (right).

While crystals are collapsing as the temperature increases
between 0°C and 10°C, increased molecular motion results in expansion. This effect is shown in the center graph in Figure 21.15.
Whether ice crystals are in the water or not, increased vibrational
motion of the molecules increases the volume of the water.

When we combine the effects of contraction and expansion, the
curve looks like the right-hand graph in Figure 21.15 (or Figure 21.13).
This behavior of water is of great importance in nature. Suppose that
the greatest density of water were at its freezing point, as is true of
most liquids. Then the coldest water would settle to the bottom,
and ponds would freeze from the bottom up. Pond organisms
would then be destroyed in winter months. Fortunately, this does
not happen. The densest water, which settles at the bottom of a

pond, is 4 degrees above the freezing temperature. Water at the freezing point, 0°C, is less dense and "floats," so ice forms at the surface while the pond remains liquid below the ice.

Let's examine this in more detail. Most of the cooling in a pond takes place at its surface, when the surface air is colder than the water. As the surface water is cooled, it becomes denser and sinks to the bottom. Water will "float" at the surface for further cooling only if it is as dense or less dense than the water below.

Consider a pond that is initially at, say, 10°C. It cannot possibly be cooled to 0°C without first being cooled to 4°C. And water at 4°C cannot remain at the surface for further cooling unless all the water below has at least an equal density—that is, unless all the water below is at 4°C. If the water below the surface is any temperature other than 4°C, any surface water at 4°C will be denser and sink before it can be further cooled. So before any ice can form, all the water in a pond must be cooled to 4°C. Only when this condition is met can the surface water be cooled to 3°, 2°, 1°, and 0°C without sinking. Then ice can form.

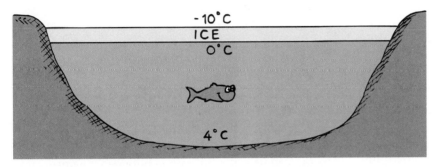

Figure 21.16 ▲
As water is cooled at the surface, it sinks until the entire lake is 4°C. Only then can the surface water cool to 0°C without sinking.

Thus, the water at the surface is first to freeze. Continued cooling of the pond results in the freezing of the water next to the ice, so a pond freezes from the surface downward. In a cold winter the ice will be thicker than in a milder winter.

Very deep bodies of water are not ice-covered even in the coldest of winters. This is because all the water in a lake must be cooled to 4°C before lower temperatures can be reached, and the winter is not long enough for all the water to be cooled to 4°C. If only some of the water is 4°C, it will lie on the bottom. Because of water's high specific heat and poor ability to conduct heat, the bottom of deep lakes in cold regions is a constant 4°C the year round. Fish should be glad that this is so.

Concept Summary

Temperature is the measurement that tells how warm or cold something is.

■ Temperature is directly proportional to the average translational kinetic energy of the molecules within an ideal gas.

Heat is energy that transfers between two things due to a temperature difference.

■ Matter does not contain heat; rather, it contains internal energy.

Specific heat is a measure of how much heat is required to raise the temperature of a unit mass of a substance by one degree.

■ Water has a much higher specific heat than other common substances.

Matter tends to expand when heated and to contract when cooled.

■ Liquids usually expand slightly more than solids.

■ Gases expand much more than liquids or solids for comparable increases in temperature (and comparable pressure).

■ Water is highly unusual in that it contracts as it warms from 0°C to 4°C and its solid form (ice) is less dense than its liquid form.

Important Terms

absolute zero (21.1)
bimetallic strip (21.8)
calorie (21.5)
Celsius scale (21.1)
Fahrenheit scale (21.1)
heat (21.2)
internal energy (21.4)
Kelvin scale (21.1)
kilocalorie (21.5)
specific heat capacity (21.6)
temperature (21.1)
thermal contact (21.2)
thermal equilibrium (21.3)
thermostat (21.8)

Review Questions

1. How is temperature commonly measured? (21.1)

2. How many degrees are between the melting point of ice and boiling point of water on the Celsius scale? Fahrenheit scale? (21.1)

3. Why is it incorrect to say that matter *contains* heat? (21.2)

4. In terms of differences in temperature between objects in thermal contact, in what direction does heat flow? (21.2)

5. What is meant by saying that a thermometer measures its own temperature? (21.3)

6. What is thermal equilibrium? (21.3)

7. What is internal energy? (21.4)

8. What is the difference between a calorie and a Calorie? (21.5)

9. What does it mean to say that a material has a high or low specific heat capacity? (21.6)

10. Do substances that heat up quickly normally have high or low specific heat capacities? (21.6)

11. How does the specific heat capacity of water compare with that of other common substances? (21.7)

12. Why is the North American west coast warmer in winter months and cooler in summer months than the east coast? (21.7)

13. Why does a bimetallic strip curve when it is heated (or cooled)? (21.8)

14. Which expands most for increases in temperature: solids, liquids, or gases? (21.8)

15. At what temperature is the density of water greatest? (21.9)

16. Ice is less dense than water because of its open crystalline structure. But why is water at 0°C less dense than water at 4°C? (21.9)

17. Why do lakes and ponds freeze from the top down rather than from the bottom up? (21.9)

18. Why do shallow lakes freeze quickly in winter, and deep lakes not at all? (21.9)

Plug and Chug

Heat transfer in calories is given by $Q = mc\Delta T$, where m is mass in grams, c is specific heat capacity in cal/g°C, and ΔT is in °C.

1. Calculate the number of calories of heat needed to change 500 grams of water by 50 Celsius degrees.

2. Calculate the number of calories given off by 500 grams of water cooling from 50°C to 20°C.

3. A 30-gram piece of iron is heated to 100°C and then dropped into cool water where the iron's temperature drops to 30°C. How many calories does it lose to the water? (The specific heat capacity of iron is 0.11 cal/g°C.)

4. Suppose the same 30-gram piece of iron is dropped into another container of water and gives off 165 calories in cooling. Calculate the iron's temperature change.

5. What mass of water will give up 240 calories when its temperature drops from 80°C to 68°C?

6. When a 50-gram piece of aluminum at 100°C is placed in water, it loses 735 calories of heat while cooling to 30°C. Calculate the specific heat capacity of the aluminum.

Think and Explain

1. If you drop a hot rock into a pail of water, the temperature of the rock and the water will change until both are equal. The rock will cool and the water will warm. Does the same principle hold true if the rock is dropped into a large lake? Explain.

2. If you stake out a plot of land with a steel tape measure using map measurements on a very hot day, will you enclose more or less land than your measurements indicate?

3. A metal ball is just able to pass through a metal ring. When the ball is heated, thermal expansion will not allow it to pass through the ring. What would happen if the ring, rather than the ball, were heated? Would the ball pass through the heated ring? Does the size of the hole in the ring increase, decrease, or stay the same?

4. A snugly fitting steel pipe circling the world would stand about 64 meters off the ground if its temperature were increased by 1°C. What would be the result if the pipe were instead cooled by 1°C?

5. After a machinist slips a hot, snugly fitting iron ring over a cold brass cylinder, the ring becomes "locked" in position and can't be removed even by subsequent heating. This procedure is called "shrink fitting." How does it occur? Can you conclude anything about the thermal expansion rates of iron and brass?

6. If you take a bite of hot pizza, the sauce may burn your mouth while the crust, at the same temperature, will not. Explain.

7. In the old days, on a cold winter night it was common to bring a hot object to bed with you. Which would be better—a 10-kilogram iron brick or a 10-kilogram jug of hot water at the same temperature? Explain.

8. On a hot day you remove from a picnic cooler a chilled watermelon and some chilled sandwiches. Which will remain cooler for a longer time? Why?

9. Iceland, so named to discourage conquest by expanding empires, is not at all ice-covered like Greenland and parts of Siberia, even though it is nearly on the Arctic Circle. The average winter temperature of Iceland is considerably higher than regions at the same latitude in eastern Greenland and central Siberia. Why is this so?

10. Why is it important to protect water pipes so they don't freeze?

11. Suppose you cut a small gap in a metal ring, as shown. If you heat the ring, will the gap become wider or narrower?

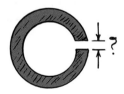

12. Would a bimetallic strip function if the two different metals happened to have the same rates of expansion? Is it important that they expand at different rates? Explain.

13. State whether water at the following temperatures will expand or contract when warmed: 0°C; 4°C; 6°C.

14. In addition to the overall motion of a molecule that is associated with temperature, some molecules can absorb large amounts of energy in the form of internal vibrations and rotations of the molecule itself. Would you expect materials composed of such molecules to have a high or a low specific heat capacity? Why?

15. If water had a lower specific heat capacity, would lakes be more likely or less likely to freeze in the winter?

Think and Solve

1. If you wished to warm 100 kg of water by 15°C for your bath, how much heat would be required? (Give your answer in calories and joules.)

2. What would be the final temperature if you mixed a liter of 20°C water with 2 liters of 40°C water?

3. What would be the final temperature if you mixed a liter of 40°C water with 2 liters of 20°C water?

4. What is the specific heat capacity of a 50-gram piece of 100°C metal that will change 400 grams of 20°C water to 22°C?

5. Suppose that a metal bar 1 m long expands 0.5 cm when it is heated. How much would it expand if it were 100 m long?

6. Steel expands 1 part in 100 000 for each 1°C increase in temperature. If the 1.5-km main span of a steel suspension bridge had no expansion joints, how much longer would it be for a temperature increase of 20°C?

22 Heat Transfer

Thick adobe walls slow heat transfer.

The spontaneous transfer of heat is always from warmer objects to cooler objects. If several objects near one another have different temperatures, then those that are warm become cooler and those that are cool become warmer, until all have a common temperature. This equalization of temperatures is brought about in three ways: by *conduction,* by *convection,* and by *radiation.*

22.1 Conduction

If you hold one end of an iron rod in a flame, before long the rod will become too hot to hold. Heat has transferred through the metal by **conduction.** Conduction of heat can take place within materials and between different materials that are in direct contact. Materials that conduct heat well are known as heat **conductors.** Metals are the best conductors. Among the common metals, silver is the most conductive, followed by copper, aluminum, and iron.

Conduction is explained by collisions between atoms or molecules, and the actions of loosely bound electrons. In the iron rod, the flame causes the atoms at the heated end to vibrate more rapidly. These atoms vibrate against neighboring atoms, which in turn do the same. More important, free electrons that can drift through the metal are made to jostle and transfer energy by colliding with atoms and other free electrons within the rod.

Materials composed of atoms with "loose" outer electrons are good conductors of heat (and electricity also). Because metals have the "loosest" outer electrons, they are the best conductors of heat and electricity.

Touch a piece of metal and a piece of wood in your immediate vicinity. Which one *feels* colder? Which is *really* colder? Your answers should be different. If the materials are in the same vicinity, they

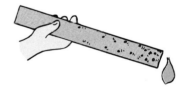

Figure 22.1 ▲
Heat from the flame causes atoms and free electrons in the end of the metal to move faster and jostle against others, which in turn do the same and increase the energy of vibrating atoms down the length of the rod.

325

Figure 22.2 ▲
The tile floor feels cold to the bare feet, while the carpet at the same temperature feels warm. This is because tile is a better conductor than carpet.

Figure 22.3 ▲
A "warm" blanket does not provide you with heat; it simply slows the transfer of your body heat to the surroundings.

should have the same temperature, room temperature. Thus neither is really colder. Yet, the metal *feels* colder because it is a better conductor; heat easily moves out of your warmer hand into the cooler metal. Wood, on the other hand, is a poor conductor. Little heat moves out of your hand into the wood, so your hand does not sense that it is touching something cooler. Wood, wool, straw, paper, cork, and polystyrene (Styrofoam) are all poor heat conductors. Instead, they are called good **insulators** because they delay the transfer of heat. A poor conductor is a good insulator.

Liquids and gases, in general, are good insulators. Air is a mixture of gases and conducts heat very poorly—air is a very good insulator. Porous materials having many small air spaces are good insulators. The good insulating properties of materials such as wool, fur, and feathers are largely due to the air spaces they contain. Birds vary their insulation by fluffing their feathers to create air spaces. Be glad that air is a poor conductor, for if it were not, you'd feel quite chilly on a 25°C (77°F) day!

Snowflakes imprison a lot of air in their crystals and are good insulators. Snow slows the escape of heat from the earth's surface, shields Eskimo dwellings from the cold, and provides protection from the cold to animals on cold winter nights. Snow, like any blanket, is not a source of heat; it simply prevents any heat from escaping too rapidly.

Heat is energy and is tangible. Cold is not; cold is simply the absence of heat. Strictly speaking, there is no "cold" that passes through a conductor or an insulator. Only heat is transferred. We don't insulate a home to keep the cold out; we insulate to keep the heat in. If the home becomes colder, it is because heat flows out.

It is important to note that no insulator can totally prevent heat from getting through it. An insulator just reduces the rate at which heat penetrates. Even the best-insulated warm homes in winter will gradually cool. Insulation delays heat transfer.

Figure 22.4 ▲
Snow lasts longest on the roof of a well-insulated house. Thus, the snow patterns reveal the conduction, or lack of conduction, of heat through the roof. Can you see how the insulation of these houses varies?

■ Questions

1. If you hold one end of a metal bar against a piece of ice, the end in your hand will soon become cold. Does cold flow from the ice to your hand?

2. Wood is a better insulator than glass. Yet fiberglass is commonly used to insulate wooden buildings. Why?

3. You can stick your hand into a hot pizza oven for several seconds without harm, whereas you'd never touch the metal insides for even a second. Why?

22.2 Convection

Recall that heat transfer by conduction involves the transfer of energy from molecule to molecule. Energy moves from one place to another, but the molecules do not. Another means of heat transfer is by movement of the hotter substance. Air in contact with a hot stove ascends and warms the region above. Water heated in a boiler in the basement rises to warm the radiators in the upper floors. This is **convection,** where heating occurs by currents in a fluid.

A simple demonstration illustrates the difference between conduction and convection. With a bit of steel wool, trap a piece of ice at the bottom of a test tube nearly filled with water. Hold the tube by the bottom with your bare hand and place the top in the flame of a Bunsen burner. (See Figure 22.5.) The water at the top will come to a vigorous boil while the ice below remains unmelted. The hot water at the top is less dense and remains at the top. Any heat that reaches the ice must be transferred by conduction, and we see that water is a poor conductor of heat. If you repeat the experiment, only this time holding the test tube at the top by means of tongs and heating the water from below while the ice floats at the surface, the ice will melt quickly. Heat gets to the top by convection, for the hot water rises to the surface, carrying its energy with it to the ice.

Convection occurs in all fluids, whether liquid or gas. Whether we heat water in a pan or heat air in a room, the process is the same. When the fluid is heated, it expands, becomes less dense, and rises. Warm air or warm water rises for the same reason that a block of wood floats in water and a helium-filled balloon rises in air. In effect,

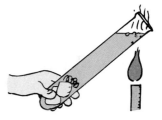

Figure 22.5 ▲
When the test tube is heated at the top, convection is prevented and heat can reach the ice by conduction only. Since water is a poor conductor, the top water will boil without melting the ice.

■ Answers

1. Cold does not flow from the ice to your hand. Heat flows from your hand to the ice. The metal is cold to your touch because you are transferring heat to the metal.

2. Fiberglass is a good insulator, many times better than glass, because of the air that is trapped between its fibers.

3. Air is a poor conductor, so the rate of heat flow from the hot air to your relatively cool hand is low. But touching the metal parts is a different story. Metal conducts heat very well, and a lot of heat in a short time is conducted into your hand when thermal contact is made.

convection is an application of Archimedes' principle, for the warmer fluid is buoyed upward by denser surrounding fluid. Cooler fluid then moves to the bottom, and the process continues. In this way, convection currents keep a fluid stirred up as it heats.

Winds

Convection currents stirring the atmosphere produce winds. Some parts of the earth's surface absorb heat from the sun more readily than others. The uneven absorption causes uneven heating of the air near the surface and creates convection currents. This phenomenon is often evident at the seashore. In the daytime the shore warms more easily than the water. Air over the shore rises, and cooler air from above the water takes its place. The result is a sea breeze (Figure 22.7).

At night the process reverses as the shore cools off more quickly than the water—the warmer air is now over the sea. If you build a fire on the beach you'll notice that the smoke sweeps inward in the day and seaward at night.

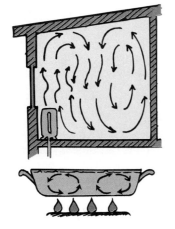

Figure 22.6 ▲
(Top) Convection currents in air. (Bottom) Convection currents in liquid.

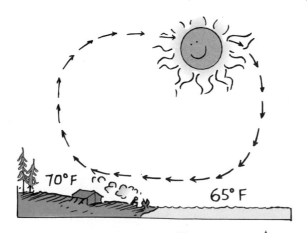

Figure 22.7 ▲
Convection currents are produced by uneven heating. The land is warmer than the water in the day and cooler than the water at night, so the direction of air flow reverses from day to night.

Cool Hand

With your mouth open wide, blow on your hand. Notice that your breath is warm. Now pucker your lips to make a small opening with your mouth and blow on your hand again. Does the temperature of the air on your hand feel the same? In which case does your exhaled breath expand more— when blowing with your mouth open wide or when blowing with your lips puckered? When did the air on your hand feel cooler? Why?

Why Rising Warm Air Cools

Rising warm air, like a rising balloon, expands. Why? Because less atmospheric pressure squeezes on it at higher altitudes. As the air expands, it cools—just the opposite of what happens when air is compressed. If you've ever compressed air with a tire pump, you probably noticed that the air and pump became quite hot. The opposite happens when air expands. Expanding air cools.

We can understand the cooling of expanding air by thinking of molecules of air as tiny balls bouncing against one another. Speed is picked up by a ball when it is hit by another that approaches with a greater speed. But when a ball collides with one that is receding, its rebound speed is reduced (Figure 22.8). Likewise for a Ping-Pong ball moving toward a paddle; it picks up speed when it hits an approaching paddle, but loses speed when it hits a receding paddle. The same idea applies to a region of air that is expanding; molecules collide, on the average, with more molecules that are receding than are approaching (Figure 22.9). Thus, in expanding air, the average speed of the molecules decreases and the air cools.*

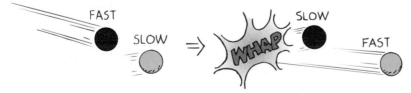

Figure 22.8 ▲
When a molecule collides with a target molecule that is receding, its rebound speed after the collision is less than it was before the collision.

---

*   Where does the energy go in this case? We will see in Chapter 24 that it goes into work done on the surrounding air as the expanding air pushes outward.

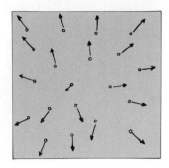

Figure 22.9 ▲
Molecules in a region of expanding air collide more often with receding molecules than with approaching ones. Their rebound speeds therefore tend to decrease and, as a result, the expanding air cools.

■ **Question**

You can hold your fingers beside the candle flame without harm, but not above the flame. Why?

22.3 Radiation

Heat from the sun is able to pass through the atmosphere and warm the earth's surface. This heat does not pass through the atmosphere by conduction, for air is one of the poorest conductors. Nor does it pass through by convection, for convection begins only after the earth is warmed. We also know that neither convection nor conduction is possible in the empty space between our atmosphere and the sun. The sun's heat is transmitted by another process—**radiation.***

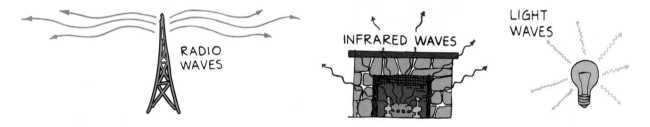

Figure 22.10 ▲
Types of radiant energy (electromagnetic waves).

Any energy, including heat, that is transmitted by radiation is called **radiant energy.** Radiant energy is in the form of *electromagnetic waves.* It includes radio waves, microwaves, infrared radiation, visible light, ultraviolet radiation, X rays, and gamma rays. These types of radiant energy are listed in order of wavelength, from longest to shortest.**

■ **Answer**

Heat travels upward by air convection. Since air is a poor conductor, very little heat travels sideways.

* The word *radiation* has more than one meaning. Do not confuse heat radiation with radioactive radiation, which is given off by the nuclei of radioactive atoms such as uranium and radium.

** Infrared (below-the-red) radiation has longer wavelengths than those of visible light. The longest visible wavelengths are for red light, and the shortest are for violet light. Ultraviolet (beyond-the-violet) radiation has shorter wavelengths. (More on wavelength in Chapter 25, and electromagnetic waves in Chapters 27 and 37.)

All objects continually emit radiant energy in a mixture of wavelengths. Objects at low temperatures emit long waves, just as long, lazy waves are produced when you shake a rope with little energy (Figure 22.11, top). Higher-temperature objects emit waves of shorter wavelengths. Objects of everyday temperatures emit waves mostly in the long-wavelength end of the infrared region, which is between radio and light waves. Shorter-wavelength infrared waves absorbed by our skin produce the sensation of heat. Thus, when we speak of heat radiation, we are speaking of infrared radiation.

If an object is hot enough, some of the radiant energy it emits is in the range of visible light. At a temperature of about 500°C an object begins to emit the longest waves we can see, red light. Higher temperatures produce a yellowish light. At about 1200°C all the different waves to which the eye is sensitive are emitted and we see an object as "white hot."

Common sources that give the sensation of heat are the burning embers in a fireplace, a lamp filament, and the sun. All of these emit both infrared radiation and visible light. When this radiant energy falls on other objects, it is partly reflected and partly absorbed. The part that is absorbed increases the internal energy of the objects.

Figure 22.11 ▲
Shorter wavelengths are produced when the rope is shaken more rapidly.

Figure 22.12 ▲
Most of the heat from a fireplace goes up the chimney by convection. The heat that warms us comes to us by radiation.

22.4 Absorption of Radiant Energy

Absorption and reflection are opposite processes. Therefore, a good absorber of radiant energy reflects very little radiant energy, including the range of radiant energy we call light. So a good absorber appears dark. A perfect absorber reflects no radiant energy and appears perfectly black. The pupil of the eye, for example, allows

Figure 22.13 ▲
Even though the interior of the box has been painted white, the hole looks black.

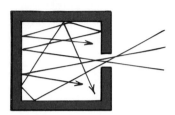

Figure 22.14 ▲
Radiant energy that enters an opening has little chance of leaving before it is completely absorbed.

Figure 22.15 ▲
Anything with a mirrorlike surface reflects most of the radiant energy it encounters. That's why it is a poor absorber of radiant energy.

radiant energy to enter with no reflection and appears perfectly black. (The pink "pupils" that appear in some flash portraits are from direct light reflected off the retina at the back of the eyeball.)

Look at the open ends of pipes in a stack. The holes appear black. Look at open doorways or windows of distant houses in the daytime, and they too look black. Openings appear black because the radiant energy that enters is reflected from the inside walls many times and is partly absorbed at each reflection until very little or none remains to come back out (Figure 22.14).

Good reflectors, on the other hand, are poor absorbers. Light-colored objects reflect more light and heat than dark-colored ones. In summer, light-colored clothing keeps people cooler.

22.5 Emission of Radiant Energy

Good absorbers are also good emitters; poor absorbers are poor emitters. For example, a radio antenna that is constructed to be a good emitter of radio waves will also, by its very design, be a good receiver of radio waves. A poorly designed transmitting antenna will also be a poor receiver. Interestingly enough, if a good absorber were not also a good emitter, then black objects would remain warmer than lighter-colored objects and never come to thermal equilibrium with them. But objects in thermal contact do come to thermal equilibrium. Each object is then emitting as much energy as it is absorbing. So a dark object that absorbs a lot must emit a lot as well.*

---

* Outdoors on a hot day, thermal equilibrium is not reached when black materials such as pavements or automobile bodies remain hotter than their surroundings—until evening, when they cool faster!

◀ **Figure 22.16**
When the containers are filled with hot water, the blackened one cools faster. If filled with cold water and exposed to radiant energy, the blackened one warms faster. Why?

To check this out, find a pair of metal containers of the same size and shape, one having a white or mirrorlike surface and the other a blackened surface (Figure 22.16). Fill the containers with hot water, and place thermometers in the water. You will find that the black container cools faster. The blackened surface is a better emitter. Coffee or tea will stay hot longer in a shiny mirrorlike pot than in a blackened one.

You can do the same experiment in reverse. This time fill each container with ice water and place the containers near a good source of radiant energy—in front of a fireplace, near a wood stove, or outside on a sunny day. You'll find that the black container warms up faster. A good emitter of radiant energy is also a good absorber.

Whether a surface plays the role of net emitter or net absorber depends on whether its temperature is above or below the surroundings. If the surface is hotter than the surroundings, for example, it will be a net emitter and will cool. If the surface is colder than the surroundings, it will be a net absorber and will become warmer. Every surface, hot or cold, both absorbs and emits radiant energy. If the surface absorbs more than it emits, it is a net absorber; if it emits more than it absorbs, it is a net emitter.

On a sunny day the earth's surface is a net absorber. At night it is a net emitter. On a cloudless night its "surroundings" are the frigid depths of space and cooling is faster than on a cloudy night, where the surroundings are nearby clouds. Record-breaking cold nights occur when the skies are clear.

■ Questions

1. If a good absorber of radiant energy were a poor emitter, how would its temperature compare with its surroundings?

2. Is it more efficient to paint a heating radiator black or silver?

■ Answers

1. If a good absorber were not also a good emitter, there would be a net absorption of radiant energy and the temperature of good absorbers would remain higher than the temperature of the surroundings. Things around us approach a common temperature only because good absorbers are, by their very nature, also good emitters.

2. Most of the heat provided by a heating radiator is accomplished by convection, so the color is not really that important. For optimum efficiency, however, the radiators should be painted a dull black so that the contribution by radiation is increased.

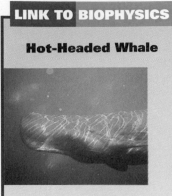
The next time you're in the direct heat of the sun, step in and out of the shade. You'll note the difference in the radiant energy you receive. Then think about the enormous amount of energy the sun emits to reach you some 150 000 000 kilometers distant. Is the sun unusually hot? Not as hot as some welding torches in auto shops. You feel the sun's heat not because it is hot (which it is), but primarily because it is *big*. Really big!

22.6 Newton's Law of Cooling

We know that an object at a temperature different from its surroundings will ultimately come to a common temperature with its surroundings. A relatively hot object cools as it warms its surroundings; a cold object warms as it cools its surroundings.

The rate of cooling of an object depends on how much hotter the object is than the surroundings. The temperature change per minute of a hot apple pie will be more if the hot pie is put in a cold freezer than if put on the kitchen table. When the pie cools in the freezer, the temperature difference is greater. A warm home will lose heat to the cold outside at a greater rate when there is a larger difference between the inside and outside temperatures. Keeping the inside of your home at a high temperature on a cold day is more costly than keeping it at a lower temperature. If you keep the temperature difference small, the rate of cooling will be correspondingly low.

The rate of cooling of an object—whether by conduction, convection, or radiation—is approximately proportional to the temperature difference ΔT between the object and its surroundings.

$$\text{rate of cooling} \sim \Delta T$$

This is known as **Newton's law of cooling.** (Guess who is credited with discovering this?)

Newton's law of cooling also holds for heating. If an object is cooler than its surroundings, its rate of warming up is also proportional to ΔT. Frozen food will warm up faster in a warm room than in a cold room.

■ Question

Since a hot cup of tea loses heat more rapidly than a lukewarm cup of tea, would it be correct to say that a hot cup of tea will cool to room temperature before a lukewarm cup of tea will?

■ Answer

No! Although the rate of cooling is greater for the hotter cup, it has farther to cool to reach thermal equilibrium. The extra time is equal to the time it takes to cool to the initial temperature of the lukewarm cup of tea. Cooling *rate* and cooling *time* are not the same thing.

22.7 Global Warming and the Greenhouse Effect

An automobile sitting in the bright sun on a hot day with its windows rolled up can get very hot inside—appreciably hotter than the outside air. This is an example of the **greenhouse effect,** so named for the same temperature-raising effect in florists' glass greenhouses. Understanding the greenhouse effect requires knowing about two concepts.

The first concept has been previously stated—that all things radiate, and the wavelength of radiation depends on the temperature of the object emitting the radiation. High-temperature objects radiate short waves; low-temperature objects radiate long waves. The second concept we need to know is that the transparency of things such as air and glass depends on the wavelength of radiation. Air is transparent to both infrared (long) waves and visible (short) waves, unless the air contains excess carbon dioxide and water vapor, in which case it is opaque to infrared. Glass is transparent to visible light waves, but is opaque to infrared waves. (The physics of transparency and opacity is discussed in Chapter 27.)

Now to why that car gets so hot in bright sunlight: Compared with the car, the sun's temperature is very high. This means the waves it radiates are very short. These short waves easily pass through both the earth's atmosphere and the glass windows of the car. So energy from the sun gets into the car interior, where, except for reflection, it is absorbed. The interior of the car warms up. Like the sun, the car interior radiates its own waves, but unlike the sun, the waves are longer. This is because the interior's temperature is much lower. The reradiated long waves encounter opaque glass windows. So reradiated energy remains in the car, which makes the car's interior even warmer. As hot as the interior gets, it won't be hot enough to radiate waves that can pass through glass (unless it glows red or white hot!).

The same effect occurs in the earth's atmosphere, which is transparent to solar radiation. The surface of the earth absorbs this energy, and reradiates part of this at longer wavelengths. Energy that the earth radiates is called **terrestrial radiation.** Atmospheric gases (mainly carbon dioxide and water vapor) absorb and re-emit much of this

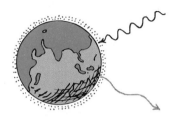

Figure 22.17 ▲
The earth's temperature depends on the energy balance between incoming solar radiation and outgoing terrestrial radiation.

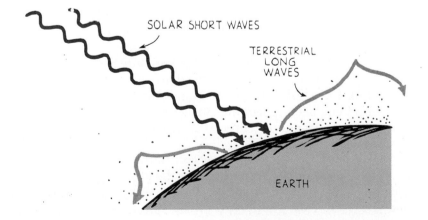

SOLAR SHORT WAVES

TERRESTRIAL LONG WAVES

EARTH

◀ **Figure 22.18**
The earth's atmosphere acts as a sort of one-way valve. It allows visible light from the sun in, but because of its water vapor and carbon dioxide content, it prevents terrestrial radiation from leaving.

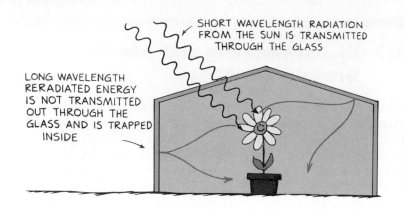

Figure 22.19 ▶
Shorter-wavelength radiant energy from the sun enters through the glass roof of the greenhouse. The soil emits long-wavelength radiant energy, which is unable to pass through the glass. Income exceeds outgo, so the interior is warmed.

SHORT WAVELENGTH RADIATION FROM THE SUN IS TRANSMITTED THROUGH THE GLASS

LONG WAVELENGTH RERADIATED ENERGY IS NOT TRANSMITTED OUT THROUGH THE GLASS AND IS TRAPPED INSIDE

long-wavelength terrestrial radiation back to Earth. So the long-wavelength radiation that cannot escape the earth's atmosphere warms the earth. This global warming process is very nice, for the earth would be a frigid –18°C otherwise. Over the last 500 000 years the average temperature of the earth has fluctuated between 19°C and 27°C and is presently at the high point, 27°C. Our present environmental concern is that increased levels of carbon dioxide and other atmospheric gases in the atmosphere may further increase the temperature and produce a new thermal balance unfavorable to the biosphere.

Interestingly enough, in the florist's greenhouse, heating is mainly due to the ability of glass to prevent convection currents from mixing the cooler outside air with the warmer inside air. So the greenhouse effect plays a bigger role in global warming than it does in the warming of greenhouses.

Averaged over a few years, the amount of solar radiation that strikes the earth exactly balances the terrestrial radiation the earth emits into space. This balance results in the average temperature of the earth—a temperature that presently supports life as we know it. Over a period of centuries, or even decades, the earth's average temperature can be changed—by natural causes and also by human activity. Adding certain materials to the atmosphere changes the absorption and reflection of solar radiation. Burning fossil fuels heats the environment. Except where the source of energy is solar, wind, or water, increased energy consumption on Earth adds heat. These activities can change the radiative balance and change the earth's average temperature.

An important credo is, "You can never change only one thing." Change one thing, and you change another. A slightly higher earth temperature means slightly warmer oceans, which means slightly increased evaporation, which means slightly increased snowfall in polar regions. The fraction of the earth presently beneath ice and snow is greater than the total area used for farmlands. These larger white areas reflect more solar radiation, which can lead to a significant drop in global temperature. So overheating the earth today may cool it tomorrow and trigger the next ice age! Or it might not. We don't know.

What we do know is that energy consumption is related to population size. We are seriously questioning the idea of continued growth. (Please take the time to read Appendix E, "Exponential Growth and Doubling Time"—very important stuff.)

22 Chapter Review

Concept Summary

Heat transfer by conduction takes place within certain materials and from one material to another when in contact.

- Metals are good conductors.
- Poor conductors, such as wood, cork, polystyrene, and most liquids and gases, are good insulators.

Heat transfer by convection takes place by the movement of heated material itself.

- Convection occurs in all fluids (both liquids and gases).
- Winds result from convection currents that stir the atmosphere.

Heat transfer by radiation takes place from everything to everything, even in empty space.

- Energy transmitted by radiation is called radiant energy.
- A good absorber of radiant energy reflects very little radiant energy, including visible light, and thus appears dark.
- Good absorbers of radiant energy are good emitters.
- According to Newton's law of cooling, the rate of cooling (or warming) of an object is approximately proportional to the temperature difference between the object and its surroundings.

Important Terms

conduction (22.1)
conductor (22.1)
convection (22.2)
greenhouse effect (22.7)
insulator (22.1)
Newton's law of cooling (22.6)
radiant energy (22.3)
radiation (22.3)
terrestrial radiation (22.7)

Review Questions

1. What is the role of "loose" electrons in heat conductors? (22.1)

2. Why does a piece of room-temperature metal feel cooler to the touch than paper, wood, or cloth? (22.1)

3. What is the difference between a conductor and an insulator? (22.1)

4. Why are materials such as wood, fur, feathers, and even snow good insulators? (22.1)

5. What is meant by saying that cold is not a tangible thing? (22.1)

6. How does Archimedes' principle relate to convection? (22.2)

7. Why does the direction of coastal winds change from day to night? (22.2)

8. How does the temperature of a gas change when it is compressed? expands? (22.2)

9. A row of dominoes is placed upright, one next to the other. When one is tipped over, it knocks against its neighbor, which does the same in cascade fashion until the whole row collapses. Which of the three types of heat transfer is this most similar to? (22.1–22.3)

10. What is radiant energy? (22.3)

11. How do the wavelengths of radiant energy vary with the temperature of the radiating source? (22.3)

12. Why does a good absorber of radiant energy appear black? (22.4)

13. Why do eye pupils appear black? (22.4)

14. Is a good absorber of radiation a good emitter or a poor emitter? (22.5)

15. Which will normally cool faster, a black pot of hot tea or a silvered pot of hot tea? (22.5)

16. Which will undergo the greater rate of cooling, a red-hot poker in a warm oven or a red-hot poker in a cold room (or do both cool at the same rate)? (22.6)

17. Does Newton's law of cooling apply to warming as well as to cooling? (22.6)

18. What is terrestrial radiation? (22.7)

19. Solar radiant energy is composed of short waves, yet terrestrial radiation is composed of relatively longer waves. Why? (22.7)

20. **a.** What does it mean to say that the greenhouse effect is like a one-way valve?

    b. Is the greenhouse effect more pronounced for florists' greenhouses or for the earth's surface? (22.7)

Activities

1. If you live where there is snow, do as Benjamin Franklin did nearly two centuries ago and lay samples of light and dark cloth on the snow. Note the differences in the rate of melting beneath the cloths.

2. Wrap a piece of paper around a thick metal bar and place it in a flame. Note that the paper will not catch fire. Can you figure out why? (*Hint:* Paper generally will not ignite until its temperature reaches about 230°C.)

Think and Explain

1. At what common temperature will both a block of wood and a piece of metal feel neither hot nor cool when you touch them with your hand?

2. If you stick a metal rod in a snowbank, the end in your hand will soon become cold. Does cold flow from the snow to your hand?

3. Wood is a poor conductor, which means that heat is slow to transfer—even when wood is very hot. Why can firewalkers safely walk barefoot on red-hot wooden coals, but not safely walk barefoot on red-hot pieces of iron?

4. Notice that a desk lamp often has small holes near the top of the metal lampshade. How do these holes keep the lamp cool?

5. When a space shuttle is in orbit and there appears to be no gravity in the cabin, why can a candle not stay lit?

6. In Montana, the state highway department spreads coal dust on top of snow. When the sun comes out, the snow rapidly melts. Why?

7. Suppose that a person at a restaurant is served coffee before he or she is ready to drink it. In order that the coffee be hottest when the person is ready for it, should cream be added to it right away or just before it is drunk?

8. Will a can of beverage cool just as fast in the regular part of the refrigerator as it will in the freezer compartment? (What physical law do you think about in answering this?)

9. If you wish to save fuel on a cold day, and you're going to leave your warm house for a half hour or so, should you turn your thermostat down a few degrees, down all the way, or leave it at room temperature?

10. If the composition of the upper atmosphere were changed so that it permitted a greater amount of terrestrial radiation to escape, what effect would this have on the earth's climate? Conversely, what would be the effect if the upper atmosphere reduced the escape of terrestrial radiation?

23 Change of Phase

Matter around us exists in three common **phases**—solid, liquid, and gas. Matter can change from one phase (or *state*, as it is sometimes called) to another. Ice, for example, is the solid phase of H_2O. Add energy, and the rigid molecular structure breaks down to the liquid phase, water. Add more energy, and the liquid changes to the gaseous phase as the water boils to become steam.

The phase of matter depends upon its temperature and the pressure that is exerted upon it. Changes of phase usually involve a transfer of energy.

Phase changes transfer energy.

23.1 Evaporation

Water in an open container will eventually evaporate, or dry up. The liquid that disappears becomes water vapor in the air. **Evaporation** is a change of phase from liquid to gas that takes place at the surface of a liquid.

The temperature of anything is related to the average kinetic energy of its molecules. Molecules in the liquid phase continuously move about in all directions and bump into one another at different speeds. Some of the molecules gain kinetic energy while others lose kinetic energy. Those molecules at the surface of the liquid that gain kinetic energy by being bumped from below may have enough energy to break free of the liquid. They can leave the surface and fly into the space above the liquid. They now comprise a *vapor,* molecules in the gaseous phase.

The increased kinetic energy of molecules bumped free of the liquid comes from molecules remaining in the liquid. This is

Figure 23.1 ▲
The cloth covering on the sides of the canteen promotes cooling when it is wet.

Figure 23.2 ▲
Dogs have no sweat glands (except between the toes). They cool themselves by panting. In this way evaporation occurs in the mouth and within the bronchial tract.

"billiard-ball physics": When balls bump into one another and some gain kinetic energy, the other balls lose this same amount of kinetic energy. So the average kinetic energy of the molecules remaining behind in the liquid is lowered. Thus, evaporation is a cooling process.

A canteen (Figure 23.1) keeps cool by evaporation when the cloth covering on the sides is kept wet. As the faster-moving water molecules leave the cloth, the temperature of the cloth decreases. The cool cloth in turn cools the metal canteen by conduction, which in turn cools the water inside.

When the human body overheats, sweat glands produce perspiration. Evaporation of perspiration cools us and helps us maintain a stable body temperature. Animals that lack sweat glands must cool themselves in other ways (Figures 23.2 and 23.3).

Figure 23.3 ▲
Pigs lack sweat glands. They wallow in mud to cool themselves.

■ Question

Would evaporation be a cooling process if each molecule at the surface of a liquid had the same kinetic energy before and after bumping into other molecules?

■ Answer

No. If there were no change in kinetic energy during molecular bumping, there would be no change in temperature. The liquid cools only when there is a lowering of average kinetic energy of molecules in the liquid. This occurs when some molecules (like billiard balls) gain speed at the expense of others that lose speed. Those that leave (evaporate) are gainers while losers remain behind in the liquid and effectively lower the average kinetic energy.

23.2 Condensation

The process opposite to evaporation is **condensation**—the changing of a gas to a liquid. The formation of droplets of water on the outside of a cold soda can is an example. Water vapor molecules collide with the slower-moving molecules of the cold can surface. The vapor molecules give up so much kinetic energy that they can't stay in the gaseous phase. They condense.

Condensation also occurs when gas molecules are captured by liquids. In their random motion, gas molecules may strike the surface of a liquid and so lose kinetic energy. Then, attractive forces exerted on them by the liquid may hold them in the liquid. Gas molecules thus become liquid molecules.*

Condensation is a warming process. Kinetic energy lost by condensing gas molecules warms the surface they strike. A steam burn, for example, is more damaging than a burn from boiling water of the same temperature. Steam gives up energy when it condenses to the liquid that wets the skin.

Figure 23.4 ▲
Heat is given up by steam when it condenses inside the radiator.

Condensation in the Atmosphere

The air always contains some water vapor. At any given temperature, however, there is a limit to the amount of it in the air. When this limit is reached, the air is said to be **saturated.** In weather reports, the **relative humidity** indicates how much water vapor is in the air, compared with the limit for that temperature. It is important to note that relative humidity is *not* a measure of how much water vapor is in the air. On a balmy summer day with a low relative humidity, for example, there may be more water vapor in the air than on a cold winter day with a high relative humidity.

At a relative humidity of 100%, the air is saturated. More water vapor is required to saturate high-temperature air than low-temperature air. The warm air of tropical regions is capable of containing much more moisture than cold Arctic air.

For saturation, there must be water vapor molecules in the air undergoing condensation. When slow-moving molecules collide, some stick together—they condense. To understand this, think of a fly making grazing contact with flypaper. At low speed it would surely get stuck, whereas at high speed it is more able to rebound into the air. Similarly, when water vapor molecules collide, they are more likely to stick together and become part of a liquid if they are moving slowly (Figure 23.5). At higher speeds, they can bounce apart and remain in the gaseous phase. The faster water molecules move, the less able they are to condense to form droplets.

---

* Some solids such as solid carbon dioxide (dry ice) and naphthalene crystals (moth balls) go directly to the gaseous phase in a process called sublimation. Snow and ice in dry, direct sunlight do the same. The reverse happens as H_2O molecules in cold air form snow.

Figure 23.5 ▶
Molecules of water vapor are
more likely to stick together and
form a liquid at lower speeds
than at higher speeds.

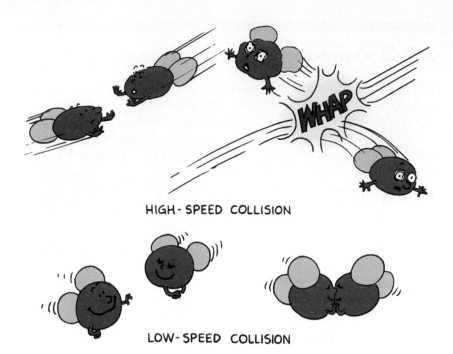

HIGH-SPEED COLLISION

LOW-SPEED COLLISION

Although condensation in the air occurs more readily at low
temperatures, it can occur at high temperatures also. Recall that
temperature is a measure of *average* kinetic energy. There are always
some molecules moving faster than average, and some moving
slower. Even at high temperature, there will be enough slow molecules to cause condensation—provided there is enough water
vapor present. Whatever the temperature, it is the slower molecules
that are more likely to stick.

Fog and Clouds

Warm air rises. As it rises, it expands. As it expands, it cools. As it
cools, water vapor molecules begin sticking together after colliding,
rather than bouncing off one another. If there are larger and slower-
moving particles or ions present, water vapor condenses upon these
particles, and we have a cloud.

Fog is basically a cloud that forms near the ground. Flying
through a cloud is much like driving through fog. Fog occurs in areas
where moist air near the ground cools. For example, moist air that
has blown in from over an ocean or lake may pass over cooler land.
Some of the water vapor condenses out of the air as it cools, and we
have fog.*

---

* What keeps the droplets of water in a cloud or fog from falling to Earth? If they're very
small, like dust, they have small terminal velocities, typically about 1 cm/s. This means
it takes 100 s to fall 1 m. A slow 1-cm/s updraft keeps the droplets in suspension. As
drops grow, their terminal velocity increases, and when terminal velocity is more than
updraft velocity, drops do fall. This is rain!

23.3 Evaporation and Condensation Rates

When you emerge from a shower and step into a dry room, you often feel chilly because evaporation is taking place quickly. If you stay in the shower stall, even with the water off, you will not feel as chilly. When you are in a moist environment, moisture from the air condenses on your skin and produces a warming effect that counteracts the cooling effect of evaporation. If as much moisture condenses as evaporates, you will feel no change in body temperature. That's why you can dry off with a towel more comfortably if you remain in the shower area.

If you leave a dish of water on a table for several days and no apparent evaporation takes place, you might conclude that nothing is happening in the water. You'd be mistaken, for much activity is taking place at the molecular level. Evaporation *and* condensation are occurring continuously at equal rates. The molecules and energy leaving the liquid's surface by evaporation are counteracted by as many molecules and as much energy returning by condensation. The liquid is in **equilibrium**—that is, in a state of balance—since evaporation and condensation have canceling effects.

Evaporation and condensation normally take place at the same time. If evaporation exceeds condensation, a liquid is cooled. If condensation exceeds evaporation, a liquid is warmed. Most often heat is transferred from *and* to the surroundings, so we do not notice cooling and warming due to evaporation and condensation.

Figure 23.6 ▲
If you feel chilly outside the shower stall, step back inside and be warmed by the condensation of the excess water vapor there.

23.4 Boiling

Evaporation takes place at the surface of a liquid. A change of phase from liquid to gas can also take place beneath the surface of a liquid under the proper conditions. The gas that forms beneath the surface causes bubbles. The bubbles are buoyed upward to the surface, where they escape into the surrounding air. This change of phase is called **boiling.**

The pressure of the vapor within the bubbles in a boiling liquid must be great enough to resist the pressure of the surrounding water. Unless the vapor pressure is great enough, the surrounding pressures will collapse any bubbles that may form. At temperatures below the boiling point, vapor pressure is not great enough, so bubbles do not form until the boiling point is reached.

As the atmospheric pressure is increased, the molecules in the vapor are required to move faster to exert increased pressure within the bubble in order to counteract the additional atmospheric pressure. So increasing the pressure on the surface of a liquid raises the boiling point of the liquid. Conversely, lowered pressure (as at high altitudes) decreases the boiling point. Thus, boiling depends not only on temperature but on pressure also.

PRESSURE OF ATMOSPHERE

Figure 23.7 ▲
The motion of molecules in the bubble of steam (much enlarged) creates a gas pressure that counteracts the water pressure against the bubble.

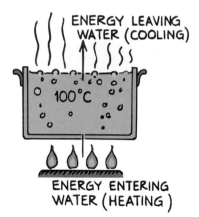

Figure 23.8 ▲
Heating and boiling are two distinct processes. Heating warms the water, and boiling cools it.

A pressure cooker is based on this fact. A pressure cooker has a tight-fitting lid that does not allow vapor to escape until it reaches a certain pressure greater than normal air pressure. As the evaporating vapor builds up inside the sealed pressure cooker, pressure on the surface of the liquid is increased, which prevents boiling. This raises the boiling point. The increased temperature of the water cooks the food faster.

It is important to note that it is the high temperature of the water that cooks the food, not the boiling process itself. At high altitudes, water boils at a lower temperature. In Denver, Colorado, the "mile-high city," for example, water boils at 95°C, instead of the 100°C boiling temperature characteristic of sea level. If you try to cook food in boiling water of a lower temperature, you must wait a longer time for proper cooking. A "three-minute" boiled egg in Denver is runny. If the temperature of the boiling water were very low, food would not cook at all.

Boiling, like evaporation, is a cooling process. This is surprising to some people because they associate boiling with heating. Heating water is one thing; boiling is another. When 100°C water at atmospheric pressure is boiling, it is in thermal equilibrium. Figure 23.8

■ Question

Since boiling is a cooling process, would it be a good idea to cool your hot and sticky hands by dipping them into boiling water?

■ Answer

No, no, no! When we say boiling is a cooling process, we mean that the water (not your hands!) is being cooled relative to the higher temperature it would attain otherwise. Because of cooling, it remains at 100°C instead of getting hotter. A dip in 100°C water would be most uncomfortable for your hands!

Physics can help with even the simplest of all cooked creations—the boiled egg. Test the egg for freshness by placing it in water. If it sinks and lies on its side, it's fresh. If it floats, it's rotten. An egg loses density as it ages because it loses moisture through pores in its shell, eventually becoming less dense than water. To test that the egg is raw, spin it on a tabletop. If it wobbles, it's uncooked. The wobbling indicates that the yolk is moving within the egg, thus changing the egg's center of gravity. Eggs sometimes crack while boiling due to an air pocket inside. With heat, the air pressure in the pocket increases enough to crack the shell. If you carefully pierce the egg's big end with a small, clean pin before boiling, it won't crack. Finally be sure you actually boil the water. You can heat an egg indefinitely at lower temperatures, but it doesn't cook. Cooking requires exceeding a threshold temperature so that the long-stranded molecules of the egg become cross-linked. That's why an egg won't cook by boiling at very high altitudes—the boiling water is not hot enough to cook the egg.

shows the water is being cooled by boiling as fast as it is being heated by energy from the heat source. If cooling did not take place, continued application of heat to a pot of boiling water would result in a continued increase in temperature. A pressure cooker reaches higher temperatures because it prevents boiling, which also prevents cooling.

23.5 Freezing

When energy is continually withdrawn from a liquid, molecular motion slows until the forces of attraction between the molecules cause them to fuse. The molecules then vibrate about fixed positions and form a solid. Water provides a good example of this process. When energy is extracted from water at a temperature of 0°C and at atmospheric pressure, ice is formed. The liquid water gives way to the solid ice phase. The change in phase from liquid to solid is called **freezing.**

Interestingly enough, if sugar or salt is dissolved in the water, the freezing temperature will be lowered. These "foreign" molecules or ions get in the way of water molecules that ordinarily would join together into the six-sided ice-crystal structure. As ice crystals do form, the hindrance is intensified, for the proportion of foreign molecules or ions among nonfused water molecules increases. Connections become more and more difficult. In general, dissolving anything in water has this result. Antifreeze is a practical application of this process.

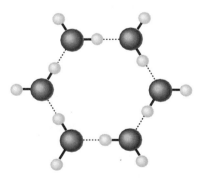

Figure 23.9 ▲
The open structure of pure ice crystals that normally fuse at 0°C. When other kinds of molecules or ions are introduced, crystal formation is interrupted, and the freezing temperature is lowered.

23.6 Boiling and Freezing at the Same Time

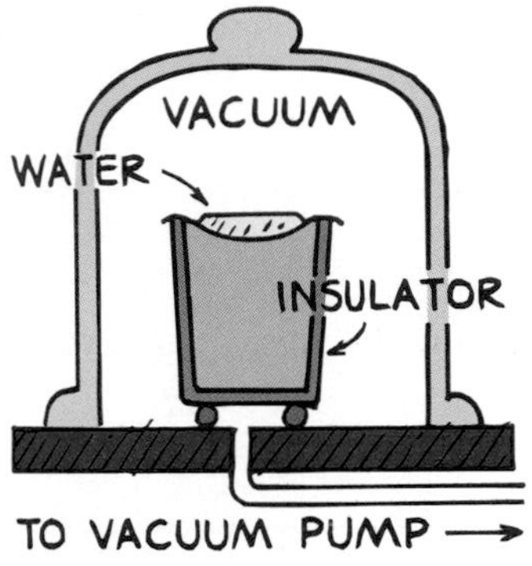

Suppose that a dish of water at room temperature is placed in a vacuum jar (Figure 23.10). If the pressure in the jar is slowly reduced by a vacuum pump, the water will start to boil. The boiling process takes heat away from the water left in the dish, which cools to a lower temperature. As the pressure is further reduced, more and more of the slower-moving molecules boil away. Continued boiling results in a lowering of temperature until the freezing point of approximately 0°C is reached. Continued cooling by boiling causes ice to form over the surface of the bubbling water. Boiling and freezing are taking place at the same time! This must be witnessed to be appreciated. Frozen bubbles of boiling water are a remarkable sight.

If some drops of coffee are sprayed into a vacuum chamber, they too will boil until they freeze. Even after they are frozen, the water molecules will continue to evaporate into the vacuum until little crystals of coffee solids are left. This is how freeze-dried coffee is made. The low temperature of this process tends to keep the chemical structure of coffee solids from changing. When hot water is added, much of the original flavor of the coffee is restored.

Figure 23.10 ▲
Apparatus to demonstrate that water will freeze and boil at the same time in a vacuum. A gram or two of water is placed in a dish that is insulated from the base by a polystyrene cup.

23.7 Regelation

The open-structured crystals of ice (Figure 23.9) can be crushed by the application of pressure. Whereas ice normally melts at 0°C, the application of pressure lowers the melting point. The crystals are simply crushed to the liquid phase. At twice standard atmospheric pressure, the melting point is lowered to −0.007°C. Quite a bit more pressure must be applied for an observable effect.

When the pressure is removed, refreezing occurs. This phenomenon of melting under pressure and freezing again when the pressure is reduced is called **regelation.** It is one of the properties of water that make it different from other substances.

You can see regelation in operation if you suspend a fine wire that supports heavy weights over an ice cube, as shown in Figure 23.11. The wire will slowly cut its way through the ice, but its track will refill with ice. You'll see the wire and weights fall to the floor, leaving the ice in a single solid piece!

To make a snowball, you use regelation. When you compress the snow with your hands, you cause a slight melting, which helps to bind the snow into a ball. Making snowballs is difficult in very cold weather, because the pressure you can apply may not be enough to melt the snow.

An ice skater skates on a thin film of water between the blade and the ice, which is produced by the blade pressure and friction. As soon as the pressure is released, the water refreezes.

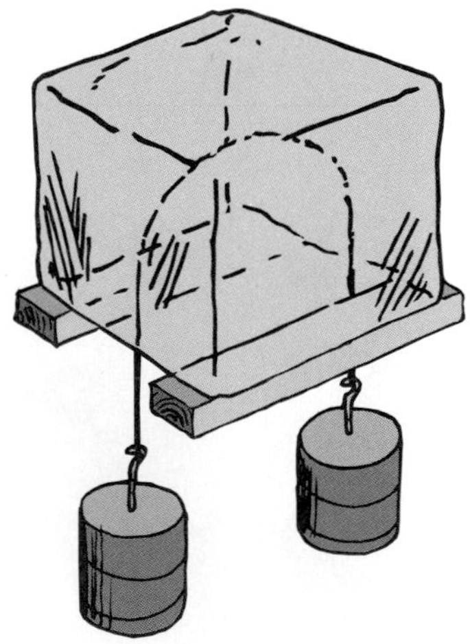

Figure 23.11 ▲
Regelation.

23.8 Energy and Changes of Phase

If you heat a solid sufficiently, it will melt and become a liquid. If you heat the liquid, it will vaporize and become a gas. Energy must be put into a substance to change its phase in the direction from solid to liquid to gas. Conversely, energy must be extracted from a substance to change its phase in the direction from gas to liquid to solid (Figure 23.12).

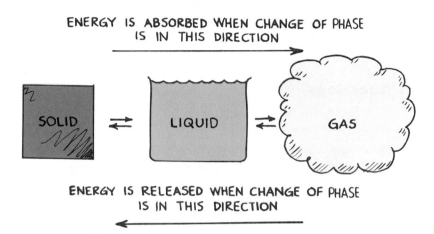

ENERGY IS ABSORBED WHEN CHANGE OF PHASE IS IN THIS DIRECTION

SOLID ⇌ LIQUID ⇌ GAS

ENERGY IS RELEASED WHEN CHANGE OF PHASE IS IN THIS DIRECTION

◄ **Figure 23.12**
Energy changes with change of phase.

The general behavior of many substances can be illustrated with a description of the changes of phase of H_2O. To make the numbers simple, suppose we have a 1-gram piece of ice at a temperature of –50°C in a closed container, and it is put on a stove to heat. A thermometer in the container reveals a slow increase in temperature up to 0°C. (It takes about half of a calorie to raise the temperature of ice by 1°C.) At 0°C, the temperature stops rising, yet heat is continually added. This heat melts the ice.

In order for the whole gram of ice to melt, 80 calories of heat energy must be absorbed by the ice. Not until all the ice melts does the temperature again begin to rise. Each additional calorie absorbed by the gram increases its temperature by 1°C until it reaches its boiling temperature, 100°C. Again, as heat is added, the temperature remains constant while more and more of the gram of water is boiled away and becomes steam. The water must absorb 540 calories of heat energy to vaporize the whole gram.* Finally, when all the water has become steam at 100°C, the temperature begins to rise once more. It continues to rise as long as heat is added (again taking about a half calorie per gram for each 1°C rise in temperature). This process is shown graphed in Figure 23.13.

---

* The value 540 calories per gram (in SI units, 2.26 megajoules/kilogram) required for vaporization or condensation is known as the *heat of vaporization* of water. The 80 calories per gram (in SI units, 0.335 mJ/kg) required for melting or freezing is known as the *heat of fusion* for water. These are energies per mass required to break intermolecular bonds during vaporization or melting, or equivalently, are energies released when bonds are formed during condensation or freezing. They vary with temperature and pressure.

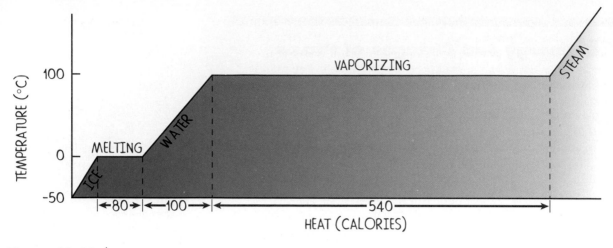

Figure 23.13 ▲
A graph showing the energy involved in the heating and the change of phase of 1 gram of H₂O. (See the same graph expressed in joules instead of calories in the lab manual.)

■ **Questions**

1. How much energy is released when a gram of steam at 100°C condenses to water at 100°C?

2. How much energy is released when a gram of steam at 100°C condenses and then cools to ice water at 0°C?

The phase change sequence is reversible. When the molecules in a gram of steam condense to form boiling water, they liberate 540 calories of heat to the environment. When the water is cooled from 100°C to 0°C, 100 additional calories are liberated to the environment. When ice water fuses to become solid ice, 80 more calories of energy are released by the water.

The 540 calories required to vaporize a gram of water is a relatively large amount of energy—much more than is required to change a gram of ice at absolute zero to boiling water at 100°C. Although the molecules in steam and boiling water at 100°C have the same average kinetic energy, steam has more potential energy, because the molecules are free of each other and are not held together in the liquid. Steam contains a vast amount of energy that can be released during condensation.

The large value of 540 calories per gram explains why under some conditions hot water will freeze faster than warm water.* This occurs for water hotter than 80°C. It is evident when the surface area

■ **Answers**

1. One gram of steam at 100°C releases 540 calories of energy when it condenses to become water at the same temperature.

2. The same steam releases 640 calories to reach ice water. That's 540 calories to change phase to water, and 100 more calories at the rate of 1 calorie per degree to cool to 0°C.

* Hot water will not freeze before cold water does, but it will freeze before lukewarm water does. Water at 100°C, for example, will freeze before water warmer than 60°C, but not before water cooler than 60°C. What's the "trick"? The hot water freezes first because more of it evaporates leaving less of it to freeze. Try it and see!

that cools by rapid evaporation is large compared with the amount of water involved. Examples are a car washed with hot water on a cold winter day, and a skating rink flooded with hot water to melt and smooth out the rough spots and refreeze quickly. The rate of cooling by rapid evaporation is very high because each gram of water that evaporates draws at least 540 calories from the water left behind. This is an enormous quantity of energy compared with the 1 calorie per Celsius degree that is drawn for each gram of water that cools by thermal conduction. Evaporation truly is a cooling process.

Figure 23.14 ▲
When a car is washed on a cold day, hot water will freeze more readily than warm water because of the energy that the rapidly evaporating water takes with it.

■ Question

Consider 10 grams of water at 100°C. What will be the temperature of the remaining 9 grams of water if 1 gram rapidly evaporates?

A refrigerator's cooling cycle is a good example of the energy interchanges that occur with the changes of phase of the refrigeration fluid. The liquid is pumped into the cooling unit, where it is forced through a tiny opening to evaporate and draw heat from the things stored in the food compartment. The gas is then directed outside the cooling unit to coils located in the back. As the gas condenses in the coils, appropriately called condensation coils, heat is given off to the surrounding air. The liquid returns to the cooling unit, and the cycle continues. A motor pumps the fluid through the system, where it enters the cyclic processes of vaporization and condensation. The

◀ **Figure 23.15**
The refrigeration cycle in a common refrigerator.

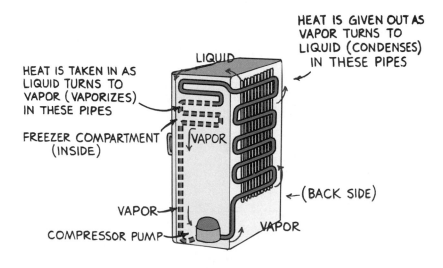

HEAT IS GIVEN OUT AS VAPOR TURNS TO LIQUID (CONDENSES) IN THESE PIPES

LIQUID

HEAT IS TAKEN IN AS LIQUID TURNS TO VAPOR (VAPORIZES) IN THESE PIPES

FREEZER COMPARTMENT (INSIDE)

VAPOR

(BACK SIDE)

VAPOR

VAPOR

COMPRESSOR PUMP

■ Answer

40°C, if we assume all the energy for evaporation is supplied by the remaining water. Why? Because 540 calories were taken away by the 1 gram that evaporated. This means that each of the remaining 9 grams gives up 60 calories (since (540 cal) ÷ 9 = 60 cal). Water cools at a rate of 1 degree Celsius per calorie, which means it drops 60 degrees to 40°C (since 100°C − 60°C = 40°C).

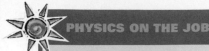

Fire Fighting

Firefighters regularly enter burning buildings to save lives and property. In order to perform their job effectively and safely, firefighters must be knowledgeable about the physics of heat. The most common fire control is dousing a flame with water. In some cases a fine mist is more effective in quenching a fire. Why? Because the fine mist readily turns to steam, and in so doing quickly absorbs energy and cools the burning material. Properly dealing with flames saves lives, including their own. To firefighters, the physics of heat is much more than a classroom assignment. It's a matter of staying alive. Job opportunities exist for firefighters with city or country fire departments and the national forest service.

next time you're near a refrigerator, place your hand near the condensation coils in the back, and you will feel the heat that has been extracted from within the cooling unit.

An air conditioner employs the same principles. It simply pumps heat from one part of the unit to another. When the roles of vaporization and condensation are reversed, the air conditioner becomes a heater. This is appropriately called a heat pump.

A way that some people judge the hotness of a clothes iron is to touch it briefly with a finger. This is also a way to burn the finger—unless it is first moistened. Energy that ordinarily would go into burning the finger goes, instead, into changing the phase of the moisture on it. The energy converts the moisture to a vapor, which additionally provides an insulating layer between the finger and the hot surface.

■ Question

When H_2O in the vapor phase condenses, is the surrounding air warmed or cooled?

Similarly, you may have seen news photos or heard stories about people walking barefoot without harm over red-hot coals from firewood. (CAUTION: Never try this on your own; even experienced "firewalkers" have received bad burns when the conditions were not just right.) The primary factor here is the low conductivity of wood—even red-hot wood. Although its temperature is high, relatively little heat is conducted to the feet, just as little heat is conducted by air when you put your hand briefly into a hot pizza oven. If you touch metal in the hot oven, OUCH! Similarly, a firewalker who steps on a hot piece of metal or another good conductor will be burned. Evaporation can play a role in firewalking too. A secondary factor is skin moisture. Perspiration on the soles of the feet decreases heat transfer to the feet. Much of the heat that would go to the feet instead goes to vaporizing the moisture—again, like touching a hot clothes iron with a wetted finger. Temperature is one thing; heat transfer is another.

In brief, a solid absorbs energy when it melts; a liquid absorbs energy when it vaporizes. Conversely, a gas emits energy when it liquefies; a liquid releases energy when it solidifies.

■ Answer

The surrounding air is warmed. Energy is released by the vapor when it turns into liquid (Figure 23.12). Another way to see that the air is warmed is to return to Figure 23.5 and our model of molecules in the air (a mixture of gases, including H_2O) as tiny billiard balls bouncing off one another. The total kinetic energy before and after collisions always remains the same. If one molecule gains kinetic energy in a collision, the other loses the same amount. Some molecules gain speed; some lose speed. What happens to losers of kinetic energy when they get near each other? They stick together. They condense from the air. But before they condense, they will have transferred much of their kinetic energy to other molecules. So the air is warmed. How much? By about 540 calories for each gram of H_2O that condenses.

23 Chapter Review

Concept Summary

During evaporation, a liquid changes phase at its surface and becomes a gas.

- Evaporation is a cooling process.

During condensation, a gas changes phase and becomes a liquid.

- Condensation is a warming process.
- At the same relative humidity, there is more water vapor in warm air than in cold air.
- Clouds and fog form when air cools and is unable to contain as much water vapor.

When evaporation and condensation occur at the same rate, the liquid is in equilibrium and there is no change in the liquid's volume.

- A liquid is in equilibrium when the surrounding air is saturated with its vapor.
- In dry air, water evaporates much faster than it condenses; in humid air, it evaporates only slightly faster than it condenses.

During boiling, a liquid changes phase at any place within the liquid, and gas bubbles form.

- The boiling temperature of a liquid depends on the pressure on its surface.
- Boiling, like evaporation, is a cooling process.

During freezing, a liquid changes phase and becomes a solid.

- The freezing temperature of a liquid is lowered by adding other substances to it.
- During regelation, ice melts under pressure and refreezes when the pressure is removed.

During phase changes, energy is given off or taken in.

- While a substance changes phase, its temperature does not change.
- Much more energy is given off when water vapor condenses than when an equal mass of water freezes.

Important Terms

boiling (23.4)
condensation (23.2)
equilibrium (23.3)
evaporation (23.1)
freezing (23.5)
phase (23.0)
regelation (23.7)
relative humidity (23.2)
saturated (23.2)

Activity

1. Boil some water in a pan and note that bubbles form at particular regions of the pan. These are nucleation sites—scratched or defected regions of the pan, or simply bits of crud. When water reaches the boiling point these sites provide havens where microscopic bubbles can collect long enough to become big bubbles. Nucleation sites are also important for phase changes of condensation and solidification. Snowflakes and raindrops typically form around dust particles, for example.

Review Questions

1. Do all the molecules or atoms in a liquid have about the same speed, or much different speeds? (23.1)

2. What is evaporation, and why is it also a cooling process? (23.1)

3. Why does a dog pant on a hot day? (23.1)

4. What is condensation, and why is it also a warming process? (23.2)

5. Why is being burned by steam more damaging than being burned by boiling water of the same temperature? (23.2)

6. Which usually contains more water vapor—warm air or cool air? (23.2)

7. Why does warm moist air form clouds when it rises? (23.2)

8. Why do you feel less chilly if you dry yourself inside the shower stall after taking a shower? (23.3)

9. How can you tell if the rate of evaporation equals the rate of condensation? (23.3)

10. What is the difference between evaporation and boiling? (23.4)

11. Why does the temperature at which a liquid boils depend on atmospheric pressure? (23.4)

12. Why is a pressure cooker even more useful when cooking food in the mountains than when cooking at sea level? (23.4)

13. Why does antifreeze or any soluble substance put in water lower its freezing temperature? (23.5)

14. How can water be made to both boil and freeze at the same time? (23.6)

15. What is regelation, and what does it have to do with the open-structured crystals in ice? (23.7)

16. a. How many calories are needed to raise the temperature of 1 gram of water by 1°C?

b. How many calories are needed to melt 1 gram of ice at 0°C?

c. How many calories are needed to vaporize 1 gram of boiling water at 100°C? (23.8)

17. Does a vapor give off or absorb energy when it turns into a liquid? (23.8)

18. What is the effect of rapid evaporation on the temperature of water? (23.8)

19. In a refrigerator, does the food cool when a vapor turns to a liquid, or vice versa? (23.8)

20. Why is it important that a finger be wet before it is touched briefly to a hot clothes` iron? (23.8)

Plug and Chug

Quantity of heat energy required for change of phase = (mass) × (heat of fusion or heat of vaporization), or in equation form,
$Q = mL$

Quantity of heat energy responsible for a temperature change = (mass) × (specific heat) × (change in temperature), or in equation form,
$Q = mc\Delta T$

For water, heat of fusion = 80 cal/g; heat of vaporization = 540 cal/g.

1. Calculate the energy (in calories) absorbed by 20 grams of water that warms from 30°C to 90°C.

2. Calculate the energy needed to melt 50 grams of 0°C ice.

3. Calculate the energy needed to melt 100 grams of 0°C ice and then heat it to 30°C.

4. Calculate the energy absorbed by 20 grams of 100°C water that is turned into 100°C steam.

5. Calculate the energy released by 20 grams of 100°C steam that condenses and then cools to 0°C?

Think and Explain

1. a. Evaporation is a cooling process. What cools and what warms during evaporation?

b. Condensation is a warming process. What warms and what cools during condensation?

2. You can determine wind direction if you wet your finger and hold it up into the air. Explain.

3. Give two reasons why pouring a hot cup of coffee into a saucer results in faster cooling.

4. At a picnic, why would wrapping a bottle in a wet cloth be a better method of cooling than placing the bottle in a bucket of cold water?

5. Why is the constant temperature of boiling water on a hot stove evidence that boiling is a cooling process? (What would happen to its temperature if boiling were not a cooling process?)

6. Will potatoes boiling in a pot of water cook faster if the water is boiling vigorously than if the water is boiling gently?

7. People who live where snowfall is common will attest to the fact that air temperatures are always higher on snowy days than on clear days. Some people get cause and effect mixed up when they say that snowfall cannot occur on very cold days. Explain.

8. If a large tub of water is kept in a small unheated room, even on a very cold day the temperature of the room will not go below 0°C. Why not?

9. On cold winter days the windows of your warm home sometimes get wet on the inside. Why is this so?

10. On a clear night, why does more dew form in an open field than under a tree or beneath a park bench?

Think and Solve

1. How much steam at 100°C must be condensed in order to melt 1 gram of 0°C ice and have the resulting ice water remain at 0°C? (The answer is *not* 0.148 grams!)

2. Calculate the energy released by 1 gram of 100°C boiling water that cools to form ice, and then continues releasing energy until it reaches absolute zero. (Absolute zero is −273°C, and the specific heat capacity of ice over this broad temperature range averages about 0.3 cal/g°C.)

3. Calculate the energy released by 1 gram of 100°C steam condensing to 1 gram of boiling water of the same temperature. How does this energy compare with the energy released in the previous problem?

4. If 20 grams of hot water at 80°C is poured into a cavity in a very large block of ice at 0°C, what will be the final temperature of the water in the cavity? How much ice must melt in order to cool the hot water down to this temperature?

5. If a 100-g piece of iron is heated to 100°C and then dropped into a cavity in a large block of ice at 0°C, how much ice will melt? (The specific heat capacity of iron is 0.11 cal/g°C.)

24 Thermodynamics

Tapping the earth's internal energy.

Thermodynamics is the study of heat and its transformation into mechanical energy. The word *thermodynamics* stems from Greek words meaning "movement of heat." The science of thermodynamics was developed in the mid-1800s, before the atomic and molecular nature of matter was understood. So far, our study of heat has been concerned with the microscopic behavior of jiggling atoms and molecules. Now we will see that thermodynamics bypasses the molecular details of systems and focuses on the macroscopic level—mechanical work, pressure, temperature, and their roles in energy transformation. The foundation of thermodynamics is the conservation of energy and the fact that heat flows from hot to cold, and not the other way around. It provides the basic theory of heat engines, from steam turbines to fusion reactors, and the basic theory of refrigerators and heat pumps. We begin our study of thermodynamics with a look at one of its early concepts—a lowest limit of temperature.

24.1 Absolute Zero

As thermal motion of atoms increases, temperature increases. There seems to be no upper limit of temperature. In contrast, there is a definite limit at the other end of the temperature scale. If we continually decrease the thermal motion of atoms in a substance, the temperature will drop. As the thermal motion approaches zero, the kinetic energy of atoms approaches zero, and the temperature of the substance approaches a lower limit. This limit is the **absolute zero** of temperature. At absolute zero, no more energy can be extracted from a substance and no further lowering of its temperature is possible. This limiting temperature is 273 degrees below zero on the Celsius scale.*

Absolute zero corresponds to zero degrees on the Kelvin, or thermodynamic, scale and is written 0 K (short for "zero kelvin"). Unlike the Celsius scale, there are no minus numbers on the thermodynamic scale. Degrees on the Kelvin scale are the same size as those on the Celsius scale. Thus, ice melts at 0°C, or 273 K, and water boils at 100°C, or 373 K. The Kelvin scale was named after the British physicist Lord Kelvin, who coined the word *thermodynamics* and first suggested such a scale.

■ Questions

1. Which is larger, a Celsius degree or a Kelvin?

2. A chunk of iron has a temperature of 0°C. If a second, identical chunk of iron is twice as hot (has twice the absolute temperature), what is its temperature in degrees Celsius?

24.2 First Law of Thermodynamics

In the eighteenth century, heat was thought to be an invisible fluid called *caloric*, which flowed like water from hot objects to cold objects. Caloric was conserved in its interactions, a discovery that led to the law of conservation of energy. In the 1840s, it became

■ Answers

1. Neither. They are equal.

2. The iron twice as hot is 273°C, because the 0°C chunk has an absolute temperature of 273 K, which when doubled is 546 K. To convert to Celsius, simply subtract 273 from Kelvin. Can you see why?

* This value was found in the 1800s by experimenters who discovered that all gases contract by the same proportion when temperature is decreased. It was found that any gas at 0°C, regardless of its initial pressure or volume, changes by 1/273 of its initial volume for each 1°C change in temperature, when pressure is held constant. For example, when the temperature is reduced to –100°C, the volume of gas is reduced by 100/273. More striking, if a gas at 0°C were cooled to –273°C, its volume would be reduced by 273/273 and become zero. Clearly, we cannot have a substance with zero volume. It was also found that the pressure of any gas in any container of fixed volume would change by 1/273 for each 1°C change. So gas in a container of fixed volume cooled to –273°C would have no pressure whatsoever. In practice, every gas liquefies before it gets this cold. Nevertheless, these decreases of volume and pressure by increments of 1/273 suggested the idea of the lowest temperature: –273°C (more precisely, –273.15°C, and –459.69° on the Fahrenheit scale). Interestingly enough, even at absolute zero, atoms still have a small kinetic energy, called the *zero-point energy*. Helium, for example, has enough motion at absolute zero to keep it from freezing. The explanation involves quantum theory.

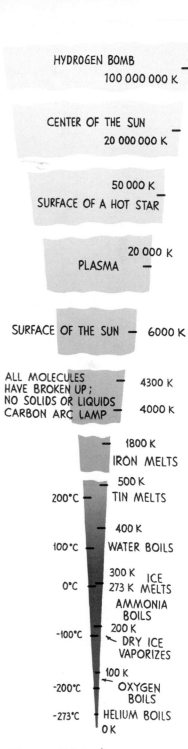

Figure 24.1 ▲
Some absolute temperatures.

apparent that the flow of heat was nothing more than the flow of energy itself. The caloric theory of heat was gradually abandoned.* Today we view heat as a form of energy. Energy can neither be created nor destroyed.

When the law of energy conservation is applied to thermal systems, we call it the **first law of thermodynamics.** We state it generally in the following form:

> Whenever heat is added to a system, it transforms to an equal amount of some other form of energy.

By *system,* we mean any group of atoms, molecules, particles, or objects we wish to deal with. The system may be the steam in a steam engine, the whole earth's atmosphere, or even the body of a living creature. It is important to define what is contained within the system as well as what is outside of it. If we add heat energy to the steam in a steam engine, to the earth's atmosphere, or to the body of a living creature, these systems will be able to do work on external things. This added energy does one or both of two things: (1) increases the internal energy of the system if it remains in the system and (2) does external work if it leaves the system. So, more specifically, the first law of thermodynamics states

$$\text{Heat added} = \frac{\text{increase in}}{\text{internal energy}} + \frac{\text{external work done}}{\text{by the system}}$$

Let's say you put an airtight can filled with air on a hot stove and heated it up. Warning: Do not actually do this. Since the can has a fixed volume, nothing moves, so no work is done. All the heat going into the can increases the internal energy of the enclosed air, so its temperature rises. This makes sense, for if heat is added to a system that does no external work, then the amount of heat added will be equal to the increase in the internal energy of the system. But if the system does external work, then the increase in internal energy will be correspondingly less. For example, if the can is fitted with a movable piston, then the heated air can do work as it expands—it can push the piston outward. Can you see that the temperature of the enclosed air will be less than if no work were done on the piston? The first law of thermodynamics makes good sense.

So we see that if a given quantity of heat is supplied to a steam engine, some of this heat increases the internal energy of the steam and the rest is transformed into mechanical work. That is, heat input equals the increase in internal energy plus the work output. The first law of thermodynamics is simply the thermal version of the law of conservation of energy.

Adding heat is not the only way to increase the internal energy of a system. If we set the "heat added" part of the first law to zero, we will see that changes in internal energy are equal to the work done

Figure 24.2 ▲
Paddle-wheel apparatus first used to compare heat energy with mechanical energy. As the weights fall, they give up potential energy and warm the water accordingly. This was first demonstrated by James Joule, for whom the unit of energy is named.

---

* Popular ideas, when proven wrong, are seldom suddenly discarded. People tend to identify with the ideas that characterize their time; hence, it is often the young who are more prone to discover and accept new ideas and push the human adventure forward.

Work Your Palms

Briskly rub your palms together. You're doing work on your skin.

What is the effect of the work done on the temperature of your palms? You can see that work can easily be converted to thermal energy. Is thermal energy as easily converted into work?

Activity

on or by the system.* If work is done on a system—compressing it, for example—the internal energy will increase. We have therefore raised the temperature of the system with no heat input. On the other hand, if work is done by the system—expanding against its surroundings, for example—the internal energy will decrease. With no heat extracted, the system cools.

Consider a bicycle pump. When we pump on the handle, the pump becomes hot. Why? Because we are putting mechanical work into the system and raising its internal energy. If the process happens quickly enough, so that very little heat is conducted from the system during compression, then nearly all of the work input will go into increasing internal energy, significantly raising temperature.

■ Questions

1. If 10 J of energy is added to a system that does no external work, by how much will the internal energy of that system be raised?

2. If 10 J of energy is added to a system that does 4 J of external work, by how much will the internal energy of that system be raised?

■ Answers

1. 10 J.

2. 6 J. We see from the first law that 10 J = 6 J + 4 J.

* Δ Heat = Δ internal energy + work

   $0 = \Delta$ internal energy + work

Then we can say

   –Work = Δ internal energy

Figure 24.3 ▲
Do work on the pump by pressing down on the piston and you compress the air inside. Adiabatic compression—the air is warmed.

24.3 Adiabatic Processes

The process of compression or expansion of a gas so that no heat enters or leaves a system is said to be **adiabatic** (Greek for "impassible"). Adiabatic changes of volume can be achieved by performing the process rapidly so that heat has little time to enter or leave (as with a bicycle pump), or by thermally insulating a system from its surroundings (with Styrofoam, for example).

A common example of a near adiabatic process is the compression and expansion of gases in the cylinders of an automobile engine (Figure 24.4). Compression and expansion occur in only a few hundredths of a second, too short a time for appreciable heat energy to leave the combustion chamber. For very high compressions, like those in a diesel engine, the temperatures achieved are high enough to ignite a fuel mixture without the use of a spark plug. Diesel engines have no spark plugs.

So when work is done on a gas by adiabatically compressing it, the gas gains internal energy and becomes warmer. When a gas adiabatically expands, it does work on its surroundings and gives up

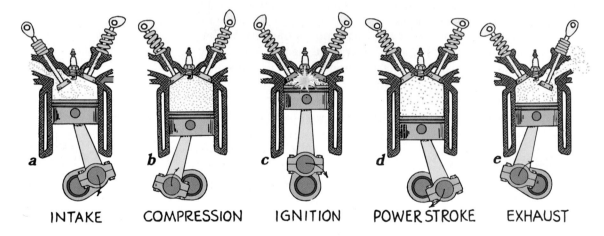

INTAKE   COMPRESSION   IGNITION   POWER STROKE   EXHAUST

Figure 24.4 ▲
One cycle of a four-cycle internal combustion engine. (a) A fuel-air mixture fills the cylinder as the piston moves down. (b) The piston moves up and compresses the mixture—adiabatically, since no heat transfer occurs. (c) The spark plug fires, ignites the mixture, and raises its temperature. (d) Adiabatic expansion pushes the piston downward—the power stroke. (e) The burned gases are pushed out the exhaust valve, and the cycle repeats.

internal energy, and thus becomes cooler. Recall the activity in Chapter 22 of blowing on your hand with puckered lips so your breath expands as it leaves your mouth (repeated here in Figure 24.5). Your breath is considerably cooler than when blown from your wide-open mouth without expanding.

Air temperature may be changed by adding or subtracting heat, by changing the pressure of the air, or by both. Heat may be added by solar radiation, by long-wave earth radiation, by moisture condensation, or by contact with the warm ground. Heat may be subtracted by radiation to space, by evaporation of rain falling through dry air, or by contact with cold surfaces.

There are many atmospheric processes, usually involving time scales of a day or less, in which the amount of heat added or subtracted is very small—small enough so that the process is nearly adiabatic. We then have the adiabatic form of the first law:

Change in air temperature   ~   pressure change

Adiabatic processes in the atmosphere occur in large masses of air that have dimensions on the order of kilometers. We'll call these large masses of air, *blobs*. Due to their large size, mixing of different temperatures or pressures of air occur only at their edges and don't appreciably alter the overall composition of the blobs. A blob behaves as if it were enclosed in a giant, tissue-light garment bag. As a blob of air flows up the side of a mountain, its pressure lessens, allowing it to expand and cool. The reduced pressure results in reduced temperature. Measurements show that the temperature of a blob of dry air drops by 10°C for each 1-kilometer increase in altitude (or for a decrease in pressure due to a 1-kilometer increase in altitude). So dry air cools 10°C for each kilometer it rises (Figure 24.6). Air flowing over tall mountains or rising in thunderstorms or cyclones may change elevation by several kilometers. So if a blob of dry air at ground level with a comfortable temperature of 25°C rose to 6 kilometers, its temperature would be a frigid –35°C. On the other hand, if air at a typical temperature of –20°C at an altitude of 6 kilometers descended to the ground, its temperature would be a roasting 40°C.

A dramatic example of this adiabatic warming is the *chinook*—a wind that blows down from the Rocky Mountains across the Great Plains. Cold air moving down the slopes of the mountains is compressed by the atmosphere into a smaller volume and is appreciably

Figure 24.5 ▲
Blow warm air onto your hand from your wide-open mouth. Now reduce the opening between your lips so the air expands as you blow. Adiabatic expansion—the air is cooled.

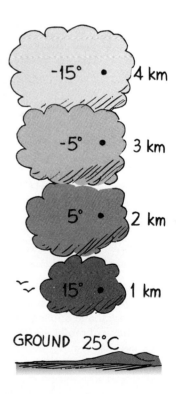

Figure 24.6 ▲
The temperature of a blob of dry air that expands adiabatically changes by about 10°C for each kilometer of elevation.

■ Questions

1. If a blob of air initially at 0°C expands adiabatically while flowing upward alongside a mountain a vertical distance of 1 km, what will its temperature be? When it has risen 5 km?

2. Imagine a giant dry-cleaner's garment bag full of air at a temperature of –10°C floating like a balloon with a string hanging from it 6 km above the ground. If you were able to yank it suddenly to the ground, what would its approximate temperature be?

■ Answers

1. At 1-km elevation, its temperature will be –10°C; at 5 km, –50°C.

2. If it were pulled down so quickly that heat conduction was negligible, it would be adiabatically compressed by the atmosphere and its temperature would rise to a piping hot 50°C (122°F), just as compressed air gets hot in a bicycle pump.

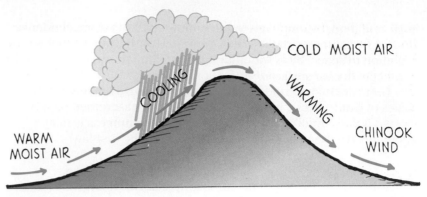

Figure 24.7 ▲
Chinooks, warm dry winds, occur when high-altitude air descends and is adiabatically warmed.

warmed. In this way communities in the paths of chinooks experience relatively warm weather in midwinter. The effect of expansion or compression on gases is quite impressive.*

24.4 Second Law of Thermodynamics

If we place a hot brick next to a cold brick, the hot brick will cool as heat flows to the cold brick. The cold brick will warm and the hot brick will cool until both bricks arrive at a common temperature: thermal equilibrium. No energy will be destroyed, in accord with the first law of thermodynamics. But pretend the hot brick takes heat from the cold brick and becomes hotter. Would this violate the first law of thermodynamics? Not if the cold brick becomes correspondingly colder so that the total energy of both bricks remains the same. This would not violate the first law, but it would violate the second law of thermodynamics. The second law tells us the direction of heat flow in natural processes. The **second law of thermodynamics** can be stated in many ways, but most simply it is this:

> Heat will never of itself flow from a cold object to a hot object.

Heat flows one way, downhill from hot to cold. In winter, heat flows from inside a warm heated home to the cold air outside. In summer, heat flows from the hot air outside into the cooler interior. The direction of heat flow is from hot to cold. Heat can be made to flow the other way, but only by imposing external effort—as occurs

Figure 24.8 ▲
A thunderhead is the result of the rapid adiabatic cooling of a rising mass of moist air. Its energy comes from condensation and freezing of water vapor.

---

* Interestingly enough, when you're flying at high altitudes where outside air temperature is typically –35°C, you're quite comfortable in your warm cabin—but not because of heaters. The process of compressing outside air to near sea-level cabin pressure would normally heat the air to a roasting 55°C (131°F). So air conditioners must be used to extract heat from the pressurized air.

with heat pumps that increase the temperature of air, or air conditioners that reduce air temperature. Without external effort, the direction of heat flow is from hot to cold.

There is a huge amount of internal energy in the ocean, but all this energy cannot be used to light a single flashlight lamp without external effort. Energy will not of itself flow from the lower-temperature ocean to the higher-temperature lamp filament.

24.5 Heat Engines and the Second Law

It is easy to change work completely into heat—simply rub your hands together briskly. Or push a crate at constant speed along a floor. All the work you do in overcoming friction is completely converted to heat. But the reverse process, changing heat completely into work, can never occur. The best that can be done is the conversion of some heat to mechanical work. The first heat engine to do this was the steam engine, invented in about 1700.

A **heat engine** is any device that changes internal energy into mechanical work. The basic idea behind a heat engine, whether a steam engine, internal combustion engine, or jet engine, is that mechanical work can be obtained only when heat flows from a high temperature to a low temperature. In every heat engine only some of the heat can be transformed into work.

In considering heat engines, we talk about *reservoirs*. Heat flows out of a high-temperature reservoir and into a low-temperature reservoir. Every heat engine will (1) absorb heat from a reservoir of higher temperature, increasing its internal energy, (2) convert some of this energy into mechanical work, and (3) expel the remaining energy as heat to some lower-temperature reservoir, usually called a sink (Figure 24.9). In a gasoline engine, for example, (1) the burning fuel in the combustion chamber is the high-temperature reservoir, (2) mechanical work is done on the piston, and (3) the expelled energy goes out as exhaust. The second law tells us that no heat engine can convert all the heat input to mechanical energy output. Only some of the heat can be transformed into work, with the remainder expelled in the process. Applied to heat engines, the second law may be stated:

> When work is done by a heat engine running between two temperatures, T_{hot} and T_{cold}, only some of the input heat at T_{hot} can be converted to work, and the rest is expelled as heat at T_{cold}.

There is always heat exhaust, which may be desirable or undesirable. Hot steam expelled in a laundry on a cold winter day may be quite desirable, while the same steam on a hot summer day is something else. When expelled heat is undesirable, we call it *thermal pollution*.

Before the second law was understood, it was thought that a very low friction heat engine could convert nearly all the input energy to

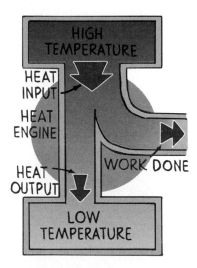

Figure 24.9 ▲
When heat energy flows in any heat engine from a high-temperature place to a low-temperature place, part of this energy is transformed into work output. (If work is put into a heat engine, the flow of energy may go from a low-temperature to a high-temperature place, as in a refrigerator or air conditioner.)

useful work. But not so. In 1824 the French engineer Sadi Carnot carefully analyzed the cycles of compression and expansion in a heat engine and made a fundamental discovery. He showed that the upper fraction of heat that can be converted to useful work, even under ideal conditions, depends on the temperature difference between the hot reservoir and the cold sink. His equation gives the ideal efficiency, or **Carnot efficiency,** of a heat engine.

$$\text{Ideal efficiency} = \frac{T_{\text{hot}} - T_{\text{cold}}}{T_{\text{hot}}}$$

T_{hot} is the temperature of the hot reservoir and T_{cold} is the temperature of the cold. Ideal efficiency depends only on the temperature difference between input and exhaust. Whenever ratios of temperatures are involved, the absolute temperature scale must be used. So T_{hot} and T_{cold} are expressed in kelvins. For example, when the hot reservoir in a steam turbine is 400 K (127°C) and the sink is 300 K (27°C), the ideal efficiency is

$$\frac{(400 - 300)}{400} = \frac{1}{4}$$

This means that even under *ideal* conditions, only 25% of the internal energy of the steam can be converted into work, while the remaining 75% is expelled as waste. This is why steam is super-heated to high temperatures in steam engines and power plants. The higher the steam temperature driving a motor or turbogenerator, the higher the efficiency of power production. (Increasing operating temperature in the example to 600 K yields an efficiency of (600 × 300)/600 = 1/2; twice the efficiency at 400 K.)

We can see the role of temperature difference between heat reservoir and sink in the operation of the steam-turbine engine in Figure 24.10. Steam from the boiler is the hot reservoir while the sink

Figure 24.10 ▶
A simplified steam turbine. The turbine turns because high-temperature steam from the boiler exerts more pressure on the front side of the turbine blades than the low-temperature steam exerts on the back side of the blades. Without a pressure difference, the turbine would not turn and deliver energy to an external load (an electric generator, for example). The presence of steam pressure on the back side of the blades, even in the absence of friction, prevents an engine from being perfectly efficient.

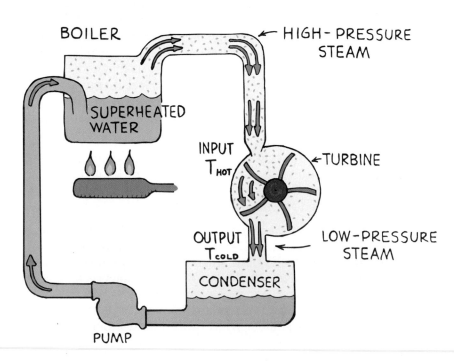

is the exhaust region after the steam passes through the turbine. The hot steam exerts pressure and does work on the turbine blades when it pushes on their front sides. This is nice. But steam pressure is not confined to the front sides of the blades; steam pressure is also exerted on the back sides of the blades—countereffective and not so nice. Pressure on the back sides is reduced most importantly because steam cools after giving much of its energy to the blades (and in practice, pressure is further reduced by lowering exhaust temperature by condensation outside the turbine). Even if friction were absent, the turbine's net work output would be the difference in work done on the blades by hot steam and work done by the blades on the cooler steam in exhausting it. We know that with confined steam, temperature and pressure go hand in hand—increase temperature and you increase pressure; decrease temperature and you decrease pressure. So the pressure difference necessary for the operation of a heat engine is directly related to the temperature difference between source and sink. The greater the temperature difference, the greater the efficiency.

Carnot's equation states the upper limit of efficiency for all heat engines. The higher the operating temperature (compared with exhaust temperature) of any heat engine, whether in an ordinary automobile, a nuclear-powered ship, or a jet aircraft, the higher the efficiency of that engine. In practice, friction is always present in all engines, and efficiency is always less than ideal.* So whereas friction is solely responsible for the inefficiencies of many devices, in the case of heat engines, the overriding concept is the second law of thermodynamics; only some of the heat input can be converted to work—even without friction.

■ Questions

1. What is the ideal efficiency of an engine if both its hot reservoir and exhaust are the same temperature—say 400 K?

2. What is the ideal efficiency of a machine having a hot reservoir at 400 K and a cold reservoir at absolute zero, 0 K?

■ Answers

1. Zero efficiency; (400 – 400)/400 = 0. This means no work output is possible for any heat engine unless a temperature difference exists between the reservoir and the sink.

2. (400 – 0)/400 = 1; only in this idealized case is an ideal efficiency of 100% possible.

* The ideal efficiency of an ordinary automobile engine is somewhat more than 50%, but in practice the actual efficiency is about 25%. Engines of higher operating temperatures (compared with sink temperatures) would be more efficient, but the melting point of engine materials limits the upper temperatures at which they can operate. Higher efficiencies await engines made with new materials with higher melting points. Watch for ceramic engines!

24.6 Order Tends to Disorder

The first law of thermodynamics states that energy can be neither created nor destroyed. The second law adds that whenever energy transforms, some of it degenerates into waste. The wasted energy is unavailable and is lost. Another way to say this is that organized energy (concentrated and therefore usable energy) degenerates into disorganized energy (nonusable energy). The energy of gasoline is organized and usable energy. When gasoline burns in an automobile engine, part of its energy does useful work such as moving the pistons, part of the energy heats the engine and surroundings, and part of the energy goes out the exhaust. Useful energy degenerates to nonuseful forms and is unavailable for doing the same work again, such as driving another automobile.

Figure 24.11 ▲
The Transamerica® Pyramid and some other buildings are heated by electric lighting, which is why the lights are on most of the time.

Organized energy in the form of electricity that goes into electric lights in homes and office buildings degenerates to heat energy. This is a principal source of heating in many office buildings in moderate climates, such as the Transamerica Pyramid in San Francisco. All of the electrical energy in the lamps, even the part that briefly exists in the form of light, turns into heat energy, which is used to warm the buildings (that explains why the lights are on most of the time). This energy is degenerated and has no further use. All the heat energy in the buildings cannot be reused to light a single lamp without some outside organizational effort.

We see that the quality of energy is lowered with each transformation. Energy of an organized form tends to disorganized forms. In this broader regard, the second law can be stated another way:

> Natural systems tend to proceed toward a state of greater disorder.

Imagine an orderly stream of molecules injected into a closed empty bottle. Molecules of gas all moving in harmony make up an orderly state—and also an unlikely state. The orderly state of molecules soon becomes disordered. Molecules of gas moving in haphazard directions and speeds make up a disorderly state—a more random and more likely state. So, order tends toward disorder. If we remove the lid of the bottle, the gas molecules will escape into the room and become even more disordered.

You would not expect the reverse to happen; that is, you would not expect the gas molecules to spontaneously order themselves back into the bottle to return to the more ordered containment. This is because compared with the infinite number of ways the molecules can randomly move, the chance of them returning to such an ordered state is practically zero. Such processes in which disorderly states tend toward orderly states are simply not observed.

Disordered energy can be changed to ordered energy only at the expense of some organizational effort or work input. For example, air can be ordered into a small region by using a compressor. But without some imposed work input, no increase in order occurs.

In the broadest sense, the message of the second law of thermodynamics is that the tendency of the universe, and all that is in it, tends to disorder.

Figure 24.12 ▲
Try to push a heavy crate across a rough floor and all your work will go into heating the floor and crate. Work against friction turns into disorganized energy.

Figure 24.13 ▲
Molecules of gas go from the bottle to the air and not the other way around.

24.7 Entropy

The idea of ordered energy tending to disordered energy is embodied in the concept of **entropy.*** Entropy is the measure of the amount of disorder. Disorder increases; entropy increases. The second law states that for natural processes, in the long run, entropy always increases. Gas molecules escaping from a bottle move from a relatively orderly state to a disorderly state. Organized structures in time become disorganized messes. Things left to themselves run down. Whenever a physical system is allowed to distribute its energy freely,

---

*   Entropy can be expressed as a mathematical equation, stating that the increase in entropy, ΔS, in an ideal thermodynamic system is equal to the amount of heat added to a system, ΔQ, divided by the temperature, T, of the system: $\Delta S = \Delta Q/T$.

Figure 24.14 ▶
Entropy.

it always does so in a manner such that entropy increases while the available energy of the system for doing work decreases.

Entropy normally increases in physical systems. However, when there is work input, as in living organisms, entropy decreases. All living things, from bacteria to trees to human beings, extract energy from their surroundings and use it to increase their own organization. This order in life forms is maintained by increasing entropy elsewhere, so life forms plus their waste products have a net increase in

Figure 24.15 ▲
Why is the motto of this contractor—"Increasing entropy is our business"— so appropriate?

SCIENCE, TECHNOLOGY, AND SOCIETY

Thermodynamics and Thermal Pollution

A modern electric power plant, though large and complex, can be approximated as a simple heat engine. The power plant uses heat from the burning of coal, oil, gas, or heat from nuclear fission to produce energy that does work by turning electric generators. In this process, it also produces waste heat as an inevitable consequence of the second law of thermodynamics. This waste heat is sometimes called *thermal pollution* because, like chemical wastes, it pollutes the environment.

Waste heat discharged into waterways can raise temperatures of aquatic environments enough to kill organisms and disrupt ecosystems. Waste heat discharged into the air can contribute to weather changes. Thermal pollution is unlike chemical pollution, since chemical pollution can be reduced by various methods. The only way to manage thermal pollution is to spread waste heat over areas large enough to absorb it without significantly increasing temperatures.

Try as we might, the second law of thermodynamics tells us that it is impossible to produce usable energy with zero environmental impact. Conservation and efficient technology are absolutely crucial to the health of our planet.

entropy.* Energy must be transformed into the living system to support life. When it is not, the organism soon dies and tends toward disorder.

The first law of thermodynamics is a universal law of nature for which no exceptions have been observed. The second law, however, is a probability statement. Given enough time, even the most improbable states may occur; entropy may sometimes spon–taneously decrease. For example, the haphazard motions of air molecules could momentarily become harmonious in a corner of the room, just as a barrelful of pennies dumped on the floor could all come up heads. These situations are possible—but not probable. The second law tells us the most probable course of events—not the only possible one.

The laws of thermodynamics are sometimes put this way: You can't win (because you can't get any more energy out of a system than you put in), you can't break even (because you can't even get as much energy out as you put in), and you can't get out of the game (entropy in the universe is always increasing).

---

* Interestingly enough, the American writer Ralph Waldo Emerson, who lived when the second law of thermodynamics was the new science topic of the day, philosophically speculated that not everything becomes more disordered with time. He cited the example of human thought. Ideas about the nature of things grow increasingly refined and organized as they pass through the minds of succeeding generations. Human thought is evolving toward more order.

24 Chapter Review

Concept Summary

Thermodynamics is the study of heat and work.

- Absolute zero is the lowest possible temperature that a substance may have; where molecules of a substance have minimum kinetic energy.

First law of thermodynamics: The heat added to a system equals the sum of the increase in internal energy plus the external work done by the system. This is a restatement of the law of energy conservation applied to heat.

- An adiabatic process is one usually of expansion or compression, wherein no heat enters or leaves a system.

Second law of thermodynamics: Heat does not spontaneously flow from a cold object to a hot object. No machine can be completely efficient in converting energy to work; some input energy is dissipated as heat. All systems tend to become more and more disordered as time goes by.

Entropy is a measure of the disorder of a system. Whenever energy freely transforms from one form to another, the direction of transformation is toward a state of greater disorder (greater entropy).

Important Terms

absolute zero (24.1)
adiabatic (24.3)
Carnot efficiency (24.5)
entropy (24.7)
first law of thermodynamics (24.2)
heat engine (24.5)
second law of thermodynamics (24.4)
thermodynamics (24.0)

Review Questions

1. What is the meaning of the Greek words from which we get the word *thermodynamics?* (24.0)

2. Is the study of thermodynamics concerned primarily with microscopic or macroscopic processes? (24.0)

3. What is the lowest possible temperature on the Celsius scale? On the Kelvin scale? (24.1)

4. What is the temperature of melting ice in kelvins? Of boiling water? (24.1)

5. How does the law of the conservation of energy relate to the first law of thermodynamics? (24.2)

6. What happens to the internal energy of a system when work is done on it? What happens to its temperature? (24.2)

7. What is the relationship between heat added to a system and the internal energy and external work done by the system? (24.2)

8. If work is done on a system, will the internal energy of the system increase or decrease? If work is done by a system, will the internal energy of the system increase or decrease? (24.2)

9. What condition is necessary for a process to be adiabatic? (24.3)

10. What happens to the temperature of air when it is adiabatically compressed? When it adiabatically expands? (24.3)

11. What generally happens to the temperature of rising air? (24.3)

12. What generally happens to the temperature of sinking air? (24.3)

13. How does the second law of thermodynamics relate to the direction of heat flow? (24.4)

14. What three processes occur in every heat engine? (24.5)

15. What is thermal pollution? (24.5)

16. If all friction could be removed from a heat engine, would it be 100% efficient? Explain. (24.5)

17. What is the ideal efficiency of a heat engine that operates with its hot reservoir at 500 K and its sink at 300 K? (24.5)

18. Why are heat engines intentionally run at high operating temperatures? (24.5)

19. Give at least two examples to distinguish between organized energy and disorganized energy. (24.6)

20. How much of the electrical energy transformed by a common lightbulb becomes heat energy? (24.6)

21. With respect to orderly and disorderly states, what do natural systems tend to do? Can a disorderly state ever transform to an orderly state? Explain. (24.6)

22. What is the physicist's term for a measure of messiness? (24.7)

23. Under what condition can entropy decrease in a system? (24.7)

24. What is the relationship between the second law of thermodynamics and entropy? (24.7)

25. Distinguish between the first and second laws of thermodynamics in terms of whether or not exceptions occur. (24.7)

Plug and Chug

1. Calculate the ideal efficiency of a heat engine that takes in energy at 800 K and expels heat to a reservoir at 300 K.

2. Calculate the ideal efficiency of a ship's boiler when steam comes out at 530 K, pushes through a steam turbine, and exits into a condenser that is kept at 290 K by circulating seawater.

3. Calculate the ideal efficiency of a steam turbine that has a hot reservoir of 112°C high-pressure steam and a sink at 27°C.

4. In a heat engine driven by ocean temperature differences, the heat source (water near the surface) is at 293 K and the heat sink (deeper water) is at 283 K. Calculate the ideal efficiency of the engine.

Think and Explain

1. A friend said the temperature inside a certain oven is 600 and the temperature inside a certain star is 60 000. You're unsure about whether your friend meant kelvins or degrees Celsius. How much difference does it make in each case?

2. When you pump a tire with a bicycle pump, the cylinder of the pump becomes hot. Give two reasons why this is so.

3. Is it possible to entirely convert a given amount of heat into mechanical energy? Is it possible to entirely convert a given amount of mechanical energy into heat? Cite examples to illustrate your answers.

4. We know that warm air rises. So it might seem that the air temperature should be higher at the top of mountains than down below. But the opposite is most often the case. Why?

5. Will the efficiency of a car engine increase, decrease, or remain the same if the muffler is removed? If the car is driven on a very cold day? Defend your answers.

6. The combined molecular kinetic energies of molecules in a very large container of cold water are greater than the combined molecular kinetic energies in a cup of hot tea. Pretend you partially immerse the teacup in the cold water and that the tea absorbs 10 joules of energy from the water and becomes hotter, while the water that gives up 10 joules of energy becomes cooler. Would this energy transfer violate the first law of thermodynamics? The second law of thermodynamics? Explain.

7. A mixture of fuel and air is burned rapidly in a combustion engine to push a piston in the engine that in turn propels the vehicle. In a jet engine a mixture of fuel and air is burned rapidly and, instead of pushing pistons, pushes the aircraft itself. Which do you suppose is more efficient?

8. Suppose one wishes to cool a kitchen by leaving the refrigerator door open and closing the kitchen door and windows. What will happen to the room temperature? Why?

9. In buildings that are being heated electrically, is it wasteful to turn on all the lights? Is turning on all the lights wasteful if the building is being cooled by air conditioning? Defend your answers.

10. Water put into a freezer compartment in your refrigerator goes to a state of less molecular disorder when it freezes. Is this an exception to the entropy principle? Explain.

Think and Solve

1. Helium has the special property that its internal energy is directly proportional to its absolute temperature. Consider a flask of helium with a temperature of 10°C. If it is heated until it has twice the internal energy, what will its temperature be?

2. A heat engine takes in 100 kJ of energy from a source at 800 K and expels 50 kJ to a reservoir at 300 K. Calculate the ideal efficiency and the actual efficiency of the engine.

3. Which heat engine has greater ideal efficiency, one that operates between the temperatures 600 K and 400 K or one that operates between 500 K and 400 K? Explain how your answer conforms to the idea that a higher operating temperature yields higher efficiency.

4. Imagine a giant dry-cleaner's bag full of air at a temperature of –35°C floating like a balloon with a string hanging from it 10 km above the ground. Estimate its temperature if you were able to yank it suddenly to the earth's surface.

IV Sound and Light

Isn't this disc the pits? I mean, there are billions of them, carefully inscribed in an array that is scanned at millions of pits per second by a laser beam. Digitized music! Or a whole encyclopedia! But the beauty of a CD is more than what it holds—just look at the brilliant spectrum of colors diffracted by the evenly spaced rows of pits. I find it even more beautiful when I know **why** it's so colorful and **why** it holds so much music or information. That's the physics of it all!

25 Vibrations and Waves

Vibrations carry energy.

A ll around us we see things that wiggle and jiggle. Even things too small to see, such as atoms, are constantly wiggling and jiggling. A wiggle in time is a **vibration.** A vibration cannot exist in one instant, but needs time to move back and forth. Strike a bell and the vibrations will continue for some time before they die down.

A wiggle in space and time is a **wave.** A wave cannot exist in one place but must extend from one place to another. Light and sound are both forms of energy that move through space as waves. This chapter is about vibrations and waves, and the following chapters continue with the study of sound and light.

25.1 Vibration of a Pendulum

Suspend a stone at the end of a string and you have a simple pendulum. Pendulums swing back and forth with such regularity that they have long been used to control the motion of clocks. Galileo discovered that the time a pendulum takes to swing back and forth through small angles does not depend on the mass of the pendulum or on the distance through which it swings. The time of a back-and-forth swing—called the **period**—depends only on *the length of the pendulum* and *the acceleration of gravity.**

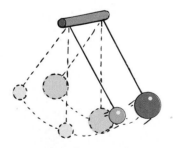

Figure 25.1 ▲
Two pendulums of the same length have the same period regardless of mass.

---

* The exact relationship for the period *T* of a simple pendulum is

$$T = 2\pi \sqrt{\frac{L}{g}}$$

where *L* is the length of the pendulum, and *g* is the acceleration of gravity.

A long pendulum has a longer period than a shorter pendulum; that is, it swings back and forth more slowly—less frequently—than a short pendulum. When walking, we allow our legs to swing with the help of gravity, like a pendulum. In the same way that a long pendulum has a greater period, a person with long legs tends to walk with a slower stride than a person with short legs. This is most noticeable in long-legged animals such as giraffes, horses, and ostriches, which run with a slower gait than do short-legged animals such as dachshunds, hamsters, and mice.

25.2 Wave Description

The back-and-forth vibratory motion (often called oscillatory motion) of a swinging pendulum is called **simple harmonic motion.*** The pendulum bob filled with sand in Figure 25.2 exhibits simple harmonic motion above a conveyor belt. When the conveyor belt is stationary (left), the sand traces out a straight line. More interestingly, when the conveyor belt is moving at constant speed (right), the sand traces out a special curve known as a **sine curve.**

Figure 25.2 ▲
Frank Oppenheimer, founder of the Exploratorium® science museum in San Francisco, demonstrates that a pendulum swinging back and forth traces out a straight line over a stationary surface, and a sine curve when the surface moves at constant speed.

---

* The condition for simple harmonic motion, nearly always met for most vibrations, is that the restoring force is proportional to the displacement from equilibrium. The component of weight that restores a displaced pendulum to its equilibrium position is directly proportional to the pendulum's displacement (for small angles)—likewise for a weight attached to a spring. Recall from Section 18.3, Hooke's law for a spring: $F = k\Delta x$, where the force that stretches (or compresses) a spring is directly proportional to the distance the spring is stretched (or compressed).

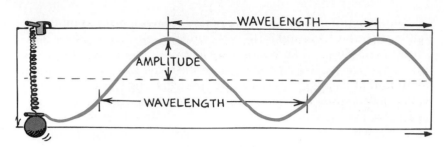

Figure 25.3 ▲
A sine curve.

The same can be done with a weight attached to a spring undergoing vertical simple harmonic motion, shown in Figure 25.3. A marking pen attached to the bob traces a sine curve on a sheet of paper that is moving horizontally at constant speed. A sine curve is a pictorial representation of a wave. Like a water wave, the high points are called **crests,** and the low points are called **troughs.** The straight dashed line represents the "home" position, or midpoint of the vibration. The term **amplitude** refers to the distance from the midpoint to the crest (or trough) of the wave. So the amplitude equals the maximum displacement from equilibrium.

The **wavelength** of a wave is the distance from the top of one crest to the top of the next one. Or equivalently, the wavelength is the distance between successive identical parts of the wave. The wavelengths of waves at the beach are measured in meters, the wavelengths of ripples in a pond in centimeters, and the wavelengths of light in billionths of a meter (nanometers).

How frequently a vibration occurs is described by its **frequency.** The frequency of a vibrating pendulum, or object on a spring, specifies the number of back-and-forth vibrations it makes in a given time (usually one second). A complete back-and-forth vibration is one cycle. If it occurs in one second, the frequency is one vibration per second or one cycle per second. If two vibrations occur in one second, the frequency is two vibrations or two cycles per second.

The unit of frequency is called the **hertz** (Hz), after Heinrich Hertz, who demonstrated radio waves in 1886. One cycle per second is 1 hertz, two cycles per second is 2 hertz, and so on. Higher frequencies are measured in kilohertz (kHz—thousands of hertz), and still higher frequencies in megahertz (MHz—millions of hertz) or gigahertz (GHz—billions of hertz). AM radio waves are broadcast in kilohertz, while FM radio waves are broadcast in megahertz; radar and microwave ovens operate at gigahertz. A station at 960 kHz on the AM radio dial, for example, broadcasts radio waves that have a frequency of 960 000 vibrations per second. A station at 101 MHz on the FM dial broadcasts radio waves with a frequency of 101 000 000 hertz. These radio-wave frequencies are the frequencies at which electrons are forced to vibrate in the antenna of a radio station's transmitting tower.

The source of all waves is something that vibrates. The frequency of the vibrating source and the frequency of the wave it produces is the same.

Figure 25.4 ▲
Electrons in the transmitting antenna of a radio station at 960 kHz on the AM dial vibrate 960 000 times each second and produce 960-kHz radio waves.

If the frequency of a vibrating object is known, its period can be calculated, and vice versa. Suppose, for example, that a pendulum makes two vibrations in one second. Its frequency is 2 Hz. The time needed to complete one vibration—that is, the period of vibration—is 1/2 second. Or if the vibration period is 3 Hz, then the period is 1/3 second. As you can see below, frequency and period are inverses of each other:

$$\text{frequency} = \frac{1}{\text{period}}$$

and vice versa,

$$\text{period} = \frac{1}{\text{frequency}}$$

■ Questions

1. What is the frequency in vibrations per second of a 100-Hz wave?

2. The Sears® Building in Chicago sways back and forth at a frequency of about 0.1 Hz. What is its period of vibration?

25.3 Wave Motion

Most of the information around us gets to us in some form of wave. Sound is energy that travels to our ears in the form of one kind of wave. Light is energy that comes to our eyes in the form of a different kind of wave (an electromagnetic wave). The signals that reach our radio and television sets also travel in the form of electromagnetic waves.

When energy is transferred by a wave from a vibrating source to a distant receiver, there is no transfer of matter between the two points. To see this, think about the very simple wave produced when one end of a horizontally stretched string is shaken up and down (Figure 25.5). After the end of the string is shaken, a rhythmic disturbance travels along the string. Each part of the string moves up and down while the disturbance moves horizontally along the length of the string. It is the disturbance that moves along the length of the string, not parts of the string itself.

■ Answers

1. A 100-Hz wave vibrates 100 times per second.

2. The period is

$$\frac{1}{\text{frequency}} = \frac{1 \text{ vib}}{0.1 \text{ Hz}} = \frac{1 \text{ vib}}{0.1 \text{ vib/s}} = 10 \text{ s.}$$

Thus, each vibration takes 10 seconds.

Figure 25.5 ▲
When the string is shaken up and down, a disturbance moves along the length of the string.

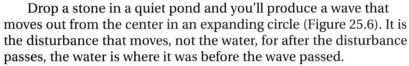

DOING PHYSICS

Making Waves

Oscillate a marking pen back and forth across a piece of paper as you pull the paper in a direction perpendicular to your oscillation. You'll get a curve that may resemble a sine curve, and will have a certain wavelength. What happens to the wavelength when you pull the paper faster? Next, repeatedly dip your finger into a wide pan of water to make circular waves on the surface. Will the wavelength of the waves increase, decrease, or remain the same when you dip your finger more frequently?

Activity

Drop a stone in a quiet pond and you'll produce a wave that moves out from the center in an expanding circle (Figure 25.6). It is the disturbance that moves, not the water, for after the disturbance passes, the water is where it was before the wave passed.

When someone speaks to you from across the room, the sound wave is a disturbance in the air that travels across the room. The air molecules themselves do not move along, as they would in a wind. The air, like the rope and the water in the previous examples, is the medium through which wave energy travels. The energy transferred from a vibrating source to a receiver is carried by a *disturbance* in a medium, not by matter moving from one place to another within the medium.

Figure 25.6 ▲
A circular water wave in a still pond.

25.4 Wave Speed

The speed of a wave depends on the medium through which the wave moves. Sound waves, for example, move at speeds of about 330 m/s to 350 m/s in air (depending on temperature), and about four times faster in water. Whatever the medium, the speed, frequency, and wavelength of the wave are related. Consider the simple case of water waves. Imagine that you fix your eyes at a stationary point on the surface of water and observe the waves passing by this point. If you count the number of crests that pass each second (the frequency) and also observe the distance between crests (the wavelength), you can then calculate the horizontal distance a particular crest moves each second.

Figure 25.7 ▶
If the wavelength is 1 meter, and one wavelength per second passes the pole, then the speed of the wave is 1 m/s.

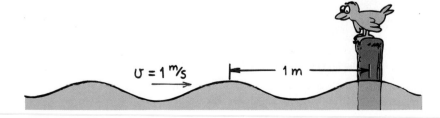

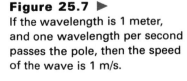

For example, if two crests pass a stationary point each second, and if the wavelength is 3 meters, then 2 × 3 meters of waves pass by in 1 second. The waves therefore move at 6 meters per second. We can say the same thing this way:

$$\text{wave speed} = \text{frequency} \times \text{wavelength}$$

Or in equation form

$$v = f\lambda$$

where v is wave speed, f is wave frequency, and λ (Greek letter lambda) is wavelength. This relationship holds for all kinds of waves, whether they are water waves, sound waves, radio waves, or light waves.

Table 25.1 shows some frequencies and corresponding wavelengths of sound in air at the same temperature. Notice that the product of frequency and wavelength is the same for each example—340 m/s in this case. During a concert, you do not hear the high notes in a chord before you hear the low notes. The sounds of all instruments reach you at the same time. Notice that low frequencies have long wavelengths, and high frequencies have shorter wavelengths. Frequency and wavelength vary inversely to produce the same wave speed for all sounds.

Table 25.1 Sound Waves

Frequency (Hz)	Wavelength (m)	Wave Speed (m/s)
160	2.13	340
264	1.29	340
396	0.86	340
528	0.64	340

Computational Example

If a train of freight cars, each 10 m long, rolls by you at the rate of 2 cars each second, what is the speed of the train?

This can be seen in two ways, the Chapter 2 way and the Chapter 25 way.

From Chapter 2 recall that

$$v = \frac{d}{t} = \frac{2 \times 10 \text{ m}}{1 \text{ s}} = 20 \text{ m/s}$$

where d is the length of that part of the train that passes you in time t.

Here in Chapter 25 we compare the train to wave motion, where the wavelength corresponds to 10 m, and the frequency is 2 Hz. Then

$$\begin{aligned}\text{wave speed} &= \text{frequency} \times \text{wavelength} \\ &= (2 \text{ Hz}) \times (10 \text{ m}) = 20 \text{ m/s}\end{aligned}$$

One of the nice things about physics is that different ways of looking at things produce the same answer. When this doesn't happen, and there is no error in computation, then the validity of one (or both!) of those ways is suspect.

25.5 Transverse Waves

Suppose you create a wave along a rope by shaking the free end up and down as shown in Figure 25.8. In this case the motion of the rope (shown by the up and down arrows) is at right angles to the direction in which the wave is moving. Whenever the motion of the medium (the rope in this case) is at right angles to the direction in which a wave travels, the wave is a **transverse wave.**

Figure 25.8 ▲
A transverse wave.

Waves in the stretched strings of musical instruments and upon the surfaces of liquids are transverse. As Chapter 27 will show, the electromagnetic waves that make up radio waves and light are also transverse.

25.6 Longitudinal Waves

Not all waves are transverse. Sometimes the particles of the medium move back and forth in the same direction in which the wave travels. The particles move *along* the direction of the wave rather than at right angles to it. This kind of wave is a **longitudinal wave.**

Both transverse and longitudinal waves can be demonstrated with a loosely-coiled spring, or Slinky®, as shown in Figure 25.9. A

■ **Answers**

1. The frequency of the wave is 2 Hz; its wavelength is 1.5 m; and its wave speed is frequency × wavelength = (2 Hz) × (1.5 m) = 3 m/s.

2. The wavelength of the 340-Hz sound wave must be 1 m. Then wave speed = (340 Hz) × (1 m) = 340 m/s.

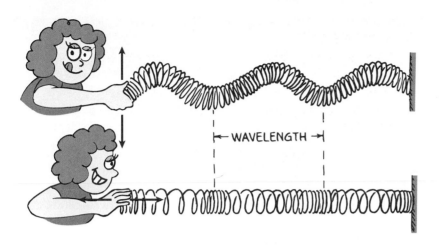

transverse wave is demonstrated by shaking the end of a Slinky up and down. A longitudinal wave is demonstrated by shaking the end of the Slinky in and out. In this case we see that the medium vibrates parallel to the direction of energy transfer. Sound waves are longitudinal waves, and will be discussed in the next chapter.

25.7 Interference

A material object such as a rock will not share its space with another rock. But more than one vibration or wave can exist at the same time in the same space. If you drop two rocks in water, the waves produced by each can overlap and form an **interference pattern.** Within the pattern, wave effects may be increased, decreased, or neutralized.

When the crest of one wave overlaps the crest of another, their individual effects add together. The result is a wave of increased amplitude. This is called **constructive interference,** or reinforcement (Figure 25.10, top). When the crest of one wave overlaps the trough of another, their individual effects are reduced. The high part of one wave simply fills in the low part of another. This is called **destructive interference,** or cancellation (Figure 25.10, bottom).

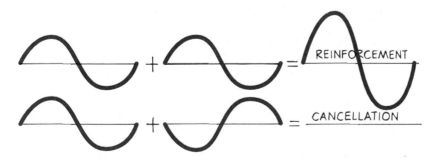

Figure 25.10 ▲
Constructive interference (top) and destructive interference (bottom) in a transverse wave.

Figure 25.11 ▶
Two overlapping water waves
produce an interference pattern.

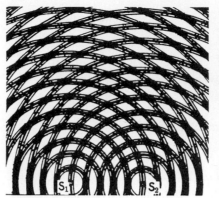

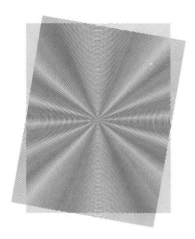

Figure 25.12 ▲
Moiré pattern.

Wave interference is easiest to see in water. Figure 25.11 (right) shows the interference pattern made when two vibrating objects touch the surface of water. The gray "spokes" are regions where a crest of one wave overlaps the trough of another to produce regions of zero amplitude. At points along these regions, the waves from the two objects arrive "out of step." We say that they are **out of phase** with one another. The dark- and light-striped regions are where the crests of one wave overlap the crests of the other, and the troughs overlap as well. In these regions, the two waves arrive "in step." They are **in phase** with each other.

Interference patterns are nicely illustrated by the overlapping of concentric circles printed on a pair of clear sheets, as shown in Figure 25.12. When the sheets overlap with their centers slightly apart, a so-called *moiré pattern* is formed that is very similar to the interference pattern of water waves (or any kind of waves). A slight shift in either of the sheets produces noticeably different patterns. If a pair of such sheets is available, be sure to try this and see the variety of patterns for yourself.

Interference is characteristic of all wave motion, whether the waves are water waves, sound waves, or light waves. The interference of sound is treated in the next chapter, and the interference of light in Chapter 31.

25.8 Standing Waves

If you tie a rope to a wall and shake the free end up and down, you will produce a wave in the rope. The wall is too rigid to shake, so the wave is reflected back along the rope to you. By shaking the rope just right, you can cause the incident (original) and reflected waves to form a **standing wave.** In a standing wave certain parts of the rope, called the **nodes,** remain stationary.

Interestingly enough, you could hold your fingers on either side of the rope at a node, and the rope would not touch them. Other parts of the rope would make contact with your fingers. The positions on a standing wave with the largest amplitudes are known as **antinodes**. Antinodes occur halfway between nodes.

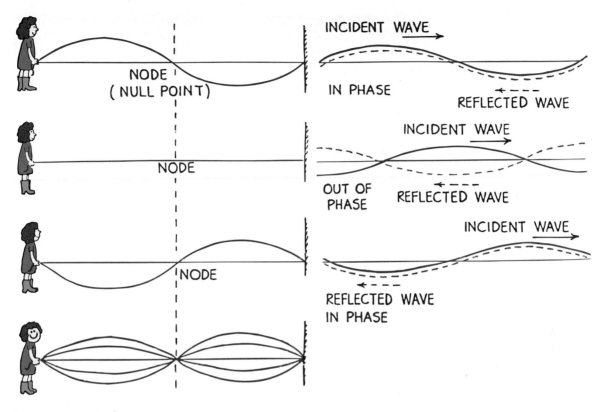

Figure 25.13 ▲
The incident and reflected waves interfere to produce a standing wave. The nodes are places that remain stationary.

Standing waves are the result of interference. When two waves of equal amplitude and wavelength pass through each other in opposite directions, the waves are always out of phase at the nodes. The nodes are stable regions of destructive interference (Figure 25.13).

You can produce a variety of standing waves by shaking the rope at different frequencies. The easiest standing wave to produce has one segment (Figure 25.14, top). If you keep doubling the frequency, you'll produce more interesting waves.

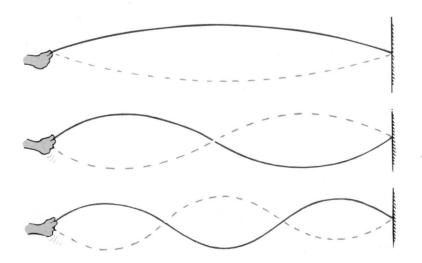

◄ **Figure 25.14**
(Top) Shake the rope until you set up a standing wave of one segment (rope length equals $\frac{1}{2}$ wavelength). (Center) Shake with twice the frequency and produce a standing wave with two segments (rope length equals 1 wavelength). (Bottom) Shake with three times the frequency and produce a standing wave with three segments (rope length equals $1\frac{1}{2}$ wavelengths).

Standing waves are set up in the strings of musical instruments that are plucked, bowed, or struck. They are set up in the air in an organ pipe and the air of a soda-pop bottle when air is blown over the top. Standing waves can be produced in either transverse or longitudinal waves.

■ Questions

1. Is it possible for one wave to cancel another wave so that the combined amplitude is zero?

2. Suppose you set up a standing wave of three segments, as shown in Figure 25.14 (bottom). If you shake with twice the frequency, how many wave segments will occur in your new standing wave? How many wavelengths will there be?

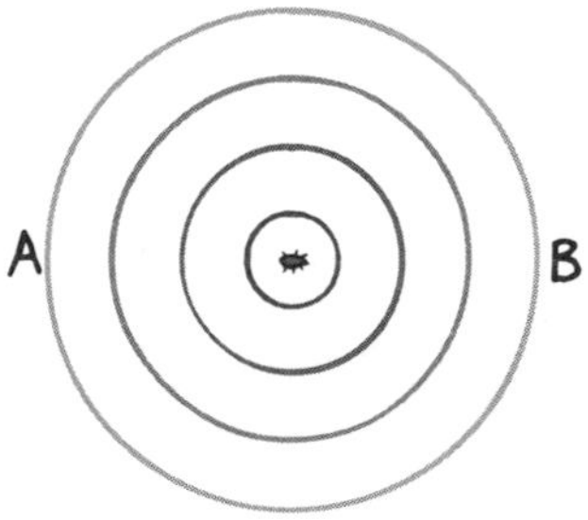

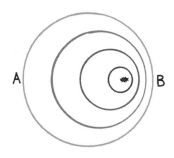

Figure 25.15 ▲
Top view of circular water wave made by a stationary bug jiggling in still water.

Figure 25.16 ▲
The wave pattern made by a bug swimming in still water.

25.9 The Doppler Effect

Imagine a bug jiggling its legs and bobbing up and down in the middle of a quiet puddle, as shown in Figure 25.15. Suppose the bug is not going anywhere but is merely treading water in a fixed position. The crests of the wave it makes are concentric circles, because the wave speed is the same in all directions. If the bug bobs in the water at a constant frequency, the distance between wave crests (the wavelength) will be the same for all successive waves. Waves encounter point A as frequently as they encounter point B. This means that the frequency of wave motion is the same at points A and B, or anywhere in the vicinity of the bug. This wave frequency is the same as the bobbing frequency of the bug.

Suppose the jiggling bug moves across the water at a speed less than the wave speed. In effect, the bug chases part of the crests it has produced. The wave pattern is distorted and is no longer concentric, as shown in Figure 25.16. The center of the outer crest was made when the bug was at the center of that circle. The center of the next smaller crest was made when the bug was at the center of that circle, and so forth. The centers of the circular crests move in the direction of the swimming bug. Although the bug maintains the same bobbing frequency as before, an observer at B would encounter the crests more often. The observer would encounter a *higher* frequency. This is because each successive crest has a shorter distance to travel so they arrive at B more frequently than if the bug were not moving toward B.

■ Answers

1. Yes. This is called destructive interference. In a standing wave in a rope, for example, parts of the rope have no amplitude—the nodes.

2. If you impart twice the frequency to the rope, you'll produce a standing wave with twice as many segments. You'll have six segments. Since a full wavelength has two segments, you'll have three complete wavelengths in your standing wave.

An observer at A, on the other hand, encounters a *lower* frequency because of the longer time between wave-crest arrivals. To reach A, each crest has to travel farther than the one ahead of it due to the bug's motion. This change in frequency due to the motion of the source (or receiver) is called the **Doppler effect** (after the Austrian scientist Christian Doppler, 1803–1853). The greater the speed of the source, the greater will be the Doppler effect.

Water waves spread over the flat surface of the water. Sound and light waves, on the other hand, travel in three-dimensional space in all directions like an expanding balloon. Just as circular wave crests are closer together in front of the swimming bug, spherical sound or light wave crests ahead of a moving source are closer together than those behind the source and encounter a receiver more frequently.

The Doppler effect is evident when you hear the changing pitch of a car horn as the car passes you. When the car approaches, the pitch is higher than normal (that is, higher on the musical scale). This occurs because the sound wave crests are encountering you more frequently. And when the car passes and moves away, you hear a drop in pitch because the wave crests are encountering you less frequently.

Figure 25.17
The pitch of sound is greater when the source moves toward you, and less when the source moves away.

Police make use of the Doppler effect of radar waves in measuring the speeds of cars on the highway. Radar waves are electromagnetic waves, lower in frequency than light and higher in frequency than radio waves. Police bounce them off moving cars, and a computer built into the radar system calculates the speed of the car relative to the radar unit by comparing the frequency of the radar emitted by the antenna with the frequency of the reflected waves (Figure 25.18).

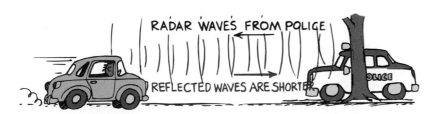

Figure 25.18
The police calculate a car's speed by measuring the Doppler effect of radar waves.

The Doppler effect also occurs for light. When a light source approaches, there is an increase in its measured frequency, and when it recedes, there is a decrease in its frequency. An increase in frequency is called a **blue shift,** because the increase is toward the high-frequency, or blue, end of the color spectrum. A decrease in frequency is called a **red shift,** referring to the low-frequency, or red, end of the color spectrum. Distant galaxies, for example, show a red shift in the light they emit. A measurement of this shift enables astronomers to calculate their speeds of recession. A rapidly spinning star shows a red shift on the side turning away from us and a blue shift on the side turning toward us. This enables a calculation of the star's spin rate.

■ Question

When a source moves toward you, do you measure an increase or decrease in wave speed?

25.10 Bow Waves

When the speed of the source in a medium is as great as the speed of the waves it produces, something interesting happens. The waves pile up. Consider the bug in the previous example when it swims as fast as the wave speed. Can you see that the bug will "keep up" with the wave crests it produces? Instead of the crests getting ahead of the bug, they pile up or superimpose on one another directly in front of the bug, as suggested in Figure 25.19. The bug moves right along with the leading edge of the waves it is producing.

The same thing happens when an aircraft travels at the speed of sound. In the early days of jet aircraft, it was believed that this pileup of sound waves in front of the airplane imposed a "sound barrier" and that to go faster than the speed of sound, the plane would have to "break the sound barrier." What actually happens is that the overlapping wave crests disrupt the flow of air over the wings, so that it is harder to control the plane when it is flying close to the speed of sound. But the barrier is not real. Just as a boat can easily travel faster than the speed of water waves, an airplane with sufficient power can easily travel faster than the speed of sound. Then we say that it is *supersonic*—faster than sound. A supersonic airplane flies into smooth, undisturbed air because no sound wave can propagate out in front of it. Similarly, a bug swimming faster than the speed of water waves finds itself always entering into water with a smooth, unrippled surface.

When the bug swims faster than wave speed, ideally it produces a wave pattern as shown in Figure 25.20. It outruns the wave crests it produces. The crests overlap at the edges, and the pattern made by these overlapping crests is a V shape, called a **bow wave,** which appears to be dragging behind the bug. The familiar bow wave generated by a speedboat knifing through the water is produced by the overlapping of many circular wave crests.

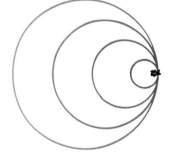

Figure 25.19 ▲
The wave pattern made by a bug swimming at the wave speed.

■ Answer

Neither! It is the *frequency* of a wave that undergoes a change where there is motion of the source, not the *wave speed.* Be clear about the distinction between frequency and speed. How frequently a wave vibrates is altogether different from how fast it moves from one place to another.

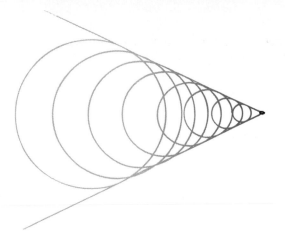

Figure 25.21 shows some wave patterns made by sources moving at various speeds. Note that after the speed of the source exceeds the wave speed, increased speed produces a narrower V shape.

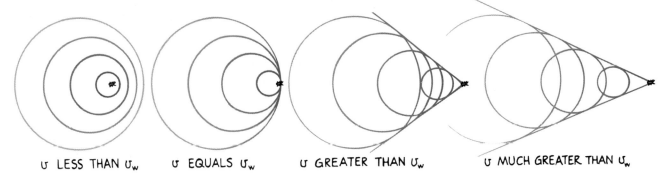

U LESS THAN U$_w$ U EQUALS U$_w$ U GREATER THAN U$_w$ U MUCH GREATER THAN U$_w$

Figure 25.21 ▲
Patterns made by a bug swimming at successively greater speeds. Overlapping at the edges occurs only when the source travels faster than wave speed.

25.11 Shock Waves

A speedboat knifing through the water generates a two-dimensional bow wave. A supersonic aircraft similarly generates a three-dimensional **shock wave.** Just as a bow wave is produced by overlapping circles that form a V, a shock wave is produced by overlapping spheres that form a cone. And just as the bow wave of a speedboat spreads until it reaches the shore of a lake, the conical shock wave generated by a supersonic craft spreads until it reaches the ground.

The bow wave of a speedboat that passes by can splash and douse you if you are at the water's edge. In a sense, you can say that you are hit by a "water boom." In the same way, when the conical shell of compressed air that sweeps behind a supersonic aircraft reaches listeners on the ground below, the sharp crack they hear is described as a **sonic boom.**

We don't hear a sonic boom from a slower-than-sound, or subsonic, aircraft, because the sound wave crests reach our ears one at a time and are perceived as a continuous tone. Only when the craft

Figure 25.22 ▶
A shock wave from a supersonic aircraft.

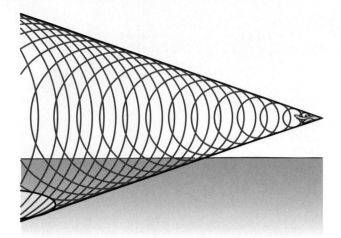

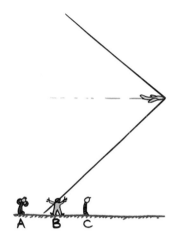

moves faster than sound do the crests overlap and encounter the listener in a single burst. The sudden increase in pressure has much the same effect as the sudden expansion of air produced by an explosion. Both processes direct a burst of high-pressure air to the listener. The ear cannot distinguish between the high pressure from an explosion and the high pressure from many overlapping wave crests.

A common misconception is that sonic booms are produced at the moment that an aircraft flies through the "sound barrier"—that is, just as the aircraft surpasses the speed of sound. This is equivalent to saying that a boat produces a bow wave only when it first overtakes its own waves. This is not so. The fact is that a shock wave and its resulting sonic boom are swept continuously behind an aircraft traveling faster than sound, just as a bow wave is swept continuously behind a speedboat. In Figure 25.23, listener B is in the process of hearing a sonic boom. Listener A has already heard it, and listener C will hear it shortly. The aircraft that generated this shock wave may have broken through the sound barrier hours ago!

It is not necessary that the moving source emit sound for it to produce a shock wave. Once an object is moving faster than the speed of sound, it will *make* sound. A supersonic bullet passing overhead produces a crack, which is a small sonic boom. If the bullet were larger and disturbed more air in its path, the crack would be more boomlike. When a lion tamer cracks a circus whip, the cracking sound is actually a sonic boom produced by the tip of the whip when it travels faster than the speed of sound. Snap a towel and the end can exceed the speed of sound and produce a mini sonic boom. The bullet, whip, and towel are not in themselves sound sources, but when traveling at supersonic speeds they produce their own sound as waves of air are generated to the sides of the moving objects.

On the matter of sound in general: You know that you'll damage your eyes if you stare at the sun. What many people don't know is that you'll similarly damage your ears if you overexpose them to loud sounds. Do as your author does when in a room with very loud music—leave. If for any reason you don't want to leave—really enjoyable music or good camaraderie with friends—stay, but use ear plugs of some kind! You're not being a wimp when you give the same care to your ears that you give to your eyes.

Concept Summary

A vibration is a wiggle in time, and a wave is a wiggle in time and space.

- The period of a wave is the time it takes for one complete back-and-forth vibration.

- The wavelength is the distance between successive identical parts of the wave.

- A wave carries energy from a vibrating source to a receiver without transferring matter from one to the other.

- The frequency, or the number of vibrations in a given time, multiplied by the wavelength equals the speed of the wave.

In a transverse wave, the medium moves at right angles to the direction in which the wave travels.

- Electromagnetic waves, such as light and radio waves, are transverse.

In a longitudinal wave, the medium moves back and forth parallel to the direction in which the wave travels.

- Sound waves are longitudinal.

Interference patterns occur when waves from different sources arrive at the same point at the same time.

- In constructive interference, crest overlaps crest, or trough overlaps trough.

- In destructive interference, a crest overlaps a trough.

- In a standing wave, points of complete destructive interference (at which the medium does not move) remain at the same location.

The Doppler effect is a shift in frequency received due to motion of a vibrating source toward or away from a receiver.

When an object moves through a medium faster than the speed of waves in the medium, a bow wave or shock wave spreads out behind it.

Important Terms

amplitude (25.2)
antinodes (25.8)
blue shift (25.9)
bow wave (25.10)
constructive interference (25.7)
crest (25.2)
destructive interference (25.7)
Doppler effect (25.9)
frequency (25.2)
hertz (25.2)
in phase (25.7)
interference pattern (25.7)
longitudinal wave (25.6)
node (25.8)
out of phase (25.7)
period (25.1)
red shift (25.9)
shock wave (25.11)
simple harmonic motion (25.2)
sine curve (25.2)
sonic boom (25.11)
standing wave (25.8)
transverse wave (25.5)
trough (25.2)
vibration (25.0)
wave (25.0)
wavelength (25.2)

Review Questions

1. **a.** What is a wiggle in time called?

   b. What is a wiggle in space and time called? (25.0)

2. What is the period of a pendulum? (25.1)

3. What is the period of a pendulum that takes one second to make a complete back-and-forth vibration? (25.1)

4. Suppose that a pendulum has a period of 1.5 seconds. How long does it take to make a complete back-and-forth vibration? Is this 1.5-second period pendulum longer or shorter in length than a 1-second period pendulum?

5. How is a sine curve related to a wave? (25.2)

6. Distinguish among these different parts of a wave: amplitude, crest, trough, and wavelength. (25.2)

7. Distinguish between the *period* and the *frequency* of a vibration or a wave. How do they relate to one another? (25.2)

8. Does the medium in which a wave travels move along with the wave itself? Defend your answer. (25.3)

9. How does the speed of a wave relate to its frequency and wavelength? (25.4)

10. As the frequency of sound is increased, does the wavelength increase or decrease? Give an example. (25.4)

11. Distinguish between a *transverse* wave and a *longitudinal* wave. (25.5–25.6)

12. Distinguish between *constructive* interference and *destructive* interference. (25.7)

13. Is interference a property of only some types of waves or of all types of waves? (25.7)

14. What causes a standing wave? (25.8)

15. When a wave source moves toward a receiver, does the receiver encounter an increase in wave frequency, wave speed, or both? (25.9)

16. Does the Doppler effect occur for only some types of waves or all types of waves? (25.9)

17. How fast must a bug swim to keep up with the waves it is producing? How fast must a boat move to produce a bow wave? (25.10)

18. Distinguish between a *bow* wave and a *shock* wave. (25.10–25.11)

19. a. What is a sonic boom?

    b. How fast must an aircraft fly in order to produce a sonic boom? (25.11)

20. If you encounter a sonic boom, is that evidence that an aircraft of some sort exceeded the speed of sound moments ago to become supersonic? Defend your answer. (25.11)

Activity

1. Tie a rubber tube, a spring, or a rope to a fixed support and produce standing waves, as Figure 25.14 suggests. See how many nodes you can produce.

Plug and Chug

1. A nurse counts 76 heartbeats in one minute. What are the period and frequency of the heart's oscillations?

2. New York's 300-m high Citicorp® Tower oscillates in the wind with a period of 6.80 s. Calculate its frequency of vibration.

3. Calculate the speed of waves in a puddle that are 0.15 m apart and made by tapping the water surface twice each second.

4. Calculate the speed of waves in water that are 0.4 m apart and have a frequency of 2 Hz.

5. The lowest frequency we can hear is 20 Hz. Calculate the wavelength associated with this frequency for sound that travels at 340 m/s. How long is this in feet?

Think and Explain

1. Red light has a longer wavelength than violet light. Which has the greater frequency?

2. If you triple the frequency of a vibrating object, what will happen to its period?

3. How far, in terms of wavelength, does a wave travel in one period?

4. The wave patterns seen in Figure 25.6 are composed of circles. What does this tell you about the speed of the waves in different directions?

5. If a wave vibrates up and down twice each second and travels a distance of 20 m each second, what is its frequency? Its wave speed? (Why is this question best answered by careful reading of the question rather than searching for a formula?)

6. Astronomers find that light coming from point A at the edge of the sun has a slightly higher frequency than light from point B at the opposite side. What do these measurements tell us about the sun's motion?

7. Would it be correct to say that the Doppler effect is the apparent change in the speed of a wave due to motion of the source? (Why is this question a test of reading comprehension as well as a test of physics knowledge?)

8. Whenever you watch a high-flying aircraft overhead, it seems that its sound comes from behind the craft rather than from where you see it. Why is this?

9. As a supersonic aircraft gains speed, does the conical angle of its shock wave become wider, narrower, or remain constant?

10. Why is it that a subsonic aircraft, no matter how loud it may be, cannot produce a sonic boom?

Think and Solve

1. While watching ocean waves at the dock of the bay, Otis notices that 10 waves pass beneath him in 30 seconds. He also notices that the crests of successive waves exactly coincide with the posts that are 5 meters apart. What are the period, frequency, wavelength, and speed of the ocean waves?

2. If a wave vibrates back-and-forth three times each second, and its wavelength is 2 meters, what is its frequency? Its period? Its speed?

3. Radio waves are electromagnetic waves that travel at the speed of light, 300 000 kilometers per second. What is the wavelength of FM radio waves received at 100 megahertz on your radio dial?

4. The wavelength of red light is about 700 nanometers, or 7×10^{-7} m. The frequency of the red light reflected from a metal surface and the frequency of the vibrating electron that produces it are the same. What is this frequency?

26 Sound

Vibrations carry energy.

Figure 26.1 ▲
Vibrate a Ping-Pong paddle in the midst of a lot of Ping-Pong balls, and they will transmit rhythmic pulses.

Pretend an entire room is filled with Ping-Pong table tennis balls, and in the middle of the room is a big paddle. You shake the paddle back and forth. What happens? When you move the paddle to the right, it hits some Ping-Pong balls and moves them to the right. They in turn hit others, moving them to the right, and so on. You set up a "Ping-Pong ripple" that moves across the room. The process is repeated the next time you move the paddle to the right, and another Ping-Pong ripple follows the first one. As you keep shaking the paddle back and forth, you keep creating Ping-Pong ripples that flow across the room. Can you see that what you are doing is making a longitudinal wave? At the far side of the room, Ping-Pong impulses arrive at the same frequency as the vibration of your paddle.

Molecules of air behave like tiny Ping-Pong balls. Place a tuning fork in the middle of a room and strike it with a rubber hammer. What happens? The surrounding air molecules are set into motion just like balls being hit by a paddle. Longitudinal waves flow through the air with a frequency equal to that of the vibrating prongs of the tuning fork. We hear these vibrations as sound. There is very little difference between the idea of a shaking paddle bumping into Ping-Pong balls and a vibrating tuning fork bumping into air molecules. In both cases vibrations are carried throughout the surrounding medium—the balls or the air.

26.1 The Origin of Sound

All sounds are produced by the vibrations of material objects. In a piano, violin, or guitar, a sound wave is produced by vibrating strings; in a saxophone, by a vibrating reed; in a flute, by a fluttering column of air at the mouthpiece. Your voice results from the vibration of your vocal chords.

In each of these cases, the original vibration stimulates the vibration of something larger or more massive—the sounding board of a stringed instrument, the air column within a reed or wind instrument, or the air in the throat and mouth of a singer. This vibrating material then sends a disturbance through a surrounding medium, usually air, in the form of longitudinal waves. Under ordinary conditions, the frequency of the vibrating source equals the frequency of sound waves produced.

We describe our subjective impression about the frequency of sound by the word **pitch.** A high-pitched sound like that from a piccolo has a high vibration frequency, while a low-pitched sound like that from a fog horn has a low vibration frequency.

A young person can normally hear pitches with frequencies from about 20 to 20 000 hertz. As we grow older, our hearing range shrinks, especially at the high-frequency end. Sound waves with frequencies below 20 hertz are called **infrasonic,** and those with frequencies above 20 000 hertz are called **ultrasonic.** We cannot hear infrasonic or ultrasonic sound waves.

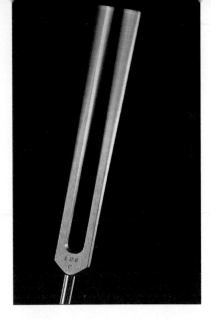

Figure 26.2 ▲
The source of all sound waves is vibration.

26.2 Sound in Air

Clap your hands and you produce a pulse that goes out in all directions. The pulse vibrates the air somewhat as a similar pulse would vibrate a coiled spring or a Slinky spring toy. Each particle moves back and forth along the direction of motion of the expanding wave.

Figure 26.3 ▲
A compression travels along the spring.

For a clearer picture of this process, consider the long room shown in Figure 26.4. At one end is an open window with a curtain over it. At the other end is a door.

When you quickly open the door (top sketch), you can imagine the door pushing the molecules next to it away from their initial positions, and into their neighbors. Neighboring molecules, in turn, push into their neighbors, and so on, like a compression wave moving along a spring, until the curtain flaps out the window. A pulse of compressed air has moved from the door to the curtain. This pulse of compressed air is called a **compression.**

When you quickly close the door (bottom sketch), the door pushes neighboring air molecules out of the room. This produces an area of low pressure next to the door. Neighboring molecules then

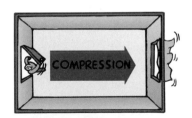

Figure 26.4 ▲
(Top) When the door is opened, a compression travels across the room. (Bottom) When the door is closed, a rarefaction travels across the room.

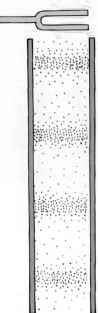

Figure 26.5 ▲
Compressions and rarefactions traveling from the tuning fork through the tube.

Figure 26.6 ▲
Sound can be heard from the ringing bell when air is inside the jar, but not when the air is removed.

move into it, leaving a zone of lower pressure behind them. We say the air in this zone of lower pressure is *rarefied.* Other molecules farther from the door, in turn, move into these rarefied regions, resulting in a pulse of rarefied air moving from the door to the curtain. This is evident when the lower-pressure air reaches the curtain, which flaps inward. This time the disturbance is a **rarefaction.**

For all wave motion, it is not the medium that travels across the room, but a *pulse* that travels. In both cases the pulse travels from the door to the curtain. We know this because in both cases the curtain moves *after* the door is opened or closed.

If you swing the door open and closed in periodic fashion, you can set up a wave of periodic compressions and rarefactions that will make the curtain swing in and out of the window. On a much smaller but more rapid scale, this is what happens when a tuning fork is struck. The vibrations of the tuning fork and the waves it produces are considerably higher in frequency and lower in amplitude than in the case of the swinging door. You don't notice the effect of sound waves on the curtain, but you are well aware of them when they meet your sensitive eardrums.

Consider sound waves in the tube shown in Figure 26.5. For simplicity, only the waves that travel in the tube are shown. When the prong of the tuning fork next to the tube moves toward the tube, a compression enters the tube. When the prong swings away, in the opposite direction, a rarefaction follows the compression. It is like the Ping-Pong paddle moving back and forth in a room packed with Ping-Pong balls. As the source vibrates, a series of compressions and rarefactions is produced.

26.3 Media That Transmit Sound

Most sounds you hear are transmitted through the air. But sound also travels in solids and liquids. Put your ear to the ground as Native Americans did, and you can hear the hoofbeats of distant horses through the ground before you can hear them through the air. More practically, put your ear to a metal fence and have a friend tap it far away. The sound is transmitted louder and faster by the metal than by the air.

Or click two rocks together under water while your ear is submerged. You'll hear the clicking sound very clearly. If you've ever been swimming in the presence of motorized boats, you've probably noticed that you can hear the boats' motors much more clearly under water than above water. Solids and liquids are generally good conductors of sound—much better than air. The speed of sound differs in different materials. In general, sound is transmitted faster in liquids than in gases, and still faster in solids.

Sound cannot travel in a vacuum (Figure 26.6). The transmission of sound requires a medium. If there is nothing to compress and expand, there can be no sound. There may still be vibration, but without a medium there is no sound.

26.4 Speed of Sound

Have you ever watched a distant person chopping wood or hammering, and noticed that the sound of the blow takes time to reach your ears? You see the blow before you hear it. This is most noticeable in the case of lightning. You hear thunder *after* you see a flash of lightning (unless you're at the source). These experiences are evidence that sound is much slower than light.

The speed of sound in dry air at 0°C is about 330 meters per second, or about 1200 kilometers per hour, about one-millionth the speed of light. Water vapor in the air increases this speed slightly. Increased temperature increases the speed of sound also. A little thought will show that this makes sense, for the faster-moving molecules in warm air bump into each other more often and therefore can transmit a pulse in less time. For each degree increase in air temperature above 0°C, the speed of sound in air increases by 0.60 m/s. So in air at a normal room temperature of about 20°C, sound travels at about 340 m/s.

The speed of sound in a material depends not on the material's density, but on its elasticity. Elasticity is the ability of a material to change shape in response to an applied force, then resume its initial shape once the distorting force is removed. Steel is very elastic; putty is inelastic.* In elastic materials, the atoms are relatively close together and respond quickly to each other's motions, transmitting energy with little loss. Sound travels about fifteen times faster in steel than in air, and about four times faster in water than in air.

■ Question

How far away is a storm if you note a 3-second delay between a lightning flash and the sound of thunder?

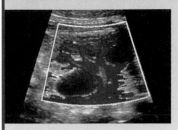
26.5 Loudness

The intensity of a sound is proportional to the square of the amplitude of a sound wave. Sound intensity is objective and is measured by instruments such as the oscilloscope shown in Figure 26.7.

■ Answer

For a speed of sound in air at 340 m/s, the distance is (340 m/s) × (3 s) = 1020 m. Time for the light is negligible, so the storm is slightly more than 1 km away.

---

* You may be surprised that steel is considered elastic and putty, inelastic. After all, that stretchy material that keeps our socks up is called *elastic,* and putty is more stretchy than steel. But elasticity is not "stretchability"; it's the tendency of a material to resume its initial shape after having been exposed to a distorting force. Some very stiff materials are elastic!

Table 26.1	
Source of Sound	**Level (dB)**
Jet engine, at 30 m	140
Threshold of pain	120
Loud rock music	115
Old subway train	100
Average factory	90
Busy street traffic	70
Normal speech	60
Library	40
Close whisper	20
Normal breathing	10
Hearing threshold	0

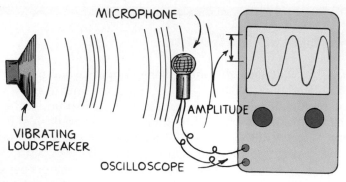

Figure 26.7 ▲

The radio loudspeaker at the left is a paper cone that vibrates in rhythm with an electric signal. The sound that is produced sets up similar vibrations in the microphone (center), which are displayed on the screen of an oscilloscope (right). The shape of the waveform on the oscilloscope reveals information about the sound.

Loudness, on the other hand, is a physiological sensation sensed in the brain. It differs for different people. Loudness is subjective but is related to sound intensity. Despite subjective variations, loudness varies nearly as the logarithm of intensity (powers of ten). The unit of intensity for sound is the decibel (dB), after Alexander Graham Bell, inventor of the telephone. Some common sources and sound levels are given in Table 26.1.

Starting with zero at the threshold of hearing for a normal ear, an increase of each 10 dB means that sound intensity increases by a factor of 10. A sound of 10 dB is 10 times as intense as sound of 0 dB; 20 dB is not twice but 10 times as intense as 10 dB, or 100 times as intense as the threshold of hearing. A 60-dB sound is 100 times as intense as a 40-dB sound.

Roughly, the sensation of loudness follows this decibel scale. We hear a 100-dB sound to be about as much louder than a 70-dB sound as the 70-dB sound is louder than a 40-dB sound. Because of this, we say that human hearing is approximately logarithmic.

26.6 Forced Vibration

When you strike an unmounted tuning fork, the sound it makes is faint. Strike a tuning fork while holding its base on a tabletop, and the sound is relatively loud. Why? This is because the table is forced to vibrate, and its larger surface sets more air in motion. The tabletop becomes a sounding board, and can be forced into vibration with forks of various frequencies. This is a case of **forced vibration.**

The mechanism in a music box is mounted on a sounding board. Without the sounding board, the sound the music box mechanism makes is barely audible. The vibration of guitar strings in an acoustical guitar would be faint if they weren't transmitted to the guitar's wooden body. Sounding boards are important in all stringed musical instruments.

Figure 26.8 ▲

When the string is plucked, the washtub is set into forced vibration and serves as a sounding board.

26.7 Natural Frequency

Drop a wrench and a baseball bat on the floor, and you hear distinctly different sounds. Objects vibrate differently when they strike the floor. Tap a wrench, and the vibrations it makes are different from the vibrations of a baseball bat, or of anything else.

When any object composed of an elastic material is disturbed, it vibrates at its own special set of frequencies, which together form its special sound. We speak of an object's **natural frequency,** which depends on factors such as the elasticity and shape of the object. Bells and tuning forks vibrate at their own characteristic frequencies. Interestingly enough, most things—from planets to atoms and almost everything else in between—have a springiness to them and vibrate at one or more natural frequencies. A natural frequency is one at which minimum energy is required to produce forced vibrations. It is also the frequency that requires the least amount of energy to continue this vibration.

Figure 26.9 ▲
The natural frequency of the smaller bell is higher than that of the big bell, and it rings at a higher pitch.

26.8 Resonance

When the frequency of a forced vibration on an object matches the object's natural frequency, a dramatic increase in amplitude occurs. This phenomenon is called **resonance.** Resonance means to re-sound, or sound again. Putty doesn't resonate because it isn't elastic, and a dropped handkerchief is too limp. In order for something to resonate, it needs a force to pull it back to its starting position and enough energy to keep it vibrating.

A common experience illustrating resonance occurs on a swing. When pumping a swing, you pump in rhythm with the natural frequency of the swing. More important than the force with which you pump is the timing. Even small pumps, or even small pushes from someone else, if delivered in rhythm with the natural frequency of the swinging motion, produce large amplitudes.

Figure 26.10 ▲
Pumping a swing in rhythm with its natural frequency produces larger amplitudes.

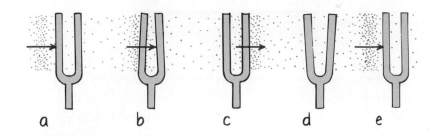

Figure 26.11 ▶
Stages of resonance. (a) The first compression meets the fork and gives it a tiny and momentary push. The fork bends (b) and then returns to its initial position (c) just at the time a rarefaction arrives. It keeps moving and (d) overshoots in the opposite direction. Just when it returns to its initial position (e), the next compression arrives to repeat the cycle. Now it bends farther because it is already moving.

a   b   c   d   e

A common classroom demonstration of resonance uses a pair of tuning forks adjusted to the same frequency and spaced about a meter apart. When one of the forks is struck, it sets the other fork into vibration. This is a small-scale version of pushing a friend on a swing—it's the timing that's important. When a sound wave impinges on the fork, each compression gives the prong a tiny push. Since the frequency of these pushes corresponds to the natural frequency of the fork, the pushes successively increase the amplitude of vibration. This is because the pushes occur at the right time and are repeatedly in the same direction as the instantaneous motion of the fork.

If the forks are not adjusted for matched frequencies, the timing of pushes will be off and resonance will not occur. When you tune your radio set, you are similarly adjusting the natural frequency of the electronics in the set to match one of the many incoming signals. The set then resonates to one station at a time, instead of playing all the stations at once.

Resonance is not restricted to wave motion. It occurs whenever successive impulses are applied to a vibrating object in rhythm with its natural frequency. English infantry troops marching across a footbridge in 1831 inadvertently caused the bridge to collapse when they marched in rhythm with the bridge's natural frequency. Since then, it is customary for troops to "break step" when crossing bridges. The Tacoma Narrows Bridge disaster in this century, Figure 26.12, is attributed to wind-generated resonance!

Figure 26.12 ▲
In 1940, four months after being completed, the Tacoma Narrows Bridge in the state of Washington was destroyed by a 40-mile-per-hour wind. The mild gale produced a fluctuating force that is said to have resonated with the natural frequency of the bridge, steadily increasing the amplitude over several hours until the bridge collapsed.

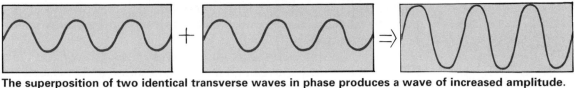

The superposition of two identical transverse waves in phase produces a wave of increased amplitude.

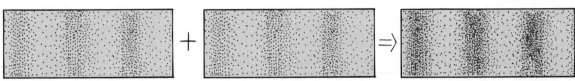

The superposition of two identical longitudinal waves in phase produces a wave of increased amplitude.

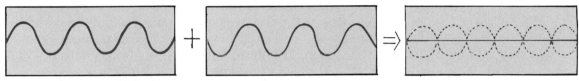

Two identical transverse waves that are out of phase destroy each other when they are superimposed.

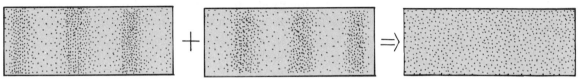

Two identical longitudinal waves that are out of phase destroy each other when they are superimposed.

Figure 26.13 ▲
Wave interference for transverse and longitudinal waves.

26.9 Interference

Sound waves, like any waves, can be made to interfere. Recall that wave interference was discussed in the previous chapter. A comparison of interference for transverse waves and longitudinal waves is shown in Figure 26.13. In either case, when the crests of one wave overlap the crests of another wave, there is constructive interference and an increase in amplitude. Or when the crests of one wave overlap the troughs of another wave, there is destructive interference and a decrease in amplitude. For sound, the crest of a wave corresponds to a compression, and the trough of a wave corresponds to a rarefaction. Interference occurs for both transverse and longitudinal waves.

Interference affects the loudness of sounds. If you are equally distant from two sound speakers that simultaneously trigger identical sound waves of constant frequency (see Figure 26.14, top), the sound is louder because the waves add. The compressions and rarefactions arrive in phase, that is, in step.

If you move to the side so that paths from the speakers differ by a half-wavelength (see Figure 26.14, bottom), rarefactions from one speaker reach you at the same time as compressions from the other. It's like the crest of one water wave exactly filling in the trough of

Figure 26.14 ▲
Interference of sound waves. (Top) Waves arrive in phase. (Bottom) Waves arrive out of phase.

Figure 26.15 ▲
Ken Ford tows gliders in quiet comfort when he wears his noise-canceling earphones.

another water wave—destructive interference. (If the speakers emit many frequencies, not all wavelengths destructively interfere for a given difference in path lengths.)

Destructive interference of sound waves is usually not a problem because there is usually enough reflection of sound to fill in canceled spots. Nevertheless, "dead spots" are sometimes evident in poorly designed theaters and gymnasiums, where sound waves reflected off walls interfere with unreflected waves to form zones of low amplitude. Often, moving your head a few centimeters in either direction can make a noticeable difference.

Destructive sound interference is a useful property in antinoise technology. Noisy devices such as jackhammers are being equipped with microphones that send the sound of the device to electronic microchips. The microchips create mirror-image wave patterns of the sound signals. For the jackhammer, this mirror-image sound signal is fed to earphones worn by the operator. Sound compressions (or rarefactions) from the hammer are neutralized by mirror-image rarefactions (or compressions) in the earphones. The combination of signals neutralizes the jackhammer noise. Noise-canceling earphones are already common for pilots. Watch for the antinoise principle applied to electronic mufflers in cars, where antinoise is blasted through loudspeakers, canceling about 95% of the original noise.

26.10 Beats

An interesting and special case of interference occurs when two tones of slightly different frequency are sounded together. A fluctuation in the loudness of the combined sounds is heard; the sound is loud, then faint, then loud, then faint, and so on. This periodic variation in the loudness of sound is called **beats.**

Beats can be heard when two slightly mismatched tuning forks are sounded together. Because one fork vibrates at a frequency different from the other, the vibrations of the forks will be momentarily in step, then out of step, then in again, and so on. When the combined waves reach your ears in step—say when a compression from one fork overlaps a compression from the other—the sound is a maximum.

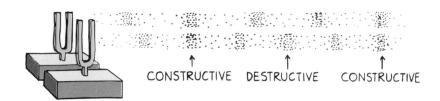

Figure 26.16 ▲
The interference of two sound sources of slightly different frequencies produces beats.

A moment later, when the forks are out of step, a compression from one fork is met with a rarefaction from the other, resulting in a minimum. The sound that reaches your ears throbs between maximum and minimum loudness and produces a tremolo effect.

If you walk side by side with someone who has a different stride, there will be times when you are both in step, and times when you are both out of step. Suppose, for example, that you take exactly 70 steps in one minute and your friend takes 72 steps in the same time. Your friend gains two steps per minute on you. A little thought will show that you two will be momentarily in step twice each minute. In general, when two people with different strides walk together, the number of times they are in step in each unit of time is equal to the difference in the frequencies of their steps. This applies also to a pair of tuning forks. When one fork vibrates 264 times per second, and the other fork vibrates 262 times per second, they are in step twice each second. A beat frequency of 2 hertz is heard.

Beats can be nicely displayed on an oscilloscope. When sound signals of slightly different frequencies are fed into an oscilloscope, graphical representations of their pressure patterns can be displayed both individually and when the sounds overlap. Figure 26.17 shows the wave forms for two waves separately, and superposed. Although the separate waves are of constant amplitude, we see amplitude variations in the superposed wave form. Careful inspection of the figure shows this variation is produced by the interference of the two superposed waves. Maximum amplitude of the composite wave occurs when both waves are in phase, and minimum amplitude

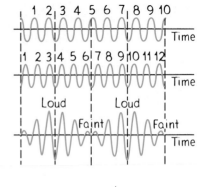

Figure 26.17 ▲
Sinusoidal representations of a 10-Hz sound wave and a 12-Hz sound wave during a 1-second time interval. When the two waves overlap, they produce a composite wave with a beat frequency of 2 Hz.

◀ **Figure 26.18**
The unequal spacings of the combs produce a moiré pattern that is similar to beats.

■ **Question**

What is the beat frequency when a 262-Hz and a 266-Hz tuning fork are sounded together? A 262-Hz and a 272-Hz?

■ **Answer**

The 262-Hz and 266-Hz forks will produce 4 beats per second, that is, 4 Hz (266 Hz minus 262 Hz). The tone heard will be halfway between, at 264 Hz, as the ear averages the frequencies. The 262-Hz and 272-Hz forks will sound like a tone at 267 Hz beating 10 times per second, or 10 Hz, which some people cannot hear. Beat frequencies greater than 10 Hz are normally too rapid to be heard.

Noise and Your Health

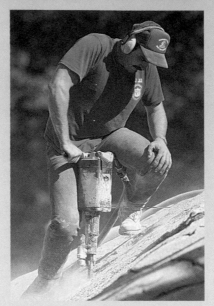

Most of us try to protect our eyes from excess light, but few give the same care to our ears. Near loudspeakers during her first time at a concert, Meidor was alarmed at the pain in her ears. Her friends meant to reassure her when they told her she'd get used to it. But what they didn't tell her was that after the fine tuning of her ears was blasted, she wouldn't know the difference.

Industrial noise is even more damaging to the ears than amplified music because of its sudden high-energy peaks. Loud motorcycles, jackhammers, chain saws, and power tools not only produce steady high-volume sound, but also produce sporadic peaks of energy that can destroy tiny hair cells in the inner ear. When these tiny sensory cells in the inner ear are destroyed they can *never* be restored. Noise-induced hearing loss is insidious.

Fortunately for music devotees, damage caused by energetic peaks is somewhat limited by an inadequate response of electronic amplifiers and loudspeakers. Similarly for live music where most of the sound comes from amplifying equipment. If amplifying equipment were more responsive to sudden sound bursts, hearing loss at concerts would be more severe.

The impact of hearing loss isn't fully apparent until compounded by age. Today's young people will be tomorrow's old people—probably the hardest of hearing ever. Start now to care for your ears and prevent further hearing loss!

---

occurs when both waves are out of phase. Like the walkers in the previous example, the waves are in step twice each second, producing a beat frequency of 2 Hz. The 10- and 12-Hz waves, chosen for convenience here, are infrasonic, so they and their beats are inaudible. Higher-frequency audible waves behave exactly the same way and can produce audible beats.

If you overlap two combs of different teeth spacings, you'll see a moiré pattern that is related to beats. The number of beats per length will equal the difference in the number of teeth per length for the two combs (Figure 26.18).

Beats can occur with any kind of wave and are a practical way to compare frequencies. To tune a piano, a piano tuner listens for beats produced between a standard tuning fork and a particular string on the piano. When the frequencies are identical, the beats disappear. The members of an orchestra tune up by listening for beats between their instruments and a standard tone produced by an oboe or some other instrument.

26 Chapter Review

Concept Summary

Sound waves are produced by the vibrations of material objects.

- A disturbance in the form of a longitudinal wave travels away from the vibrating source.
- High-pitched sounds are produced by sources vibrating at high frequency, while low-pitched sounds are produced by low-frequency sources.

Sound waves consist of traveling pulses of high-pressure zones, or compressions, alternating with pulses of low-pressure zones, or rarefactions.

- Sound can travel through gases, liquids, and solids, but not through a vacuum.
- Sound travels fastest through very elastic materials, such as steel.

Every object vibrates at its own set of natural frequencies.

- When an object such as a sounding board is forced to vibrate by a sound source, the sound becomes louder.
- When an object is forced to vibrate at one of its own natural frequencies, resonance occurs and the sound becomes much louder.

Like any waves, two sound waves can exhibit interference and make sound louder or softer.

- Rapid changes in loudness, known as beats, occur when two tones very close in frequency are heard at the same time.

Important Terms

beats (26.10)
compression (26.2)
forced vibration (26.6)
infrasonic (26.1)
natural frequency (26.7)
pitch (26.1)
rarefaction (26.2)
resonance (26.8)
ultrasonic (26.1)

Review Questions

1. What is the source of all sounds? (26.1)

2. How does pitch relate to frequency? (26.1)

3. What is the average frequency range of a young person's hearing? (26.1)

4. Distinguish between *infrasonic* and *ultrasonic* sound. (26.1)

5. **a.** Distinguish between *compressions* and *rarefactions* of a sound wave.

   b. How are compressions and rarefactions produced? (26.2)

6. Light can travel through a vacuum, as is evidenced when you see the sun or the moon. Can sound travel through a vacuum also? Explain why or why not. (26.3)

7. **a.** How fast does sound travel in dry air at room temperature?

   b. How does air temperature affect the speed of sound? (26.4)

8. How does the speed of sound in air compare with its speed in water and in steel? (26.4)

9. Why does sound travel faster in solids and liquids than in gases? (26.4)

10. Why is sound louder when a vibrating source is held to a sounding board? (26.6)

11. Why do different objects make different sounds when dropped on a floor? (26.7)

12. What does it mean to say that everything has a natural frequency of vibration? (26.7)

13. What is the relationship between forced vibration and resonance? (26.8)

14. Why can a tuning fork or bell be set into resonance, while tissue paper cannot? (26.8)

15. How is resonance produced in a vibrating object? (26.8)

16. What does tuning in a radio station have to do with resonance? (26.8)

17. Is it possible for one sound wave to cancel another? Explain. (26.9)

18. Why does destructive interference occur when the path lengths from two identical sources differ by half a wavelength? (26.9)

19. How does interference of sound relate to beats? (26.10)

20. What is the beat frequency when a 494-Hz tuning fork and a 496-Hz tuning fork are sounded together? (26.10)

Activities

1. Suspend the wire grill of a refrigerator or oven shelf from a string, the ends of which you hold to your ears. Let a friend gently stroke the grill with pieces of broom straw and other objects. The effect is best appreciated if you are in a relaxed condition with your eyes closed. Be sure to try this.

2. If you blow air across the top of a pop bottle, a puff of air (compression) travels downward, bounces from the bottom, and travels back to the opening. When it arrives (less than a thousandth of a second later), it disturbs the flow of air that you are still producing across the top. This causes a slightly bigger puff of air to start again on its way down the bottle. This happens repeatedly until a very large (and loud) vibration is built up and you hear it as sound. The pitch of the sound depends on the time taken for the back-and-forth trip, which depends on the depth of the bottle. If the bottle is empty, a long wave is reinforced and a relatively low tone is produced. With liquid in the bottle, the bottom of the air space is closer to the top and the pitch is higher. With a series of bottles properly filled, you can make your own music.

3. Wet your finger and rub it slowly around the rim of a thin-rimmed stemmed glass while you hold its base firmly to a tabletop with your other hand. The friction of your finger will produce standing waves in the glass, much like the wave the friction from a violin bow makes on the strings of a violin.

4. If you are ever in a room with a ventilation fan, try to hum at the frequency of the fan. As you approach its frequency, you will hear beats. How nice that physics is everywhere!

Think and Explain

1. When watching a baseball game, we often hear the bat hitting the ball after we actually see the hit. Why?

2. Why will marchers at the end of a long parade following a band be out of step with marchers nearer the band?

3. You watch a distant farmer driving a stake into the ground with a sledgehammer. He hits the stake at a regular rate of one stroke per second. You hear the sound of the blows

exactly synchronized with the blows you see. And then you hear one more blow after you see him stop hammering. How far away is the farmer?

4. What two physics mistakes occur in a science fiction movie when you see and hear at the same time a distant explosion in outer space?

5. When a sound wave propagates past a point in the air, what are the changes that occur in the pressure of air at this point?

6. How much more intense is (a) a close whisper than the threshold of hearing? (b) a close whisper than normal breathing?

7. The signal-to-noise ratio for a tape recorder is listed at 50 dB, meaning that when music is played back, the intensity level of the music is 50 dB greater than that of the noise from tape hiss and so forth. By what factor is the sound intensity of the music greater than that of the noise?

8. If the handle of a tuning fork is held solidly against a table, the sound becomes louder. Why? How will this affect the length of the time the fork keeps vibrating? Explain, using the law of energy conservation.

9. The sitar, an Indian musical instrument, has a set of strings that vibrate and produce music, even though they are never plucked by the player. These "sympathetic strings" are identical to the plucked strings and are mounted below them. What is your explanation?

10. Suppose three tuning forks of frequencies 260 Hz, 262 Hz, and 266 Hz are available. What beat frequencies are possible for pairs of these forks sounded together?

11. Suppose a piano tuner hears 2 beats per second when listening to the combined sound from her tuning fork and the piano note being tuned. After slightly tightening the string, she hears 1 beat per second. Should she loosen or should she further tighten the string?

12. Do all people in a group hear the same music when they listen to it attentively? Do all see the same sight when looking at a painting? Do all taste the same flavor when sampling the same cheddar cheese? Do all perceive the same aroma when smelling the same flower? Do all feel the same texture when touching the same fabric? Do all come to the same conclusion when listening to a logical presentation of ideas? Explain.

Think and Solve

1. Sound waves travel at approximately 340 m/s. What is the wavelength of a sound with a frequency of 20 Hz (the lowest note we can hear as a sound)? What is the wavelength of a sound with a frequency of 20 kHz (the highest note we can hear)?

2. Suppose you wish to produce a sound wave that has a wavelength of 1 m in room-temperature air. What would its frequency be?

3. An oceanic depth-sounding vessel surveys the ocean bottom with ultrasonic sound that travels 1530 m/s in seawater. Find the depth of the water if the time delay of the echo to the ocean floor and back is 8 seconds.

4. Two sounds, one at 240 Hz and the other at 243 Hz, occur at the same time. What beat frequency do you hear?

5. Two notes are sounding, one of which is 440 Hz. If a beat frequency of 5 Hz is heard, what is the other note's frequency?

27 Light

Light—the only thing we see.

The only thing we can *really* see is light. But what *is* light? We know that during the day the primary source of light is the sun, and the secondary source is the brightness of the sky. Other common sources are flames, white-hot filaments in lamps, and glowing gases in glass tubes. Almost everything we see, such as this page, is made visible by the light it reflects from such sources. Some materials, such as air, water, or window glass, allow light to pass straight through. Other materials, such as thin paper or frosted glass, allow the passage of light in diffused directions so that we can't see objects through them. Most materials do not allow the passage of any light, except through a very thin layer.

Why do things such as water and glass allow light to go straight through, while things such as wood and steel block it? To answer these questions, you must know something about light itself.

27.1 Early Concepts of Light

Light has been studied for thousands of years. Some of the ancient Greek philosophers thought that light consisted of tiny particles, which could enter the eye to create the sensation of vision. Others, including Socrates and Plato, thought that vision resulted from streamers or filaments emitted by the eye making contact with an object. This view was supported by Euclid, when he asked how else we can explain why we do not see a needle on the floor until our eyes fall upon it.

Up until the time of Newton and beyond, most philosophers and scientists thought that light consisted of particles. However, one Greek, Empedocles, taught that light traveled in waves. One of

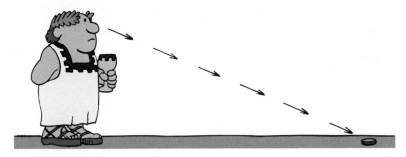

Figure 27.1 ▲
Some ancients believed that light traveled from our eyes to the objects we look at, rather than from the objects to our eyes.

Newton's contemporaries, the Dutch scientist Christian Huygens, also argued that light was a wave.

The particle theory was supported by the fact that light seemed to move in straight lines instead of spreading out as waves do. Huygens provided evidence that under some circumstances light *does* spread out (this is *diffraction*, which is covered in Chapter 31). Other scientists later found more evidence to support the wave theory. The wave theory became the accepted theory in the nineteenth century.

Then in 1905 Einstein published a theory explaining the *photo-electric effect*. According to this theory, light consists of particles—massless bundles of concentrated electromagnetic energy—later called **photons.**

Scientists now agree that light has a dual nature, part particle and part wave. This chapter discusses only the wave nature of light, and leaves the particle nature of light to Chapter 38.

27.2 The Speed of Light

It was not known whether light travels instantaneously or with finite speed until the latter part of the seventeenth century. Galileo had tried to measure the time a light beam takes to travel to a distant mirror and back, but the time was so short he couldn't begin to measure it. Others tried the experiment at longer distances with lanterns they blinked on and off between distant mountaintops. All they succeeded in doing was measuring their own reaction times.

The first demonstration that light travels at a finite speed was supplied by the Danish astronomer Olaus Roemer about 1675. Roemer made very careful measurements of the periods of Jupiter's moons. The innermost moon, Io, is visible through a small telescope and was measured to revolve around Jupiter in 42.5 hours. Io disappears periodically into Jupiter's shadow, so this period could be measured with great precision. Roemer was puzzled to find an irregularity in the measurements of Io's observed period. He found that while the earth was moving away from Jupiter, say from position B to C in Figure 27.2, the measured periods of Io were all somewhat

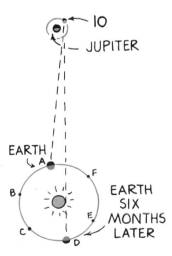

Figure 27.2 ▲
Roemer's method of measuring the speed of light. Light coming from Jupiter's moon Io takes a longer time to reach the earth at position D than at position A. The extra distance that the light travels divided by the extra time it takes gives the speed of light.

longer than average. When the earth was moving toward Jupiter, say from position E to F, the measured periods were shorter than average. Roemer estimated that the cumulative discrepancy between positions A and D amounted to about 22 minutes. That is, when the earth was at position D, Io would pass into Jupiter's shadow 22 minutes late compared with observations at position A.*

The Dutch physicist Christian Huygens correctly interpreted this discrepancy. When the earth was farther away from Jupiter, it was the *light* that was late, not the *moon*. Io passed into Jupiter's shadow at the predicted time, but the light carrying the message did not reach Roemer until it had traveled the extra distance across the diameter of the earth's orbit. There is some doubt as to whether Huygens knew the value of this distance. In any event, this distance is now known to be 300 000 000 km. Using the correct travel time of 1000 s for light to move across the earth's orbit makes the calculation of the speed of light quite simple:

$$\text{speed of light} = \frac{\text{extra distance traveled}}{\text{extra time measured}}$$

$$= \frac{300\ 000\ 000\ \text{km}}{1000\ \text{s}} = 300\ 000\ \text{km/s}$$

The most famous experiment measuring the speed of light was performed by the American physicist Albert Michelson in 1880. Figure 27.3 is a simplified diagram of his experiment. Light from an

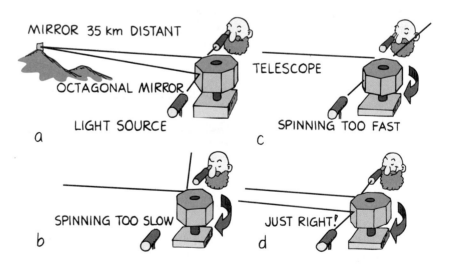

Figure 27.3 ▲
The mirror arrangement used by Michelson to measure the speed of light. Light is reflected back to the eyepiece when the mirror is at rest (a). Reflected light fails to enter the eyepiece when the mirror spins too slowly (b) or too fast (c). When it rotates at the correct speed (d), light reaches the eyepiece.

---

* Roemer's estimate was not quite correct. The correct value is 17 minutes, or about 1000 seconds.

intense source was directed by a lens to an octagonal mirror initially at rest. The mirror was carefully adjusted so that a beam of light was reflected to a stationary mirror located on a mountain 35 km away, and then reflected back to the octagonal mirror and into the eye of an observer. The distance the light had to travel to the distant mountain was carefully surveyed, so Michelson had only to find the time it took to make a round-trip. He accomplished this by spinning the octagonal mirror at a high rate.

When the mirror was spun, the light beam scanned across the horizon with only short bursts of light reaching the mountain mirror to be reflected back to the spinning octagonal mirror. If the rotating mirror made exactly one-eighth rotation in the time the light made the trip to the distant mountain and back, the mirror would be in a position to reflect light into the eyepiece of the observer. If the mirror was rotated too slowly or too quickly, it would not be in a position to reflect light into the eyepiece. When the speed of rotation of the mirror was adjusted so that the light entered the eyepiece, Michelson knew that the time for the light to make the round-trip and the time for the octagonal mirror to make one-eighth of a rotation was the same. He divided the 70-km round-trip distance by this time. Michelson's experimental value for the speed of light was 299 920 km/s, which we round to 300 000 km/s. Michelson received the 1907 Nobel Prize in physics for this experiment. He was the first American scientist to receive this prize.

We now know that the speed of light in a vacuum is a universal constant. Light is so fast that if a beam of light could travel around the earth, it would make 7.5 trips in one second. Light takes 8 minutes to travel from the sun to the earth, and 4 years from the next nearest star, Alpha Centauri. The distance light travels in one year is called a **light-year.**

So Alpha Centauri is 4 light-years away. Our galaxy has a diameter of 100 000 light-years, which means that light takes 100 000 years just to travel across the galaxy. Some galaxies are 10 billion light-years from Earth. If one of those galaxies had exploded 5 billion years ago, this information would not reach Earth for another 5 billion years to come. Light is fast and the universe is big!

■ Question

Light entered the eyepiece when Michelson's octagonal mirror made exactly one-eighth of a rotation during the time light reflected to the distant mountain and back. Would light enter the eyepiece if the mirror turned one-quarter of a rotation in this time?

■ Answer

Yes, light would enter the eyepiece whenever the octagonal mirror turned in multiples of $\frac{1}{8}$ rotation—$\frac{1}{4}$, $\frac{1}{2}$, 1, etc.—in the time the light made its round-trip. What is required is that any of the eight faces be in place when the reflected flash returns from the mountain. Michelson did not spin the mirror fast enough, however, for these other possibilities to occur.

Cinematography

Why do wagon wheels in movies generally act weird? Sometimes they appear still as the wagon moves, or go backward instead of forward. This illusion occurs because we don't see a continuous flow of action in a movie, but instead see a series of still shots at 24 frames per second. At this rapid rate, the human eye can't detect the gaps between the frames. If a wheel spun 24 times per second, the spokes would return to the same position for every frame and the wheel would appear stationary. The wheel doesn't have to spin this fast for the stationary effect. Consider a wheel with six identical spokes. Suppose it turns four times each second. That's 24 spokes each second, and that's exactly the number of pictures the camera takes each second. So a picture of the spoke is always at the same place on the screen. What happens when a spoke doesn't quite make it to the position of the spoke ahead in 1/24 second? Being a little further *behind,* how does the wheel appear to turn?

How far, in kilometers, would a beam of uninterrupted light travel in one year?

The speed of light is constant, so its instantaneous speed and average speed are the same—c. From the equation for speed, $v = d/t$, or in this case, $c = d/t$, we can say

$$d = ct$$
$$= (300\ 000 \text{ km/s}) \times (1 \text{ yr})$$

Introducing conversion factors for the time units, we find

$$d = \left(\frac{300\ 000 \text{ km}}{1 \text{ s}}\right) \times (1 \text{ yr}) \times \left(\frac{365 \text{ d}}{1 \text{ yr}}\right) \times \left(\frac{24 \text{ h}}{1 \text{ d}}\right) \times \left(\frac{3600 \text{ s}}{1 \text{ h}}\right)$$
$$= 9.5 \times 10^{12} \text{ km}$$

This distance is one light-year.

27.3 Electromagnetic Waves

Light is energy that is emitted by accelerating electric charges—often electrons in atoms. This energy travels in a wave that is partly electric and partly magnetic. Such a wave is an **electromagnetic wave.** Light is a small portion of the broad family of electromagnetic waves that includes such familiar forms as radio waves, microwaves, and X rays. The range of electromagnetic waves, or the **electromagnetic spectrum,** as it is called, is shown in Figure 27.4.

The lowest frequency of light we can see with our eyes appears red. The highest visible frequencies are nearly twice the frequency of red and appear violet. Electromagnetic waves of frequencies lower than the red of visible light are called **infrared.** Heat lamps give off infrared waves. Electromagnetic waves of frequencies higher than those of violet are called **ultraviolet.** These higher-frequency waves are responsible for sunburns.

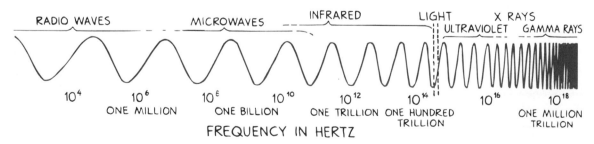

Figure 27.4 ▲

The electromagnetic spectrum is a continuous range of waves extending from radio waves to gamma rays. The descriptive names of the sections are merely a historical classification, for all waves are the same in nature, differing principally in frequency and wavelength; all have the same speed.

27.4 Light and Transparent Materials

Light is energy carried in an electromagnetic wave that is generated by vibrating electric charges. When light is incident upon matter, electrons in the matter are forced into vibration. In effect, vibrations in an emitter are transferred to vibrations in a receiver. This is similar to, but in other ways different from, the way sound is received by a receiver (see Figure 27.5).

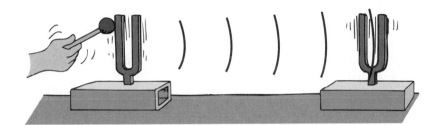

◄ **Figure 27.5**
Just as a sound wave can force a sound receiver into vibration, a light wave can force charged particles in materials into vibration.

Exactly how a receiving material responds when light is incident upon it depends on the frequency of the light and the natural frequency of electrons in the material. Visible light vibrates at a very high rate, more than 100 trillion times per second (10^{14} hertz). If a charged object is to respond to these ultrafast vibrations, it must have very little inertia. Electrons have a small enough mass to vibrate this fast.

Glass and water are two materials that allow light to pass through in straight lines. They are **transparent** to light. To understand how light goes through a transparent material such as glass, visualize the electrons in an atom as connected by imaginary springs (Figure 27.6). When a light wave hits them, they vibrate.

All materials that are springy (elastic) respond more to vibrations at some frequencies than others. Bells ring at a particular frequency, tuning forks vibrate at a particular frequency, and so do the electrons in matter. The natural vibration frequencies of an electron depend on how strongly it is attached to a nearby nucleus. Different materials have different electric "spring strengths."

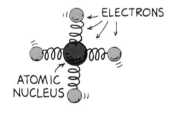

Figure 27.6 ▲
The electrons of atoms in glass can be imagined to be bound to the atomic nucleus as if connected by springs.

Electrons in glass have a natural vibration frequency in the ultraviolet range. When ultraviolet light shines on glass, resonance occurs as the wave builds and maintains a large vibration between the electron and the atomic nucleus, just as a large vibration is built when pushing someone at the resonant frequency on a swing. The energy received by the atom can be either passed on to neighboring atoms by collisions, or reemitted as light. If ultraviolet light interacts with an atom that has the same natural frequency, the vibration amplitude of its electrons becomes unusually large. The atom typically holds on to this energy for quite a long time (about 1 million vibrations or 100 millionths of a second). During this time the atom makes many collisions with other atoms and gives up its energy in the form of heat. That's why glass is not transparent to ultraviolet.

But when the electromagnetic wave has a lower frequency than ultraviolet, as visible light does, the electrons are forced into vibration with smaller amplitudes. The atom holds the energy for less time, with less chance of collision with neighboring atoms, and less energy is transferred as heat. The energy of the vibrating electrons is reemitted as transmitted light. Glass is transparent to all the frequencies of visible light. The frequency of the reemitted light passed from atom to atom is identical to that of the light that produced the vibration to begin with. The main difference is a slight time delay between absorption and reemission.

This time delay results in a lower average speed of light through a transparent material. See Figure 27.7. Light travels at different average speeds through different materials. In a vacuum the speed of light is a constant 300 000 km/s; we call this speed of light c. Light travels very slightly more slowly than this in the atmosphere, but its speed there is usually rounded off to c. In water light travels at 75% of its speed in a vacuum, or $0.75c$. In glass light travels at about $0.67c$, depending on the type of glass. In a diamond light travels at only $0.40c$, less than half its speed in a vacuum. When light emerges from these materials into the air, it travels at its original speed, c.

Infrared waves, which have frequencies lower than visible light, vibrate not only the electrons, but also the entire structure of the

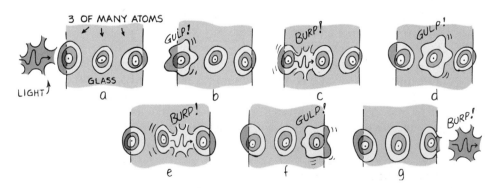

Figure 27.7 ▲

A light wave incident upon a pane of glass sets up vibrations in the atoms that produce a chain of absorptions and reemissions that pass the light energy through the material and out the other side. Because of the time delay between absorptions and reemissions, the average speed of light in glass is less than c.

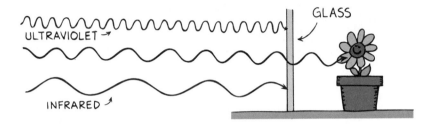

◄ **Figure 27.8**
Glass blocks both infrared and ultraviolet, but is transparent to all the frequencies of visible light.

glass. This vibration of the structure increases the internal energy of the glass and makes it warmer. In sum, glass is transparent to visible light, but not to ultraviolet and infrared light.

27.5 Opaque Materials

Most materials absorb light without reemission and thus allow no light through them; they are **opaque.** Wood, stone, and people are opaque to visible light. In opaque materials, any coordinated vibrations given by light to the atoms and molecules are turned into random kinetic energy—that is, into internal energy. The materials become slightly warmer.

Metals are also opaque. Interestingly enough, in metals the outer electrons of atoms are not bound to any particular atom. They are free to wander with very little restraint throughout the material. That's why metal conducts electricity and heat so well. When light shines on metal and sets these free electrons into vibration, their energy does not "spring" from atom to atom in the material, but is reemitted as visible light. This reemitted light is seen as a reflection. That's why metals are shiny.

Our atmosphere is transparent to visible light and some infrared, but fortunately, almost opaque to high-frequency ultraviolet waves. The small amount of ultraviolet that does get through is responsible for sunburns. If it all got through, we wouldn't dare go out in the sun without protection. Clouds are semitransparent to ultraviolet, which is why you can get a sunburn on a cloudy day. Ultraviolet also reflects

Figure 27.9 ▲
Metals are shiny because light that shines on them forces free electrons into vibration. These electrons then emit their "own" light waves as a reflection.

from sand and water, which is why you can sometimes get a sunburn while in the shade of a beach umbrella.

■ Question

Why is glass transparent to visible light, but opaque to ultraviolet and infrared?

27.6 Shadows

A thin beam of light is often called a **ray.** Any beam of light—no matter how wide—can be thought of as made of a bundle of rays. When light shines on an object, some of the rays may be stopped while others pass on in a straight-line path. A **shadow** is formed where light rays cannot reach. Sharp shadows are produced by a small light source nearby or by a larger source farther away. However, most shadows are somewhat blurry. There is usually a dark part on the inside and a lighter part around the edges. A total shadow is called an **umbra,** and a partial shadow a **penumbra.** A penumbra appears where some of the light is blocked, but where other light fills in. This can happen where light from one source is blocked and light from another source fills in (Figure 27.11). Or a penumbra occurs where light from a broad source is only partially blocked.

Figure 27.11 ▲
An object held close to a wall casts a sharp shadow because light coming from slightly different directions does not spread much behind the object. As the object is moved farther away, penumbras are formed and cut down on the umbra. When it is very far away, no shadow is evident because all the penumbras mix together into a big blur.

Figure 27.10 ▲
A large light source produces a softer shadow than a smaller source.

■ Answer

The natural frequency of vibration for electrons in glass matches the frequency of ultraviolet light, so resonance in the glass occurs when ultraviolet waves shine on it. These energetic vibrations of electrons generate heat instead of wave reemission, so the glass is opaque to ultraviolet. In the range of visible light, the forced vibrations of electrons in the glass are more subtle, and reemission of light rather than the generation of heat occurs, so that the glass is transparent. Lower-frequency infrared causes entire atomic structures, not just electrons, to resonate, so heat is generated and the glass is opaque to infrared.

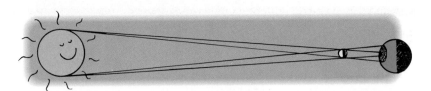

A dramatic example of this occurs when the moon passes between the earth and the sun—during a solar eclipse. Because of the large size of the sun, the rays taper to provide an umbra and a surrounding penumbra (Figure 27.12). The moon's shadow barely reaches the earth. If you stand in the umbra part of the shadow, you experience brief darkness during the day. If you stand in the penumbra, you experience a partial eclipse. The sunlight is dimmed and the sun appears as a crescent.*

The earth, like most objects in sunlight, casts a shadow. This shadow extends into space, and sometimes the moon passes into it. When this happens, we have a lunar eclipse. Whereas a solar eclipse can be observed only in a small region of the earth at a given time, a lunar eclipse can be seen by all observers on the nighttime half of the earth (Figure 27.13).

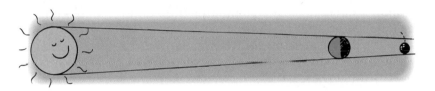

◀ **Figure 27.13**
An eclipse of the moon.

Shadows occur when light is bent in passing through a transparent material such as water. In Figure 27.14 shadows are cast by turbulent, rising warm water. Light travels at slightly different speeds in warm and in cold water. The difference bends light, just as layers of warm and cool air in the night sky bend starlight and cause the twinkling of stars. Some of the light gets deflected a bit and leaves darker places on the wall. The shapes of the shadows depend on how the light is bent. Chapter 29 returns to the bending of light.

■ Question

Why are lunar eclipses more commonly seen than solar eclipses?

■ Answer

There are usually two of each every year. However, the shadow of the moon on the earth is very small compared with the shadow of the larger earth on the smaller moon. Only a relatively few people are in the shadow of the moon (solar eclipse), while everybody who views the nighttime sky can see the shadow of the earth on the moon (lunar eclipse).

* People are cautioned not to look at the sun at the time of a solar eclipse because the brightness and ultraviolet radiation of direct sunlight is damaging to the eyes. This good advice is often misunderstood by those who then think that sunlight is more damaging at this special time. But staring at the sun when it is high in the sky is harmful whether or not an eclipse occurs. In fact, staring at the bare sun is more harmful than when part of the moon blocks it! The reason for special caution at the time of an eclipse is simply that more people are interested in looking at the sun during an eclipse.

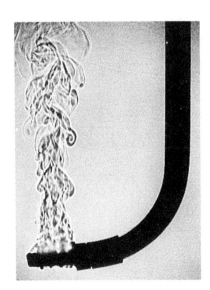

Figure 27.14 ▲
A heater at the tip of the submerged J-tube produces convection currents in the water. They are revealed by shadows cast by light that is deflected differently by the water of different temperatures.

Figure 27.15 ▲
A vertically polarized wave (left) and a horizontally polarized wave (right).

Figure 27.16 ▲
Polarized light lies along the same plane as that of the vibrations of the electron that emits it.

27.7 Polarization

Light travels in waves. The fact that the waves are transverse—and not longitudinal—is demonstrated by the phenomenon of **polarization.** If you shake the end of a horizontal rope, as in Figure 27.15, a transverse wave travels along the rope. The vibrations are back and forth in one direction, and the wave is said to be *polarized*. If the rope is shaken up and down, a vertically polarized wave is produced; that is, the waves traveling along the rope are confined to a vertical plane. If the rope is shaken from side to side, a horizontally polarized wave is produced.

A single vibrating electron emits an electromagnetic wave that is polarized. A vertically vibrating electron emits light that is vertically polarized, while a horizontally vibrating electron emits light that is horizontally polarized (Figure 27.16).

A common light source, such as an incandescent or fluorescent lamp, a candle flame, or the sun, emits light that is not polarized. This is because the vibrating electrons that produce the light vibrate in random directions. When light from these sources shines on a polarizing filter, such as that from which Polaroid® sunglasses are made, the light that is transmitted *is* polarized. The filter is said to have a *polarization axis* that is in the direction of the vibrations of the polarized light wave.

Figure 27.17 ▲
Polaroid sunglasses block out horizontally vibrating light. When the lenses overlap at right angles, no light gets through.

Light will pass through a pair of polarizing filters when their polarization axes are aligned, but not when they are crossed at right angles. This behavior is very much like the filtering of a vibrating rope that passes through a pair of picket fences, as illustrated in Figure 27.18.

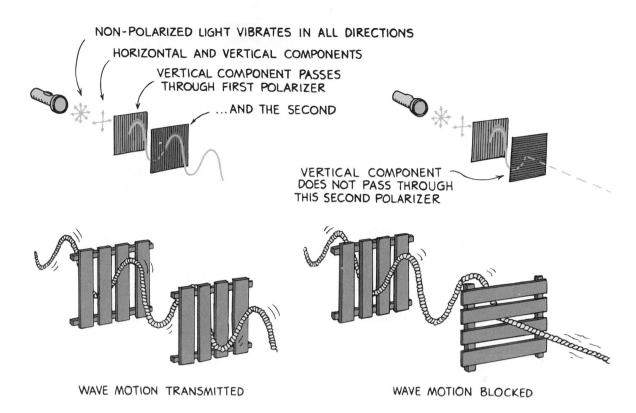

NON-POLARIZED LIGHT VIBRATES IN ALL DIRECTIONS

HORIZONTAL AND VERTICAL COMPONENTS

VERTICAL COMPONENT PASSES
THROUGH FIRST POLARIZER

...AND THE SECOND

VERTICAL COMPONENT
DOES NOT PASS THROUGH
THIS SECOND POLARIZER

WAVE MOTION TRANSMITTED

WAVE MOTION BLOCKED

When you skip flat stones across the surface of a pond, stones with flat sides parallel to the water bounce ("reflect"), but stones with flat sides at right angles to the surface penetrate the water ("refract"). Light behaves similarly. The flat side of a stone is like the plane of vibration of polarized light. Light that reflects at glancing angles from nonmetallic surfaces, such as glass, water, or roads, vibrates mainly in the plane of the reflecting surface. So glare from a horizontal surface is horizontally polarized. Do you see why the axes of Polaroid sunglasses are vertical? In this way glare from horizontal surfaces is eliminated.

Figure 27.18 ▲
A rope analogy illustrates the effect of crossed sheets of polarizing material.

Figure 27.19 ▲
Light is transmitted when the axes of the polarizing filters are aligned (left), but absorbed when they are at right angles to each other (center). Interestingly enough, when a third filter is sandwiched between the crossed ones (right), light is transmitted. Why? (To answer, you'll have to know more about vectors. See Appendix D, Vector Applications.)

27.8 Polarized Light and 3-D Viewing

Vision in three dimensions depends on the fact that both eyes give impressions simultaneously (or nearly so), each eye viewing a scene from a slightly different angle. To convince yourself of this, hold an upright finger at arm's length and see how it switches position relative to the background as you alternately close each eye. The view seen by each eye is different. The combination of views in the eye-brain system gives depth (Figure 27.20).

Figure 27.20 ▶
When your left eye looks at the left view of the statement while your right eye looks at the right view, your eye-brain system combines them to produce depth. The second and fourth lines appear farther away. To see this, place your face to the book with your nose touching the page. Now *very slowly,* without trying to focus your eyes at any one point, move away from the figure. If you've moved 30 centimeters and still haven't seen the stereo effect, start over. It may take a few tries. (If you view the overlapped images with crossed eyes, the second and fourth lines will appear closer!)

> Life success is not acquiring all the things you want, but becoming the kind of person you'd like to be.

> Life success is not acquiring all the things you want, but becoming the kind of person you'd like to be.

A pair of photographs or movie frames, taken a short distance apart (about average eye spacing), can be seen in 3-D when the left eye sees only the left view and the right eye sees only the right view. Slide shows or movies accomplish this by projecting the pair of views through polarization filters onto a screen. Their polarization axes are at right angles to each other (Figure 27.21). The overlapping pictures look blurry to the naked eye. To see in 3-D, the viewer wears polarizing eyeglasses with the lens axes also at right angles. In this way each eye sees a separate picture, just as in real life. The brain interprets the two pictures as a single picture with a feeling of depth. (Hand-held stereo viewers produce the same effect.)

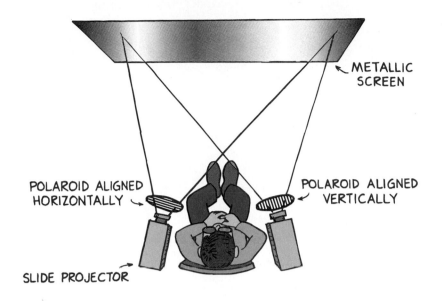

◀ **Figure 27.21**
A 3-D slide show using polarizing filters. The left eye sees only polarized light from the left projector; the right eye sees only polarized light from the right projector. Both views merge in the brain to produce an image with depth.

Depth is also seen in computer-generated stereograms, as in Figure 27.22. Here the slightly different patterns are hidden from a casual view. You can view the message of the figure (what this book is about!) with the procedure for viewing Figure 27.20. Once you've mastered the viewing technique, head for the local mall and check the variety of stereograms in posters and books.

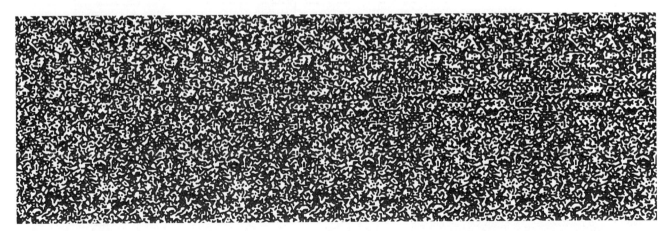

Figure 27.22 ▲
A computer-generated stereogram.

■ Question

Which pair of glasses is best suited for automobile drivers? (The polarization axes are shown by the straight lines.)

■ Answer

Pair A is best suited because the vertical axis blocks horizontally polarized light that composes much of the glare from horizontal surfaces. (Pair C is suited for viewing 3-D movies.)

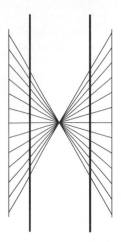

Are the vertical lines parallel?

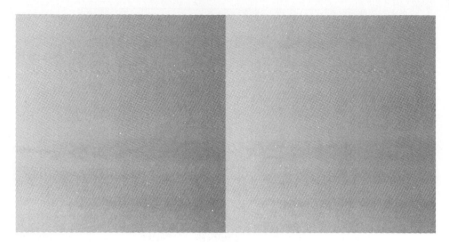

Both rectangles are equally bright. Cover the boundary between them with a pencil and see.

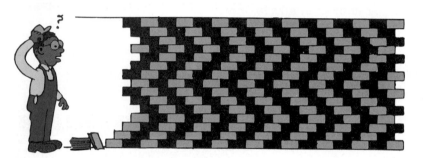

Are the tiles really crooked?

Is the hat taller than the brim is wide?

Could you make this in the shop?

Do these lines move?

PARIS
IN THE
THE SPRING

What does this sign read?

Figure 27.23 ▲
Optical illusions.

27 Chapter Review

Concept Summary

Light has a speed of 300 000 km/s in a vacuum, and lower average speeds in matter.

Light is energy that travels in electromagnetic waves within a certain range of frequencies.

- Light is produced by vibrating electric charges in atoms.

- Light passes through materials whose atoms absorb the energy and immediately reemit it as light.

- Light doesn't pass through a material when the energy is changed to random kinetic energy of the atoms.

Light waves are transverse, so they can be polarized (with vibrations all in the same direction).

- Polarizing filters transmit components of incident nonpolarized light that are parallel to the polarization axis, and block components vibrating at right angles to the polarization axis. The result is the emergence of polarized light.

Important Terms

electromagnetic spectrum (27.3)
electromagnetic wave (27.3)
infrared (27.3)
light-year (27.2)
opaque (27.5)
penumbra (27.6)
photon (27.1)
polarization (27.7)
ray (27.6)
shadow (27.6)
transparent (27.4)
ultraviolet (27.3)
umbra (27.6)

Review Questions

1. **a.** What is a photon?

   b. Which theory of light is the photon more consistent with—the wave theory or the particle theory? (27.1)

2. How long does it take for light to travel across the diameter of the earth's orbit around the sun? (27.2)

3. How did Michelson know the time that light took to make the round-trip to the distant mountain? (27.2)

4. How long does light take to travel from the sun to the earth? From the star Alpha Centauri to the earth? (27.2)

5. How long does light take to travel a distance of one light-year? (27.2)

6. What is the source of electromagnetic waves? (27.3)

7. Is the color spectrum simply a small segment of the electromagnetic spectrum? Defend your answer. (27.3)

8. How do the frequencies of infrared, visible, and ultraviolet light compare? (27.3)

9. How does the role of inertia relate to the rate at which electric charges can be forced into vibration? (27.4)

10. Different bells and tuning forks have their own natural vibrations, and emit their own tones when struck. How is this analogous to atoms, molecules, and light? (27.4)

11. Light incident upon a pane of glass slows down in passing through the glass. Does it emerge at a slower speed or at its initial speed? Explain. (27.4)

12. Will glass be transparent to frequencies of light that match its own natural frequencies? (27.4)

13. Does the time delay between the absorption and reemission of light affect the average speed of light in a material? Explain. (27.4)

14. Why would you expect the speed of light to be slightly less in the atmosphere than in a vacuum? (27.4)

15. When light encounters a material, it can build up vibrations in the electrons of certain atoms that may be intense enough to last over a long period of time. Will the energy of these vibrations tend to be absorbed and turned into heat, or absorbed and reemitted as light? (27.4)

16. What determines whether or not a material is transparent or opaque? (27.4–27.5)

17. Why are metals shiny in appearance? (27.5)

18. Distinguish between an umbra and a penumbra. (27.6)

19. a. Distinguish between a solar eclipse and a lunar eclipse.

b. Which type of eclipse is dangerous to your eyes if viewed directly? (27.6)

20. What is the difference between light that is polarized and light that is not? (27.7)

21. Why is light from a common lamp or from a candle flame nonpolarized? (27.7)

22. In what direction is the polarization of the glare that reflects from a horizontal surface? (27.7)

23. How do polarizing filters allow each eye to see separate images in the projection of three-dimensional slides or movies? (27.8)

Think and Explain

1. What evidence can you cite to support the idea that light can travel through a vacuum?

2. If the octagonal mirror in the Michelson apparatus were spun at twice the speed that produced light in the eyepiece, would light still be seen? At 2.1 times the speed? Explain.

3. If the mirror in Michelson's apparatus had had six sides instead of eight, would it have had to spin faster or more slowly to measure the speed of light? Explain.

4. You can get a sunburn on a sunny day and on an overcast day. But you cannot get a sunburn if you are behind glass. Explain.

5. If you fire a bullet through a tree, it will slow down in the tree and emerge at less than its initial speed. But when light shines on a pane of glass, even though it slows down inside, its speed upon emerging is the same as its initial speed. Explain.

6. Short wavelengths of visible light interact more frequently with the atoms in glass than do longer wavelengths. Which do you suppose takes the longer time to get through glass—red light or blue light?

7. Suppose that sunlight is incident upon both a pair of reading glasses and a pair of sunglasses. Which pair would you expect to be warmer, and why?

8. Why does a high-flying plane cast little or no shadow on the ground, while a low-flying plane casts a sharp shadow?

9. An ideal polarizing filter transmits 50% of the incident nonpolarized light. Why is this so?

10. What percentage of light would be transmitted by two ideal polarizing filters, one atop the other, with their axes aligned? With their axes crossed at right angles?

28 Color

Roses are red and violets are blue; colors intrigue artists and physics types too. To the physicist, the colors of things are not in the substances of the things themselves. Color is in the eye of the beholder and is provoked by the frequencies of light emitted or reflected by things. We see red in a rose when light of certain frequencies reaches our eyes. Other frequencies will provoke the sensation of other colors. Whether or not these frequencies of light are actually perceived as colors depends on the eye-brain system. Many organisms, including people with defective color vision, see no red in a rose.

Color is in the eye of the beholder.

28.1 The Color Spectrum

Isaac Newton was the first to make a systematic study of color. By passing a narrow beam of sunlight through a triangular-shaped glass prism, he showed that sunlight is composed of a mixture of all the colors of the rainbow. The prism cast the sunlight into an elongated patch of colors on a sheet of white paper (Figure 28.1). Newton called this spread of colors a **spectrum,** and noted that the colors were formed in the order red, orange, yellow, green, blue, and violet.

Sunlight is an example of what is called **white light.** Under white light, white objects appear white and colored objects appear in their individual colors. Newton showed that the colors in the spectrum were a property not of the prism but of white light itself. He demonstrated this when he recombined the colors with a second prism to produce white light again. In other words, all the colors, one atop the other, combine to produce white light. Strictly speaking, white is not a color but a combination of all colors.

Black is similarly not a color itself, but is the absence of light. Objects appear black when they absorb light of all visible frequencies.

Figure 28.1 ▲
Newton passed sunlight through a glass prism to form the color spectrum.

421

Figure 28.2 ▲
When sunlight passes through a prism, it separates into a spectrum of all the colors of the rainbow.

Figure 28.3 ▲
When a stack of razor blades bolted together is viewed end on, the edges appear black. Light that enters the wedge-shaped spaces between the blades is reflected so many times that most of it is absorbed.

Carbon soot is an excellent absorber of light and looks very black. The dull finish of black velvet is an excellent absorber also. But even a polished surface may look black under some conditions. For example, highly polished razor blades are not black, but when stacked together and viewed end on, they appear quite black (Figure 28.3). Most of the light that gets between the closely spaced edges of the blades gets trapped and is absorbed after being reflected many times.

Black objects that you can see do not absorb all light that falls on them, for there is always some reflection at the surface. If not, you wouldn't be able to see them.

28.2 Color by Reflection

The colors of most objects around you are due to the way the objects reflect light. Light is reflected from objects in a manner similar to the way sound is "reflected" from a tuning fork when another that is nearby sets it into vibration. A tuning fork can be made to vibrate even when the frequencies are not matched, although at significantly reduced amplitudes. The same is true of atoms and molecules. We can think of atoms and molecules as three-dimensional tuning forks with electrons that behave as tiny oscillators that whirl in orbits around the nuclei. Electrons can be forced temporarily into larger orbits by the vibrations of electromagnetic waves (such as light). Like acoustical tuning forks, once excited to more vigorous motion, electrons send out their own energy waves in all directions.

Different materials have different natural frequencies for absorbing and emitting radiation. In one material, electrons oscillate readily at certain frequencies; in another material, they oscillate readily at different frequencies. At the resonant frequencies where the amplitudes of oscillation are large, light is absorbed (recall from the previous chapter that glass absorbs ultraviolet light for this reason). But at frequencies below and above the resonant frequencies, light is reemitted. If the material is transparent, the reemitted light passes through it. If the material is opaque, the light passes back into the medium from which it came. This is reflection.

Figure 28.4 ▲
The square on the left *reflects* all the colors illuminating it. In sunlight it is white. When illuminated with blue light, it is blue. The square on the right *absorbs* all the colors illuminating it. In sunlight it is warmer than the white square.

Most materials absorb light of some frequencies and reflect the rest. If a material absorbs light of most visible frequencies and reflects red, for example, the material appears red. If it reflects light of all the visible frequencies, like the white part of this page, it will be the same color as the light that shines on it. If a material absorbs all the light that shines on it, it reflects none and is black.

When white light falls on a flower, light of some frequencies is absorbed by the cells in the flower and some light is reflected. Cells that contain chlorophyll absorb light of most frequencies incident upon them and reflect the green part, so they appear green. The petals of a red rose, on the other hand, reflect primarily red light, with a lesser amount of blue. Interestingly enough, the petals of most yellow flowers, such as daffodils, reflect red and green as well as yellow. Yellow daffodils reflect light of a broad band of frequencies. The reflected colors of most objects are not pure single-frequency colors, but are composed of a spread of frequencies. So something yellow, for example, may simply be a mixture of colors without blue and violet—or it can be built of red and green together.

It is important to note that an object can reflect only light of frequencies present in the illuminating light. The appearance of a colored object therefore depends on the kind of light used. A candle flame emits light that is deficient in the higher frequencies; it emits a yellowish light. Things look yellowish in candlelight. An incandescent lamp emits light of all the visible frequencies, but is richer toward the lower frequencies, enhancing the reds. A fluorescent lamp is richer in

the higher frequencies, so blues are enhanced when illuminated with fluorescent lamps. In a fabric with a little bit of red, for example, the red will be more apparent when illuminated with an incandescent lamp than with a fluorescent lamp. Colors in daylight appear different from the way they appear when illuminated with either of these lamps (Figure 28.5). The perceived color of an object is subjective and depends on the light source, although color differences between two objects are most easily detected in bright sunlight.

Figure 28.5 ▲
Color depends on the light source.

28.3 Color by Transmission

The color of a transparent object depends on the color of the light it transmits. A red piece of glass appears red because it absorbs all the colors that compose white light, except red, which it transmits. Similarly, a blue piece of glass appears blue because it transmits primarily blue and absorbs the other colors that illuminate it.

The material in the glass that selectively absorbs colored light is known as a **pigment.** From an atomic point of view, electrons in the pigment atoms selectively absorb light of certain frequencies in the illuminating light. Light of other frequencies is reemitted from atom to atom in the glass. The energy of the absorbed light increases the kinetic energy of the atoms, and the glass is warmed. Ordinary window glass is colorless because it transmits light of all visible frequencies equally well.

Figure 28.6 ▶
Blue glass transmits only energy of the frequency of blue light; energy of the other frequencies is absorbed and warms the glass.

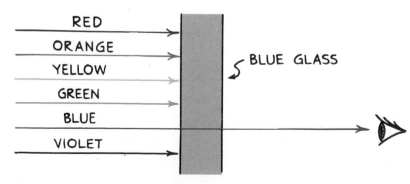

1. When red light shines on a red rose, why do the leaves become warmer than the petals?

2. When green light shines on a red rose, why do the petals look black?

3. What color does a ripe banana appear when illuminated with red light? With yellow light? With green light? With blue light?

▶ **DOING PHYSICS**

Reflections

Hold a candle flame, match flame, or any small source of white light in between you and a piece of colored glass. You'll see two reflections from the glass; one from the front surface and one from the back surface. What is the color of the flame reflected from the front surface? From the back surface? Is there a difference, and if so, why?

Activity

28.4 Sunlight

White light from the sun is a composite of all the visible frequencies. The brightness of solar frequencies is uneven, as indicated in the graph of brightness versus frequency in Figure 28.7. The graph indicates that the lowest frequencies of sunlight, in the red region, are not as bright as those in the middle-range yellow and green region. Yellow-green light is the brightest part of sunlight. Since humans evolved in the presence of sunlight, it is not surprising that we are most sensitive to yellow green. That is why it is more and more common for new fire engines to be painted yellow green, particularly at airports where visibility is vital. This also explains why at night we

---

■ **Answers**

1. The petals appear red because they reflect red light. The leaves absorb rather than reflect red light, so the leaves become warmer.

2. The petals absorb rather than reflect the green light. Since green is the only color illuminating the rose, and green contains no red to be reflected, the rose appears to have no color at all—black.

3. A banana reflects red, yellow, and green light, so when illuminated with any of these colors it reflects that color and appears that color. A banana does not reflect blue, so when illuminated with blue light it appears black.

are able to see better under the illumination of yellow sodium-vapor lamps than we are under tungsten lamps of the same brightness. The blue portion of sunlight is not as bright, and the violet portion is even less bright.

Figure 28.7 ▶
The radiation curve of sunlight is a graph of brightness versus frequency. Sunlight is brightest in the yellow-green region, in the middle of the visible range.

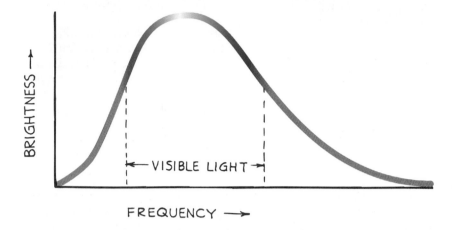

The graphical distribution of brightness versus frequency in Figure 28.7 is called the *radiation curve* of sunlight. Most whites produced from reflected sunlight have this frequency distribution.

28.5 Mixing Colored Light

Light of all the visible frequencies mixed together produces white. Interestingly enough, white also results from the combination of only red, green, and blue light. When a combination of only red, green, and blue light of equal brightness is overlapped on a screen, as shown in Figure 28.8, it appears white. Where red and green light alone overlap, the screen appears yellow. Red and blue light alone produce the bluish red color called *magenta*. Green and blue light alone produce the greenish blue color called *cyan*.

Figure 28.8 ▼
When red light, green light, and blue light of equal brightness are projected on a white screen, the overlapping areas appear different colors. Where all three overlap, white is produced.

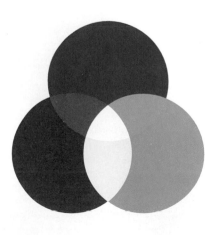

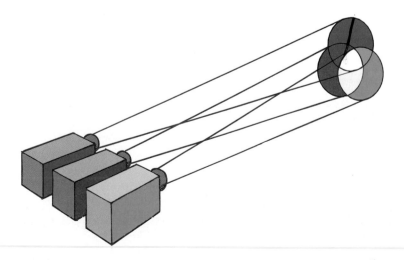

This can be understood if the frequencies of white light are divided into three regions: the lower-frequency red end, the middle-frequency green part, and the higher-frequency blue end (Figure 28.9). The low and middle frequencies combined appear yellow to the human eye. The middle and high frequencies combined appear greenish blue (cyan). The low and high frequencies combined appear bluish red (magenta).

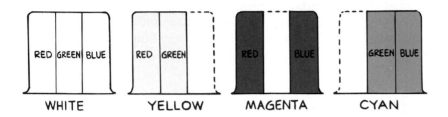

◀ **Figure 28.9**
The low-frequency, middle-frequency, and high-frequency parts of white light appear *red, green,* and *blue.* To the human eye, red + green = yellow; red + blue = magenta; green + blue = cyan.

In fact, almost any color at all can be made by overlapping light of three colors and adjusting the brightness of each color of light. This amazing phenomenon is due to the way the human eye works. The three colors do not have to be red, green, and blue, although those three produce the highest number of different colors. For this reason red, green, and blue are called the **additive primary colors.**

Color television is based on the ability of the human eye to see combinations of three colors as a variety of different colors. A close examination of the picture on most color television tubes will reveal that the picture is made up of an assemblage of tiny spots, each less than a millimeter across. When the screen is lit, some of the spots are red, some green, and some blue. At a distance the mixtures of these colors provide a complete range of colors, plus white.*

28.6 Complementary Colors

What happens when two of the three additive primary colors are combined?

red + green = yellow

red + blue = magenta

blue + green = cyan

Now, a little thought and inspection of Figure 28.8 will show that when we add in the third color, we get white.

---

* On a black-and-white television set, the black we see in the darkest scenes is simply the color of the tube face itself, which is more a light gray than black. Our eyes are sensitive to the contrast with the illuminated parts of the screen, and we see the light gray as black. In our minds, we make it black.

$$yellow + blue = white$$

$$magenta + green = white$$

$$cyan + red = white$$

When two colors are added together to produce white, they are called **complementary colors.** For example, we see that yellow and blue are complementary because yellow, after all, is the combination of red and green. And red, green, and blue light together appear white. By similar reasoning we see that magenta and green are complementary colors, as are cyan and red. Every hue has some complementary color that when added will produce white.

Figure 28.10 ▲
(Top left) Under white light, the colored blocks appear red, orange, yellow, green, blue, and violet, and the shadows are gray. (Top right) When the blocks are lit by red light from the right and green light from the left, the blocks reflect no blue light; the shadow on the left, where red light is nearly absent, appears green; similarly the shadow on the right appears red; the background reflects both red and green light and appears yellow. (Bottom left) The blocks are lit by blue light. (Bottom right) The blocks are lit by red light and blue light. Can you explain the colors of the shadows and backgrounds in the bottom photos?

Now, if you begin with white light and *subtract* some color from it, the resulting color will appear as the complement of the one subtracted. Not all the light incident upon an object is reflected. Some is absorbed. The part that is absorbed is in effect subtracted from the

■ Questions

1. What are the complementary colors of (a) magenta, (b) blue, and (c) cyan?
2. What color does red light plus blue light appear?
3. What color does white light *minus* yellow light appear?
4. What color does white light *minus* green light appear?

■ Answers

1. (a) Green; (b) yellow; (c) red
2. Magenta
3. Blue
4. Magenta

Retinal Fatigue

When you stare at a colored object for a while the color receptors in your eyeballs become fatigued. Try the following activity to see what happens. Stare at the flag for a minute or so. Now look at a white area. The afterimage you see is comprised of the complementary colors! This occurs because the fatigued receptors send a weaker signal to the brain. White minus a color produces the complementary to the missing color. Try this with other colors also.

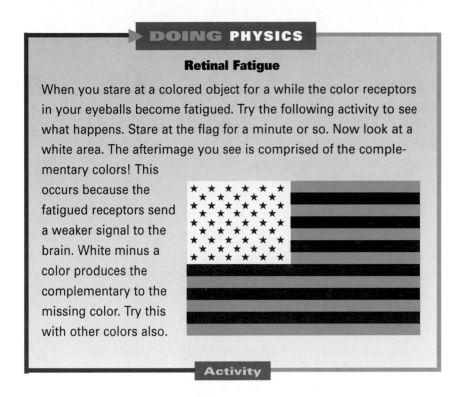

Activity

incident light. For example, if white light falls on a pigment that absorbs red light, the light reflected appears cyan. A pigment that absorbs blue light will appear yellow; similarly, a pigment that absorbs yellow light will appear blue. Whenever you subtract a color from white light, you end up with the complementary color.

28.7 Mixing Colored Pigments

Artists know that if we mix red, green, and blue paint, the result will be not white but a muddy dark brown. Red and green paint certainly do not combine to form yellow as red and green light do. The mixing of paints and dyes is an entirely different process from the mixing of colored light.

Paints and dyes contain finely divided solid particles of pigment that produce their colors by absorbing light of certain frequencies and reflecting light of other frequencies. Pigments absorb light of a relatively wide range of frequencies and reflect a wide range as well. In this sense, pigments reflect a mixture of colors.

Blue paint, for example, reflects mostly blue light, but also violet and green; it absorbs red, orange, and yellow light. Yellow paint reflects mostly yellow light, but also red, orange, and green; it absorbs blue and violet light. When blue and yellow paints are mixed, then between them they absorb all the colors except green.

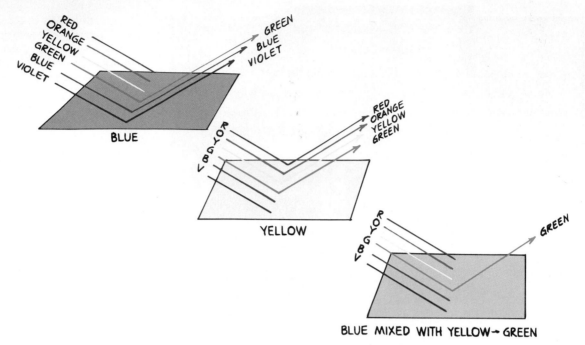

Figure 28.11 ▲

(Left) Blue pigment reflects not only blue light, but also colors to either side of blue—namely, green and violet. It absorbs red, orange, and yellow light. (Center) Yellow pigment reflects not only yellow light, but also red, orange, and green. It absorbs blue and violet light. (Right) When blue and yellow pigments are mixed, the only common color reflected is green. The other colors have been *subtracted* from the incident white light.

The only color they both reflect is green (Figure 28.11), which is why the mixture looks green. This process is called *color mixing by subtraction,* to distinguish it from the effect of mixing colored light, which is called *color mixing by addition.*

So when you cast lights on the stage at a school play, you use the rules of color addition to produce various colors. But when you mix paint, you use the rules of color subtraction.

You may have learned as a child that you can make any color with crayons or paints of three so-called primary colors: red, yellow, and blue. Actually, the three paint or dye colors that are most useful in color mixing by subtraction are magenta (bluish red), yellow, and cyan (greenish blue). These are the **subtractive primary colors,** used in printing illustrations in full color.*

Color printing is done on a press that prints each page with four differently colored inks (magenta, yellow, cyan, and black) in succession. Each color of ink comes from a different plate, which transfers

---

* Note that magenta, yellow, and cyan are used to make other colors by *subtraction,* as when colored paints or dyes are mixed. When colors are mixed by *addition,* as when colored light is mixed, red, green, and blue are the most useful colors to mix.

the ink to the paper. The ink deposits are regulated on different parts of the plate by tiny dots. Examine the colored pictures in this book, or in any magazine, with a magnifying glass and see how the overlapping dots of three colors plus black give the appearance of many colors. Figure 28.12 shows the images made by the four plates, separately and combined.

Table 28.1 Color Subtraction		
Pigment	**Absorbs**	**Reflects**
red	blue, green	red
green	blue, red	green
blue	red, green	blue
yellow	blue	red, green
cyan	red	green, blue
magenta	green	red, blue

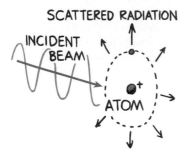

Figure 28.13 ▲
A beam of light falls on an atom and causes the electrons in the atom to move temporarily in larger orbits. The more vigorously oscillating electrons reemit light in various directions. Light is scattered.

28.8 Why the Sky Is Blue

If a sound beam of a particular frequency is directed to a tuning fork of similar frequency, the tuning fork will be set into vibration and effectively redirect the beam in multiple directions. The tuning fork **scatters** the sound. A similar process occurs with the scattering of light from molecules and larger specks of matter that are far apart from one another—as in the atmosphere.

We know that atoms and molecules behave like tiny optical tuning forks and reemit light waves that shine on them. Very tiny particles do the same. The tinier the particle, the higher the frequency of light it will scatter. This is similar to small bells ringing with higher notes than larger bells. The nitrogen and oxygen molecules and the tiny particles that make up the atmosphere are like tiny bells that "ring" with high frequencies when energized by sunlight. Like the sound from bells, the reemitted light is sent in all directions. It is scattered.

Figure 28.14 ▲
The sky is blue because its tiny particles scatter high-frequency light. The blue "sky" between the viewer and the distant mountains produces bluish mountains.

Most of the ultraviolet light from the sun is absorbed by a protective layer of ozone gas in the upper atmosphere. The remaining ultraviolet sunlight passing through the atmosphere is scattered by atmospheric particles and molecules. Of the visible frequencies, violet light is scattered the most, followed by blue, green, yellow, orange, and red, in that order. Red light is scattered only a tenth as much as violet. Although violet light is scattered more than blue, our eyes are not very sensitive to violet light. Our eyes are more sensitive to blue, so we see a blue sky.

The blue of the sky varies in different places under different conditions. Where there are a lot of particles of dust and other particles larger than oxygen and nitrogen molecules, the lower frequencies of light are scattered more. This makes the sky less blue, and it takes on a whitish appearance. After a heavy rainstorm, when the particles have been washed away, the sky becomes a deeper blue.

The higher that one goes into the atmosphere, the fewer molecules there are in the air to scatter light. The sky appears darker. When there are no molecules, as on the moon for example, the "sky" is black.

Water droplets in a variety of sizes—some of them microscopic—make up clouds. The different-size droplets result in a variety of frequencies for scattered light: low frequencies from larger droplets and high frequencies from tinier droplets of water molecules. The overall result is a white cloud. The electrons in a tiny droplet vibrate together and in step, which results in the scattering of a greater amount of energy than when the same number of electrons vibrate separately. Hence, clouds are bright!

Figure 28.15 ▲
The droplets that compose a cloud come in a wide variety of sizes. Hence a wide variety of colors are scattered, which is why the cloud is white.

28.9 Why Sunsets Are Red

The lower frequencies of light are scattered the least by nitrogen and oxygen molecules. Therefore red, orange, and yellow light are transmitted through the atmosphere more readily than violet and blue. Red light, which is scattered the least, passes through more atmosphere without interacting with matter than light of any other color. Therefore, when light passes through a thick atmosphere, light of the lower frequencies is transmitted while light of the higher frequencies is scattered. At dawn and at sunset, sunlight reaches us through a longer path through the atmosphere than at noon.

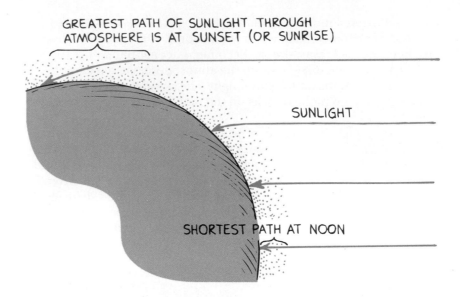

GREATEST PATH OF SUNLIGHT THROUGH ATMOSPHERE IS AT SUNSET (OR SUNRISE)

SUNLIGHT

SHORTEST PATH AT NOON

At noon sunlight travels through the least amount of atmosphere to reach the earth's surface (Figure 28.16). Then a relatively small amount of light is scattered from sunlight. As the day progresses and the sun is lower in the sky, the path through the atmosphere is longer, and more blue is scattered from the sunlight. Less and less blue remains in the sunlight that reaches the earth. The sun appears progressively redder, going from yellow to orange and finally to a reddish orange at sunset. (The sequence is reversed between dawn and noon.)

The colors of the sun and sky are consistent with our rules for color mixing. When blue is subtracted from white light, the complementary color that is left is yellow. The subtraction of violet leaves orange. When green is subtracted, magenta is left. The relative amounts of scattering depend on atmospheric conditions, which change from day to day and give us a variety of sunsets.

The next time you find yourself admiring a crisp blue sky, or delighting in the shapes of bright clouds, or watching a beautiful sunset, think about all those ultra-tiny optical tuning forks vibrating; you'll appreciate these everyday wonders of nature even more!

Figure 28.17 ▶
The sunset sky is red because of the absence of high-frequency light.

1. If molecules in the sky scattered low-frequency light more than high-frequency light, how would the colors of the sky and sunsets appear?

2. Distant dark mountains are bluish in color. What is the source of this blueness? (*Hint:* What is between you and the mountains you see?)

3. Distant snow-covered mountains reflect a lot of light and are bright. But they sometimes look yellowish, depending on how far away they are. Why are they yellow? (*Hint:* What happens to the reflected white light as it travels from the mountain to you?)

Why do you see the scattered blue when the background is dark, but not when the background is bright? Because the scattered blue is faint. A faint color will show itself against a dark background, but not against a bright background. For example, when you look from the earth's surface at the atmosphere against the darkness of space, the atmosphere is sky blue. But astronauts above who look below through the same atmosphere to the bright surface of the earth do not see the same blueness.

28.10 Why Water Is Greenish Blue

We often see a beautiful deep blue when we look at the surface of a lake or the ocean. But that is not the color of water. It is the reflected color of the sky. The color of water itself, as you can see by looking at a piece of white material under water, is a pale greenish blue.

Water is transparent to nearly all the visible frequencies of light. Water molecules absorb infrared waves because they resonate to the frequencies of infrared. The energy of the infrared waves is transformed into kinetic energy of the water molecules. Infrared is a strong component of the sunlight that warms water.

■ **Answers**

1. If low frequencies were scattered more, red light would be scattered out of the sunlight on its long path through the atmosphere at sunset, and the sunlight to reach your eye would be predominantly blue and violet. So sunsets would appear blue!

2. If you look at distant dark mountains, very little light from them reaches you, and the blueness of the atmosphere between you and the mountains predominates. The blueness is of the low-altitude "sky" between you and the mountains. That's why distant mountains look blue!

3. The reason that distant snow-covered mountains often appear a pale yellow is because the blue in the white light from the snowy mountains is scattered on its way to you. What happens to white when blue is scattered from it? The complementary color left is yellow.

Figure 28.18 ▲
Ocean water is cyan because it absorbs red. The froth in the waves is white because its droplets of many sizes scatter many colors.

Water molecules resonate somewhat to the visible-red frequencies. This causes a gradual absorption of red light by water. A 15-m layer of water reduces red light to a quarter of its initial brightness. There is very little red light in the sunlight that penetrates below 30 m of water. When red is taken away from white light, what color remains? This question can be asked in another way: What is the complementary color of red? The complementary color of red is cyan—a greenish blue color. In seawater, the color of everything at these depths looks greenish blue.

It is interesting to note that many crabs and other sea animals that appear black in deep water are found to be red when they are raised to the surface. At great depths, black and red look the same. So both black and red sea animals are hardly seen by predators and prey in deep water. They have survived an evolutionary history while more visible varieties have not.

In summary, the sky is blue because blue from sunlight is reemitted in all directions by molecules in the atmosphere. Water is greenish blue because red is absorbed by molecules in the water. The colors of things depend on what colors are reflected by molecules, and also by what colors are absorbed by molecules.

28.11 The Atomic Color Code— Atomic Spectra

Every element has its own characteristic color when made to emit light. If the atoms are far enough apart so that their vibrations are not interrupted by neighboring atoms, their true colors are emitted. This occurs when atoms are made to glow in the gaseous state. (In the solid state, as in a lamp filament, where atoms are crowded together, the characteristic colors of the atoms are smudged to produce a continuous spectrum.) Neon gas, for example, glows a brilliant red; mercury vapor glows a bluish violet; and helium glows a pink. The glow of each element is unlike the glow of any other element.

The light from glowing elements can be analyzed with an instrument called a **spectroscope.** This chapter began with a brief account of Newton's investigation of light passing through a prism. The spectrum formed in Newton's first experiment was impure, because it was formed by overlapping circular images of the circular hole in his window shutter. He later produced a better spectrum by first passing light through a thin slit and then focusing it with lenses through the prism and onto a white screen (Figure 28.19). If the slit is made narrow, overlapping is reduced and the colors in the resulting spectrum are much clearer.

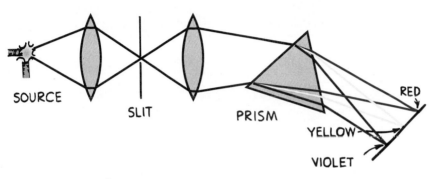

Figure 28.19 ▲
A fairly pure spectrum is produced by passing white light through a thin slit, two lenses, and a prism.

This arrangement of thin slit, lenses, and a prism (or a diffraction grating) is the basis for the spectroscope.* A simple spectroscope with a diffraction grating is shown in Figure 28.20. A spectroscope displays the spectra of the light from hot gases and other light sources. (*Spectra* is the plural of *spectrum.*) The spectra of light sources are viewed through a magnifying eyepiece.

When light from a glowing element is analyzed through a spectroscope, it is found that the colors are the composite of a variety of different frequencies of light. The spectrum of an element appears not as a continuous band of color but as a series of lines, as shown in Figure 28.21. Each line corresponds to a distinct frequency of light. Such a spectrum is known as a **line spectrum.** The spectral lines seen in the spectroscope are images of the slit through which the light passes. Note that each colored line appears in the same position as that color in the continuous spectrum.

Figure 28.20 ▲
A spectroscope. Light to be analyzed illuminates the thin slit at the left, where it is focused by lenses onto either a diffraction grating (shown) or a prism on the rotating table in the middle, and then viewed through the eyepiece on the right.

---

* Although a *diffraction grating* works differently from a prism, it too spreads light into a spectrum. It is more commonly used than a prism in spectroscopes. A *spectrometer* is similar to a spectroscope except that it also measures the wavelengths of a spectrum and records the spectrum (on film for example).

Figure 28.21 ▲
(From top to bottom) The continuous spectrum of an incandescent lamp and the line spectra of three elements: hydrogen, sodium, and mercury.

The light from each different element produces its own characteristic pattern of lines. This is because each element has its own distinct configuration of electrons, and these emit distinct frequencies of light when electrons change from one energy state to another in the atom.* The frequencies of light emitted by atoms in the gaseous state are the "fingerprints" of the elements. Much of the information that physicists have about atomic structure is from the study of atomic spectra. The atomic composition of common materials, the sun, and distant galaxies is revealed in the spectra of these sources. Even the element helium, the second most common element in the universe, was discovered through its "fingerprint" in sunlight. The spectrometer is a very useful and powerful tool.

After the French Emperor Napoleon Bonaparte died in 1821, scientists analyzed samples of his hair with a spectroscope. Murder was suspected, for they found traces of arsenic—a poison. More recently, however, it has been discovered that the coloring in Napoleon's wallpaper contained high levels of arsenic. It is now speculated that dampness and mold may have turned the arsenic into a deadly gas.

---

* Electrons in atoms behave differently when materials glow and are sources of illumination than when materials simply reflect light that shines on them. This distinction won't be treated in detail here, except to say that when materials are made to glow, the electrons in their atoms jump to orbits of higher energy in a process called *excitation*.

Concept Summary

White light is a combination of light of all visible frequencies.

■ Black is the absence of light; objects that appear black absorb all visible frequencies.

The color of an object is due to the color of the light it reflects (if opaque) or transmits (if transparent).

■ Light is absorbed when its frequency matches a natural vibration frequency of electrons in the material illuminated by the light.

Color mixing by addition is the mixing of light of different frequencies.

■ The eye sees a combination of red, green, and blue light of equal brightness as white.

■ Red, green, and blue are the additive primary colors.

Color mixing by subtraction is the mixing of colored paints or dyes, which absorb most frequencies except for the ones that give them their characteristic color.

■ When paints or dyes are mixed, the mixture absorbs all the frequencies each paint or dye absorbs.

■ Magenta, cyan, and yellow are the subtractive primary colors.

Scattering of violet and blue frequencies of sunlight in all directions is what gives the sky its blue color.

■ When sunlight travels a long path through the atmosphere, as at dawn or sunset, only the lower frequencies of light are transmitted; the higher ones are scattered out.

Atoms of each element have characteristic line spectra that can be used to identify the element.

Important Terms

additive primary colors (28.5)
complementary colors (28.6)
line spectrum (28.11)
pigment (28.3)
scatter (28.8)
spectroscope (28.11)
spectrum (28.1)
subtractive primary colors (28.7)
white light (28.1)

Review Questions

1. List the order of colors in the color spectrum. (28.1)

2. Are black and white real colors, in the sense that red and green are? Explain. (28.1)

3. A vibrating tuning fork emits sound. What is emitted by the vibrating electrons of atoms? (28.2)

4. What happens to light of a certain frequency that encounters atoms of the same resonant frequency? (28.2)

5. Why does the color of an object look different under a fluorescent lamp from the way it looks under an incandescent lamp? (28.2)

6. **a.** What color(s) of light does a transparent red object *transmit*?

   b. What color(s) does it *absorb*? (28.3)

7. What is the function of a pigment? (28.3)

8. Why are more and more fire engines being painted yellow green instead of red? (28.4)

9. How can yellow be produced on a screen if only red light and green light are available? (28.5)

10. What is the name of the color produced by a mixture of green and blue light? (28.5)

11. What colors of spots are lit on a television tube to give full color? (28.5)

12. What are complementary colors? (28.6)

13. What color is the complement of blue? (28.6)

14. The process of producing a color by mixing pigments is called *color mixing by subtraction*. Why do we say "subtraction" instead of "addition" in this case? (28.7)

15. What colors of ink are used to print full-color pictures in books and magazines? (28.7)

16. What is light scattering? (28.8)

17. a. Do tiny particles in the air scatter high or low frequencies of light?

b. What frequencies do large particles scatter? (28.8)

18. Why is the sky blue? (28.8)

19. Why is the sky sometimes whitish? (28.8)

20. Why are clouds white? (28.8)

21. Why are sunsets red? (28.9)

22. Why is water greenish blue? (28.10)

23. What is a spectroscope, and what is its function? (28.11)

24. Does the red light from glowing neon gas have only one frequency or a mixture of frequencies? (28.11)

25. Why might atomic spectra be considered the "fingerprints" of atoms? (28.11)

Activities

1. Cover the ends of a cardboard tube with metal foil. Punch a hole in each end with a pencil, one about 3 or so millimeters and the other twice as big. Put your eye to the smaller hole and look through the tube at the colors of things against the black background of the tube. Colors will appear very different from how they appear against ordinary backgrounds.

2. If you have a computer with a color monitor and a color-controlled program available, try the following. Add full-strength red and full-strength green. Note that you produce yellow. Add about two-thirds strength blue and you get a lighter (not darker) yellow. Try full-strength red, blue, and green, and you get white. Can you see that the more light you shed on something, the brighter (that is, closer to white) it gets?

3. Simulate your own sunset: Add a few drops of milk to a glass of water and look through it to a lit incandescent bulb. The bulb appears to be red or pale orange, while light scattered to the side appears blue. Try it and see.

Think and Explain

1. What is the color of common tennis balls, and why?

2. Shine red light on a rose. Why will the temperature of the leaves increase more than the temperature of the red petals?

3. Why are the interiors of optical instruments painted black?

4. Suppose two beams of white light are shone on a white screen, one beam through a pane of red glass and the other through a pane of green glass. What color appears on the screen where the two beams overlap? What occurs if instead the two panes of glass are placed in the path of a single beam?

5. In a dress shop that has only fluorescent lighting, a customer insists on taking a garment into the daylight at the doorway. Is she being reasonable? Explain.

6. What color would a yellow cloth appear if illuminated with sunlight? With yellow light? With blue light?

7. A spotlight is coated so that it won't transmit blue from its white-hot filament. What color is the emerging beam of light?

8. How could you use the spotlights at a play to make the yellow clothes of the performers suddenly change to black?

9. A stage performer stands where beams of red and green light cross.
   a. What is the color of her white shirt under this illumination?

   b. What are the colors of the shadows she casts on the stage floor?

10. What colors of ink do color ink-jet printers use to produce the colors you see? Do the inks form colors by color addition or by color subtraction?

11. On a photographic print, your dearest friend is seen wearing a red sweater. What color is the sweater on the negative?

12. Why can't we see a laser beam going across the room unless there is fog, chalk dust, or a mist in the air?

13. Very big particles, such as droplets of water, absorb more radiation than they scatter. How does this fact help to explain why rain clouds appear dark?

14. The only light to reach very far beneath the surface of the ocean is greenish blue. Objects at these depths either reflect greenish blue or reflect no color at all. If a ship that is painted red, green, and white sinks to the bottom of the ocean, how will these colors appear?

15. A lamp filament is made of tungsten. When made to glow, it emits a continuous spectrum—all the colors of the rainbow. When tungsten gas is made to glow, however, the light is a composite of very discrete colors. Why is there a difference in spectra?

29 Reflection and Refraction

Changes in light speed produce refraction.

When you shine a beam of light on a mirror, the light doesn't travel through the mirror, but is returned by the mirror's surface back into the air. When sound waves strike a canyon wall, they return to you as an echo. When a transverse wave transmitted along a spring reaches a wall, it reverses direction. In all these situations, waves remain in one medium rather than enter a new medium. These waves are *reflected*.

In other situations, as when light passes from air into water, waves travel from one medium into another. When waves strike the surface of a medium at an angle, their direction changes as they enter the second medium. These waves are *refracted*. This is evident when a pencil in a glass of water appears to be bent.

Usually waves are partly reflected and partly refracted when they fall on a transparent medium. When light shines on water, for example, some of the light is reflected and some is refracted. To understand this, let's see how reflection occurs.

29.1 Reflection

When a wave reaches a boundary between two media, some or all of the wave bounces back into the first medium. This is **reflection.** For example, suppose you fasten a spring to a wall and send a pulse along the spring's length (Figure 29.1). The wall is a very rigid medium compared with the spring. As a result, all the wave energy is reflected back along the spring rather than being transmitted into the wall. Waves that travel along the spring are almost *totally reflected* at the wall.

If the wall is replaced with a less rigid medium, such as the heavy spring shown in Figure 29.2, some energy is transmitted into

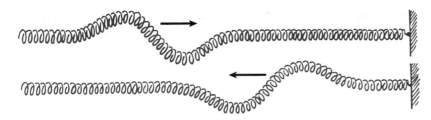

Figure 29.1 ▲
A wave is totally reflected when it reaches a completely rigid boundary.

the new medium. Some of the wave energy is still reflected. These waves are *partially reflected.*

A metal surface is rigid to light waves that shine upon it. Light energy does not propagate into the metal, and instead is returned in a reflected wave. The wave reflected from a metal surface has almost the full intensity of the incoming wave, apart from small energy losses due to the friction of the vibrating electrons in the surface. This is why metals such as silver and aluminum are so shiny. They reflect almost all the frequencies of visible light. Smooth surfaces of these metals are therefore used as mirrors.

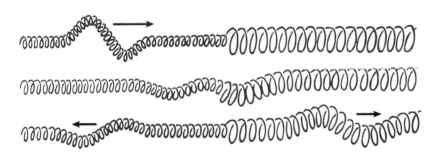

◀ **Figure 29.2**
When the wave reaches the heavy spring, it is partially reflected. Part of the wave energy bounces back along the first spring, while the other part travels along the heavy spring.

Other materials such as glass and water are not as rigid to light waves. Like the different springs of Figure 29.2, wave energy is both reflected and transmitted at the boundary. When light shines perpendicularly on the surface of still water, about 2% of its energy is reflected and the rest is transmitted. When light strikes glass perpendicularly, about 4% of its energy is reflected. Except for slight losses, the rest is transmitted.

29.2 The Law of Reflection

In one dimension, reflected waves simply travel back in the direction from which they came. Let a ball drop to the floor, and it bounces straight up along its initial path. In two dimensions, the situation is a little different. Toss a ball at an angle to the floor, and it normally bounces at the same angle in a new direction. Likewise with light.

The direction of incident and reflected waves is best described by straight-line *rays*. Incident rays and reflected rays make equal angles with a line perpendicular to the surface, called the **normal,** as shown in Figure 29.3. The angle made by the incident ray and the normal, called the **angle of incidence,** is equal to the angle made by the reflected ray and the normal, called the **angle of reflection.** That is,

angle of incidence = angle of reflection

Figure 29.3 ▶
In reflection, the angle between the incident ray and the normal is equal to the angle between the reflected ray and the normal.

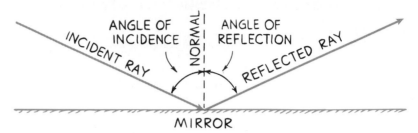

This relationship is called the **law of reflection.** The incident ray, the normal, and the reflected ray all lie in the same plane. The law of reflection applies to both partially reflected and totally reflected waves.

29.3 Mirrors

Consider a candle flame placed in front of a plane (flat) mirror. Rays of light are reflected from the mirror surface in all directions. The number of rays is infinite, and every one obeys the law of reflection.

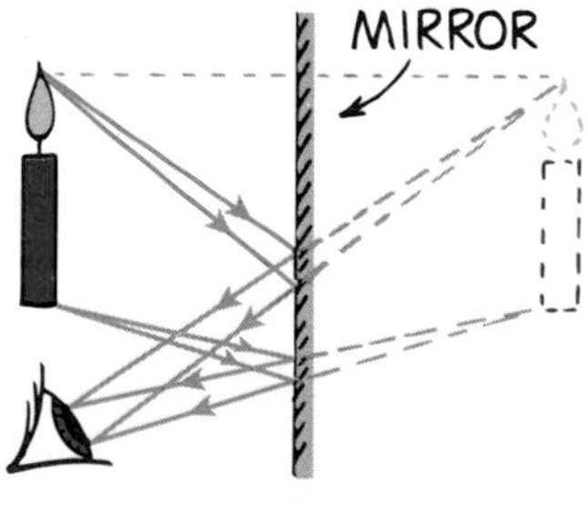

Figure 29.4 ▲
A virtual image is formed behind the plane mirror and is located at the position where the extended reflected rays (broken lines) converge.

> ▶ **DOING PHYSICS**
>
> ### Mirror Image
>
> Look at your face in a mirror. Then look at something on the surface of the mirror, such as a dust speck. Notice that you have to adjust your eyes as you refocus from looking at your image to looking at the mirror surface. Is it apparent that your image is farther away than the mirror surface? How much farther? Use a manual focus camera to help you determine these distances. To do this, focus on the dust speck on the mirror surface and then read the distance off the scale on the camera's lens. Repeat this process using your image in the mirror.
>
> When the mirror is curved, the sizes and distances of object and image are no longer equal. This text will not treat curved mirrors, except to say that the law of reflection still holds for curved mirrors. At every part of the surface, the angle of incidence is equal to the angle of reflection, as shown in Figure 29.6. Note that for a curved mirror, unlike a plane mirror, the normals (shown as dashed black lines) at different points on the surface are not parallel to each other.
>
> **Activity**

Figure 29.4 shows only two rays that originate at the tip of the candle flame and reflect from the mirror to your eye. Note that the rays diverge (spread apart) from the tip of the flame, and continue diverging from the mirror upon reflection. These divergent rays *appear* to originate from a point located behind the mirror. So you see an *image* of the candle in the mirror (actually *behind* the mirror). The image is called a **virtual image,** because light does not actually start there.

Your eye cannot ordinarily tell the difference between an object and its virtual image. This is because the light that enters your eye is entering in exactly the same manner, physically, as it would without the mirror if there really were an object there. Notice that the image is as far behind the mirror as the object is in front of the mirror. Notice also that the image and object are the same size. When you view yourself in a mirror, your image is the same size your identical twin would appear if located as far behind the mirror as you are in front—as long as the mirror is flat.

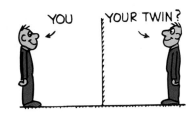

Figure 29.5 ▲
For reflection in a plane mirror, object size equals image size and object distance equals image distance.

◀ **Figure 29.6**
(a) The virtual image formed by a *convex* mirror (a mirror that curves outward) is smaller and closer to the mirror than the object is. (b) When the object is close to a *concave* mirror (a mirror that curves inward like a "cave"), the virtual image is larger and farther away than the object is. In any case, the law of reflection holds for each ray.

■ Questions

1. If you look at your blue shirt in a mirror, what is the color of its image? What does this tell you about the frequency of light incident upon a mirror compared with the frequency of the light after it is reflected?

2. If you wish to take a picture of your image while standing 2 m in front of a plane mirror, for what distance should you set your camera to provide sharpest focus?

■ Answers

1. The color of the image will be the same as the color of the object. This is evidence that the frequency of light is not changed by reflection.

2. You should set your camera for a distance of 4 m. The situation is equivalent to your standing 2 m in front of an open window and viewing your twin standing 2 m in back of the window.

29.4 Diffuse Reflection

When light is incident on a rough surface, it is reflected in many directions. This is **diffuse reflection** (Figure 29.7). Although the reflection of each single ray obeys the law of reflection, the many different angles that incident light rays encounter cause reflection in many directions.

What constitutes a rough surface for some rays may be a polished surface for others. If the differences in elevations in a surface are small (less than about one-eighth the wavelength of the light that falls on it), the surface is considered polished. A surface therefore may be polished for long wavelengths, but not polished for short wavelengths. The wire-mesh "dish" shown in Figure 29.8 is very rough for light waves, not mirrorlike at all. Yet for long-wavelength radio waves, it is polished. It acts as a mirror to radio waves and is an

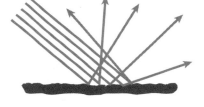

Figure 29.7 ▲
Diffuse reflection from a rough surface.

Figure 29.8 ▶
The open-mesh parabolic dish acts like a diffuse reflector for light waves but like a polished reflector for long-wavelength radio waves.

excellent reflector. Whether a surface is a diffuse reflector or a polished reflector depends on the length of the waves it reflects.

Light that reflects from this page is diffuse. The page may be smooth to a long radio wave, but to the short wavelengths of visible light it is rough. This roughness is evident in the microscopic view of an ordinary paper surface (Figure 29.9). Rays of light incident on this page encounter millions of tiny flat surfaces facing in all directions, so they are reflected in all directions. This is very nice, for it allows us to read the page from any direction or position. We see most of the things around us by diffuse reflection.

Figure 29.9 ▲
A microscopic view of the surface of ordinary paper.

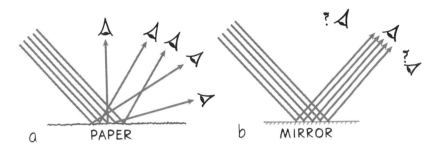

Figure 29.10 ▲
(a) If you shine a beam of light on paper, you can see diffusely reflected light at any position. (b) However, your eye must be at the right place to see a reflected beam from a small mirror.

29.5 Reflection of Sound

An echo is reflected sound. The fraction of sound energy reflected from a surface is more when the surface is rigid and smooth, and less when the surface is soft and irregular. Sound energy not reflected is absorbed or transmitted.

Sound reflects from all surfaces—the walls, ceiling, floor, furniture, and people—of a room. Designers of interiors of buildings, whether office buildings, factories, or auditoriums, need to understand the reflective properties of surfaces. The study of these properties is *acoustics*.

When the walls of a room, auditorium, or concert hall are too reflective, the sound becomes garbled. This is due to multiple reflections called **reverberations.** But when the reflective surfaces are more absorbent, the sound level is lower, and the hall sounds dull and lifeless. Reflection of sound in a room makes it sound lively and full, as you have probably found out while singing in the shower. In the design of an auditorium or concert hall, a balance between reverberation and absorption is desired.

The walls of concert halls are often designed with grooves so that the sound waves are diffused. This is illustrated in Figure 29.11 (top). In this way a person in the audience receives a small amount of reflected sound from many parts of the wall, rather than a larger amount of sound from one part of the wall.

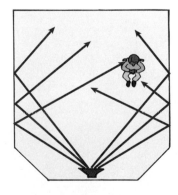

Figure 29.11 ▲
(Top) With grooved walls, sound reflects from many small sections of the wall to a listener. (Bottom) With flat walls, an intense reflected sound comes from only one part of the wall.

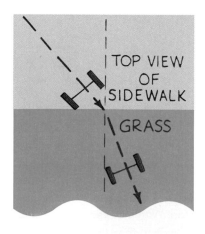

Figure 29.12 ▶

The shiny plates above the orchestra in Davies Symphony Hall in San Francisco reflect both light and sound. Adjusting them is quite simple: What you see is what you hear.

Highly reflective surfaces are often placed behind and above the stage to direct sound out to an audience. The large shiny plastic plates in Figure 29.12 also reflect light. A listener can look up at these reflectors and see the reflected images of the members of the orchestra. (The plastic reflectors are somewhat curved, which increases the field of view.) Both sound and light obey the same law of reflection, so if a reflector is oriented so that you can *see* a particular musical instrument, rest assured that you will *hear* it also. Sound from the instrument will follow the line of sight to the reflector and then to you.

29.6 Refraction

Suppose you take a rear axle with its wheels attached off an old toy cart and let it roll along a pavement that slopes gently downward and onto a downward-sloping mowed lawn. It rolls more slowly on the lawn because of the interaction of the wheels with the blades of grass. If you roll it at an angle as shown in Figure 29.13, it will be deflected from its straight-line course. The direction of the axle and rolling wheels is shown in the illustration. Note that the wheel that first meets the lawn slows down first—because it interacts with the grass while the opposite wheel is still rolling on the pavement. The axle pivots, and the path is bent toward the normal (the thin dashed line perpendicular to the grass-pavement boundary). The axle then continues across the lawn in a straight line at reduced speed.

Water waves similarly bend when one part of each wave is made to travel slower (or faster) than another part. This is **refraction.** Waves travel faster in deep water than in shallow water. Figure 29.14 (left) shows a view from above of straight wave crests (the bright lines) moving toward the right edge of the photo. They are moving from deep water across a diagonal boundary into shallow water. At the boundary,

Figure 29.13 ▲

The direction of the rolling wheels changes when one wheel slows down before the other one.

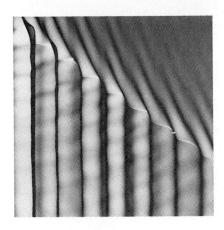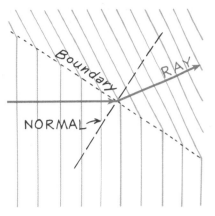

Figure 29.14 ▲

(Left) Photograph of the refraction of a water wave at a boundary where the wave speed changes because the water depth changes. (Right) Diagram of wave fronts and a sample ray. The ray is perpendicular to the wave front it intersects.

the wave speed and direction of travel are abruptly altered. Since the wave moves more slowly in shallow water, the crests are closer together. If you look carefully, you'll see some reflection from the boundary.

In drawing a diagram of a wave, as in Figure 29.14 (right), it is convenient to draw lines that represent the positions of different crests. Such lines are called **wave fronts.*** At each point along a wave front, the wave is moving perpendicular to the wave front. The direction of motion of the wave can thus be represented by rays that are perpendicular to the wave fronts. The ray in Figure 29.14 (right) shows how the water wave changes direction after it crosses the boundary between deep and shallow water. Sometimes we analyze waves in terms of wave fronts, and at other times in terms of rays. Both are useful models for understanding wave behavior.

29.7 Refraction of Sound

Sound waves are refracted when parts of a wave front travel at different speeds. This happens in uneven winds or when sound is traveling through air of uneven temperature. On a warm day, for example, the air near the ground may be appreciably warmer than the air above. Since sound travels faster in warmer air, the speed of sound near the ground is increased. The refraction is not abrupt but gradual (Figure 29.15). Sound waves therefore tend to bend away from warm ground, making it appear that the sound does not carry well.

---

*   Wave fronts can also represent the positions of different troughs—or any continuous portions of the wave that are all vibrating the same way at the same time.

Figure 29.15 ▲
The wave fronts of sound are bent in air of uneven temperature.

On a cold day or at night, when the layer of air near the ground is colder than the air above, the speed of sound near the ground is reduced. The higher speed of the wave fronts above cause a bending of the sound toward the earth. When this happens, sound can be heard over considerably longer distances.

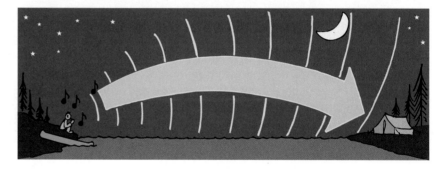

Figure 29.16 ▲
At night, when the air is cooler over the surface of the lake, sound is refracted toward the ground and carries unusually well.

■ Question

Suppose you are downwind from a factory whistle. In which case will the whistle sound louder—if the wind speed near the ground is more than the wind speed several meters above the ground, or if it is less?

■ Answer

You'll hear the whistle better if the wind speed near the ground is less than the wind speed higher up. For this condition, the sound will be refracted toward the ground. If the wind speed were greater near the ground, the refraction would be upward.

29.8 Refraction of Light

A pond or swimming pool both appear shallower than they actually are. A pencil in a glass of water appears bent, the air above a hot stove seems to shimmer, and stars twinkle. These effects are caused by changes in the speed of light as it passes from one medium to another, or through varying temperatures and densities of the same medium—which changes the directions of light rays. In short, these effects are due to the refraction of light.*

Figure 29.17 shows rays and wave fronts of light refracted as it passes from air into water. (The wave fronts would be curved if the source of light were close, just as the wave fronts of water waves near a stone thrown into the water are curved. If we assume that the source of light is the sun, then it is so far away that the wave fronts are practically straight lines.) Note that the left portions of the wave fronts are the first to slow down when they enter the water (or right portion if you look along the direction of travel). The refracted ray of light, which is at right angles to the refracted wave fronts, is closer to the normal than is the incident ray.

Compare the refraction in this case to the bending of the axle's path in Figure 29.13. When light rays enter a medium in which their speed decreases, as when passing from air into water, the rays bend toward the normal. But when light rays enter a medium in which their speed increases, as when passing from water into air, the rays bend away from the normal.

Figure 29.18 shows a laser beam entering a container of water at the left and exiting at the right. The path would be the same if the light entered from the right and exited at the left. The light paths are reversible for both reflection and refraction. If you can see somebody by way of a reflective or refractive device, such as a mirror or a prism, then that person can see you (or your eyes) by looking through the device also.

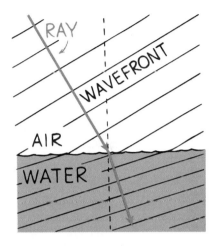

Figure 29.17 ▲
As a light wave passes from air into water, its speed decreases. Note that the refracted ray is closer to the normal than is the incident ray.

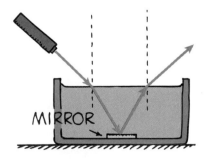

Figure 29.18 ▲
The laser beam bends toward the normal when it enters the water, and away from the normal when it leaves.

---

* The ratio n of the speed of light in a vacuum to the speed in a given material is called the *index of refraction* of that material.

$$\text{index of refraction } n = \frac{\text{speed of light in vacuum}}{\text{speed of light in material}}$$

The quantitative law of refraction, called *Snell's law*, was first worked out in 1621 by W. Snell, a Dutch astronomer and mathematician. According to Snell's law,

$$n \sin \theta = n' \sin \theta'$$

where n and n' are the indices of refraction of the media on either side of the boundary, and θ and θ' are the respective angles of incidence and refraction. If three of these values are known, the fourth can be calculated from this relationship.

Figure 29.19 ▲
Because of refraction, the apparent depth of the glass block is less than the real depth (left), the fish appears to be nearer than it actually is (center), and the full glass mug appears to hold more root beer than it actually does (right).

As Figure 29.19 (left) shows, a thick pane of glass appears to be only two-thirds its real thickness when viewed straight on. (For clarity, the diameter of the eye pupil is made larger than true scale.) Similarly, water in a pond or pool appears to be only three-quarters its true depth. Look at a fish in water from a bank, and the fish appears to be nearer the surface than it really is (Figure 29.19, center). It also seems closer. Another illusion is shown in the right of the figure. Light from the root beer is refracted through the sides of the thick glass, making the glass appear thinner than it is. The eye, accustomed to perceiving light traveling along straight lines, perceives the root beer to be at the outer edge of the glass, along the broken lines. These effects are due to the refraction of light whenever it crosses a boundary between air and another transparent medium.

29.9 Atmospheric Refraction

Although the speed of light in air is only 0.03% less than its speed in a vacuum, in some situations atmospheric refraction is quite noticeable. One interesting example is the **mirage.** On hot days there may be a layer of very hot air in contact with the ground. Since molecules in hot air are farther apart, light travels faster through it than through cooler air above. The speeding up of the part of the wave nearest the ground produces a gradual bending of the light rays. This can produce an image, say, of the tree in Figure 29.20. The image appears upside down to an observer at the right, just as if it were reflected from a surface of water. But the light is not reflected; it is refracted.

Figure 29.20 ▶
The refraction of light in air produces a mirage.

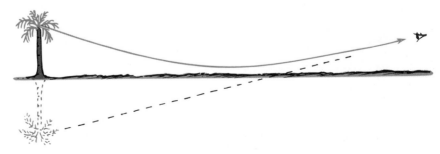

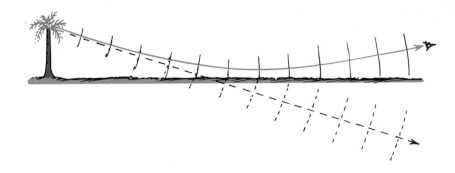

◀ **Figure 29.21**
Wave fronts of light travel faster
in the hot air near the ground,
thereby bending the rays of light
upward.

Wave fronts of light are shown in Figure 29.21. The refraction of light in air in this case is very much like the refraction of sound in Figure 29.15. Undeflected wave fronts would travel at one speed and in the direction shown by the broken lines. Their greater speed near the ground, however, causes the light ray to bend upward as shown.

A motorist experiences a similar situation when driving along a hot road that appears to be wet ahead. The sky appears to be reflected from a wet surface but, in fact, light from the sky is being refracted through a layer of hot air. A mirage is not, as some people mistakenly believe, a "trick of the mind." A mirage is formed by real light and can be photographed (see Figure 29.22).

◀ **Figure 29.22**
A mirage.

> ### DOING PHYSICS
>
> #### Refraction in Air
>
> Look across a hot stove or hot pavement. Those shimmering images, or "heat waves," you see are the effects of atmospheric refraction. The speed of light changes as it travels through varying temperatures and densities of air. Similarly, the twinkling of stars in the nighttime sky is produced by refractions of light as it passes through unstable layers in the atmosphere. Do you see one reason why many observatories are located atop mountains?
>
> **Activity**

Figure 29.23 ▶
When the sun is already below
the horizon, you can still see it.

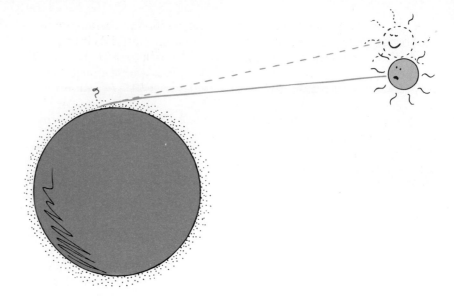

Figure 29.24 ▲
Atmospheric refraction produces
a "pumpkin" sun.

When you watch the sun set, you see the sun for several minutes after it has really sunk below the horizon. This is because light is refracted by the earth's atmosphere (Figure 29.23). Since the density of the atmosphere changes gradually, the refracted rays bend gradually to produce a curved path. The same thing occurs at sunrise, so our day-times are about 5 minutes longer because of atmospheric refraction.

When the sun (or moon) is near the horizon, the rays from the lower edge are bent more than the rays from the upper edge. This produces a shortening of the vertical diameter and makes the sun (or moon) look elliptical instead of round (Figure 29.24).

■ Question

If the speed of light were the same for the various temperatures and densities of air, would there still be mirages, slightly longer daytimes, and a "pumpkin" sun at sunset?

29.10 Dispersion in a Prism

Chapter 27 discussed how the average speed of light is less than c in a transparent medium. How much less depends on the medium and the frequency of the light. Light of frequencies closer to the natural frequency of the electron oscillators in a medium travels more slowly in the medium. This is because there are more interactions with the

■ Answer

No! There would be no refraction if light traveled at the same speed in air of different temperatures and densities.

medium in the process of absorption and reemission. Since the natural or resonant frequency of most transparent materials is in the ultraviolet part of the spectrum, visible light of higher frequencies travels more slowly than light of lower frequencies. Violet light travels about 1% slower in ordinary glass than red light. The colors between red and violet travel at their own intermediate speeds.

Since different frequencies of light travel at different speeds in transparent materials, they will refract differently and bend at different angles. When light is bent twice at nonparallel boundaries, as in a prism, the separation of the different colors of light is quite apparent. This separation of light into colors arranged according to their frequency is called **dispersion** (Figure 29.25). Dispersion is what enabled Isaac Newton to produce a spectrum.

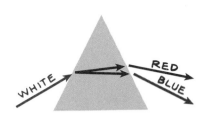

Figure 29.25 ▲
Dispersion through a prism.

29.11 The Rainbow

A spectacular illustration of dispersion is the rainbow. The conditions for seeing a rainbow are that the sun be shining in one part of the sky and that water droplets in a cloud or in falling rain be in the opposite part of the sky. When you turn your back to the sun, you see the spectrum of colors in a bow. Seen high enough from an airplane, the bow forms a complete circle. All rainbows would be completely round if the ground were not in the way.

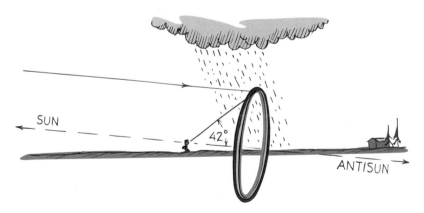

Figure 29.26 ▲
The rainbow is seen in a part of the sky opposite the sun and is centered on the imaginary line extending from the sun to the observer.

To understand how light is dispersed by raindrops, consider an individual spherical raindrop, as shown in Figure 29.27. Follow the ray of sunlight as it enters the drop near its top surface. Some of the light here is reflected (not shown), and the rest is refracted into the water. At this first refraction, the light is dispersed into its spectral colors. Violet is bent the most and red the least.

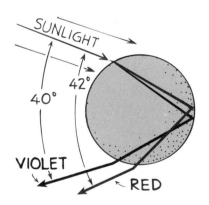

Figure 29.27 ▲
Dispersion of sunlight by a single drop, which produces a rainbow. Light is concentrated at the angles shown.

The rays reach the opposite part of the drop to be partly refracted out into the air (not shown) and partly reflected back into the water. Part of the rays that arrive at the lower surface of the drop are refracted into the air. This second refraction is similar to that of a prism, where refraction at the second surface increases the dispersion already produced at the first surface. This twice-refracted, once-reflected light is concentrated in a narrow range of angles.

Each drop disperses a full spectrum of colors. An observer, however, is in a position to see only a single color from any one drop (see Figure 29.28). If violet light from a single drop enters your eye, red light from the same drop falls below your eye. To see red light you have to look at a drop higher in the sky. You'll see the color red when the angle between a beam of sunlight and the dispersed light is 42°. The color violet is seen when the angle between the sunbeam and dispersed light is 40°.

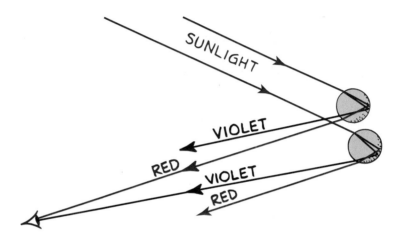

Figure 29.28 ▲
Sunlight strikes two sample drops and emerges as dispersed light. The observer sees red from the upper drop and violet from the lower drop. Millions of drops produce the whole spectrum.

Figure 29.29 ▲
Only raindrops along the dashed arc disperse red light to the observer at a 42° angle.

You don't need to look only upward at 42° to see dispersed red light. You can see red by looking sideways at the same angle or anywhere along a circular arc swept out at a 42° angle (see Figure 29.29). The dispersed light of other colors is along similar arcs, each at their own slightly different angle. Altogether, the arcs for each color form the familiar rainbow shape.

If you rotate the triangle in Figure 29.29 you sweep out the portion of a cone, with your eye at the apex. The raindrops that disperse light to you lie at the far edges of such a cone. The thicker the region of water drops, the thicker the conical edge you look through, and the more vivid the rainbow.

Your cone of vision that intersects the raindrops creating your rainbow is different from that of a person next to you. So when a friend says, "Look at the beautiful rainbow," you can reply, "Okay, move aside so I can see it too." Everybody sees his or her own personal rainbow.

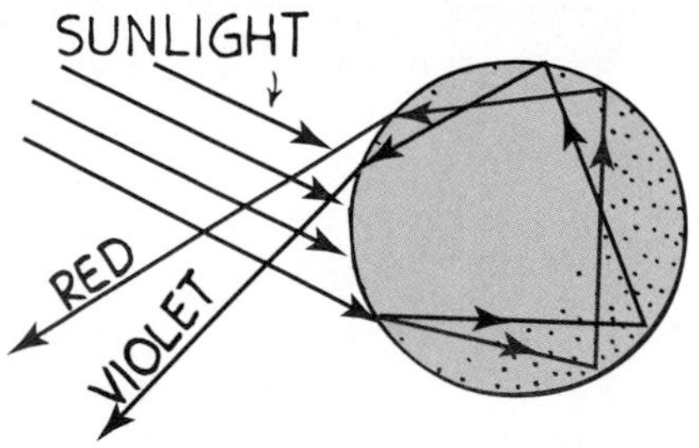

◀ **Figure 29.30**
Double reflection in a drop produces a secondary bow.

So when you move, your rainbow moves with you. This means you can never approach the side of a rainbow, or see it end-on as in the exaggerated view of Figure 29.26. You *can't* get to its end. Hence the expression "looking for the pot of gold at the end of the rainbow" means pursuing something you can never reach.

Often a larger, secondary bow with colors reversed can be seen arching at a greater angle around the primary bow. The secondary bow is formed by similar circumstances and is a result of double reflection within the raindrops (Figure 29.30). Because some light is refracted out the back during the extra reflection, the secondary bow is much dimmer.

■ **Question**

If light traveled at the same speed in raindrops as it does in air, would we still have rainbows?

29.12 Total Internal Reflection

When you're in a physics mood and you're going to take a bath, fill the tub extra deep and bring a waterproof flashlight into the tub with you. Turn the bathroom light off. Shine the submerged light straight up and then slowly tip it and note how the intensity of the emerging beam diminishes and how more light is reflected from the water surface to the bottom of the tub.

At a certain angle, called the **critical angle,** you'll notice that the beam no longer emerges into the air above the surface. The intensity of the emerging beam reduces to zero where it tends to graze the surface. When the flashlight is tipped beyond the critical angle (48° from the normal in water), you'll notice that the beam cannot enter the air; it is only reflected. The beam is experiencing **total internal reflection.** The only light emerging from the water surface is that which is diffusely reflected from the bottom of the bathtub.

■ **Answer**

No.

457

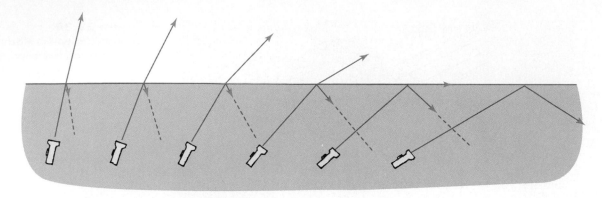

Figure 29.31 ▲
Light emitted in the water at angles below the critical angle is partly refracted and partly reflected at the surface. At the critical angle (second sketch from right), the emerging beam skims the surface. Past the critical angle (far right), there is total internal reflection.

This procedure is shown in Figure 29.31. The proportions of light refracted and reflected are indicated by the relative lengths of the solid arrows. Note that the light reflected beneath the surface obeys the law of reflection: The angle of incidence is equal to the angle of reflection.

The critical angle for glass is about 43°, depending on the type of glass. This means that within the glass, rays of light that are more than 43° from the normal to a surface will be totally internally reflected at that surface. Rays of light in the glass prisms shown in Figure 29.32, for example, meet the back surface at 45° and are totally internally reflected. They will stay inside the glass until they meet a surface at an angle between 0° (straight on) and 43° to the normal.

Total internal reflection is as the name implies: total—100%. Silvered or aluminized mirrors reflect only 90 to 95% of incident light, and are marred by dust and dirt; prisms are more efficient. This is the main reason they are used instead of mirrors in many optical instruments.

The critical angle for a diamond is 24.6°, smaller than any other known substance. This small critical angle means that light inside a

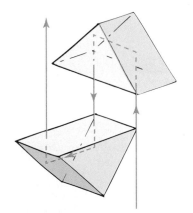

Figure 29.32 ▲
Total internal reflection in glass prisms.

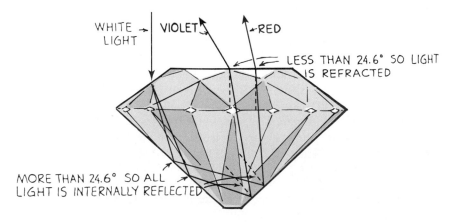

Figure 29.33 ▲
Paths of light in a diamond.

Figure 29.34 ▲
In an optical fiber, light is piped from one end to the other by a succession of total internal reflections.

diamond is more likely to be totally internally reflected than to escape. All light rays more than 24.6° from the normal to a surface in a diamond stay inside by total internal reflection. When a diamond is cut as a gemstone, light that enters at one facet is usually totally internally reflected several times, without any loss in intensity, before exiting from another facet in another direction. That's why you see unexpected flashes from a diamond. A small critical angle, plus the pronounced refraction because of the unusually low speed of light in diamond, produces wide dispersion and a wide array of colors. The colors seen in a diamond are quite brilliant.

Total internal reflection underlies the usefulness of **optical fibers,** sometimes called *light pipes.* As the name implies, these transparent fibers pipe light from one place to another. They do this by a series of total internal reflections, much like the ricocheting of a bullet inside a steel pipe. Optical fibers are useful for getting light to inaccessible places. Mechanics and machinists use them to look at the interior of engines, and physicians use them to look inside a patient's body. Light shines down some of the fibers to illuminate the scene and is reflected back along others.

Optical fibers are important in communications. In many cities, thin glass fibers have replaced thick, bulky, and expensive copper cables to carry thousands of simultaneous telephone messages between major switching centers. Undersea copper cables are also being replaced by optical fibers. More information can be carried in the high frequencies of visible light than in the lower frequencies of electric current. Optical fibers are more and more replacing electric circuits and microwave links in communications technology.

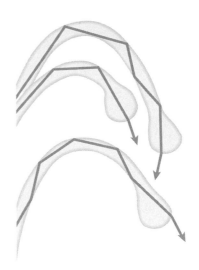

LINK TO BIOLOGY

Fiber-Optic Bears

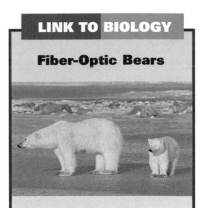

Ever wonder how a polar bear survives the extreme cold of an arctic winter? Fiber optics of course! The polar bear's fur provides more than insulation against the cold; the hairs of its fur are actually transparent optical fibers that trap ultraviolet light. A polar bear's fur appears white because visible light is reflected by the rough inner surface of each hollow hair. Higher-frequency radiant energy shines down the fibers to the bear's skin. The skin is very efficient at absorbing all the solar energy it can get—guess what color the skin is? Black!

29 Chapter Review

Concept Summary

In reflection, a wave reaches the boundary between two media and bounces back into the first medium.

- At a boundary, usually part of a wave is reflected and part passes into the second medium.

- According to the law of reflection, the angle of incidence is equal to the angle of reflection.

- A plane mirror forms a virtual image of an object; the image appears to be as far in back of the mirror as the object is in front of it, and is the same size as the object.

- Light that falls on a rough surface is reflected diffusely.

In refraction, a wave reaches the boundary between two media and changes direction as it passes into the second medium.

- Refraction is caused by a difference in the speed of the wave in the two media.

- The speed of light in materials depends on frequency, causing the different colors that make up white light to refract differently and spread out to form a visible spectrum.

In total internal reflection, an incident wave on a boundary is at an angle such that none of the wave can be refracted, so only reflection occurs.

Important Terms

angle of incidence (29.2)
angle of reflection (29.2)
critical angle (29.12)
diffuse reflection (29.4)
dispersion (29.10)
law of reflection (29.2)
mirage (29.9)
normal (29.2)
optical fiber (29.12)
reflection (29.1)
refraction (29.6)
reverberation (29.5)
total internal reflection (29.12)
virtual image (29.3)
wave front (29.6)

Review Questions

1. What becomes of a wave's energy when the wave is totally reflected at a boundary? When it is partially reflected at a boundary? (29.1)

2. Why do smooth metal surfaces make good mirrors? (29.1)

3. When light strikes perpendicular to the surface of a pane of glass, how much light is reflected and how much is transmitted? (29.1)

4. What is meant by the normal to a surface? (29.2)

5. What is the law of reflection? (29.2)

6. When you view your image in a plane mirror, how far behind the mirror is your image compared with your distance in front of the mirror? (29.3)

7. Does the law of reflection hold for *curved* mirrors? (29.3)

8. Does the law of reflection hold for diffuse reflection? Explain. (29.4)

9. What is meant by the idea that a surface may be polished for some waves and rough for others? (29.4)

10. Distinguish between an echo and a reverberation. (29.5)

11. Does the law of reflection hold for both sound waves and light waves? (29.5)

12. Distinguish between reflection and refraction. (29.1, 29.6)

13. When a wave crosses a surface at an angle from one medium into another, why does it

"pivot" as it moves across the boundary into the new medium? (29.6)

14. What is the orientation of a ray in relation to the wave front of a wave? (29.6)

15. Give an example where refraction is abrupt, and another where refraction is gradual. (29.6–29.7)

16. Does refraction occur for both sound waves and light waves? (29.7–29.8)

17. If light had the same speed in air and in water, would light be refracted in passing from air into water? (29.8)

18. If you can see the face of a friend who is underwater, can she also see you? (29.8)

19. Does refraction tend to make objects submerged in water seem shallower or deeper than they really are? (29.8)

20. Is a mirage a result of refraction or reflection? Explain. (29.9)

21. Is daytime a bit longer or a bit shorter because of atmospheric refraction? (29.9)

22. As light passes through glass or water, do the high or low frequencies of light interact more in the process of absorption and reemission, and therefore lag behind? (29.10)

23. Why does blue light refract at greater angles than red light in transparent materials? (29.10)

24. What conditions are necessary for viewing a rainbow in the sky? (29.11)

25. How is a raindrop similar to a prism? (29.11)

26. What is the *critical angle* in terms of refraction and total internal reflection? (29.12)

27. Why are optical fibers often called *light pipes*? (29.12)

Activities

1. What must be the minimum length of a plane mirror in order for you to see a full view of yourself? To find out, stand in front of a mirror and put pieces of tape on the glass: one piece where you see the top of your head, and the other where you see the bottom of your feet. Compare the distance between the pieces of tape with your height. If a full-length mirror is not handy, use a smaller mirror and find the minimum length of mirror to see your face. Mark where you see the top of your head and the bottom of your chin. Then compare the distance between the marks with the length of your face.

2. What effect does your distance from the mirror have on the answer to Activity 1? (*Hint:* Move closer and farther from your initial position. Be sure the top of your head lines up with the top piece of tape. At greater distances, is your image smaller than, larger than, or the same size as the space between the pieces of tape?) Surprised?

3. If available, look at a diamond or similar transparent gemstone under bright light. Turn the stone and note the flashes of color that refract, reflect, and refract toward you. When the flash encounters only one eye instead of two, your brain registers it differently than for both eyes. The one-eyed flash is a sparkle!

Think and Explain

1. Suppose that a mirror and three lettered cards are set up as in the figure. If a person's eye is at point P, which of the lettered cards will be seen reflected in the mirror?

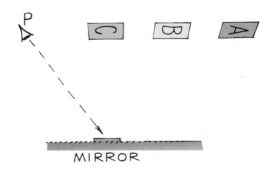

ƎƆNA⅃UᗺMA

2. Why is the lettering on the front of some vehicles "backward," as seen here?

3. Trucks often have signs on their backs that say, "If you can't see my mirrors, I can't see you." Explain the physics here.

4. Contrast the types of reflection from a rough road and from the smooth surface of a wet road to explain why it is difficult for a motorist to see the roadway ahead when driving on a rainy night.

5. Cameras with automatic focus bounce a sonar (sound) beam from the object being photographed, and compute distance from the time interval between sending and receiving the signal. Why will these cameras not focus properly for photographs of mirror images?

6. Why is an echo weaker than the original sound?

7. Does the reflection of a scene in calm water look exactly the same as the scene itself only upside down? (*Hint:* Place a mirror on the floor between you and a table. Do you see the top of the table in the reflected image?)

8. If you were spearing a fish with a spear, would you aim above, below, or directly at the observed fish to make a direct hit? Would your answer be the same if you used laser light to "spear" the fish? Defend your answer.

9. A rainbow viewed from an airplane may form a complete circle. Will the shadow of the airplane appear at the center of the circle? Explain with the help of Figure 29.26.

10. The photo below shows two identical cola bottles, each with the *same* amount of cola. The right bottle is in air, and the left bottle is encased in solid plastic that has nearly the same "index of refraction" as glass (the speed of light in the plastic and in glass are nearly the same). Which bottle shows an illusion of the amount of cola? How does the other bottle give a truer view of its contents?

Think and Solve

1. When light strikes glass perpendicularly, about 4% is reflected at each surface. How much light is transmitted through a pane of window glass?

2. Suppose you walk toward a mirror at 1 m/s. How fast do you and your image approach each other? (The answer is *not* 1 m/s.)

3. A bat flying in a cave emits a sound and receives its echo in one second. How far away is the cave wall?

4. An oceanic depth-sounding vessel surveys the ocean bottom with ultrasonic sound that travels 1530 m/s in seawater. Find the depth of the water if the time delay of the echo to the ocean floor and back is 6 s.

30 Lenses

Lenses manipulate light.

A light ray bends as it enters glass and bends again as it leaves. The bending (refraction) is due to the difference in the average speed of light in glass and in air. Glass of a certain shape can form an image that appears larger, smaller, closer, or farther than the object being viewed. For example, magnifying glasses have been used for centuries and were well known to the early Greeks and medieval Arabs. Today, eyeglasses allow millions of people to read in comfort, and cameras, projectors, telescopes, and microscopes widen our view of the world.

30.1 Converging and Diverging Lenses

When a piece of glass has just the right shape, it bends parallel rays of light so that they cross and form an image. Such a piece of glass is a **lens.**

The shape of a lens can be understood by considering a lens to be a large number of portions of triangular prisms, as shown in Figure 30.1. When arranged in certain positions, the prisms bend incoming parallel rays so they converge to (or diverge from) a single point. The arrangement shown at the left is thicker in the middle; it

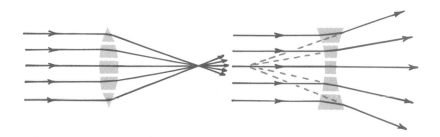

◀ **Figure 30.1**
A lens may be thought of as a set of prisms that converge light (left) or diverge light (right).

converges the light. The arrangement at the right is thinner in the middle; it diverges the light.

In both arrangements, the most net bending of rays occurs at the outermost prisms, for they have the greatest angle between the two refracting surfaces. No net bending occurs in the middle "prism," for its glass faces are parallel and rays emerge in their original direction.

Real lenses are made not of prisms, of course, but of solid pieces of glass with surfaces that are usually ground to a spherical shape. Figure 30.2 shows how smooth lenses refract rays of light and form wave fronts. The lens at the left is thicker in the middle and rays of light that are initially parallel (straight wave fronts) are made to converge. This is a **converging lens.** The lens at the right is thinner in the middle and the rays of light are made to diverge. This is a **diverging lens.**

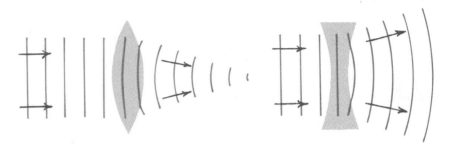

Figure 30.2 ▲
Wave fronts travel more slowly in glass than in air. In the converging lens (left), the wave fronts are retarded more through the center of the lens, and the light converges. In the diverging lens (right), the waves are retarded more at the edges, and the light diverges.

Figure 30.3 illustrates some important terms for a lens. The **principal axis** of a lens is the line joining the centers of curvature of its surfaces. For a converging lens, the **focal point** is the point at which a beam of parallel light, parallel to the principal axis, converges. Incident parallel beams that are not parallel to the principal axis focus at points above or below the focal point. All such possible points make up a **focal plane.** A lens affects light coming from the right in the same way as light coming from the left (*or* has the same effect on light incident from either side). Therefore, a lens has two focal points and two focal planes. When the lens of a camera is set for distant objects, the film is in the focal plane behind the lens in the camera.

Figure 30.3 ▶
Key features of a converging lens.

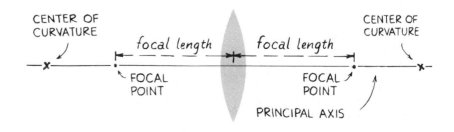

For a diverging lens, an incident beam of light parallel to the principal axis is not converged to a point, but is diverged so that the light appears to come from a point. The **focal length** of a lens, whether converging or diverging, is the distance between the center of the lens and its focal point. When the lens is thin, the focal lengths on either side are equal, even when the curvatures on the two sides are not.

30.2 Image Formation by a Lens

With unaided vision, an object far away is seen through a relatively small angle of view (Figure 30.4a). When you are closer, the same object is seen through a larger angle of view (Figure 30.4b). This wider angle enables the perception of more detail. Magnification occurs when an image is observed through a wider angle with the use of a lens than without the lens and allows more detail to be seen. A magnifying glass is simply a converging lens that increases the angle of view and allows more detail to be seen.

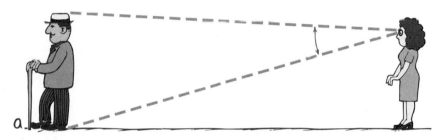

Figure 30.4 ▲
(a) A distant object is viewed through a narrow angle. (b) When the same object is viewed through a wide angle, more detail is seen.

When you use a magnifying glass, you hold it close to the object you wish to see magnified. This is because a converging lens will magnify only when the object is between the focal point and the lens. The magnified image will be farther from the lens than the object, and it will be right-side up. If a screen were placed at the image distance, no image would appear on the screen because no light is actually directed to the image position. The rays that reach your eye, however, behave *as if* they came from the image position, so the image is a **virtual image**.

When the object is far enough away to be beyond the focal point of a converging lens, light from the object does converge and can be focused on a screen (Figure 30.6). An image formed in this way by converging light is called a **real image.** A real image formed by a single converging lens is upside down (inverted). Converging lenses are used for projecting slides and motion pictures on a screen, and for projecting a real image on the film of a camera.

Figure 30.5 ▲
A converging lens can be used as a magnifying glass to produce a virtual image of a nearby object. The image appears larger and farther from the lens than the object.

Figure 30.6 ▶
A converging lens forms a real,
upside-down image of a more
distant object.

When a diverging lens is used alone, the image is always virtual,
right-side up, and smaller than the object. It makes no difference
how far or how near the object is. A diverging lens is often used for
the viewfinder on a camera. When you look at the object to be
photographed through the viewfinder, you see a right-side up virtual
image that approximates the same proportions as the photograph to
be taken.

Figure 30.7 ▶
The moving pattern of bright
lines on the bottom of a swimming pool results from the
uneven surface of water, which
behaves as a moving blanket of
lenses.

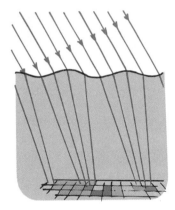

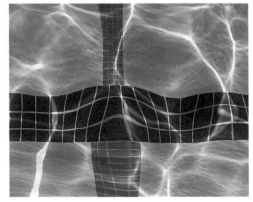

Figure 30.8 ▲
A virtual image produced by a
diverging lens.

■ Question

Why is the greater part of the photograph in Figure 30.8 out of
focus?

■ Answer

Both Jamie and his cat and the virtual image of Jamie and his cat are "objects" for the
lens of the camera that took this photograph. Since the objects are at different distances
from the camera lens, their respective images are at different distances with respect to the
film in the camera. So only one can be brought into focus. The same is true of your eyes.
You cannot focus on near and far objects at the same time.

30.3 Constructing Images Through Ray Diagrams

Ray diagrams, like the one in Figure 30.9, show the principal rays that can be used to determine the size and location of an image. To construct a ray diagram the size and location of the object, its distance from the center of the lens, and the focal length of the lens must be known.* An arrow is used to represent the object (which may be anything from a microbe viewed in a microscope to a galaxy viewed through a telescope). For simplicity, one end of the object is placed right on the principal axis.

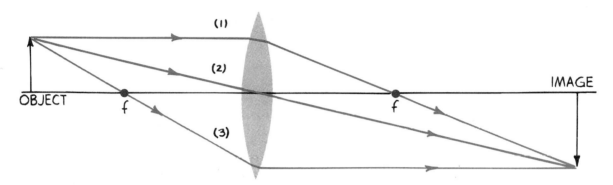

Figure 30.9 ▲
Ray diagram. Three useful rays from the object that converge on the image.

To locate the position of the image, you only have to know the paths of two rays from a point on the object. Any point except for the point on the principal axis will work, but it is customary to choose a point at the tip of the arrow.

The path of one refracted ray is known from the definition of the focal point. A ray parallel to the principal axis will be refracted by the lens to the focal point, as shown in Figure 30.9.

Another path is known: through the center of the lens where the faces are parallel to each other. A ray of light will pass through the center with no appreciable change in direction. Therefore, a ray from the tip of the arrowhead proceeds in a straight line through the center of the lens.

---

* The mathematical relationship between object distance o, image distance i, and focal length f is given by

$$\frac{1}{o} + \frac{1}{i} = \frac{1}{f}$$

This is called the *thin-lens equation*.

A third path is known: A ray of light that passes through the focal point in front of the lens emerges from the lens and proceeds parallel to the principal axis.

All three paths are shown in Figure 30.9, which is a typical ray diagram. The image is located where the three rays intersect. Any two of these three rays is sufficient to locate the relative size and location of the image.

The ray diagram for a converging lens used as a magnifying glass is shown in Figure 30.10. In this case, where the distance from the lens to the object is less than the focal length, the rays diverge as they leave the lens. The rays of light appear to come from a point in front of the lens (same side of the lens as the object). The location of the image is found by extending the rays backward to the point where they converge. The virtual image that is formed is magnified and right-side up.

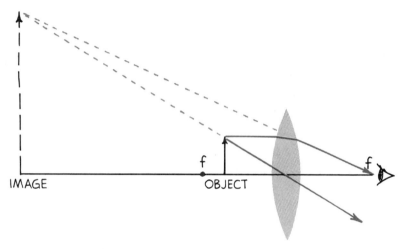

Figure 30.10 ▲
Ray diagram for a magnifying glass. The object is less than one focal length from the lens, so the image is virtual, right-side up, and magnified.

The three rays useful for the construction of a ray diagram are summarized:

1. A ray parallel to the principal axis that passes through the focal point after refraction by the lens.

2. A ray through the center of the lens that does not change direction.

3. A ray through the focal point in front of the lens that emerges parallel to the principal axis after refraction by the lens.

Any two rays are sufficient to locate an image; which particular pair is chosen is merely a matter of convenience.

The ray diagrams in Figure 30.11 show image formation by a converging lens as an object initially at the focal point is moved away from the lens along the principal axis. Since the object is not located between the focal point and the lens, all the images that are formed are real and inverted.

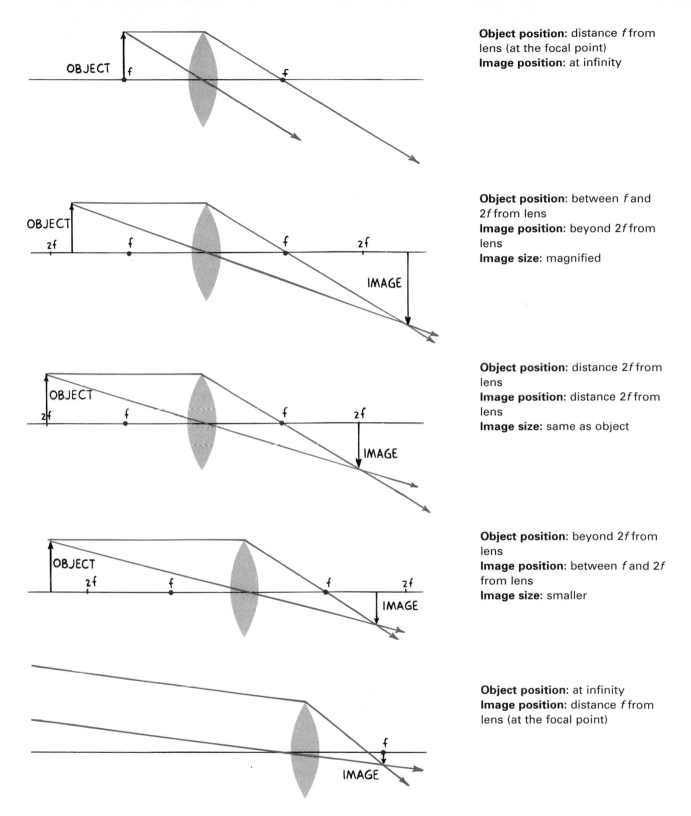

Object position: distance f from lens (at the focal point)
Image position: at infinity

Object position: between f and $2f$ from lens
Image position: beyond $2f$ from lens
Image size: magnified

Object position: distance $2f$ from lens
Image position: distance $2f$ from lens
Image size: same as object

Object position: beyond $2f$ from lens
Image position: between f and $2f$ from lens
Image size: smaller

Object position: at infinity
Image position: distance f from lens (at the focal point)

Figure 30.11 ▲
Ray diagrams for different positions of an object in relation to a converging lens of focal length f.

The method of drawing ray diagrams applies also to diverging lenses (Figure 30.12). A ray parallel to the principal axis from the tip of the arrow will be bent by the lens in the same direction as if it had come from the focal point. A ray through the center goes straight through. A ray that is heading for the focal point on the far side of the lens is bent so that it emerges parallel to the axis of the lens.

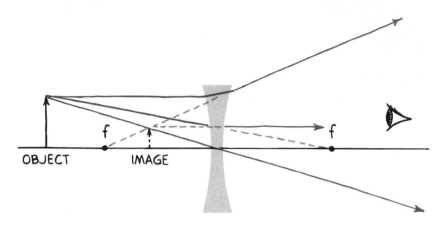

Figure 30.12 ▲
Ray diagram for a diverging lens.

On emerging from the lens, the three rays appear to come from a point on the same side of the lens as the object. This point defines the position of the virtual image. The image is nearer the lens than the object. It is smaller than the object and right-side up. Regardless of the object's position, the image formed by a diverging lens is always virtual, reduced, and right-side up.

30.4 Image Formation Summarized

A converging lens is a simple magnifying glass when the object is within one focal length of the lens. The image is then virtual, magnified, and right-side up.

When the object is beyond one focal length, a converging lens produces a real, inverted image. The location of the image depends on how close the object is to the focal point. If it is close to the focal point, the image is far away (as with a slide projector or movie projector). If the object is far from the focal point, the image is nearer (as with a camera). In all cases where a real image is formed, the object and the image are on opposite sides of the lens.

When the object is viewed with a diverging lens, the image is virtual, reduced, and right-side up. This is true for all locations of the object. In all cases where a virtual image is formed, the object and the image are on the same side of the lens.

Where must an object be located so that the image formed by a converging lens will be (a) at infinity? (b) as near the object as possible? (c) right-side up? (d) the same size? (e) inverted and enlarged?

30.5 Some Common Optical Instruments

The advent of eyeglasses probably occurred in Italy in the late 1200s. If anybody viewed objects through a pair of lenses held far apart, one in front of the other, there is no record of it, for curiously enough, the telescope wasn't invented until some 300 years later. Today, lenses are used in many optical instruments. Among these are the camera, telescope (and binoculars), compound microscope, and projector.

The Camera

A camera consists of a lens and sensitive film mounted in a lighttight box. In many cameras, the lens is mounted so that it can be moved back and forth to adjust the distance between the lens and film. The lens forms a real, inverted image on the film.

Figure 30.13 shows a camera with a single simple lens. In practice, most cameras make use of compound lenses to minimize distortions called *aberrations.*

The amount of light that gets to the film is regulated by a shutter and a diaphragm. The shutter controls the length of time that the film is exposed to light. The diaphragm controls the opening that light passes through to reach the film. Varying the size of the opening (aperture) varies the amount of light that reaches the film at any instant.

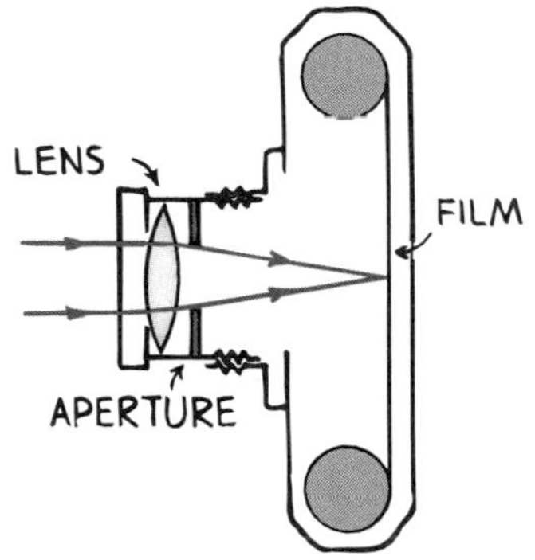

Figure 30.13 ▲
A simple camera.

The Telescope

A simple telescope uses a lens to form a real image of a distant object. The real image is not caught on film but is projected in space to be examined by another lens used as a magnifying glass. The second lens, called the **eyepiece,** is positioned so that the image produced by the first lens is within one focal length of the eyepiece. The eyepiece

---

■ **Answer**

The object should be (a) one focal length from the lens (at the focal point) (see Figure 30.11); (b) and (c) within one focal length of the lens (see Figure 30.10); (d) at two focal lengths from the lens (see Figure 30.11); (e) between one and two focal lengths from the lens (see Figure 30.11).

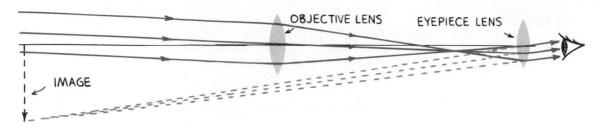

Figure 30.14 ▲
Lens arrangement for an astronomical telescope. (For simplification, the image is shown close here; it is actually located at infinity.)

Figure 30.15 ▲
The arrangement of prisms in binoculars.

forms an enlarged virtual image of the real image. When you look through a telescope, you are looking at an image of an image.

Figure 30.14 shows the lens arrangement for an *astronomical telescope.* The image is inverted, which explains why maps of the moon are printed with the moon upside down.

A third lens or a pair of reflecting prisms is used in the *terrestrial telescope,* which produces an image that is right-side up. A pair of these telescopes side by side, each with a pair of prisms to provide four reflecting surfaces to turn images right-side up, makes up a pair of *binoculars* (Figure 30.15).

Since no lens transmits 100% of the light incident upon it, astronomers prefer the brighter, inverted images of a two-lens telescope to the less bright, right-side-up images that a third lens or prisms would provide. For nonastronomical uses, such as viewing distant landscapes or sporting events, right-side-up images are more important than brightness, so the additional lens or prisms are used.

Telescopes that use lenses are *refracting telescopes.* Larger astronomical telescopes use mirrors instead of lenses.

The Compound Microscope

A compound microscope uses two converging lenses of short focal length, arranged as shown in Figure 30.16. The first lens, called the **objective lens,** produces a real image of a close object. Since the image is farther from the lens than the object, it is enlarged. A second lens, the eyepiece, forms a virtual image of the first image, further enlarged. The instrument is called a compound microscope because it enlarges an already enlarged image.

Figure 30.16 ▶
Lens arrangement for a compound microscope.

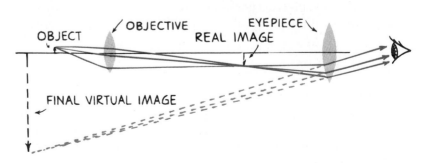

The Projector

The arrangement of converging lenses for a slide or movie projector is shown in Figure 30.17. A concave mirror reflects light from an intense source back onto a pair of *condenser lenses.* The condenser lenses direct the light through the slide or movie frame to a *projection lens.* The projection lens is mounted in a sliding tube so that it can be positioned back and forth to focus a sharp image on the screen.

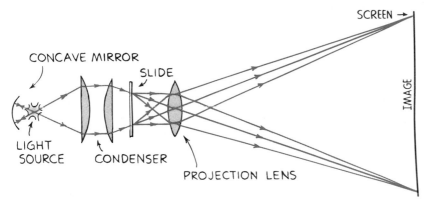

Figure 30.17 ▲
Lens arrangement for a projector.

30.6 The Eye

In many respects the human eye is similar to the camera. The amount of light that enters is regulated by the **iris,** the colored part of the eye that surrounds the opening called the **pupil.*** Light enters through the transparent covering called the **cornea,** passes through the pupil and lens, and is focused on a layer of tissue at the back of the eye—the **retina**—extremely sensitive to light. Different parts of the retina receive light from different directions.

The retina is not uniform. There is a small region in the center of our field of view where we have the most distinct vision. This spot is called the *fovea.* Much greater detail can be seen here than at the side parts of the eye.

There is also a spot in the retina where the nerves carrying all the information leave the eye in a narrow bundle. This is the *blind spot.* You can demonstrate that you have a blind spot in each eye if you hold this book at arm's length, close your left eye, and look at the circle in Figure 30.19 with only your right eye. You can see both the circle and the X at this distance. If you now move the book slowly

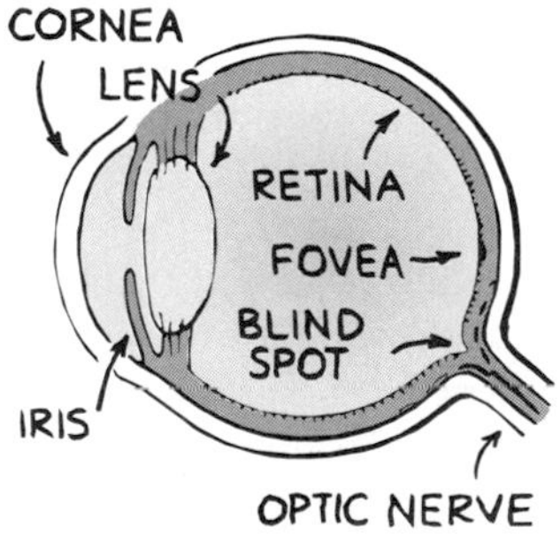

Figure 30.18 ▲
The human eye.

---

* The hole of the pupil usually looks black because light is going in but not coming out. Sometimes in flash photos, the light from the flashbulb enters the eye at just the right angle to reflect off the retina at the back of the eye. That's why flash photographs sometimes show the pupils to be pinkish.

toward your face, with your right eye still fixed upon the circle, you'll reach a position about 20 to 25 cm from your eye where the X disappears. To establish the blind spot in your left eye, close your right eye and similarly look at the X with your left eye so that the circle disappears. With both eyes opened, you'll find no position where either the X or the circle disappears because one eye "fills in" the part of the object to which the other eye is blind. It's nice to have two eyes.

Figure 30.19 ▲
For the blind spot experiment.

In both the camera and the eye, the image is upside down, and this is compensated for in both cases. You simply turn the camera film around to look at it. Your brain has learned to turn around images it receives from your retina!

A principal difference between a camera and the human eye has to do with focusing. In a camera, focusing is accomplished by altering the distance between the lens and the film. In the human eye, most of the focusing is done by the cornea, the transparent membrane at the outside of the eye. Adjustments in focusing of the image on the retina are made by changing the thickness and shape of the lens to regulate its focal length. This is called *accommodation* and is brought about by the action of the *ciliary muscle,* which surrounds the lens.

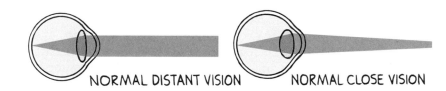

Figure 30.20 ▲
The shape of the lens changes to focus light on the retina.

30.7 Some Defects in Vision

If you have what is called normal vision, your eye can accommodate to clearly see objects from infinity (the *far point*) down to 25 cm (the *near point,* which normally recedes for all people with advancing age).

The eyes of a **farsighted** person form images behind the retina (Figure 30.21). The eyeball is too short. Farsighted people have to hold things more than 25 cm away to be able to focus them. The remedy is to increase the converging effect of the eye. This is done by wearing eyeglasses or contact lenses with converging lenses. Converging lenses will converge the rays that enter the eye sufficiently to focus them on the retina instead of behind the retina.

Figure 30.21 ▲
The eyeball of the farsighted eye is too short. A converging lens moves the image closer and onto the retina.

A **nearsighted** person can see nearby objects clearly, but does not see distant objects clearly because they are focused too near the lens, in front of the retina (Figure 30.22). The eyeball is too long. A remedy is to wear corrective lenses that diverge the rays from distant objects so that they focus on the retina instead of in front of it.

Figure 30.22 ▲
The eyeball of the nearsighted eye is too long. A diverging lens moves the image farther away and onto the retina.

Astigmatism of the eye is a defect that results when the cornea is curved more in one direction than the other, somewhat like the side of a barrel. Because of this defect, the eye does not form sharp images. The remedy is cylindrical corrective lenses that have more curvature in one direction than in another.

30.8 Some Defects of Lenses

No lens gives a perfect image. The distortions in an image are called **aberrations.** By combining lenses in certain ways, aberrations can be minimized. For this reason, most optical instruments use compound lenses, each consisting of several simple lenses, instead of single lenses.

Spherical aberration results when light passes through the edges of a lens and focuses at a slightly different place from light passing through the center of the lens (Figure 30.23). This can be remedied by covering the edges of a lens, as with a diaphragm in a camera. Spherical aberration is corrected in good optical instruments by a combination of lenses.

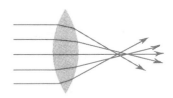

Figure 30.23 ▲
Spherical aberration.

> ▶ **DOING PHYSICS**
>
> ### Pinhole Image
>
> Poke a tiny hole in a piece of paper or card. Hold it in front of your eye close to this page. Whether or not you normally wear glasses, you'll see the print clearly. Because you're close, the print will seem magnified. Why is bright light needed? What advice do you have for someone who wears glasses and misplaces them, and can't see the small print in a telephone book?
>
> **Activity**

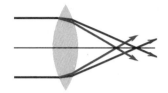

Figure 30.24 ▲
Chromatic aberration.

Chromatic aberration is the result of the different speeds of light of various colors and hence the different refractions they undergo. In a simple lens red light and blue light bend by different amounts (as in a prism), so they do not come to focus in the same place. *Achromatic lenses,* which combine simple lenses of different kinds of glass, correct this defect.

In the eye, vision is sharpest when the pupil is smallest because light then passes through only the center of the eye's lens, where spherical and chromatic aberrations are minimal. Also, light bends the least through the center of a lens, so minimal focusing is required for a sharp image. You see better in bright light because your pupils are smaller.

■ Question

Why is there chromatic aberration in light that passes through a lens, but no chromatic aberration in light that reflects from a mirror?

Figure 30.25 ▲
Bifocal eyeglasses have two sets of lenses with different focal lengths. As Charlie Spiegel shows here, the smaller lenses have a shorter focal length and arc for close-up viewing.

An option for those with poor sight in the last five hundred years has been to wear spectacles, and in more recent times another option has been to wear contact lenses. It is interesting to note that at the present time there is an alternative to both spectacles and contact lenses for people with poor eyesight. Experimental and controversial techniques today allow eye surgeons to reshape the cornea of the eye for normal vision. In tomorrow's world, the wearing of eyeglasses and contact lenses may be a thing of the past. We really do live in a rapidly changing world. And that can be nice.

■ Answer

Different frequencies travel at different speeds in a transparent medium, and therefore refract at different angles, which produces chromatic aberration. The angles at which light *reflects,* on the other hand, have nothing to do with the frequency of light. One color reflects the same as any other. Mirrors are therefore preferable to lenses in telescopes because there is no chromatic aberration with reflection.

30 Chapter Review

Concept Summary

A lens refracts parallel rays of light so that they cross—or appear to cross—at a focal point.

- A converging lens is thicker in the middle; a diverging lens is thinner in the middle.

- A converging lens forms virtual, magnified images when the object is within one focal length of the lens.

- A converging lens forms real images when the object is beyond one focal length from the lens.

- A diverging lens always forms virtual, reduced images.

- Optical instruments that use lenses include the camera, telescope, compound microscope, and projector.

- The human eye refracts light and focuses it on the retina (with the help of corrective lenses if necessary).

Important Terms

aberration (30.8)
astigmatism (30.7)
converging lens (30.1)
cornea (30.6)
diverging lens (30.1)
eyepiece (30.5)
farsighted (30.7)
focal length (30.1)
focal plane (30.1)
focal point (30.1)
iris (30.6)
lens (30.1)
nearsighted (30.7)
objective lens (30.5)
principal axis (30.1)
pupil (30.6)
ray diagram (30.3)
real image (30.2)
retina (30.6)
virtual image (30.2)

Review Questions

1. Distinguish between a *converging* lens and a *diverging* lens. (30.1)

2. Distinguish between the focal *point* and focal *plane* of a lens. (30.1)

3. Distinguish between a *virtual* image and a *real* image. (30.2)

4. There are three convenient rays commonly used in ray diagrams to estimate the position of an image. Describe these three rays in terms of their orientation with respect to the principal axis and focal points. (30.3)

5. How many of the rays in Question 4 are necessary for estimating the position of an image? (30.3)

6. Do ray diagrams apply only to converging lenses, or to diverging lenses also? (30.3)

7. Explain what is meant by saying that in a telescope one looks at the image of an image. (30.5)

8. In what two ways does an astronomical telescope differ from a terrestrial telescope? (30.5)

9. How does a compound microscope differ from a telescope? (30.5)

10. Which instrument—a telescope, a compound microscope, or a camera—is most similar to the eye? (30.5–30.6)

11. Why do you not normally notice a blind spot when you look at your surroundings? (30.6)

12. Distinguish between *farsighted* and *nearsighted* vision. (30.7)

13. What is astigmatism, and how can it be corrected? (30.7)

14. Distinguish between *spherical* aberration and *chromatic* aberration, and cite a remedy for each. (30.8)

Activities

1. Make a pinhole camera, as illustrated in the figure. Cut out one end of a small cardboard box, and cover the end with tissue or onion-skin paper. Make a clean-cut pinhole at the other end. (If the cardboard is thick, place a piece of metal foil over an opening in the cardboard, and make the hole in the foil.) Aim the camera at a bright object in a darkened room, and you will see an upside-down image on the translucent tissue paper. If in a dark, windowless room you replace the tissue paper with unexposed photographic film, cover the back so it is lighttight, and cover the pinhole with a removable flap, you will be ready to take a picture. Exposure times differ, depending mostly on the kind of film and the amount of light. Try different exposure times, starting with about 3 seconds. Also try boxes of various lengths. You'll find everything in focus in your photographs, but the pictures will not have clear-cut, sharp outlines. The principal difference between your pinhole camera and a commercial one is the glass lens, which is larger than the pinhole and therefore admits more light in less time. It is because a lens camera is so fast that the pictures it takes are called "snapshots."

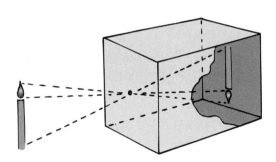

2. Note the shapes of light spots that reach the ground in the shade of a tree. Most of them are circular, or elliptical if the sun is low in the sky. The spots are "pinhole" images of the sun, which occur when the opening in the leaves above is small compared with the distance to the ground below. This is dramatic at the time of a partial solar eclipse, when the spots take the form of crescents. Physics is truly everywhere!

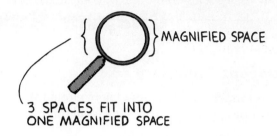

MAGNIFIED SPACE

3 SPACES FIT INTO ONE MAGNIFIED SPACE

3. Determine the magnification power of a lens by focusing on the lines of a ruled piece of paper. Count the spaces between the lines that fit into one magnified space, and you have the magnification power of the lens. For example, if three spaces fit into one magnified space, then the magnification power of the lens is 3. You can do the same with binoculars and a distant brick wall. Hold the binoculars so that only one eye looks at the bricks through the eyepiece while the other eye looks directly at the bricks. The number of bricks, as seen with the unaided eye, that will fit into one magnified brick gives the magnification of the instrument.

Think and Explain

1. **a.** What condition must exist for a converging lens to produce a virtual image?

   b. What condition must exist for a diverging lens to produce a real image?

2. How could you demonstrate that an image was indeed a real image?

3. Why do you suppose that a magnifying glass has often been called a "burning glass"?

4. In terms of focal length, how far is the camera lens from the film when very distant objects are being photographed?

5. Can you photograph yourself in a mirror and focus the camera on both your image and the mirror frame? Explain.

6. If you take a photograph of your image in a plane mirror, how many meters away should you set your focus if you are 2 meters in front of the mirror?

7. Copy the three drawings in the figure. Then use ray diagrams to find the image of each arrow.

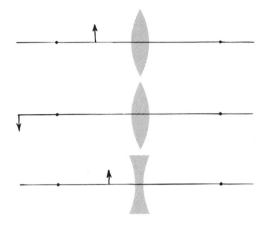

8. Why do you have to put slides into a slide projector upside down?

9. Maps of the moon are actually upside down. Why is this so?

10. What is responsible for the rainbow-colored fringe commonly seen at the edges of a spot of white light from the beam of a slide projector?

11. Would telescopes and microscopes magnify if light had the same speed in glass as in air? Explain.

12. Consider a simple magnifying glass under water. Will it magnify more or less? Explain why.

31 Diffraction and Interference

Interference colors.

Today Isaac Newton is most famous for his accomplishments in mechanics—his laws of motion and universal gravitation. In his own lifetime, however, he was originally famous for his work on light. Newton pictured light as a beam of ultra-tiny material particles. With this model he could explain reflection as a bouncing of the particles from a surface, and he could explain refraction as the result of deflecting forces by the surface on the light particles. In the eighteenth and nineteenth centuries, this particle model gave way to a wave model of light because waves could explain not only reflection and refraction, but everything else that was known about light at that time. In this chapter we will investigate the wave aspects of light, which are needed to explain two important phenomena—diffraction and interference.

31.1 Huygens' Principle

In the late 1600s a Dutch mathematician-scientist, Christian Huygens, proposed a very interesting idea about waves. Huygens stated that light waves spreading out from a point source may be regarded as the overlapping of tiny secondary wavelets, and that every point on any wave front may be regarded as a new point source of secondary waves. We see this in Figure 31.1. In other words, wave fronts are made up of tinier wave fronts. This idea is called **Huygens' principle.**

Look at the spherical wave front in Figure 31.2. Each point along the wave front AA' is the source of a new wavelet that spreads out in a sphere from that point. Only a few of the infinite number of wavelets are shown in the figure. The new wave front BB' can be regarded as a smooth surface enclosing the infinite number of overlapping wavelets that started from AA' a short time earlier.

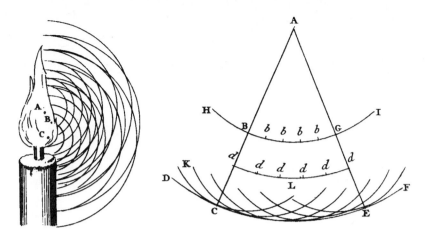

Figure 31.1 ▲

These drawings are from Huygens' book *Treatise on Light*. Light from A (left) expands in wave fronts, every point of which (right) behaves as if it were a new source of waves. Secondary wavelets starting at *b,b,b,b* form a new wave front (*d,d,d,d*); secondary wavelets starting at *d,d,d,d* form still another new wave front (DCEF).

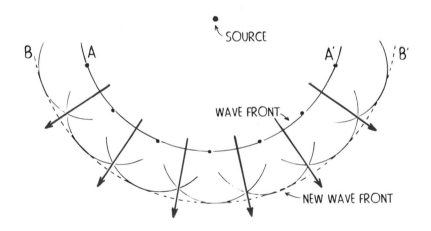

◄ **Figure 31.2**
Huygens' principle applied to a spherical wave front.

As a wave front spreads, it appears less curved. Very far from the original source, the wave fronts seem to form a plane. A good example is the plane waves that arrive from the sun. A Huygens' wavelet construction for plane waves is shown in Figure 31.3. (In a two-dimensional drawing, the planes are shown as straight lines.)

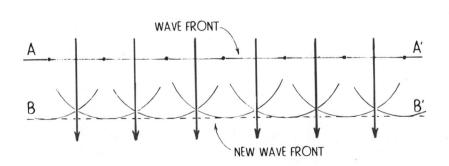

◄ **Figure 31.3**
Huygens' principle applied to a plane wave front.

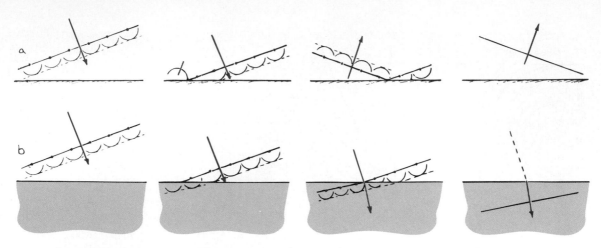

Figure 31.4 ▲
Huygens' principle applied to (a) reflection and (b) refraction.

The laws of reflection and refraction are illustrated via Huygens' principle in Figure 31.4.

You can observe Huygens' principle in water waves that are made to pass through a narrow opening. A wave with straight wave fronts can be generated in water by successively dipping a stick lengthwise into the water (Figure 31.5). A ruler works well. When the straight wave fronts pass through the opening in a barrier, interesting wave patterns result.

Figure 31.5 ▶
Making plane waves in a tank of water and watching the pattern they produce when they pass though an opening in a barrier.

When the opening is wide, you'll see straight wave fronts pass through without change—except at the corners, where the wave fronts are bent into the "shadow region" in accord with Huygens' principle. If you narrow the width of the opening, less of the wave gets through, and the spreading into the shadow region is more pronounced. When the opening is small compared with the wavelength of the waves, Huygens' idea that every part of a wave front can be regarded as a source of new wavelets becomes quite apparent. As the waves move into the narrow opening, the water sloshing up and down in the opening is easily seen to act as a point source of circular waves that fan out on the other side of the barrier. The photos in Figure 31.6 are top views of water waves generated by a vibrating stick. Note how the waves fan out more as the hole through which they pass becomes smaller.

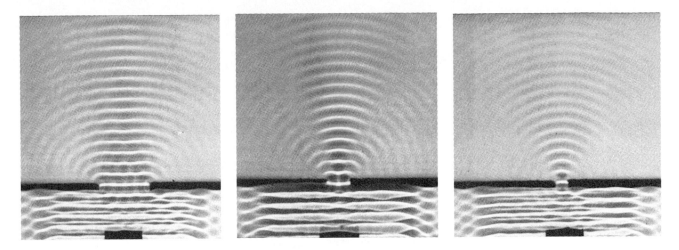

Figure 31.6 ▲
Straight waves passing through openings of various sizes. The smaller the opening, the greater the bending of the waves at the edges.

31.2 Diffraction

Any bending of a wave by means other than reflection or refraction is called **diffraction.** The photos of Figure 31.6 show the diffraction of straight water waves through various openings. When the opening is wide compared with the wavelength, the spreading effect is small. As the opening becomes narrower, the spreading of waves becomes more pronounced. The same occurs for all kinds of waves, including light waves.

When light passes though an opening that is large compared with the wavelength of light, it casts a rather sharp shadow (Figure 31.7). When light passes through a small opening, such as a thin razor slit in a piece of opaque material, it casts a fuzzy shadow, for the light fans out like the water through the narrow opening in Figure 31.6. The light is diffracted by the thin slit.

Figure 31.7 ▲
Light casts a sharp shadow with some fuzziness at its edges when the opening is large compared with the wavelength of the light. Because of diffraction, it casts a fuzzier shadow when the opening is extremely narrow.

483

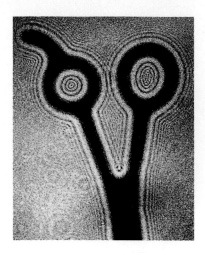

Figure 31.8 ▲
Diffraction fringes around the scissors are evident in the shadows of laser light, which is of a single frequency. These fringes would be filled in by multitudes of other fringes if the source were white light.

Diffraction is not confined to the spreading of light through narrow slits or other openings. Diffraction occurs to some degree for all shadows. On close examination, even the sharpest shadow is blurred at the edge. When light is of a single color, diffraction can produce *diffraction fringes* at the edge of the shadow, as shown in Figure 31.8. In white light, the fringes merge together to create a fuzzy blur at the edge of a shadow.

The amount of diffraction depends on the size of the wavelength compared with the size of the obstruction that casts the shadow (Figure 31.9). The longer the wave compared with the obstruction, the greater the diffraction is. Long waves are better at filling in shadows. This is why foghorns emit low-frequency sound waves—to fill in "blind spots." Likewise for radio waves of the standard AM broadcast band. These are very long compared with the size of most objects in their path. Long waves don't "see" relatively small buildings in their path. They diffract, or bend, readily around buildings and reach more places than shorter waves do.*

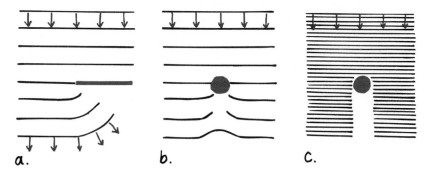

a.　　　　　　b.　　　　　　c.

Figure 31.9 ▲
(a) Waves tend to spread into the shadow region. (b) When the wavelength is about the size of the object, the shadow is soon filled in. (c) When the wavelength is short compared with the width of the object, a sharper shadow is cast.

FM radio waves are shorter, don't diffract as much around buildings, and aren't received as well as AM radio waves are in mountain canyons or city "canyons." This is why many localities have poor FM reception while AM reception comes in loud and clear. TV waves, which are also electromagnetic waves, behave much like FM waves.** TV antennas are often put on rooftops to improve reception. Both FM and TV transmission is "line of sight," meaning that diffraction is not significant, whereas AM transmission diffracts around hills and buildings to reach places that would otherwise be shadowed. Diffraction can be nice.

---

* On the other hand, a framework of connected steel girders in a building or bridge can act as a "polished" surface for long-wavelength radio waves, reflecting them away from the structure. That's why you lose reception when you drive onto a steel bridge while listening to an AM station. So although long waves get *around* a steel structure, they don't get *into* it.

** Most TV channels have wavelengths shorter than FM wavelengths, and therefore even less diffraction than FM, but some (channels 2–6) have wavelengths longer than FM wavelengths.

Diffraction is not so nice when we wish to see very small objects with microscopes. If the size of the object is the same as the wavelength of light, the image of the object will be blurred by diffraction. If the object is smaller than the wavelength of light, no structure can be seen. The entire image is lost due to diffraction. No amount of magnification or perfection of microscope design can defeat this fundamental diffraction limit.

To minimize this problem, microscopists illuminate tiny objects with shorter wavelengths. It turns out that a beam of electrons has a wavelength associated with it. This wavelength is very much shorter than the wavelengths of visible light. Microscopes that use beams of electrons to illuminate tiny things are called *electron microscopes*. The diffraction limit of an electron microscope is much less than that of an optical microscope.

The fact that smaller details can be better seen with smaller wavelengths is cleverly employed by the dolphin in scanning its environment with high-frequency sound—ultrasound.* The echoes of long-wavelength sound give the dolphin an overall image of objects in its surroundings. To examine more detail, the dolphin emits sounds of shorter wavelengths. Skin, muscle, and fat are almost transparent to dolphins, so they "see" a thin outline of the body. Bones, teeth, and gas-filled cavities are clearly apparent. Physical evidence of cancers, tumors, heart attacks, and even emotional states can all be "seen" by the dolphins. The dolphin has always done naturally what humans in the medical field have only recently been able to do with ultrasound devices.

◀ **Figure 31.10**
A dolphin emits ultrashort-wavelength sounds to locate and identify objects in its environment. Distance is sensed by the time delay between sending sound and receiving its echo, and direction is sensed by differences in time or phase for echoes reaching its two ears. A dolphin's main diet is fish and, since hearing in fish is limited to fairly long wavelength sound, they are not alerted to the fact they are being hunted.

* The primary sense of the dolphin is acoustic, for vision is not a very useful sense in the often murky and dark depths of the ocean. Whereas sound is a passive sense for us, it is an active sense for the dolphin, who sends out sounds and then perceives its surroundings on the basis of the echoes that come back. What's more interesting, the dolphin can reproduce the sonic signals that paint the image of its surroundings; therefore, the dolphin probably communicates its experience to other dolphins by communicating the full acoustic image of what is "seen," placing it directly in the minds of other dolphins. The dolphin needs no word or symbol for "fish," for example, but communicates an image of the real thing—maybe with emphasis highlighted by selective filtering, as we similarly communicate a musical concert to others through various means of sound reproduction. Is it any wonder that the language of the dolphin is very different from ours?

DOING PHYSICS

Seeing Diffraction Fringes

Hold two fingers close together between your eye and an illuminated source such as the sky. Look carefully at the narrow opening. Do you see diffraction fringes? Vary the width of the opening, and the distance of your fingers from your eye and note the effects. Go a step further and cut a thin slit in a piece of cardboard. Then look at a light source through it. You can vary the size of the opening by bending the card slightly. Diffraction fringes should be clearly evident!

Activity

Question

Why is blue light used to view tiny objects in an optical microscope?

31.3 Interference

The idea of wave interference was introduced in Chapter 25, and applied to sound in Chapter 26. The idea is important enough to summarize here before applying it to light waves.

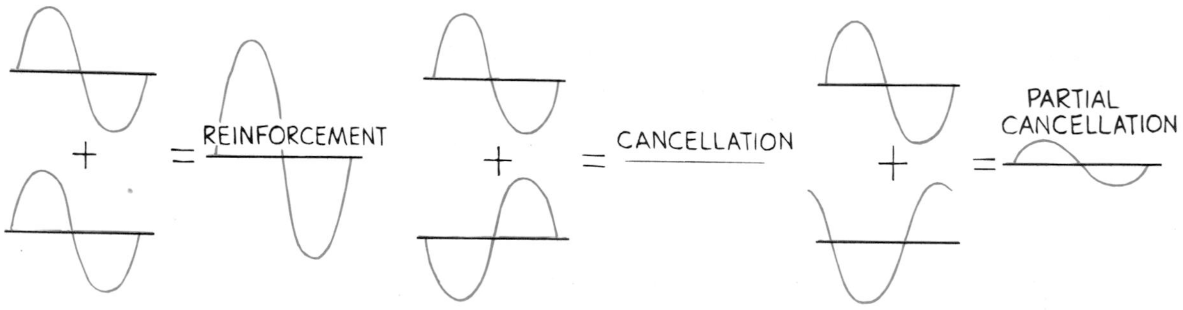

Figure 31.11 ▲
Interference.

If you drop a couple of stones into water at the same time, the two sets of waves that result cross each other and produce what is called an *interference pattern.* Within the pattern, wave effects may be increased, decreased, or neutralized. When the crest of one wave

Answer

Less diffraction results from the short wavelengths of blue light relative to other longer wavelengths.

Figure 31.12 ▲
Interference patterns of overlapping water waves from two vibrating sources.

overlaps the crest of another, their individual effects add together; this is *constructive interference*. When the crest of one wave overlaps the trough of another, their individual effects are reduced; this is *destructive interference*.

Water waves can be produced in shallow tanks of water known as *ripple tanks* under more carefully controlled conditions. Interesting patterns are produced when two sources of waves are placed side by side. Small spheres are made to vibrate at a controlled frequency in the water while the wave patterns are photographed from above, as shown in Figure 31.12. The gray "spokes" are regions of destructive interference. The dark and light striped regions are regions of constructive interference. The greater the frequency of the vibrating spheres, the closer together the stripes (and the shorter the wavelength). Note how the number of regions of destructive interference depends on the wavelength and on the distance between the wave sources.

31.4 Young's Interference Experiment

In 1801 the British physicist and physician Thomas Young performed an experiment that was to make him famous.* Young discovered that when **monochromatic** light—light of a single color—was directed through two closely spaced pinholes, fringes of brightness and darkness were produced on a screen behind. He realized that the bright fringes of light resulted from light waves from both holes arriving

---

* Young read fluently at the age of two; by four, he had read the Bible twice; by fourteen, he knew eight languages. During his adult life he contributed to an understanding of fluids, work and energy, and the elastic properties of materials. He was also the first person to make progress in deciphering Egyptian hieroglyphics. No doubt about it: Thomas Young was smart—very smart.

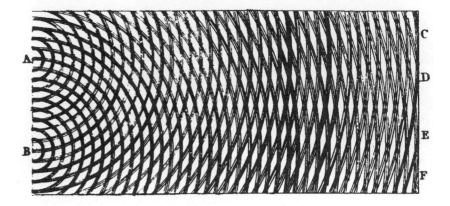

Figure 31.13 ▲
Thomas Young's original drawing of a two-source interference pattern. The dark circles represent wave crests; the white spaces between the crests represent troughs. Constructive interference occurs where crests overlap crests or troughs overlap troughs. Letters *C, D, E,* and *F* mark regions of destructive interference.

crest to crest (constructive interference—more light). Similarly, the dark areas resulted from light waves arriving trough to crest (destructive interference—no light). Young had convincingly demonstrated the wave nature of light that Huygens had proposed earlier.

Young's experiment is now done with two closely spaced slits instead of pinholes, so the fringes are straight lines. A sodium vapor lamp provides a good source of monochromatic light, and a laser is even better. The arrangement is shown in Figure 31.14. Note the similarity of this to the arrangement of sound speakers in Figure 26.14 (Chapter 26). The effects are similar. The pattern of fringes that results is shown in Figure 31.15.

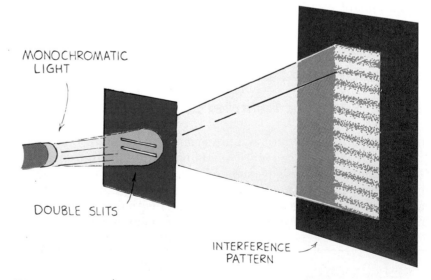

MONOCHROMATIC LIGHT

DOUBLE SLITS

INTERFERENCE PATTERN

Figure 31.14 ▲
Monochromatic light passing through two closely spaced slits diffracts. The screen is illuminated where light waves arrive in phase. The screen is dark where light waves arrive out of phase.

Figure 31.16 shows how the series of bright and dark lines results from the different path lengths from the slits to the screen. A bright fringe occurs when waves from both slits arrive in phase. Dark regions occur when waves arrive out of phase.

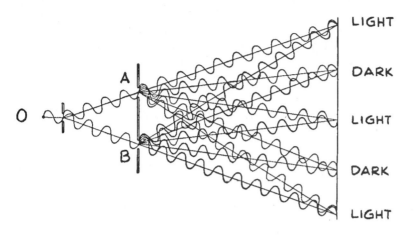

Figure 31.16 ▲
Light from *O* passes through slits *A* and *B* and produces an interference pattern on the screen at the right.

Figure 31.15 ▲
An interference pattern produced by a double slit, as illustrated in Figure 31.14.

■ Questions

1. Why is it important that monochromatic (single frequency) light be used in Young's interference experiment?

2. If the double slits were illuminated with monochromatic blue light, would the fringes be closer together or farther apart than those produced when monochromatic red light is used?

Interference patterns are not limited to double-slits arrangements. A multitude of closely spaced parallel slits makes up what is called a **diffraction grating.** Many spectrometers use diffraction gratings rather than prisms to disperse light into colors. Whereas a prism separates the colors of light by refraction, a diffraction grating separates colors by interference.

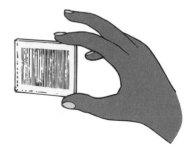

Figure 31.17 ▲
A diffraction grating disperses light into colors by interference among light beams diffracted by many slits or grooves. It may be used in place of a prism in a spectrometer.

Answers

1. If light of a variety of wavelengths were diffracted by the slits, dark fringes for one wavelength would be filled in with bright fringes for another, resulting in no distinct fringe pattern. This is similar to the person listening to the pair of speakers back in Chapter 26 (Figure 26.14). If the path difference equals one-half wavelength for one frequency, it cannot also equal one-half wavelength for any other frequency. Different frequencies will "fill in" the fringes.

2. The wavelength of blue light is shorter than (nearly half) that of red light. Investigate the differences in the number of fringes for the water waves in Figure 31.12. The fringes of shorter wavelengths are closer together than those of longer wavelengths. So blue fringes would be closer together than red fringes.

Figure 31.18 ▲
Arrays of tiny pits beneath the surface of the compact disc (CD) nicely diffract light.

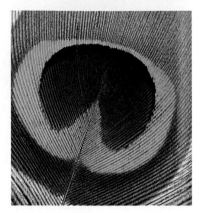

Figure 31.19 ▲
Diffraction from ridges in a peacock's feathers produce beautiful iridescent colors.

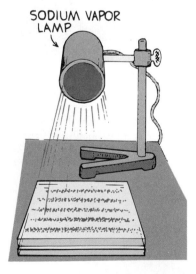

Figure 31.20 ▲
Interference fringes produced when monochromatic light is reflected from two plates of glass with an air wedge between them.

More common diffraction gratings are seen in reflective materials used in items such as costume jewelry and automobile bumper stickers. These materials are ruled with tiny grooves that diffract light into a brilliant spectrum of colors. The pits on the reflective surface of an audio compact disc not only provide high-fidelity music but also diffract light spectacularly into its component colors. But long before the advent of these high-tech items, the feathers of birds have been nature's diffraction gratings, and the striking colors of opals come from layers of tiny silica spheres that act as diffraction gratings.

31.5 Single-Color Interference from Thin Films

Interference fringes can be produced by the reflection of light from two surfaces that are very close together. If you shine monochromatic light onto two plates of glass, one atop the other as shown in Figure 31.20, you'll see dark and bright bands.

The cause of these bands is the interference between the waves reflected from the glass on the top and bottom surfaces of the air space between the plates. This is shown in Figure 31.21. The reflected light comes to the eye by two different paths. The light that hits the lower glass surface has slightly farther to go to reach your eye. If this extra distance results in light from the upper and lower reflections getting to your eye one-half wavelength out of phase, then destructive interference will occur and a dark region will be seen. Nearby, the path differences will not result in destructive interference, and a light region will be seen.

A practical use of interference fringes is the testing of precision lenses. When a lens to be tested is placed on a perfectly flat piece of glass and illuminated from above with monochromatic light, light and dark fringes are seen (Figure 31.22). Irregular fringes indicate an irregular surface. When a lens is polished until smooth and concentric, the interference fringes will be concentric and regularly spaced.

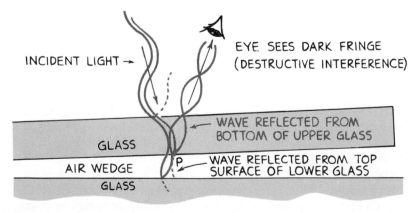

Figure 31.21 ▲
Reflection from the upper and lower surfaces of a thin air space results in constructive and destructive interference.

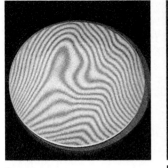

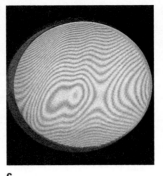

a   b   c   d

Figure 31.22 ▲

The flatness or curvature of a surface can be tested by placing the surface on a very flat piece of glass and observing the interference pattern. (a) An irregular surface; (b) a flat surface; (c) a poorly polished lens; (d) a precision lens.

31.6 Iridescence from Thin Films

Everyone who has seen soap bubbles or gasoline spilled on a wet street has noticed the beautiful spectrum of colors reflected from them. Some types of bird feathers have colors that seem to change hue as the bird moves. All these colors are produced by the interference of light waves of mixed frequencies in thin films, a phenomenon known as **iridescence.**

A thin film, such as a soap bubble, has two closely spaced surfaces. Light that reflects from one surface may cancel light that reflects from the other surface. For example, the film may be just the right thickness in one place to cause the destructive interference of, say, blue light. If the film is illuminated with white light, then the light that reflects to your eye will have no blue in it. What happens when blue is taken away from white light? The answer is, the complementary color will appear. And for the cancellation of blue, we get yellow. So the soap bubble will appear yellow wherever blue is canceled.

In a thicker part of the film, where green is canceled, the bubble will appear magenta. The different colors correspond to the cancellations of their complementary colors by different thicknesses of the film.

Figure 31.24 illustrates interference for a thin layer of gasoline on a layer of water. Light reflects from both the upper gasoline-air surface and the lower gasoline-water surface. Suppose that the incident beam is monochromatic blue, as in the illustration. If the gasoline layer is just the right thickness to cause cancellation of light of that wavelength, then the gasoline surface appears dark to the eye. On the other hand, if the incident beam is white sunlight, the gasoline surface appears yellow to the eye. Blue is subtracted from the white, leaving the complementary color, yellow.

The beautiful colors reflected from some types of seashells are produced by interference of light from their thin transparent coatings. So are the sparkling colors from fractures within opals. Interference colors can even be seen in the thin film of detergent left when dishes are not properly rinsed.

Figure 31.23 ▲

The beautiful colors of gasoline on a wet street correspond to different thicknesses of the thin film. The colors provide a vivid "contour map" of microscopic differences in surface "elevations."

Figure 31.24 ▶

The thin film of gasoline is just the right thickness so that monochromatic blue light reflected from the top surface of the gasoline is canceled by light of the same wavelength reflected from the water.

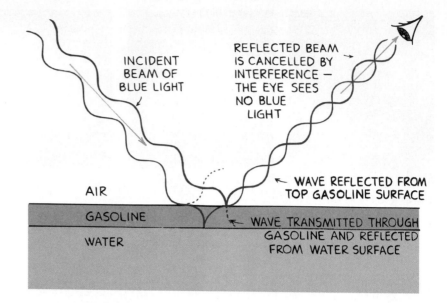

■ Questions

1. What color will reflect from a soap bubble in sunlight when its thickness is such that red light is canceled?

2. The left column lists some colored objects. Match them to the various ways that light may produce that color from the choices in the right column.

   a. yellow banana   1. interference

   b. blue sky   2. diffraction

   c. rainbow   3. selective reflection

   d. peacock feathers   4. refraction

   e. soap bubble   5. scattering

Interference provides the principal method for measuring the wavelengths of light. Wavelengths of other regions of the electromagnetic spectrum are also measured with interference techniques. Extremely small distances (millionths of a centimeter) are measured with instruments called *interferometers*, which make use of the principle of interference. These instruments are sensitive enough to detect the displacement at the end of a long, several-centimeters-thick solid steel bar when you gently apply opposite twists to opposite ends with your hands. They are among the most accurate measuring instruments known.

The next two sections describe the laser and what is perhaps the most exciting illustration of interference—the *hologram*.

■ Answers

1. You will see the color cyan, which is the complementary color of red.

2. a—3; b—5; c—4; d—2; e—1.

Swirling Colors

Dip a dark-colored coffee mug (dark is the best background for viewing interference colors) in dishwashing detergent and then hold it sideways as if you were pouring from it. Look at the reflected light from the soap film that covers its mouth. Swirling colors appear as the soap runs down to form a wedge that grows thicker at the bottom with time. The top becomes thinner— so thin it appears black. This happens when the film is thinner than $\frac{1}{4}$ the wavelength of the shortest waves of visible light. The film soon becomes so thin that it pops.

Activity

31.7 Laser Light

Light emitted by a common lamp is **incoherent.** That is, the light has many phases of vibration (as well as many frequencies). The light is as incoherent as the footsteps on an auditorium floor when a mob of people are chaotically rushing about. Incoherent light is chaotic. Interference within a beam of incoherent light is rampant, and a beam spreads out after a short distance, becoming wider and wider and less intense with increased distance.

◄ **Figure 31.25**
Incoherent white light contains waves of many frequencies and wavelengths that are out of phase with one another.

Even if a beam is filtered so that it is monochromatic (has a single frequency), it is still incoherent because the waves are out of phase and interfere with one another. The slightest differences in their directions result in a spreading with increased distance.

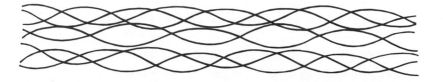

◄ **Figure 31.26**
Light of a single frequency and wavelength can still be out of phase.

Figure 31.27 ▲
Coherent light. All the waves are identical and in phase.

A beam of light that has the same frequency, phase, and direction is said to be **coherent.** There is no interference of waves within the beam. Only a beam of coherent light will not spread and diffuse.

Coherent light is produced by a **laser** (whose name comes from light amplification by stimulated emission of radiation).* Within a laser, a light wave emitted from one atom stimulates the emission of light from a neighboring atom so that the crests of each wave coincide. These waves stimulate the emission of others in cascade fashion, and a beam of coherent light is produced. This is very different from the random emission of light from atoms in common sources.

Figure 31.28 ▶
A helium-neon laser. A high voltage applied to a mixture of helium and neon gas energizes helium atoms to a prolonged energy state. Before the helium can emit light, it gives up its energy by collision with neon, which is boosted to an otherwise hard-to-come-by matched energy state. Light emitted by neon stimulates other energized neon atoms to emit matched-frequency light. The process cascades, and a coherent beam of light is produced. The output remains steady because the helium is constantly reenergized.

Figure 31.29 ▲
A product's code is read by laser light that reflects from the bar pattern and is converted to an electrical signal that is fed into a computer. The signal is high when light is reflected from the white spaces and low when reflected from a dark bar.

The laser is not a source of energy. It is simply a converter of energy, taking advantage of the process of stimulated emission to concentrate a certain fraction of the energy input (commonly much less than 1%) into a thin beam of coherent light. Like all devices, a laser can put out no more energy than it takes in.

Lasers come in many types and have broad applications in diverse fields. Surveyors and construction workers use them as "chalk lines," surgeons use them as scalpels, garment manufacturers use them as cloth-cutting saws. They are used to read product codes in cash registers, to read the music signal in CDs, and the bar codes that access the videodisc ancillary of your *Conceptual Physics* program. A most impressive product of laser light is the hologram.

---

* A word constructed from the initials of a phrase is called an *acronym.*

31.8 The Hologram

Holo- comes from the Greek word for "whole," and *gram* comes from the Greek for "message" or "information." A **hologram** is a three-dimensional version of a photograph that contains the whole message or entire picture in every portion of its surface. To the naked eye, it appears to be an imageless piece of transparent film, but on its surface is a pattern of microscopic fringes. Light diffracted from these fringes produces an image that is extremely realistic. Holograms are also difficult to reproduce—hence their use on credit cards.

A hologram is produced by the interference between two laser light beams on photographic film. The two beams are part of one beam. One part illuminates the object and is reflected from the object to the film. The second part, called the *reference beam,* is reflected from a mirror to the film, as shown in Figure 31.30. Interference between the reference beam and light reflected from the different points on the object produces a pattern of microscopic fringes on the film. Light from nearer parts of the object travels shorter paths than light from farther parts of the object. The different distances traveled will produce slightly different interference patterns with the reference beam. In this way information about the depth of an object is recorded.

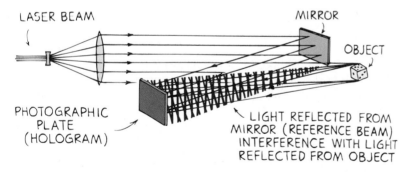

Figure 31.30 ▲

A simplified arrangement for making a hologram. The laser light that exposes the photographic film is made up of two parts; one part is reflected from the object, and one part is reflected from the mirror. The waves of these two parts interfere to produce microscopic fringes on the film. When developed, it is then a hologram.

When light falls on a hologram, it is diffracted by the fringed pattern to produce wave fronts identical in form to the original wave fronts reflected by the object. The diffracted wave fronts produce the same effect as the original reflected wave fronts. Whether you look through the hologram or see the reflections from a hologram you see a realistic three-dimensional image as though you were viewing the original object through a window (or from a mirror). Parallax is evident when you move your head to the side and see down the sides of the object, or when you lower your head and look underneath the object. Holographic pictures are extremely realistic.

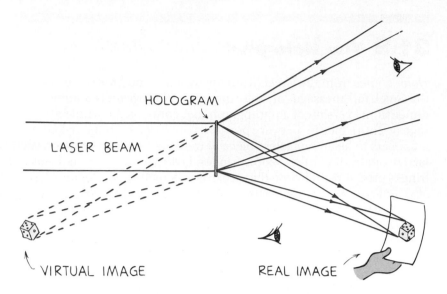

Figure 31.31 ▲
When a hologram is illuminated with coherent light, the diverging diffracted light produces a three-dimensional *virtual* image that can be seen when looking *through* the hologram, like looking through a window. You refocus your eyes to see near and far parts of the image, just as you do when viewing a real object. Converging diffracted light produces a *real* image in front of the hologram, which can be projected on a screen. Since the image has depth, you cannot see near and far parts of the image in sharp focus for any single position on a flat screen.

Interestingly enough, if the hologram is made on film, you can cut it in half and still see the entire image on each half. And you can cut one of the pieces in half again and again and see the entire image, just as you can put your eye to any part of a window to see outdoors. Every part of the hologram has received and recorded light from the entire object.

Even more interesting is holographic magnification. If holograms are made using short-wavelength light and viewed with light of a longer wavelength, the resulting image is magnified in the same proportion as the wavelengths. Holograms made with X rays would be magnified thousands of times when viewed with visible light and appropriate viewing arrangements. X-ray holograms have not been made as this book is being written. Technological growth is fast these days. Are X-ray holograms a reality as you are reading this?

Light is interesting—especially when it is diffracted through the interference fringes of that supersophisticated diffraction grating, the hologram!

31 Chapter Review

Concept Summary

Diffraction of light is the bending of light by means other than reflection or refraction.

- Huygens' principle, which states that every point on a wave front acts like a point source of secondary wavelets, can be used to understand diffraction.

- Diffraction is greatest for waves that are long compared with the size of the obstruction.

Interference of light is the combining of single-frequency light from two parts of the same beam in such a way that crest overlaps crest or crest overlaps trough.

- The colors seen in thin films (soap bubbles or thin films of gasoline on water) arc due to destructive interference of different frequencies at different thicknesses.

- Holograms are three-dimensional pictures created through the interference of two parts of a laser beam.

Important Terms

coherent (31.7)
diffraction (31.2)
diffraction grating (31.4)
hologram (31.8)
Huygens' principle (31.1)
incoherent (31.7)
iridescence (31.6)
laser (31.7)
monochromatic (31.4)

Review Questions

1. What is Huygens' principle? (31.1)

2. **a.** Waves spread out when they pass through an opening. Does spreading become more or less pronounced for narrower openings?

   b. What is this spreading called? (31.1–31.2)

3. Does diffraction aid or hinder radio reception? (31.2)

4. Does diffraction aid or hinder the viewing of images in a microscope? (31.2)

5. Is it possible for a wave to be canceled by another wave? Defend your answer.

6. Does wave interference occur for waves in general, or only for light waves? Give examples to support your answer. (31.3–31.4)

7. What was Thomas Young's discovery? (31.4)

8. What is the cause of the fringes of light in Young's experiment? (31.4)

9. What is a diffraction grating? (31.4)

10. What is required for part of the light reflected from a surface to be canceled by another part reflected from a second surface? (31.5)

11. What causes the bright and dark fringes visible in lenses that rest on flat plates of glass (as shown in Figure 31.21)? (31.5)

12. What is iridescence, and what phenomenon is it related to? (31.6)

13. If a soap bubble is thick enough to cancel yellow by interference, what color will it appear if illuminated by white light? (31.6)

14. Why is gasoline that is spilled on a wet surface so colorful? (31.6)

15. What is an interferometer, and on what physics principle is it based? (31.6)

16. How does light from a laser differ from light from an ordinary lamp? (31.7)

17. Can a laser put out more energy than is put in? (Would you have to know more about lasers to answer this question? Why?) (31.7)

18. What is a hologram, and on what physics principle is it based? (31.8)

19. How does the image of a hologram differ from that of a common photograph? (31.8)

20. What would be the advantage of making holograms with X rays? (31.8)

Activity

1. Make some slides for a slide projector by sticking crumpled cellophane onto pieces of slide-sized polarizing material. Also try strips of cellophane tape, overlapping at different angles. (Experiment with different brands of tape.) Project the slides onto a large screen or white wall and rotate a second, slightly larger piece of polarizing material in front of the projector lens. The colors are vivid! Do this in rhythm with your favorite music, and you'll have your own light and sound show.

Think and Explain

1. In our everyday environment, diffraction is much more evident for sound waves than for light waves. Why is this so?

2. Suppose the speakers at an open-air rock concert are pointed forward. You move about and notice that the sounds of female vocalists can be heard in front of the stage, but very little off to the sides. By comparison you notice that bass sounds can be heard quite well both in front and off to the sides. What is your explanation?

3. Why do radio waves diffract around buildings while light waves do not?

4. Why are TV broadcasts in the VHF range easier to receive in areas of marginal reception

than broadcasts in the UHF range? (*Hint:* UHF has higher frequencies than VHF.)

5. Suppose a pair of loudspeakers a meter or so apart emit pure tones of the same frequency and loudness. When a listener walks past in a path parallel to the line that joins the loudspeakers, the sound is heard to alternate from loud to soft. What is going on?

6. In the preceding question, suggest a path along which the listener could walk so as not to hear alternate loud and soft sounds.

7. When monochromatic light illuminates a pair of thin slits, an interference pattern is produced on a wall behind. How will the distance between the fringes of the pattern for red light differ from that for blue light?

8. When Thomas Young performed his interference experiment, monochromatic light passed through a single narrow opening before it reached the double openings. Explain why this made the fringes clear. (*Hint:* What would be the result if light reaching the double openings came from several different directions?)

9. Seashells, butterfly wings, and the feathers of some birds often change color as you look at them from different positions. Explain this phenomenon in terms of light interference.

10. If you notice the interference patterns of a thin film of oil or gasoline on water, you'll note that the colors form complete rings. How are these rings similar to the lines of equal elevation on a contour map?

11. Suppose the thickness of a soap film is just right for canceling yellow light. What color does the eye see? Why will this color change when the surface is viewed at a grazing angle?

V Electricity and Magnetism

This simple electric circuit illustrates some intriguing physics. The battery provides **voltage,** an electric pressure that pushes electrons through the wire and lamp. Electrons flow easily through the relatively thick wire, but with difficulty through the lamp filament. The filament has a **resistance** to electron flow. **Current** squeezed through it shakes the atoms so vigorously that they glow. That's why the filament emits **light** while the connecting wire doesn't. Even the light is **electrical** in nature—**magnetic** too, as Unit 5 will show. Who's more afraid of electricity—someone who has an understanding of it, or someone who doesn't?

32 Electrostatics

Enormous transfer of electrical energy.

Electricity in one form or another underlies just about everything around you. It's in the lightning from the sky; it's in the spark beneath your feet when you scuff across a rug; and it's what holds atoms together to form molecules. The control of electricity is evident in technological devices of many kinds, from lamps to computers. In this technological age it is important to have an understanding of how the basics of electricity can be manipulated to give people a prosperity that was unknown before recent times.

This chapter is about **electrostatics,** or electricity at rest. Electrostatics involves electric charges, the forces between them, and their behavior in materials. The next chapter is about the aura that surrounds electric charges—the *electric field*. Chapters 34 and 35 cover moving electric charges, or *electric currents;* the voltages that produce them; and the ways that currents can be controlled. Finally, Chapters 36 and 37 cover the relationship of electric currents to magnetism, and how electricity and magnetism can be controlled to operate motors and other electrical devices.

An understanding of electricity requires a step-by-step approach, for one concept is the building block for the next. So please study this material with extra care. It is a good idea at this time to lean more heavily on the laboratory part of your course, for *doing* physics is better than only studying physics. If you're hasty, the physics of electricity and magnetism can be difficult, confusing, and frustrating. But with careful effort, it can be comprehensible and rewarding.

32.1 Electrical Forces and Charges

You are familiar with the force of gravity. It attracts you to the earth, and you call it your weight. Now consider a force acting on you that is billions upon billions of times stronger. Such a force could compress you to a size about the thickness of a piece of paper. But suppose that in addition to this enormous force there is a repelling force that is also billions upon billions of times stronger than gravity. The two forces acting on you would balance each other and have no noticeable effect at all. It so happens that there is a pair of such forces acting on you all the time—**electrical forces.**

Electrical forces arise from particles in atoms. In the simple model of the atom proposed in the early 1900s by Ernest Rutherford and Niels Bohr, a positively charged nucleus is surrounded by electrons (Figure 32.2). The protons in the nucleus attract the electrons and hold them in orbit, just as the sun holds the planets in orbit. Electrons are attracted to protons, but electrons repel other electrons. This attracting and repelling behavior is attributed to a property called **charge.*** By convention (general agreement), electrons are *negatively* charged and protons *positively* charged. Neutrons have no charge, and are neither attracted nor repelled by charged particles.

Some important facts about atoms are

1. Every atom has a positively charged nucleus surrounded by negatively charged electrons.

2. All electrons are identical; that is, each has the same mass and the same quantity of negative charge as every other electron.

3. The nucleus is composed of protons and neutrons. (The common form of hydrogen, which has no neutrons, is the only exception.) All protons are identical; similarly, all neutrons are identical. A proton has nearly 2000 times the mass of an electron, but its positive charge is equal in magnitude to the negative charge of the electron. A neutron has slightly greater mass than a proton and has no charge.

4. Atoms usually have as many electrons as protons, so the atom has zero *net* charge.

Just *why* electrons repel electrons and are attracted to protons is beyond the scope of this book. At our level of understanding we simply say that this is nature as we find it—that this electric behavior is fundamental, or basic. The fundamental rule at the base of all electrical phenomena is

> Like charges repel; opposite charges attract.

The old saying that opposites attract, usually referring to people, was first popularized by public lecturers who traveled about by horse

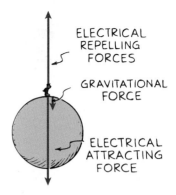

Figure 32.1 ▲
The enormous attractive and repulsive electrical forces between the charges in the earth and the charges in your body balance out, leaving the relatively weaker force of gravity, which attracts only. Hence your weight is due only to gravity.

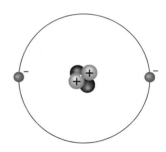

Figure 32.2 ▲
Model of a helium atom. The helium nucleus is composed of two protons and two neutrons. The positively charged protons attract two negative electrons.

---

* Why don't protons pull the oppositely charged electrons into the nucleus? Interestingly enough, the reason is *not* the same as the reason that planets orbit the sun. Within the atom, different laws of physics apply. This is the domain of *quantum physics,* which we come to in Chapter 38. According to quantum physics, an electron behaves like a wave and has to occupy a certain amount of space related to its wavelength. The size of an atom is set by the minimum amount of "elbow room" that an electron requires.

Why is it that the protons in the nucleus do not mutually repel and fly apart? What holds the nucleus together? The answer is that in addition to electrical forces in the nucleus, there are even stronger forces that are nonelectrical in nature. These are *nuclear forces* and are discussed in Chapter 39.

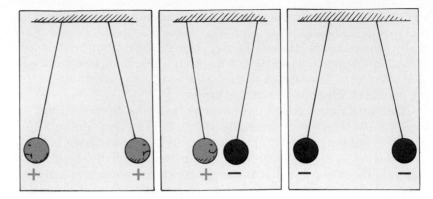

Figure 32.3 ▲
Likes repel and opposites attract.

and wagon to entertain people by demonstrating the scientific marvels of electricity. An important part of these demonstrations was the charging and discharging of pith balls. Pith is a light, spongy plant tissue that resembles Styrofoam, and balls of it were coated with aluminum paint so their surfaces would conduct electricity. When suspended from a silk thread, such a ball would be attracted to a rubber rod just rubbed with cat's fur, but when the two made contact, the force of attraction would change to a force of repulsion. Thereafter, the ball would be repelled by the rubber rod but attracted to a glass rod that had just been rubbed with silk. A pair of pith balls charged in different ways exhibited both attraction and repulsion forces (Figure 32.3). The lecturer pointed out that nature provides two kinds of charge, just as it provides two sexes.

■ Questions

1. Beneath the complexities of electrical phenomena, there lies a fundamental rule from which nearly all other effects stem. What is this fundamental rule?

2. How does the charge of an electron differ from the charge of a proton?

32.2 Conservation of Charge

Electrons and protons have electric charge. In a neutral atom, there are as many electrons as protons, so there is no net charge. The total positive charge balances the total negative charge exactly. If an electron is removed from an atom, the atom is no longer neutral. The atom has

■ Answers

1. Like charges repel; opposite charges attract.

2. The charges of both particles are equal in magnitude, but opposite in sign.

The Threat of Static Charge

Two hundred years ago young boys called "powder monkeys" ran below the decks of warships to pick up sacks of black gunpowder for the cannons above. It was ship law that they do this task barefoot. Why? Because it was important that no static charge build up on their bodies as they ran back and forth. The bare feet ensured no build up of static charge that might result in an igniting spark.

Today electronic technicians in high-technology firms that build, test, and repair electronic circuit components similarly follow procedures to guard against static charge. Not because of any danger of blowing up their buildings, but to prevent damage to delicate circuits. Some circuit components are so sensitive that they can be "fried" by static electric sparks. So electronic technicians work in environments free of high-resistant surfaces where static

charge can accumulate. They wear clothing of special fabric with ground wires between their sleeves and their socks. Some wear special wrist bands that are clipped to a grounded surface, so that any charge that builds up, by movement on a chair for example, is discharged. As electronic components become smaller and circuit elements are placed closer together, the threat of electric sparks producing short circuits becomes greater and greater. Maintaining a static-free environment is an important ongoing task for many companies.

---

one more positive charge (proton) than negative charge (electron) and is said to be positively charged.

A charged atom is called an *ion*. A *positive ion* has a net positive charge; it has lost one or more electrons. A *negative ion* has a net negative charge; it has gained one or more extra electrons.

Matter is made of atoms, and atoms are made of electrons and protons (and neutrons as well). An object that has equal numbers of electrons and protons has no net electric charge. But if there is an imbalance in the numbers, the object is then electrically charged. An imbalance comes about by adding or removing electrons.

Although the innermost electrons in an atom are bound very tightly to the oppositely charged atomic nucleus, the outermost electrons of many atoms are bound very loosely and can be easily dislodged. How much energy is required to tear an electron away from an atom varies for different substances. The electrons are held more firmly in rubber than in fur, for example. Hence, when a rubber rod is rubbed by a piece of fur, electrons transfer from the fur to the rubber rod. The rubber then has an excess of electrons and is negatively charged. The fur, in turn, has a deficiency of electrons and is positively charged. If you rub a glass or plastic rod with silk, you'll find that the rod becomes positively charged. The silk has a greater affinity

Figure 32.4 ▲

Electrons are transferred from the fur to the rod. The rod is then negatively charged. Is the fur charged? Positively or negatively?

for electrons than the glass or plastic rod. Electrons are rubbed off the rod and onto the silk. In summary:

> An object that has unequal numbers of electrons and protons is electrically charged. If it has more electrons than protons, the object is negatively charged. If it has fewer electrons than protons, then it is positively charged.

Notice that electrons are neither created nor destroyed but are simply transferred from one material to another. Charge is conserved. In every event, whether large-scale or at the atomic and nuclear level, the principle of **conservation of charge** applies. No case of the creation or destruction of net electric charge has ever been found. The conservation of charge is a cornerstone in physics, ranking with the conservation of energy and momentum.

Any object that is electrically charged has an excess or deficiency of some whole number of electrons—electrons cannot be divided into fractions of electrons. This means that the charge of the object is a whole-number multiple of the charge of an electron. It cannot have a charge equal to the charge of 1.5 or 1000.5 electrons, for example.* All charged objects to date have a charge that is a whole-number multiple of the charge of a single electron.

■ Question

If you scuff electrons onto your feet while walking across a rug, are you negatively or positively charged?

32.3 Coulomb's Law

Recall from Newton's law of gravitation that the gravitational force between two objects of mass m_1 and mass m_2 is proportional to the product of the masses and inversely proportional to the square of the distance d between them:

$$F = G\frac{m_1m_2}{d^2}$$

where G is the universal gravitational constant.

The electrical force between any two objects obeys a similar inverse-square relationship with distance. This relationship was

■ Answer

You have more electrons after you scuff your feet, so you are negatively charged (and the rug is positively charged).

---

* Within the atomic nucleus, however, elementary particles called *quarks* carry charges 1/3 and 2/3 the magnitude of the electron's charge. Each proton and each neutron is made up of three quarks. Since quarks always exist in such combinations and have never been found separated, the whole-number-multiple rule of electron charge holds for nuclear processes as well.

discovered by the French physicist Charles Coulomb (1736–1806) in the eighteenth century. **Coulomb's law** states that for charged particles or objects that are small compared with the distance between them, the force between the charges varies directly as the product of the charges and inversely as the square of the distance between them. The role that charge plays in electrical phenomena is much like the role that mass plays in gravitational phenomena. Coulomb's law can be expressed as

$$F = k \frac{q_1 q_2}{d^2}$$

where d is the distance between the charged particles; q_1 represents the quantity of charge of one particle and q_2 the quantity of charge of the other particle; and k is the proportionality constant.

The SI unit of charge is the **coulomb,** abbreviated C. Common sense might say that it is the charge of a single electron, but it isn't. For historical reasons, it turns out that a charge of 1 C is the charge of 6.24 billion billion (6.24×10^{18}) electrons. This might seem like a great number of electrons, but it represents only the amount of charge that passes through a common 100-W lightbulb in about one second.

The proportionality constant k in Coulomb's law is similar to G in Newton's law of gravitation. Instead of being a very small number like G, the electrical proportionality constant k is a very large number. Rounded off, it equals

$$k = 9\ 000\ 000\ 000\ \text{N·m}^2/\text{C}^2$$

or, in scientific notation, $k = 9.0 \times 10^9$ N·m²/C². The units N·m²/C² convert the right-hand side of the equation to the unit of force, the newton (N), when the charges are in coulombs (C) and the distance is in meters (m). Note that if a pair of charges of 1 C each were 1 m apart, the force of repulsion between the two charges would be 9 billion newtons.* That would be more than 10 times the weight of a battleship! Obviously, such amounts of *net* charge do not exist in our everyday environment.

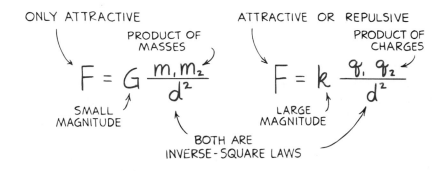

◀ **Figure 32.5**
Comparison of Newton's law of gravitation and Coulomb's law.

---

* Contrast this with the gravitational force of attraction between two masses of 1 kg, each a distance 1 m apart: 6.67×10^{-11} N. This is an extremely small force. For the force to be 1 N, two masses 1 m apart would have to be about 122 000 kilograms each! Gravitational forces between ordinary objects are much too small to be detected except in delicate experiments. Electrical forces (noncanceled) between ordinary objects are large enough to be commonly experienced.

So Newton's law of gravitation for masses is similar to Coulomb's law for electric charges.* Whereas the gravitational force of attraction between a pair of one-kilogram masses is extremely small, the electrical force between a pair of one-coulomb charges is extremely large. The greatest difference between gravitation and electrical forces is that while gravity only attracts, electrical forces may either attract or repel.

■ Questions

1. What is the chief significance of the fact that G in Newton's law of gravitation is a small number and k in Coulomb's law is a large number when both are expressed in SI units?

2. a. If an electron at a certain distance from a charged particle is attracted with a certain force, how will the force compare at twice this distance?

   b. Is the charged particle in this case positive or negative?

Because most objects have almost exactly equal numbers of electrons and protons, electrical forces usually balance out. Between the earth and the moon, for example, there is no measurable electrical force. In general, the weak gravitational force, which attracts only, is the predominant force between astronomical bodies.

Although electrical forces balance out for astronomical and everyday objects, at the atomic level this is not always true. The negative electrons of one atom may at times be closer to the positive protons of a neighboring atom than to the average location of the neighbor's electrons. Then the attractive force between these charges is greater than the repulsive force. When the net attraction is sufficiently strong, atoms combine to form molecules. The chemical bonding forces that hold atoms together to form molecules are electrical forces acting in small regions where the balances of attractive and repelling forces is not perfect. It makes good sense for anyone planning to study chemistry to know something about electricity.

■ Answers

1. The small value of G indicates that gravity is a weak force; the large value of k indicates that the electrical force is enormous in comparison.

2. a. In accord with the inverse-square law, at twice the distance the force will be one-fourth as much.

   b. Since there is a force of attraction, the charges must be opposite in sign, so the charged particle is positive.

* The similarities between these two forces have made some physicists think they may be different aspects of the same thing. Albert Einstein was one of these people; he spent the later part of his life searching with little success for a "unified field theory." In recent years, the electrical force has been unified with the nuclear *weak force*, which plays a role in radioactive decay. Physicists are still looking for a way to unify electrical and gravitational forces.

The hydrogen atom has the simplest structure of all atoms. Its nucleus is a proton (mass 1.7×10^{-27} kg), outside of which there is a single electron (mass 9.1×10^{-31} kg) at an average separation distance of 5.3×10^{-11} m. Compare the electrical and gravitational forces between the proton and the electron in a hydrogen atom.

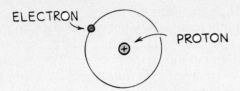

To solve for the electrical force, simply substitute the appropriate values in Coulomb's law.

$$\text{distance } d = 5.3 \times 10^{-11} \text{ m}$$
$$\text{proton charge } q_p = +1.6 \times 10^{-19} \text{ C}$$
$$\text{electron charge } q_e = -1.6 \times 10^{-19} \text{ C}$$

The electric force F_e is

$$F_e = k \frac{q_e q_p}{d^2}$$

$$= (9.0 \times 10^9 \text{ N·m}^2/\text{C}^2) \frac{(1.6 \times 10^{-19} \text{ C})^2}{(5.3 \times 10^{-11} \text{ m})^2}$$

$$= 8.2 \times 10^{-8} \text{ N}$$

The gravitational force F_g between them is

$$F_g = G \frac{m_e m_p}{d^2}$$

$$= (6.7 \times 10^{-11} \text{ N·m}^2/\text{kg}^2) \frac{(9.1 \times 10^{-31} \text{ kg})(1.7 \times 10^{-27} \text{ kg})}{(5.3 \times 10^{-11} \text{ m})^2}$$

$$= 3.7 \times 10^{-47} \text{ N}$$

A comparison of the two forces is best shown by their ratio:

$$\frac{F_e}{F_g} = \frac{8.2 \times 10^{-8} \text{ N}}{3.7 \times 10^{-47} \text{ N}} = 2.2 \times 10^{39}$$

The electrical force is more than 10^{39} times greater than the gravitational force. In other words, the electric forces that subatomic particles exert on one another are so much stronger than their mutual gravitational forces that gravitation can be completely neglected.

32.4 Conductors and Insulators

Electrons are more easily moved in some materials than in others. Outer electrons of the atoms in a metal are not anchored to the nuclei of particular atoms, but are free to roam in the material. Such materials are good **conductors.** Metals are good conductors for the motion of electric charges for the same reason they are good conductors of heat: Their electrons are "loose."

Electrons in other materials—rubber and glass, for example—are tightly bound and remain with particular atoms. They are not free to wander about to other atoms in the material. These materials are poor conductors of electricity, for the same reason they are generally poor conductors of heat. Such materials are good **insulators.**

Figure 32.6 ▶
It is easier for electric charge to flow through hundreds of kilometers of metal wire than through a few centimeters of insulating material.

All substances can be arranged in order of their ability to conduct electric charges. Those at the top of the list are the conductors, and those at the bottom are the insulators. The ends of the list are very far apart. The conductivity of a metal, for example, can be more than a million trillion times greater than the conductivity of an insulator such as glass. In a power line, charge flows much more easily through hundreds of kilometers of metal wire than through the few centimeters of insulating material that separates the wire from the supporting tower. In a common appliance cord, charges will flow through several meters of wire to the appliance, and then through its electrical network, and then back through the return wire rather than flow directly across from one wire to the other through the tiny thickness of rubber insulation.

Whether a substance is classified as a conductor or an insulator depends on how tightly the atoms of the substance hold their electrons. Some materials, such as germanium and silicon, are good insulators in their pure crystalline form but increase tremendously in conductivity when even one atom in ten million is replaced with an impurity that adds or removes an electron from the crystal structure. These materials can be made to behave sometimes as insulators and sometimes as conductors. Such materials are **semiconductors.** Thin layers of semiconducting materials sandwiched together make up *transistors,* which are used in a variety of electrical applications. How transistors and other semiconductor devices work will not be covered in this book.

At temperatures near absolute zero, certain metals acquire infinite conductivity (zero resistance to the flow of charge). These are **superconductors.** Since 1987, superconductivity at "high" temperatures (above 100 K) has been found in a variety of nonmetallic compounds. Once electric current is established in a superconductor, the electrons flow indefinitely. Explanations are presently being vigorously researched.

32.5 Charging by Friction and Contact

We are all familiar with the electrical effects produced by friction. We can stroke a cat's fur and hear the crackle of sparks that are produced, or comb our hair in front of a mirror in a dark room and see as well as hear the sparks of electricity. We can scuff our shoes across a rug and feel the tingle as we reach for the doorknob, or do the same when sliding across plastic seat covers while parked in an automobile (Figure 32.7). In all these cases electrons are being transferred by friction when one material rubs against another.

Figure 32.7 ▲
Charging by friction and then by contact while parked at a drive-in movie.

Electrons can be transferred from one material to another by simply touching. When a charged rod is placed in contact with a neutral object, some charge will transfer to the neutral object. This method of charging is simply called *charging by contact.* If the object is a good conductor, the charge will spread to all parts of its surface because the like charges repel each other. If it is a poor conductor, it may be necessary to touch the rod at several places on the object in order to get a more or less uniform distribution of charge.

32.6 Charging by Induction

If we bring a charged object *near* a conducting surface, even without physical contact, electrons will move in the conducting surface. Consider the two insulated metal spheres, A and B, in Figure 32.8. In sketch (a), the uncharged spheres touch each other, so in effect they form a single noncharged conductor. In sketch (b), a negatively charged rod is held near sphere A. Electrons in the metal are repelled by the rod, and excess negative charge has moved onto sphere B, leaving sphere A with excess positive charge. The charge on the two spheres has been redistributed. A charge is said to have been **induced** on the spheres. In sketch (c), spheres A and B are separated while the rod is still present. In sketch (d), the rod has been removed. The spheres are charged equally and oppositely. They have been charged by **induction.** Since the charged rod never touched them, it retains its initial charge.

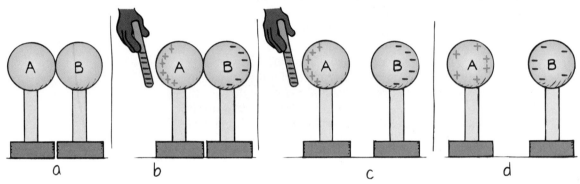

Figure 32.8 ▲
Charging by induction.

A single sphere can be charged similarly if we touch it when the charges are separated by induction. Consider a metal sphere that hangs from a nonconducting string, as shown in Figure 32.9. In sketch (a), the net charge on the metal sphere is zero. In sketch (b), a charge redistribution is induced by the presence of the charged rod. The net charge on the sphere is still zero. In sketch (c), touching the sphere

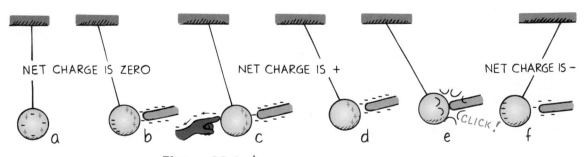

Figure 32.9 ▲
Charge induction by grounding.

removes electrons by contact. In sketch (d), the sphere is left positively charged. In sketch (e), the sphere is attracted to the negative rod; it swings over to it and touches it. Now electrons move onto the sphere from the rod. The sphere has been negatively charged by contact. In sketch (f), the negative sphere is repelled by the negative rod.

When we touch the metal surface with a finger (sketch (c)), charges that repel each other have a conducting path to a practically infinite reservoir for electric charge—the ground. When we allow charges to move off (or onto) a conductor by touching it, it is common to say that we are **grounding** it. Chapter 34 returns to this idea of grounding in the discussion of electric currents.

■ Questions

1. Would the charges induced on spheres A and B of Figure 32.8 necessarily be exactly equal and opposite?

2. Why does the negative rod in Figure 32.8 have the same charge before and after the spheres are charged, but not when charging takes place as in Figure 32.9?

Charging by induction occurs during thunderstorms. The negatively charged bottoms of clouds induce a positive charge on the surface of the earth below. Benjamin Franklin was the first to demonstrate this in his famous kite-flying experiment, in which he proved that lightning is an electrical phenomenon.* Most lightning is an electrical discharge between oppositely charged parts of clouds.

Answers

1. The charges must be equal and opposite on both spheres, because each single positive charge on sphere A is the result of a single electron being taken from A and moved to B. This is like taking bricks from the surface of a brick road and putting them all on the sidewalk. The number of bricks on the sidewalk will be exactly matched by the number of holes in the road. Similarly, the number of extra electrons on sphere B will exactly match the number of "holes" (positive charges) left in sphere A. Remember that the absence of an electron makes a positive charge.

2. In the charging process of Figure 32.8, no contact was made between the negative rod and either of the spheres. In the charging process of Figure 32.9, however, the rod touched the sphere when it was positively charged. A transfer of charge by contact reduced the negative charge on the rod.

* Benjamin Franklin was most fortunate that he was not electrocuted as were others who attempted to duplicate his experiment. In addition to being a great statesman, Franklin was a first-rate scientist. He introduced the terms *positive* and *negative* as they relate to electricity but nevertheless supported the "one-fluid theory" of electric currents. He also contributed to our understanding of grounding and insulation. As Franklin approached the height of his scientific career, a more compelling task was presented to him—helping to form the system of government of the newly independent United States. A less important undertaking would not have kept him from spending more of his energies on his favorite activity—the scientific investigation of nature.

Figure 32.10 ▲
The bottom of the negatively charged cloud induces a positive charge at the surface of the ground below.

The kind we are most familiar with is the electrical discharge between the clouds and the oppositely charged ground below.

Franklin also found that charge flows readily to or from sharp points, and fashioned the first lightning rod. If the rod is placed above a building connected to the ground, the point of the rod collects electrons from the air, preventing a large buildup of positive charge on the building by induction. This continual "leaking" of charge prevents a charge buildup that might otherwise lead to a sudden discharge between the cloud and the building. The primary purpose of the lightning rod, then, is to prevent a lightning discharge from occurring. If for any reason sufficient charge does not leak from the air to the rod, and lightning strikes anyway, it may be attracted to the rod and short-circuited to the ground, thereby sparing the building.

32.7 Charge Polarization

Charging by induction is not restricted to conductors. When a charged rod is brought near an insulator, there are no free electrons to migrate throughout the insulating material. Instead, there is a rearrangement of the positions of charges within the atoms and molecules themselves (Figure 32.11, left). One side of the atom or molecule is induced to be slightly more positive (or negative) than the opposite side. The atom or molecule is said to be **electrically polarized.** If the charged rod is negative, say, then the positive side of the atom or molecule is toward the rod, and the negative side of the atom or molecule is away from it. The atoms or molecules near the surface all become aligned this way (Figure 32.11, right).

This explains why electrically neutral bits of paper are attracted to a charged object. Molecules are polarized in the paper, with the oppositely charged sides of molecules closest to the charged object. Closeness wins, and the bits of paper experience a net attraction.

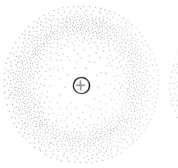
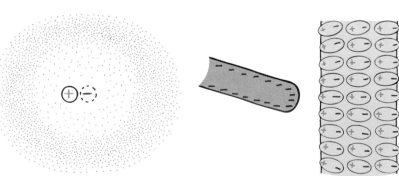

Figure 32.11 ▲
(Left) When an external negative charge is brought closer from the left, the charges within a neutral atom or molecule rearrange so that the left-hand side is slightly more positive and the right-hand side is slightly more negative. (Right) All the atoms or molecules near the surface become electrically polarized.

Charging

Charge a comb by running it through your hair. This will work especially well if the weather is dry. Now bring the comb near some tiny bits of paper. Explain your observations. Next, place the charged comb near a thin stream of running water from a faucet. Is there an electrical interaction between the comb and the stream? Does this mean the stream of water is charged? Why or why not?

Activity

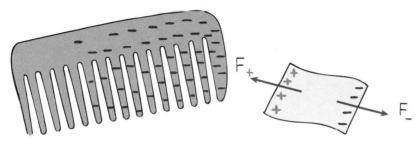

Figure 32.12 ▲
A charged comb attracts an uncharged piece of paper because the force of attraction for the closer charge is greater than the force of repulsion for the farther charge. Closeness wins, and there is a net attraction.

Sometimes they will cling to the charged object and suddenly fly off. This indicates that charging by contact has occurred; the paper bits have acquired the same sign of charge as the charged object and are then repelled.

Rub an inflated balloon on your hair and it becomes charged. Place the balloon against the wall and it sticks, because the charge on the balloon induces an opposite surface charge on the wall. Closeness wins, for the charge on the balloon is slightly closer to the opposite induced charge than to the charge of the same sign (Figure 32.13).

Many molecules—H_2O for example—are electrically polarized in their normal states. The distribution of electric charge is not perfectly even. There is a little more negative charge on one side of the molecule than on the other (Figure 32.14). Such molecules are said to be *electric dipoles*.

Figure 32.13 ▲
The negatively charged balloon polarizes molecules in the wooden wall and creates a positively charged surface, so the balloon sticks to the wall.

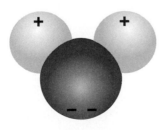

◀ **Figure 32.14**
An H_2O molecule is an electric dipole.

Microwave Cooking

Imagine an enclosure filled with Ping-Pong balls among a few batons, all at rest. Now imagine the batons suddenly flipping back and forth like semi-rotating propellers, striking neighboring Ping-Pong balls. Almost immediately most Ping-Pong balls are energized, vibrating in all directions. A microwave oven works similarly. The batons are water molecules made to flip back and forth in rhythm with microwaves in the enclosure. The Ping-Pong balls are non–water molecules that make up the bulk of material being cooked.

H_2O molecules are polar, with opposite charges on opposite sides. When an electric field is imposed on them, they align with the field like a compass aligns with a magnetic field. When the field is made to oscillate, the H_2O molecules oscillate also—and quite energetically. Food is cooked by a sort of "kinetic friction" as flip-flopping H_2O molecules impart thermal motion to surrounding food molecules. A microwave oven wouldn't work without the presence of the polar molecules in the food. That's why microwaves pass through foam, paper, or ceramic plates with no effect.

In summary, we know that objects are electrically charged in three ways.

1. By friction, when electrons are transferred by friction from one object to another.

2. By contact, when electrons are transferred from one object to another by direct contact without rubbing. A charged rod placed in contact with an uncharged piece of metal, for example, will transfer charge to the metal.

3. By induction, when electrons are caused to gather or disperse by the presence of nearby charge (even without physical contact). A charged rod held near a metal surface, for example, attracts charges of the same sign as those on the rod and repels opposite charges. The result is a redistribution of charge on the object without any change in its net charge. If the metal surface is discharged by contact, with a finger for example, then a net charge will be left.

If the object is an insulator, on the other hand, then a realignment of charge rather than a migration of charge occurs. This is charge polarization, in which the surface near the charged object becomes oppositely charged. This occurs when pieces of neutral paper are attracted to a charged object, or when you stick a charged balloon to a wall.

32 Chapter Review

Concept Summary

All electrons have the same amount of negative charge; all protons have a positive charge equal in magnitude to the negative charge on the electron.

- Electrical forces arise because of the way that like charges repel and unlike charges attract.

- Electric charge is conserved.

- According to Coulomb's law, the electrical force between two charged objects is proportional to the product of the charges and inversely proportional to the square of the distance between them.

Electrons move easily in good conductors and poorly in good insulators.

- Objects become charged when electrons move onto them or off of them.

- Charging by friction occurs when electrons are transferred by rubbing.

- Charging by contact occurs when electrons are transferred by direct contact.

- Charging by induction occurs in the presence of a charge without physical contact.

- Charge polarization occurs in insulators that are in the presence of a charged object.

Important Terms

charge (32.1)
conductor (32.4)
conservation of charge (32.2)
coulomb (32.3)
Coulomb's law (32.3)
electrical force (32.1)
electrically polarized (32.7)
electrostatics (32.0)
grounding (32.6)
induced (32.6)
induction (32.6)
insulator (32.4)
semiconductor (32.4)
superconductor (32.4)

Review Questions

1. Which force—gravitational or electrical—repels as well as attracts? (32.1)

2. Gravitational forces depend on the property called *mass*. What comparable property underlies electrical forces? (32.1)

3. How do protons and electrons differ in their electric charge? (32.1)

4. Is an electron in a hydrogen atom the same as an electron in a uranium atom? (32.1)

5. Which has more mass—a proton or an electron? (32.1)

6. In a normal atom, how many electrons are there compared with protons? (32.1)

7. **a.** How do like charges behave?

   b. How do unlike charges behave? (32.1)

8. How does a negative ion differ from a positive ion? (32.2)

9. **a.** If electrons are rubbed from cat's fur onto a rubber rod, does the rod become positively or negatively charged?

   b. How about the cat's fur? (32.2)

10. What does it mean to say that charge is conserved? (32.2)

11. **a.** How is Coulomb's law similar to Newton's law of gravitation?

    b. How are the two laws different? (32.3)

12. The SI unit of mass is the kilogram. What is the SI unit of charge? (32.3)

13. The proportionality constant k in Coulomb's law is huge in ordinary units, whereas the proportionality constant G in Newton's law of gravitation is tiny. What does this mean in terms of the relative strengths of these two forces? (32.3)

14. Why does the weaker force of gravity dominate over electrical forces for astronomical objects? (32.3)

15. Why do electrical forces dominate between atoms that are close together? (32.3)

16. What is the difference between a good conductor and a good insulator? (32.4)

17. a. Why are metals good conductors?

b. Why are materials such as rubber and glass good insulators? (32.4)

18. What is a semiconductor? (32.4)

19. What is a superconductor? (32.4)

20. a. What are the three main methods of charging objects?

b. Which method involves no touching? (32.5–32.6)

21. What is lightning? (32.6)

22. What is the function of a lightning rod? (32.6)

23. What does it mean to say an object is electrically polarized? (32.7)

24. When a charged object polarizes another, why is there an attraction between the objects? (32.7)

25. What is an electric dipole? (32.7)

Think and Explain

1. Electrical forces between charges are enormous relative to gravitational forces. Yet, we normally don't sense electrical forces between us and our environment, while we do sense our gravitational interaction with the earth. Why is this so?

2. By how much is the electrical force between a pair of ions reduced when their separation distance is doubled? Tripled?

3. If you scuff electrons from your hair onto a comb, are you positively or negatively charged? How about the comb?

4. An electroscope is a simple device. It consists of a metal ball that is attached by a conductor to two fine gold leaves that are protected from air disturbances in a jar, as shown in the sketch. When the ball is touched by a charged object, the leaves that normally hang straight down spring apart. Why? (Electroscopes are useful not only as charge detectors, but also for measuring the amount of charge: the more charge transferred to the ball, the more the leaves diverge.)

5. Would it be necessary for a charged object to actually touch the leaves of an electroscope (see Question 4) for the leaves to diverge? Defend your answer.

6. If a glass rod that is rubbed with a plastic dry cleaner's bag acquires a certain charge, why does the plastic bag have exactly the same amount of opposite charge?

7. Why is a good conductor of electricity also a good conductor of heat?

8. Explain how an object that is electrically neutral can be attracted to an object that is charged.

9. If electrons were positive and protons negative, would Coulomb's law be written the same or differently?

10. The five thousand billion billion freely moving electrons in a penny repel one another. Why don't they fly out of the penny?

33 Electric Fields and Potential

Arc discharge between electrodes and a fork.

The space around a strong magnet is different from how it would be if the magnet were not there. Put a paper clip in the space and you'll see the paper clip move. The space around a black hole is different from how it would be if the black hole were not there. Put yourself in the space, and that will be the last thing you do. Similarly, the space around a concentration of electric charge is different from how it would be if the charge were not there. If you walk by the charged dome of an electrostatic machine—a Van de Graaff generator, for example—you can sense the charge. Hair on your body stands out—just a tiny bit if you're more than a meter away, and more if you're closer. The space that surrounds each of these things—the magnet, the black hole, and the electric charge—is altered. The space is said to contain a *force field*.

33.1 Electric Fields

The force field that surrounds a mass is a gravitational field. If you throw a ball into the air, it follows a curved path. Earlier chapters showed that it curves because there is an interaction between the ball and the earth—between their centers of gravity, to be exact. Their centers of gravity are quite far apart, so this is "action at a distance."

The idea that things not in contact could exert forces on one another bothered Isaac Newton and many others. The concept of a force field eliminates the distance factor. The ball is in contact with the field all the time. We can say the ball curves because it interacts with the earth's gravitational field. It is common to think of distant rockets and space probes as interacting with gravitational fields rather than with the masses of the earth and other astronomical bodies that are responsible for the fields.

Figure 33.1 ▲
You can sense the force field that surrounds a charged Van de Graaff generator.

Just as the space around the earth and every other mass is filled with a gravitational field, the space around every electric charge is filled with an **electric field**—a kind of aura that extends through space. In Figure 33.2, a gravitational force holds a satellite in orbit about a planet, and an electrical force holds an electron in orbit about a proton. In both cases there is no contact between the objects, and the forces are "acting at a distance." Putting this in terms of the field concept, we can say that the orbiting satellite and electron interact with the force fields of the planet and the proton and are everywhere in contact with these fields. In other words, the force that one electric charge exerts on another can be described as the interaction between one charge and the electric field set up by the other.

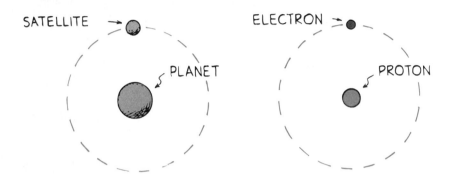

Figure 33.2 ▲
The satellite and the electron both experience forces; they are both in force fields.

An electric field has both magnitude and direction. Its magnitude (strength) can be measured by its effect on charges located in the field. Imagine a small positive "test charge" that is placed in an electric field. Where the force is greatest on the test charge, the field is strongest. Where the force on the test charge is weak, the field is small.*

The direction of the electric field at any point, by convention, is the direction of the electrical force on a small *positive* test charge placed at that point. Thus, if the charge that sets up the field is positive, the field points away from that charge. If the charge that sets up the field is negative, the field points toward that charge. (Be sure to distinguish between the hypothetical small test charge and the charge that sets up the field.)

---

* The strength of an electric field is a measure of the force exerted on a small test charge placed in the field. (The test charge must be small enough that it doesn't push the original charge around and thus alter the field we are trying to measure.) If a test charge q experiences a force F at some point in space, then the electric field E at that point is

$$E = \frac{F}{q}$$

Electric field strength can be measured in units of newtons per coulomb (N/C), or equivalently, volts per meter (V/m).

33.2 Electric Field Lines

Since an electric field has both magnitude and direction, it is a *vector quantity* and can be represented by vectors. The negatively charged particle in Figure 33.3 (left) is surrounded by vectors that point toward the particle. (If the particle were positively charged, the vectors would point away from the particle. The vectors always point in the direction of the force that would act on a positive test charge.) The magnitude of the field is indicated by the length of the vectors. The electric field is greater where the vectors are long than where the vectors are short. To represent a complete electric field by vectors, you would have to show a vector at every point in the space around the charge. Such a diagram would be totally unreadable!

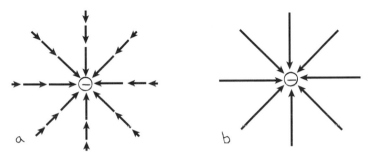

Figure 33.3 ▲
Electric field representations about a negative electric charge. (a) A vector representation; (b) a lines-of-force representation.

A more useful way to describe an electric field is with electric *field lines,* also called *lines of force* (Figure 33.3, right). Where the lines are farther apart, the field is weaker. For an isolated charge, the lines extend to infinity, while for two or more opposite charges, the lines emanate from a positive charge and terminate on a negative charge. Some electric field configurations are shown in Figure 33.4.

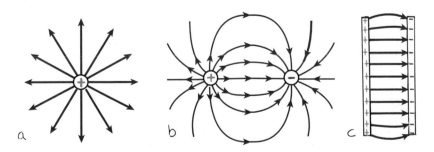

Figure 33.4 ▲
Some electric field configurations. (a) Field lines around a single positive charge. (b) Field lines for a pair of equal but opposite charges. Note that the lines emanate from the positive charge and terminate on the negative charge. (c) Evenly spaced field lines between two oppositely charged parallel plates.

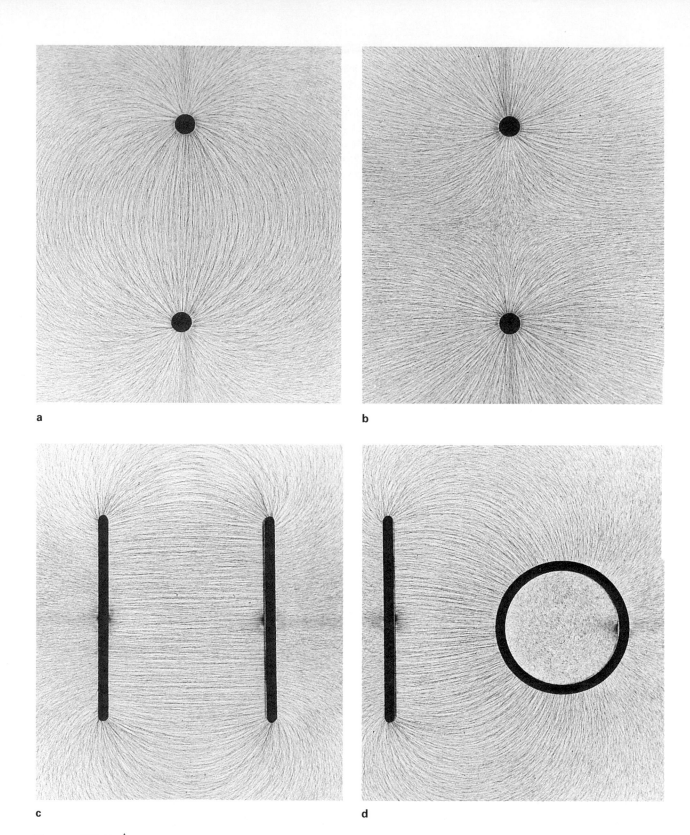

Figure 33.5 ▲
Bits of fine thread suspended in an oil bath surrounding charged conductors line up end to end along the direction of the field. (a) Equal and opposite charges. (b) Equal like charges. (c) Oppositely charged plates. (d) Oppositely charged cylinder and plate.

Photographs of field patterns are shown in Figure 33.5. The photographs show bits of thread that are suspended in an oil bath surrounding charged conductors. The ends of the bits of thread are charged by induction and tend to line up end-to-end with the field lines, like iron filings in a magnetic field. In the two top photos (a) and (b), we see the field lines are characteristic of a single pair of point charges. The oppositely charged parallel plates in (c) produce nearly parallel field lines between the plates. Their combined fields produce the resultant field lines between the plates. Except near the ends, the field between the plates has a constant strength. Notice the threads inside the cylinder (d) are unaligned. There is no electric field in the space inside a conductor. The conductor shields the space from the field outside.

If we were concerned only about the forces produced by isolated point charges, the electric field concept would be of limited use. The force between point charges is described by Coulomb's law. But charges most often are spread out over a wide variety of surfaces. Charges also move. This motion is communicated to neighboring charges by changes in the electric field.

We will see later in this and following chapters that the electric field is a storehouse of energy.

■ Question

Suppose that a beam of electrons is produced at one end of a glass tube and lights up a phosphor screen on the inner surface of the other end. When the beam is straight, it produces a spot of light in the middle of the screen. If the beam passes through the electric field of a pair of oppositely charged plates, it is deflected, say to the left. If the charge on the plates is reversed, in what direction will deflection occur?

33.3 Electric Shielding

The dramatic photo in Figure 33.6 shows a car being struck by lightning. Yet, the occupant inside the car is completely safe. This is because the electrons that shower down upon the car are mutually repelled and spread over the outer metal surface, finally discharging when additional sparks jump from the car's body to the ground. The

Figure 33.6 ▲
Electrons from the lightning bolt mutually repel and spread over the outer metal surface. Although the electric field they set up may be great *outside* the car, the overall electric field *inside* the car practically cancels to zero.

■ Answer

When the charge on the plates is reversed, the electric field will be in the opposite direction, so the electron beam will be deflected to the right. If the field is made to oscillate, the beam will be swept back and forth. With a second set of plates and further refinements it could sweep a picture onto the screen!

configuration of electrons on the car's surface at any moment is such that the electric fields inside the car practically cancel to zero. This is true of any charged conductor. In fact, if the charge on a conductor is not moving, the electric field inside the conductor is *exactly* zero.

The absence of electric field within a conductor holding static charge does not arise from the inability of an electric field to penetrate metals. It comes about because free electrons within the conductor can "settle down" and stop moving only when the electric field is zero. So the charges arrange themselves to ensure a zero field with the material.

As a simple example, consider the charged metal sphere shown in Figure 33.7. Because of mutual repulsion, the electrons have spread as far apart from one another as possible. They distribute themselves uniformly over the surface of the sphere. A positive test charge located exactly in the middle of the sphere would feel no force. The electrons on the left side of the sphere, for example, would tend to pull the test charge to the left, but the electrons on the right side of the sphere would tend to pull the test charge to the right equally hard. The net force on the test charge would be zero. Thus, the electric field is also zero. Interestingly enough, complete cancellation will occur *anywhere* inside the conducting sphere. To show why this is true involves some geometry and is beyond the scope of this text.

If the conductor is not spherical, then the charge distribution will not be uniform. If it is a cube, for example, then most of the charge is located near the corners. The remarkable thing is this: The exact charge distribution over the surfaces and corners of a conducting cube is such that the electric field everywhere inside the cube is zero. Look at it this way: If there were an electric field inside a conductor, then free electrons inside the conductor would be set in motion. How far would they move? Until equilibrium is established, which is to say, when the positions of all the electrons produce a zero field inside the conductor.

There is no way to shield gravity, because gravity only attracts. There are no repelling parts of gravity to offset attracting parts. Shielding electric fields, however, is quite simple. Surround yourself

Figure 33.7 ▲
The forces on a test charge located inside a charged hollow sphere cancel to zero.

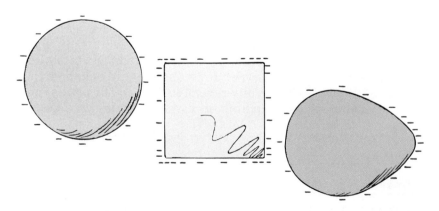

Figure 33.8 ▲
Static charges are distributed on the surface of all conductors in such a way that the electric field inside the conductors is zero.

or whatever you wish to shield with a conducting surface. Put this surface in an electric field of whatever field strength. The free charges in the conducting surface will arrange themselves on the surface of the conductor in a way such that all field contributions inside cancel one another. That's why certain electronic components are encased in metal boxes, and why certain cables have a metal covering—to shield them from all outside electrical activity.

Figure 33.9 ▲
The metal-lined cover shields the internal electrical components from external electric fields. Similarly, a metal cover shields the coaxial cable.

■ Question

It is said that a gravitational field, unlike an electric field, cannot be shielded. But the gravitational field at the center of the earth cancels to zero. Isn't this evidence that a gravitational field *can* be shielded?

33.4 Electric Potential Energy

Recall from Chapter 8 the relation between work and potential energy. Work is done when a force moves something in the direction of the force. An object has potential energy by virtue of its location, say in a force field. For example, if you lift an object, you apply a force equal to its weight. When you raise it through some distance, you are doing work on the object. You are also increasing its gravitational potential energy. The greater the distance it is raised, the greater is the increase in its gravitational potential energy. Doing work increases its gravitational potential energy (Figure 33.10).

 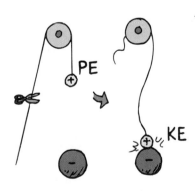

◀ **Figure 33.10**
(Left) Work is done to lift the ram of the pile driver against the gravitational field of the earth. In an elevated position, the ram has gravitational potential energy. When released, this energy is transferred to the pile below. (Right) Similar energy transfer occurs for electric charges.

■ Answer

No. Gravity can be canceled inside a planet or between planets, but it cannot be *shielded* by a planet or by any arrangement of masses. During a lunar eclipse, for example, when the earth is directly between the sun and the moon, there is no shielding of the sun's field to affect the moon's orbit. Even a very slight shielding would accumulate over a period of years and show itself in the timing of subsequent eclipses. Shielding requires a combination of repelling and attracting forces, and gravity only attracts.

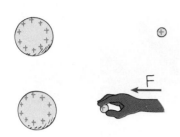

Figure 33.11 ▲
The small positive charge has more potential energy when it is closer to the positively charged sphere because work is required to move it to the closer location.

In a similar way, a charged object can have potential energy by virtue of its location in an electric field. Just as work is required to lift an object against the gravitational field of the earth, work is required to push a charged particle against the electric field of a charged body. (It may be more difficult to visualize, but the physics of both the gravitational case and the electrical case is the same.) The electric potential energy of a charged particle is increased when work is done to push it against the electric field of something else that is charged.

Figure 33.11 (top) shows a small positive charge located at some distance from a positively charged sphere. If we push the small charge closer to the sphere (Figure 33.11, bottom), we will expend energy to overcome electrical repulsion. Just as work is done in compressing a spring, work is done in pushing the charge against the electric field of the sphere. This work is equal to the energy gained by the charge. The energy the charge now possesses by virtue of its location is called **electric potential energy.** If the charge is released, it will accelerate in a direction away from the sphere, and its electric potential energy will transform into kinetic energy.

33.5 Electric Potential

If in the preceding discussion we push two charges instead, we do twice as much work. The two charges in the same location will have twice the electric potential energy as one; three charges will have three times the potential energy; a group of ten charges will have ten times the potential energy; and so on.

Rather than deal with the total potential energy of a group of charges, it is convenient when working with electricity to consider the *electric potential energy per charge.* The electric potential energy per charge is the total electric potential energy divided by the amount of charge. At any location the potential energy *per charge*—whatever the amount of charge—will be the same. For example, an object with ten units of charge at a specific location has ten times as much energy as an object with a single unit of charge. But it also has ten times as much charge, so the potential energy per charge is the same. The concept of electric potential energy per charge has a special name, **electric potential.**

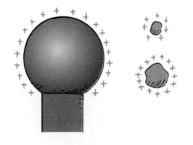

Figure 33.12 ▲
An object of greater charge has more electric potential energy in the field of the charged dome than an object of less charge, but the *electric potential* of any amount of charge at the same location is the same.

$$\text{electric potential} = \frac{\text{electric potential energy}}{\text{charge}}$$

The SI unit of measurement for electric potential is the **volt,** named after the Italian physicist Allesandro Volta (1745–1827). The symbol for volt is V. Since potential energy is measured in joules and charge is measured in coulombs,

$$1 \text{ volt} = 1 \frac{\text{joule}}{\text{coulomb}}$$

■ **Question**

If there were twice as much charge on one of the charged objects near the charged sphere in Figure 33.12, would the electric potential energy of the object in the field of the charged sphere be the same or would it be twice as great? Would the electric potential of the object be the same or would it be twice as great?

Thus, a potential of 1 volt equals 1 joule of energy per coulomb of charge; 1000 volts equals 1000 joules of energy per coulomb of charge. If a conductor has a potential of 1000 volts, it would take 1000 joules of energy per coulomb to bring a small charge from very far away and add it to the charge on the conductor.* (Since the small charge would be much less than one coulomb, the energy required would be much less than 1000 joules. For example, to add the charge of one proton to the conductor, 1.6×10^{-19} C, it would take only 1.6×10^{-16} J of energy.)

Since electric potential is measured in volts, it is commonly called **voltage.** In this book the names will be used interchangeably. The significance of voltage is that once the location of zero voltage has been specified, a definite value for it can be assigned to a location whether or not a charge exists at that location. We can speak about the voltages at different locations in an electric field whether or not any charges occupy those locations.

Rub a balloon on your hair and the balloon becomes negatively charged, perhaps to several thousand volts! If the charge on the balloon were one coulomb, it would take several thousand joules of energy to give the balloon that voltage. However, one coulomb is a very large amount of charge; the charge on a balloon rubbed on hair is typically much less than a millionth of a coulomb. Therefore, the amount of energy associated with the charged balloon is very, very small—about a thousandth of a joule. A high voltage requires great energy only if a great amount of charge is involved. This example highlights the difference between electric potential energy and electric potential.

■ **Answer**

Twice as much charge would cause the object to have twice as much electric potential energy, because it would have taken twice as much work to put the object at that location. But the electric potential would be the same, because the electric potential is total electric potential energy divided by total charge. In this case, twice the energy divided by twice the charge gives the same value as the original energy divided by the original charge. Electric potential is not the same thing as electric potential energy. Be sure you understand this before you study further.

---

* It is common practice to assign a zero electric potential to places infinitely far away from any charges. As the next chapter discusses, for electric currents the value zero is assigned to the potential of the ground.

Figure 33.13 ▲
Although the voltage of the charged balloon is high, the electric potential energy is low because of the small amount of charge.

33.6 Electric Energy Storage

Electrical energy can be stored in a common device called a **capacitor.** Capacitors are found in nearly all electronic circuits. Computer motherboards use low-energy capacitors as on-off switches. Computer keyboards have them beneath each key. Capacitors in photoflash units store larger amounts of energy slowly and release it rapidly during the short duration of the flash. Similarly, but on a grander scale, enormous amounts of energy are stored in banks of capacitors that power giant lasers in national laboratories.

The simplest capacitor is a pair of conducting plates separated by a small distance, but not touching each other. When the plates are connected to a charging device such as the battery shown in Figure 33.14, charge is transferred from one plate to the other. This occurs as the positive battery terminal pulls electrons from the plate connected to it. These electrons in effect are pumped through the battery and through the negative terminal to the opposite plate. The capacitor plates then have equal and opposite charges—the positive plate is connected to the positive battery terminal, and the negative plate is connected to the negative battery terminal. The charging process is complete when the potential difference between the plates equals the potential difference between the battery terminals—the battery voltage. The greater the battery voltage and the larger and closer the plates, the greater the charge that is stored. In practice, the plates may be thin metallic foils separated by a thin sheet of paper. This "paper sandwich" is then rolled up to save space and may be inserted into a cylinder. Such a practical capacitor is shown with others in Figure 33.15. (We will consider the role of capacitors in circuits in the next chapter.)

A charged capacitor is discharged when a conducting path is provided between the plates. Discharging a capacitor can be a

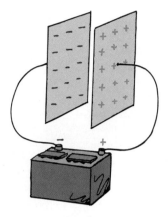

Figure 33.14 ▲
Capacitor consisting of two closely spaced metal parallel plates. When connected to a battery, the plates become equally and oppositely charged. The voltage between the plates then matches the voltage difference between the battery terminals.

Figure 33.15 ▶
Practical capacitors.

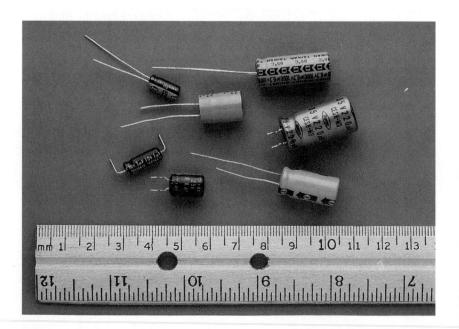

shocking experience if you happen to be the conducting path. The energy transfer that occurs can be fatal where high voltages are present, such as the power supply in a TV set—even after the set has been turned off. That's the main reason for the warning signs on such devices.

The energy stored in a capacitor comes from the work required to charge it. The energy is in the form of the electric field between its plates. Between parallel plates the electric field is uniform, as indicated in Figures 33.4c and 33.5c on previous pages. So the energy stored in a capacitor is energy stored in the electric field.

Electric fields are storehouses of energy. We will see in the next chapter that energy can be transported over long distances by electric fields, which can be directed through and guided by metal wires or directed through empty space. Then in Chapter 37 we will see how energy from the sun is radiated in the form of electric and magnetic fields. The fact that energy is contained in electric fields is truly far reaching.

33.7 The Van de Graaff Generator

A common laboratory device for building up high voltages is the *Van de Graaff generator*. This is the lightning machine often used by "evil scientists" in old science fiction movies. A simple model of the Van de Graaff generator is shown in Figure 33.16. A large hollow metal sphere is supported by a cylindrical insulating stand. A

LINK TO TECHNOLOGY

Ink-jet Printers

The printhead of an ink-jet printer typically ejects a thin, steady stream of thousands of tiny ink droplets each second as it shuttles back and forth across the paper. As the stream flows between electrodes that are controlled by the computer, selective droplets are charged. The uncharged droplets then pass undeflected in the electric field of a parallel plate capacitor and form the image on the page; the charged droplets are deflected and do not reach the page. Thus, the image produced on the paper is made from ink droplets that are *not* charged. The blank spaces correspond to deflected ink that never made it to the paper.

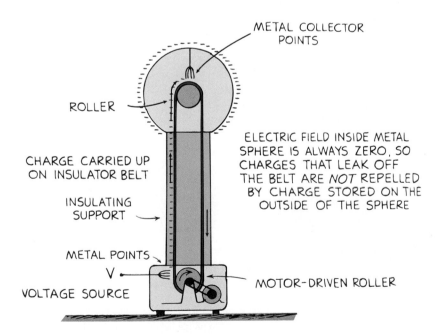

METAL COLLECTOR POINTS

ROLLER

CHARGE CARRIED UP ON INSULATOR BELT

INSULATING SUPPORT

METAL POINTS

V

VOLTAGE SOURCE

ELECTRIC FIELD INSIDE METAL SPHERE IS ALWAYS ZERO, SO CHARGES THAT LEAK OFF THE BELT ARE *NOT* REPELLED BY CHARGE STORED ON THE OUTSIDE OF THE SPHERE

MOTOR-DRIVEN ROLLER

Figure 33.16 ▲
A simple model of a Van de Graaff generator.

motor-driven rubber belt inside the support stand moves past a comblike set of metal needles that are maintained at a high electric potential. A continuous supply of electrons is deposited on the belt through electric discharge by the points of the needles and is carried up into the hollow metal sphere. The electrons leak onto metal points (which act like tiny lightning rods) attached to the inner surface of the sphere. Because of mutual repulsion, the electrons move to the outer surface of the conducting sphere. (Remember, static charge on any conductor is on the outside surface.) This leaves the inside surface uncharged and able to receive more electrons as they are brought up the belt. The process is continuous, and the charge builds up to a very high electric potential—on the order of millions of volts.

A sphere with a radius of 1 m can be raised to a potential of 3 million volts before electrical discharge occurs through the air (because breakdown occurs in air when the electric field strength is about 3×10^6 V/m).* The voltage can be further increased by increasing the radius of the sphere or by placing the entire system in a container filled with high-pressure gas. Van de Graaff generators can produce voltages as high as 20 million volts. These devices accelerate charged particles used as projectiles for penetrating the nuclei of atoms. Touching one can be a hair-raising experience (Figure 33.17).

Figure 33.17 ▶
The physics enthusiast and the dome of the Van de Graaff generator are charged to a high voltage. Why does her hair stand out?

---

* The electric field strength for arc discharge is directly proportional to the radius of the conducting sphere. Hence we see that the large-radius sphere of a Van de Graaff generator holds considerable charge before arc discharge occurs, while the sharp points of lightning rods readily leak charge. The sharp points behave as small-radius spheres. The greater the radius, the greater the charge that can be stored before arc discharge occurs.

33 Chapter Review

Concept Summary

An electric field fills the space around every electric charge.

- The field is strongest where it would exert the greatest electrical force on a test charge.

- The direction of the field at any point is the direction of the electrical force on a positive test charge.

- An electric field can be represented by electric field lines.

- Static charge occupies only the outer surface of a conductor; inside the conductor the electric field is zero.

- An electric field is a storehouse of energy.

A charged object has electric potential energy by virtue of its location in an electric field.

- The electric potential, or voltage, at any point in an electric field is the electric potential energy per charge for a charged object at that point.

- A zero of potential must be specified; it is often at infinite distance from charges.

- A capacitor is a device for storing charge and energy.

Important Terms

capacitor (33.6)
electric field (33.1)
electric potential (33.5)
electric potential energy (33.4)
volt (33.5)
voltage (33.5)

Review Questions

1. What is meant by the expression *action at a distance*? (33.1)

2. How does the concept of a field eliminate the idea of action at a distance? (33.1)

3. How are a gravitational and an electric field similar? (33.1)

4. Why is an electric field considered a vector quantity? (33.2)

5. **a.** What are electric field lines?

   b. How do their directions compare with the direction of the force that acts on a positive test charge in the same region? (33.2)

6. How is the strength of an electric field indicated with field lines? (33.2)

7. How do the electric field lines appear when the field has the same strength at all points in a region? (33.2)

8. Why are occupants inside a car struck by lightning safe? (33.3)

9. What is the size of the electric field inside any charged conductor? (33.3)

10. **a.** Can gravity be shielded?

    b. Can electric fields be shielded? (33.3)

11. What is the relationship between the amount of work you do on an object and its potential energy? (33.4)

12. How can the electric potential energy of a charged particle in an electric field be increased? (33.4)

13. What will happen to the electric potential energy of a charged particle in an electric field when the particle is released and free to move? (33.4)

14. Clearly distinguish between *electric potential energy* and *electric potential*. (33.5)

15. If you do more work to move more charge a certain distance against an electric field, and increase the electric potential energy as a result, why do you not also increase the electric potential (33.5)?

16. The SI unit for electric potential energy is the joule. What is the SI unit for electric potential? (33.5)

17. Charge must be present at a location in order for there to be electric potential energy. Must charge also be present at a location for there to be electric potential? (33.5)

18. How can electric potential be high when electric potential energy is relatively low? (33.5)

19. How does the amount of charge on the inside surface of the sphere of a charged Van de Graaff generator compare with the amount on the outside? (33.7)

20. How much voltage can be built up on a Van de Graaff generator of 1 m radius before electrical discharge occurs through the air? (33.7)

5. When a conductor is charged, the charge moves to the outer surface of the conductor. Why is this so?

6. Suppose that a metal file cabinet is charged. How will the charge concentration at the corners of the cabinet compare with the charge concentration on the flat parts of the cabinet? Defend your answer.

7. Is it correct to say that an object with twice the electric potential as another has twice the electric potential energy? Defend your answer.

8. You are not harmed by contact with a charged balloon, even though its voltage is very high. Is the reason for this similar to why you are not harmed by the greater-than-1000°C sparks from a Fourth of July–type sparkler?

9. What is the net charge of a capacitor? Explain.

10. Why does your hair stand out when you are charged by a device such as a Van de Graaff generator?

Think and Explain

1. How is an electric field different from a gravitational field?

2. The vectors for the gravitational field of the earth point *toward* the earth; the vectors for the electric field of a proton point *away* from the proton. Explain.

3. Imagine a "free" electron and "free" proton held midway between the plates of a charged parallel plate capacitor. When released, how do their accelerations and directions of travel compare? If we ignore their attraction to each other, which reaches a capacitor plate first?

4. Suppose that the strength of the electric field about an isolated point charge has a certain value at a distance of 1 m. How will the electric field strength compare at a distance of 2 m from the point charge? What law guides your answer?

Think and Solve

1. **a.** If you do 12 J of work to push 0.001 C of charge from point A to point B in an electric field, what is the voltage difference between points A and B?

   b. When the charge is released, what will be its kinetic energy as it flies back past its starting point A? What principle guides your answer?

2. **a.** Suppose that you start with a charge of 0.002 C, twice the charge of the previous example, and find that it takes 24 J of work to move it from point A to point B. Now what is the voltage difference between points A and B?

   b. If this charge is released, what will be its kinetic energy as it flies back past point A?

34 Electric Current

Negligible potential difference across the birds' feet.

The previous chapter discussed the concept of electric potential, or voltage. This chapter will show that voltage is an "electric pressure" that can produce a flow of charge, or *current,* within a conductor. The flow is restrained by the *resistance* it encounters. When the flow takes place along one direction, it is called *direct current* (dc); when it flows to and fro, it is called *alternating current* (ac). The rate at which energy is transferred by electric current is *power.* You'll note here that there are many terms to be sorted out. This is easier (and more meaningful) to do when you have some understanding of the ideas these terms represent. In turn, the ideas are better understood if you know how they relate to one another. Let's begin with the flow of electric charge.

34.1 Flow of Charge

Recall in your study of heat and temperature that heat flows through a conductor when a difference in temperature exists across its ends. Heat flows from the end of higher temperature to the end of lower temperature. When both ends reach the same temperature, the flow of heat ceases.

In a similar way, when the ends of an electric conductor are at different electric potentials, charge flows from one end to the other. Charge flows when there is a **potential difference,** or difference in potential (voltage), across the ends of a conductor. The flow of charge will continue until both ends reach a common potential. When there is no potential difference, there is no longer a flow of charge through the conductor.

As an example, if one end of a wire were connected to the ground and the other end placed in contact with the sphere of a Van de Graaff generator that is charged to a high potential, a surge of charge would flow through the wire. The flow would be brief, however, for the sphere of the generator would quickly reach a common potential with the ground.

To attain a sustained flow of charge in a conductor, some arrangement must be provided to maintain a difference in potential while charge flows from one end to the other. The situation is analogous to the flow of water from a higher reservoir to a lower one (Figure 34.1, left). Water will flow in a pipe that connects the reservoirs only as long as a difference in water level exists. (This is implied in the saying "Water seeks its own level.") The flow of water in the pipe, like the flow of charge in the wire that connects the Van de Graaff generator to the ground, will cease when the pressures at each end are equal. In order that the flow be sustained, there must be a suitable pump of some sort to maintain a difference in water levels (Figure 34.1, right). Then there will be a continual difference in water pressures and a continual flow of water. The same is true of electric current.

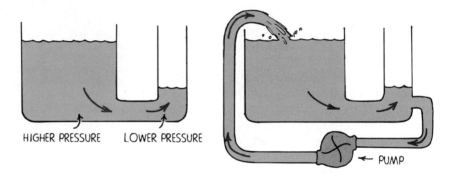

Figure 34.1 ▲
(Left) Water flows from the higher-pressure end of the pipe to the lower-pressure end. The flow will cease when the difference in pressure ceases. (Right) Water continues to flow because a difference in reservoir level is maintained with the pump.

34.2 Electric Current

Electric current is simply the flow of electric charge. In solid conductors the electrons carry the charge through the circuit because they are free to move throughout the atomic network. These electrons are called *conduction electrons*. Protons, on the other hand, are bound inside atomic nuclei that are more or less locked in fixed positions. In fluids, such as the electrolyte in a car battery, positive and negative ions as well as electrons may compose the flow of electric charge.

Electric current is measured in **amperes,** for which the SI unit is symbol A.* An ampere is the flow of 1 coulomb of charge per second. (Recall that 1 coulomb, the standard unit of charge, is the electric charge of 6.25 billion billion electrons.) In a wire that carries a current of 5 amperes, for example, 5 coulombs of charge pass any cross section in the wire each second. So that's a lot of electrons! In a wire that carries 10 amperes, twice as many electrons pass any cross section each second.

Note that a current-carrying wire does not have a *net* electric charge. While the current is flowing, negative electrons swarm through the atomic network that is composed of positively charged atomic nuclei. Under ordinary conditions, the number of electrons in the wire is equal to the number of positive protons in the atomic nuclei. When electrons flow in a wire, the number entering one end is the same as the number leaving the other. The net charge of the wire is normally zero at every moment.

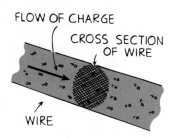

Figure 34.2 ▲
When the rate of flow of charge past any cross section is 1 coulomb (6.25 billion billion electrons) per second, the current is 1 ampere.

34.3 Voltage Sources

Charges do not flow unless there is a potential difference. A sustained current requires a suitable "electric pump" to provide a sustained potential difference. Something that provides a potential difference is known as a **voltage source.** If you charge a metal sphere positively, and another negatively, you can develop a large voltage between them. This voltage source is not a good electric pump because when the spheres are connected by a conductor, the potentials equalize in a single brief surge of moving charges. It is not practical. Dry cells, wet cells, and generators, however, are capable of maintaining a steady flow. (A battery is just two or more cells connected together.)

Dry cells, wet cells, and generators supply energy that allows charges to move. In dry cells and wet cells, energy released in a chemical reaction occurring inside the cell is converted to electric energy.** Generators—such as the alternators in automobiles—convert mechanical energy to electric energy, as discussed in Chapter 37. The electric potential energy produced by whatever means is available at the terminals of the cell or generator. The potential energy per coulomb of charge available to electrons moving between terminals is the voltage (sometimes called the *electromotive force*, or *emf*). The voltage provides the "electric pressure" to move electrons between the terminals in a circuit.

---

* The SI symbol for ampere is A. However, an older symbol still in common usage is amp. People often speak of a current of, say, "5 amps."

** A description of the chemical reactions inside dry cells and wet cells can be found in almost any chemistry textbook.

Figure 34.3 ▲
Each coulomb of charge that is made to flow in a circuit that connects the ends of this 1.5-volt flashlight cell is energized with 1.5 joules.

Power utilities use large electric generators to provide the 120 volts delivered to home outlets. The alternating potential difference between the two holes in the outlet averages 120 volts. When the prongs of a plug are inserted into the outlet, an average electric "pressure" of 120 volts is placed across the circuit connected to the prongs. This means that 120 joules of energy is supplied to each coulomb of charge that is made to flow in the circuit.

There is often some confusion between charge flowing *through* a circuit and voltage being impressed *across* a circuit. To distinguish between these ideas, consider a long pipe filled with water. Water will flow *through* the pipe if there is a difference in pressure *across* or between its ends. Water flows from the high-pressure end to the low-pressure end. Only the water flows, not the pressure. Similarly, you say that charges flow *through* a circuit because of an applied voltage *across* the circuit.* You don't say that voltage flows through a circuit. Voltage doesn't go anywhere, for it is the charges that move. Voltage causes current.

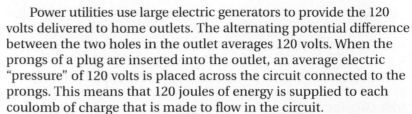

34.4 Electric Resistance

The amount of current that flows in a circuit depends on the voltage provided by the voltage source. The current flow also depends on the resistance that the conductor offers to the flow of charge—the **electric resistance.** This is similar to the rate of water flow in a pipe, which depends not only on the pressure difference between the ends of the pipe but on the resistance offered by the pipe itself. The resistance of a wire depends on the *conductivity* of the material used in the wire (that is, how well it conducts) and also on the thickness and length of the wire.

Thick wires have less resistance than thin wires. Longer wires have more resistance than short wires. In addition, electric resistance depends on temperature. The greater the jostling about of atoms within the conductor, the greater resistance the conductor offers to the flow of charge. For most conductors, increased temperature means increased resistance.** The resistance of some materials becomes zero at very low temperatures. These are the superconductors discussed briefly in Chapter 32.

Figure 34.4 ▲
For a given pressure, more water passes through a large pipe than a small one. Similarly, for a given voltage, more electric current passes through a large-diameter wire than a small-diameter one.

---

* It is conceptually simpler to say that current flows through a circuit, but don't say this around somebody who is "picky" about grammar, for the expression *current flows* is redundant. More properly, charge flows, which *is* current.

** Carbon is an interesting exception. At high temperatures, electrons are shaken from the carbon atom, which increases electric current. Carbon's resistance decreases with increasing temperature. This behavior, along with its high melting temperature, accounts for the use of carbon in arc lamps.

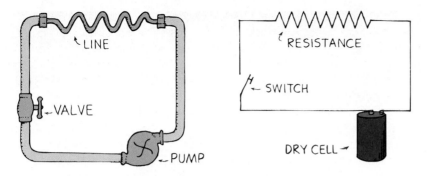

Figure 34.5 ▲
Analogy between a simple hydraulic circuit and an electric circuit.

Electric resistance is measured in units called **ohms,*** after Georg Simon Ohm, a German physicist who tested different wires in circuits to see what effect the resistance of the wire had on the current.

34.5 Ohm's Law

Ohm discovered that the current in a circuit is directly proportional to the voltage impressed across the circuit, and is inversely proportional to the resistance of the circuit. In short,

$$\text{current} = \frac{\text{voltage}}{\text{resistance}}$$

This relationship between voltage, current, and resistance is called **Ohm's law.****

The relationship between the units of measurement for these three quantities is

$$1 \text{ ampere} = 1 \frac{\text{volt}}{\text{ohm}}$$

So for a given circuit of constant resistance, current and voltage are proportional. This means that you'll get twice the current for twice the voltage. The greater the voltage, the greater the current. But if the resistance is doubled for a circuit, the current will be half what it would be otherwise. The greater the resistance, the less the current. Ohm's law makes good sense.

---

*   The Greek letter capital Ω (omega) is usually used as a symbol for ohm.

** Many texts use V for voltage, I for current, and R for resistance, and express Ohm's law as $V = IR$. It then follows that $I = V/R$, or $R = V/I$, so if any two variables are known, the third can be found.

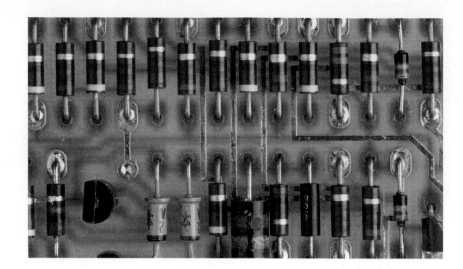

Figure 34.6 ▲
Resistors. The stripes are color coded to indicate the resistance in ohms.

Using specific values, a potential difference of 1 volt impressed across a circuit that has a resistance of 1 ohm will produce a current of 1 ampere. If a voltage of 12 volts is impressed across the same circuit, the current will be 12 amperes.

The resistance of a typical lamp cord is much less than 1 ohm, while a typical lightbulb has a resistance of about 100 ohms. An iron or electric toaster has a resistance of 15 to 20 ohms. The low resistance permits a large current, which produces considerable heat. Inside electric devices such as radio and television receivers, the current is regulated by circuit elements called *resistors*, whose resistance may range from a few ohms to millions of ohms.

■ Questions

1. What is the resistance of an electric frying pan that draws 12 amperes of current when connected to a 120-volt circuit?

2. How much current is drawn by a lamp that has a resistance of 100 ohms when a voltage of 50 volts is impressed across it?

■ Answers

1. The resistance is 10 ohms.

$$\text{resistance} = \frac{\text{voltage}}{\text{current}} = \frac{120 \text{ volts}}{12 \text{ amperes}} = 10 \text{ ohms}$$

An electric device is said to *draw* current when voltage is impressed across it, just as water is said to be drawn from a well or a faucet. In this sense, to draw is not to attract, but to *obtain*.

2. The current is 0.5 ampere.

$$\text{current} = \frac{\text{voltage}}{\text{resistance}} = \frac{50 \text{ volts}}{100 \text{ ohms}} = 0.5 \text{ ampere}$$

Electrolysis

Electrochemistry is about electric energy and chemical change. Molecules in a liquid can be broken apart and separated by the action of electric current. This is *electrolysis*. A common example is passing an electric current through water, separating water into its hydrogen and oxygen components. This common process is also at work when a car battery is recharged. Electrolysis is also used to produce metals from ores. Aluminum is a familiar metal produced by electrolysis. Aluminum is common today, but before the advent of its production by electrolysis in 1886, aluminum was much more expensive than silver and gold!

34.6 Ohm's Law and Electric Shock

What causes electric shock in the human body—current or voltage? The damaging effects of shock are the result of current passing through the body. From Ohm's law, we can see that this current depends on the voltage applied, and also on the electric resistance of the human body.

The resistance of your body depends on its condition and ranges from about 100 ohms if you're soaked with salt water to about 500 000 ohms if your skin is very dry. If you touched the two electrodes of a battery with dry fingers, the resistance your body would normally offer to the flow of charge would be about 100 000 ohms. You usually would not feel 12 volts, and 24 volts would just barely tingle. If your skin were moist, on the other hand, 24 volts could be

VOLTAGE SUPPLIES THE PUSH

RESISTANCE OPPOSES THE PUSH

CURRENT RESULTS!

Table 34.1	
Effect of Various Electric Currents on the Body	
Current in amperes	**Effect**
0.001	Can be felt
0.005	Painful
0.010	Involuntary muscle contractions (spasms)
0.015	Loss of muscle control
0.070	If through the heart, serious disruption; probably fatal if current lasts for more than 1 second

■ Questions

1. If the resistance of your body were 100 000 ohms, what would be the current in your body when you touched the terminals of a 12-volt battery?

2. If your skin were very moist so that your resistance was only 1000 ohms, and you touched the terminals of a 24-volt battery, how much current would you draw?

■ Answers

1. The current in your body, quite harmless, would be

$$\text{current} = \frac{\text{voltage}}{\text{resistance}} = \frac{12 \text{ V}}{100\ 000 \ \Omega} = 0.000\ 12 \text{ A}$$

2. You would draw $\dfrac{24 \text{ V}}{1000 \ \Omega}$, or 0.024 A, a dangerous amount of current!

Figure 34.7 ▲
Handling a wet hair dryer can be like sticking your fingers into a live socket.

Figure 34.8 ▲
The bird can stand harmlessly on one wire of high potential, but it had better not reach over and grab a neighboring wire! Why not?

Figure 34.9 ▲
The third prong connects the body of the appliance directly to ground. Any charge that builds up on an appliance is therefore conducted to the ground.

quite uncomfortable. Table 34.1 describes the effects of different amounts of current on the human body.

Many people are killed each year by current from common 120-volt electric circuits. If you touch a faulty 120-volt light fixture with your hand while you are standing on the ground, there is a 120-volt "electric pressure" between your hand and the ground. The soles of your shoes normally provide a very large resistance between your feet and the ground, so the current would probably not be enough to do serious harm. But if you are standing barefoot in a wet bathtub connected through its plumbing to the ground, the resistance between you and the ground is very small. Your overall resistance is lowered so much that the 120-volt potential difference may produce a harmful current through your body.

Drops of water that collect around the on-off switches of devices such as a hair dryer can conduct current to the user. Although distilled water is a good insulator, the ions in ordinary water greatly reduce the electric resistance. These ions are contributed by dissolved materials, especially salts. There is usually a layer of salt left from perspiration on your skin, which when wet lowers your skin resistance to a few hundred ohms or less. Handling electric devices while taking a bath is extremely dangerous.

You have seen birds perched on high-voltage wires. Every part of their bodies is at the same high potential as the wire, and they feel no ill effects. For the bird to receive a shock, there must be a *difference* in electric potential between one part of its body and another part. Most of the current will then pass along the path of least electric resistance connecting these two points.

Suppose you fell from a bridge and managed to grab onto a high-voltage power line, halting your fall. So long as you touch nothing else of different potential, you will receive no shock at all. Even if the wire is thousands of volts above ground potential and even if you hang by it with two hands, no charge will flow from one hand to the other. This is because there is no appreciable difference in electric potential between your hands. If, however, you reach over with one hand and grab onto a wire of different potential, ZAP!!

Mild shocks occur when the surfaces of electric appliances are at an electric potential different from that of the surfaces of other nearby devices. If you touch surfaces of different potentials, you become a pathway for current. Sometimes the effect is more than mild. To prevent this problem, the outsides of electric appliances are connected to a ground wire, which is connected to the round third prong of a three-wire electric plug (Figure 34.9). All ground wires in all plugs are connected together through the wiring system of the house. The two flat prongs are for the current-carrying double wire. If the live wire accidentally comes in contact with the metal surface of an appliance, the current will be directed to ground rather than shocking you if you handle it.

One effect of electric shock is to overheat tissues in the body or to disrupt normal nerve functions. It can upset the nerve center that controls breathing. In rescuing victims, the first thing to do is clear them from the electric supply with a wooden stick or some other nonconductor so that you don't get electrocuted yourself. Then apply artificial respiration.

34.7 Direct Current and Alternating Current

Electric current may be *dc* or *ac*. By *dc*, we mean **direct current,** which refers to a flow of charge that *always flows in one direction.* A battery produces direct current in a circuit because the terminals of the battery always have the same sign of charge. Electrons always move through the circuit in the same direction, from the repelling negative terminal and toward the attracting positive terminal. Even if the current moves in unsteady pulses, so long as it moves in one direction only, it is dc.

Alternating current (ac) acts as the name implies. Electrons in the circuit move first in one direction and then in the opposite direction, alternating back and forth about relatively fixed positions. This is accomplished by alternating the polarity of voltage at the generator or other voltage source. Nearly all commercial ac circuits in North America involve voltages and currents that alternate back and forth at a frequency of 60 cycles per second. This is 60-hertz current. In some places, 25-hertz, 30-hertz, or 50-hertz current is used.

Voltage of ac in North America is normally 120 volts.* In the early days of electricity, higher voltages burned out the filaments of electric lightbulbs. Tradition has it that 110 volts was settled on because it made bulbs of the day glow as brightly as a gas lamp. So the hundreds of power plants built in the United States prior to 1900 adopted 110 volts (or 115 or 120 volts) as their standard. By the time electricity became popular in Europe, engineers had figured out how to make lightbulbs that would not burn out so fast at higher voltages. Power transmission is more efficient at higher voltages, so Europe adopted 220 volts as their standard. The United States stayed with 110 volts (today officially 120 volts) because of the installed base of 110-volt equipment.

Although lamps in an American home operate on 110–120 volts, many electric stoves and other power-hungry appliances operate on 220–240 volts. How is this possible? Because most electric service in the United States is three-wire: one wire at 120 volts positive, one

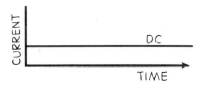

Figure 34.10 ▲
Direct current (dc) does not change direction over time. Alternating current (ac) cycles back and forth.

■ **Answer**

The initial *cause* is the voltage, but it is the current that does most of the damage.

* If you study electricity further you'll learn that 120 volts refers to the "root-mean-square" average of the voltage. The actual voltage in a 120-volt ac circuit varies between +170-volt and –170-volt peaks. It delivers the same power to an iron or a toaster as a 120-volt dc circuit.

wire at zero volts (neutral), and the other wire at a negative 120 volts. This is ac, with the positive and negative alternating at 60 hertz. A wire that is positive at one instant is negative 1/120 of a second later. Most home appliances are connected between the neutral wire and either of the other two wires, producing 120 volts. When the plus-120 is connected to the minus-120,* a 240-volt jolt is produced—just right for electric stoves, air conditioners, and clothes dryers.

The popularity of ac arises from the fact that electric energy in the form of ac can be transmitted great distances with easy voltage step-ups that result in lower heat losses in the wires. Why this is so will be discussed in Chapter 37.

The primary use of electric current, whether dc or ac, is to transfer energy quietly, flexibly, and conveniently from one place to another.

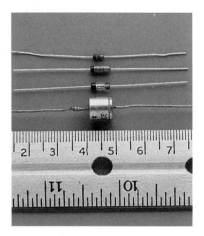

Figure 34.11 ▲
Diodes.

34.8 Converting AC to DC

The current in your home is ac. The current in a battery-operated device, such as a pocket calculator, is dc. With an ac-dc converter you can operate such a device on ac instead of batteries. In addition to a transformer to lower the voltage (Chapter 37), the converter uses a **diode,** a tiny electronic device that acts as a one-way valve to allow electron flow in only one direction. Since alternating current vibrates in two directions, only half of each cycle will pass through a diode. The output is a rough dc, off half the time. To maintain continuous current while smoothing the bumps, a capacitor is used (Figure 34.12).

Recall from the previous chapter that a capacitor acts as a storage reservoir for charge. Just as it takes time to raise or lower the water level in a reservoir, it takes time to add or remove electrons from the plates of a capacitor. A capacitor therefore produces a retarding effect on changes in current flow. It smooths the pulsed output.

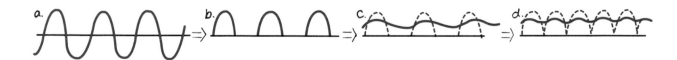

Figure 34.12 ▲
(a) When input to a diode is ac, (b) output is pulsating dc. (c) Charging and discharging of a capacitor provides continuous and smoother current. (d) In practice, a pair of diodes are used so there are no gaps in current output. The effect is to reverse the polarity of alternate half-cycles instead of eliminating them.

---

* The "plus" and "minus" are arbitrary, because they alternate. The important thing is that they are opposite.

34.9 The Speed of Electrons in a Circuit

When you flip on the light switch on your wall and the circuit is completed, the lightbulb appears to glow immediately. When you make a telephone call, the electric signal carrying your voice travels through the connecting wires at seemingly infinite speed. This signal is transmitted through the conductors at nearly the speed of light. It is *not* the electrons that move at this speed but the signal.

At room temperature, the electrons inside a metal wire have an average speed of a few million kilometers per hour due to their thermal motion. This does not produce a current because the motion is random. There is no net flow in any one direction. But when a battery or generator is connected, an electric field is established inside the wire. It is a pulsating electric field that can travel through a circuit at nearly the speed of light. The electrons continue their random motions in all directions while simultaneously being nudged along the wire by the electric field.

The conducting wire acts as a guide or "pipe" for electric field lines (Figure 34.13). In the space outside the wire, the electric field has a pattern determined by the location of electric charges, including charges in the wire. Inside the wire, the electric field is directed along the wire. If the voltage source is dc, like the battery shown in Figure 34.13, the electric field lines are maintained in one direction in the conductor.

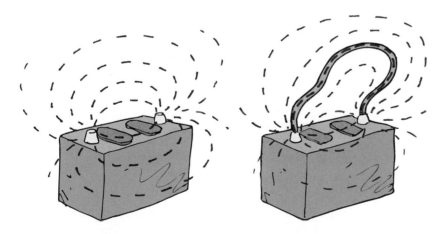

Figure 34.13 ▲
The electric field lines between the terminals of a battery are directed through a conductor, which joins the terminals.

Conduction electrons are accelerated by the field in a direction parallel to the field lines. Before they gain appreciable speed, they "bump into" the anchored metallic ions in their paths and transfer some of their kinetic energy to them. This is why current-carrying wires become hot. These collisions interrupt the motion of the electrons so that their actual *drift speed*, or *net speed* through the wire due to the field, is extremely low. In a typical dc circuit, in the electric system of an automobile for example, electrons have a net average drift speed of about 0.01 cm/s. At this rate, it would take about three hours for an electron to travel through 1 meter of wire.

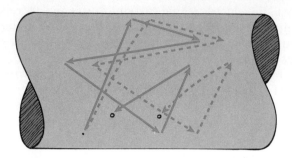

Figure 34.14 ▲
The solid lines depict a possible random path of an electron bouncing off atoms in a conductor. Instantaneous speeds are about 1/200 the speed of light. The dashed lines show an exaggerated view of how this path may be altered when an electric field is applied. The electron drifts toward the right with an average speed much less than a snail's pace.

In an ac circuit, the conduction electrons don't make any net progress in any direction. In a single cycle they drift a tiny fraction of a centimeter in one direction, then the same tiny distance in the opposite direction. Hence they oscillate rhythmically to and fro about relatively fixed positions. When you talk to your friend on the telephone, it is the *pattern* of oscillating motion that is carried across town at nearly the speed of light. The electrons already in the wires vibrate to the rhythm of the traveling pattern.

34.10 The Source of Electrons in a Circuit

In a hardware store you can buy a water hose that is empty of water. But you can't buy a piece of wire, an "electron pipe," that is empty of electrons. The source of electrons in a circuit is the conducting circuit material itself. Some people think that the electric outlets in the walls of their homes are a source of electrons. They think that electrons flow from the power utility through the power lines and into the wall outlets of their homes. This is not true. The outlets in homes are ac. Electrons do not travel appreciable distances through a wire in an ac circuit. Instead, they vibrate to and fro about relatively fixed positions.

When you plug a lamp into an ac outlet, *energy* flows from the outlet into the lamp, not electrons. Energy is carried by the electric field and causes a vibratory motion of the electrons that already exist

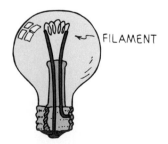

Figure 34.15 ▲
The conduction electrons that surge to and fro in the filament of the lamp do not come from the voltage source. They are in the filament to begin with. The voltage source simply provides them with surges of energy.

in the lamp filament. If 120 volts ac are impressed on a lamp, then an average of 120 joules of energy are dissipated by each coulomb of charge that is made to vibrate. Most of this electric energy appears as heat, while some of it takes the form of light. Power utilities do not sell electrons. They sell *energy.* You supply the electrons.

Thus, when you are jolted by an ac electric shock, the electrons making up the current in your body originate in your body. Electrons do not come out of the wire and through your body and into the ground; energy does. The energy simply causes free electrons in your body to vibrate in unison. Small vibrations tingle; large vibrations can be fatal.

34.11 Electric Power

Unless it is in a superconductor, a charge moving in a circuit expends energy. This may result in heating the circuit or in turning a motor. The rate at which electric energy is converted into another form such as mechanical energy, heat, or light is called **electric power.** Electric power is equal to the product of current and voltage.*

$$\text{electric power} = \text{current} \times \text{voltage}$$

If the voltage is expressed in volts and the current in amperes, then the power is expressed in watts. So in units form,

$$1 \text{ watt} = (1 \text{ ampere}) \times (1 \text{ volt})$$

If a lamp rated at 120 watts operates on a 120-volt line, you can see that it will draw a current of 1 ampere, since 120 watts = (1 ampere) × (120 volts). A 60-watt lamp draws 0.5 ampere on a 120-volt line. This relationship becomes a practical matter when you wish to know the cost of electric energy, which varies from 1 cent to 10 cents per kilowatt-hour depending on locality.

A *kilowatt* is 1000 watts, and a *kilowatt-hour* represents the amount of energy consumed in 1 hour at the rate of 1 kilowatt.**

---

* Note that this follows from the definitions of current and voltage:

$$\text{current} \times \text{voltage} = \frac{\cancel{\text{charge}}}{\text{time}} \times \frac{\text{energy}}{\cancel{\text{charge}}} = \frac{\text{energy}}{\text{time}} = \text{power}$$

** Since power = energy/time, simple rearrangement gives energy = power × time; hence, energy can be expressed in units of kilowatt-hours.

Physicists measure energy in *joules,* but utility companies customarily sell energy in units of *kilowatt-hours* (kW·h), where 1kW·h = 3.6×10^6 J. This duplication of units added to an already long list of units unfortunately makes the study of physics more difficult. It will be enough for you to become familiar with, and be able to distinguish between, the units *coulombs, volts, ohms, amperes, watts, kilowatts,* and *kilowatt-hours* here. Mastering them requires laboratory work and the help of more advanced textbooks. An understanding of electricity takes considerable time and effort, so be patient with yourself if you find this material difficult.

Figure 34.16 ▲
The power and voltage on the lightbulb read "60 W 120 V." How much current in amperes will flow through the bulb?

Therefore, in a locality where electric energy costs 5 cents per kilowatt-hour, a 100-watt electric lightbulb can be run for 10 hours at a cost of 5 cents, or a half-cent for each hour. A toaster or iron, which draws more current and therefore more power, costs several times as much to operate for the same time.

■ Questions

1. How much power is used by a calculator that operates on 8 volts and 0.1 ampere? If it is used for one hour, how much energy does it use?

2. Will a 1200-watt hair dryer operate on a 120-volt line if the current is limited to 15 amperes by a safety fuse? Can two hair dryers operate on this line?

■ Answers

1. Power = current × voltage = (0.1 A) × (8 V) = 0.8 W. If it is used for one hour, then energy = power × time = (0.8 W) × (1 h) = 0.8 watt-hour, or 0.0008 kilowatt-hour.

2. One 1200-watt hair dryer can be operated because the circuit can provide (15 A) × (120 V) = 1800 watts. But there is inadequate power to operate two hair dryers of combined power 2400 watts. This can be seen also from the amount of current involved. Since 1 watt = (1 ampere) × (1 volt), it follows that (1200 watts)/(120 volts) = 10 amperes; so the hair dryer will operate when connected to the circuit. But two hair dryers on the same plug will require 20 amperes and will blow the 15-amp fuse.

34 Chapter Review

Concept Summary

Electric current is the flow of electric charge that occurs when there is a potential difference across the ends of an electric conductor.

- The flow continues until both ends reach a common potential.

- Dry cells, wet cells, and electric generators are voltage sources that maintain a potential difference in a circuit.

The amount of current that flows in a circuit depends on the voltage and the electric resistance that the conductor offers to the flow of charge.

- An increased temperature or a longer wire increases resistance.

- A thicker wire decreases resistance.

Ohm's law states that the amount of current is directly proportional to the voltage and inversely proportional to the resistance.

- Resistors are used in many electric devices to control current.

- Electric shock is the result of an electric current passing through the body when there is a voltage difference between two parts of the body.

Direct current (dc) is electric current in which the charge flows in one direction only; electrons in an alternating current (ac) alternate their direction of flow.

- Batteries produce direct current. Power utilities produce alternating current.

- Alternating current allows low-cost, high-voltage energy transmission across great distances, with safe low-voltage use by the consumer.

Electric fields travel through circuits at nearly the speed of light, but the electrons themselves do not.

- In a dc circuit, electrons have a low drift speed within wires.

- In ac circuits, energy, not electrons, flows from the outlet; the electrons superimpose a rhythmic vibration on rapid random motion.

Electric power, the rate at which electric energy is converted into other forms of energy, is equal to the product of current and voltage.

Important Terms

alternating current (34.7)
ampere (34.2)
diode (34.8)
direct current (34.7)
electric current (34.2)
electric power (34.11)
electric resistance (34.4)
ohm (34.4)
Ohm's law (34.5)
potential difference (34.1)
voltage source (34.3)

Review Questions

1. What condition is necessary for the flow of heat? What analogous condition is necessary for the flow of charge? (34.1)

2. What is meant by the term *potential?* What is meant by *potential difference?* (34.1)

3. What condition is necessary for the sustained flow of water in a pipe? What analogous condition is necessary for the sustained flow of charge in a wire? (34.1)

4. What is electric current? (34.2)

5. What is an ampere? (34.2)

6. What is voltage? (34.3)

7. How many joules per coulomb are given to charges that flow in a 120-volt circuit? (34.3)

8. Does charge flow through a circuit or into a circuit? (34.3)

9. Does voltage flow through a circuit, or is voltage established across a circuit? (34.3)

10. What is electric resistance? (34.4)

11. Is electric resistance greater in a short fat wire or a long thin wire? (34.4)

12. What is Ohm's law? (34.5)

13. If the voltage impressed across a circuit is constant but the resistance doubles, what change occurs in the current? (34.5)

14. If the resistance of a circuit remains constant while the voltage across the circuit decreases to half its former value, what change occurs in the current? (34.5)

15. How does wetness affect the resistance of your body? (34.6)

16. Why is it that a bird can perch without harm on a high-voltage wire? (34.6)

17. What is the function of the third prong in a household electric plug? (34.6)

18. Distinguish between dc and ac. Which is produced by a battery and which is usually produced by a generator? (34.7)

19. A diode converts ac to pulsed dc. What electric device smooths the pulsed dc to a smoother dc? (34.8)

20. What are the roles of a diode and a capacitor in an ac-dc converter? (34.8)

21. What is a typical "drift" speed of electrons that make up a current in a typical dc circuit? In a typical ac circuit? (34.9)

22. From where do the electrons originate that flow in a typical electric circuit? (34.10)

23. What is power? (34.11)

24. Which of these is a unit of power and which is a unit of electric energy: a watt, a kilowatt, a kilowatt-hour? (34.11)

25. How many amperes flow through a 60-watt bulb when 120 volts are impressed across it? (34.11)

Activity

1. Batteries are made up of electric cells, which are composed of two unlike pieces of metal separated by a conducting solution. A simple 1.5-volt cell, equivalent to a flashlight cell, can be made by placing a strip of copper and a strip of zinc in a moist vegetable or piece of fruit as shown in the figure. A lemon or banana works fine. Hold the ends of the strips close together but not touching, and place the ends on your tongue. The slight tingle you feel and the metallic taste you experience result from a slight current of electricity pushed by the cell through the metal strips when your moist tongue closes the circuit. Try this and compare the results for different metals and different fruits and vegetables.

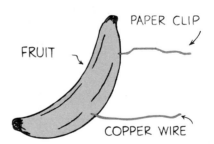

Plug and Chug

1. Calculate the current where 10 coulombs of charge pass a point in 5 seconds.

2. Calculate the current of a lightning bolt that delivers a charge of 35 coulombs to the ground in a time of 1/1000 second.

3. Calculate the current in a toaster that has a heating element of 14 ohms when connected to a 120-V outlet.

4. Calculate the current in the coiled heating element of a 240-V stove. The resistance of the element is 60 ohms at its operating temperature.

5. Electric socks, popular in cold weather, have a 90-ohm heating element that is powered by a 9-volt battery. How much current warms your feet?

6. How much current moves through your fingers (resistance: 1200 ohms) if you touch them to the terminals of a 6-volt battery?

7. Calculate the resistance of the filament in a lightbulb that carries 0.4 A when 3.0 V is impressed across it.

8. Calculate the current in a 140-W electric blanket connected to a 120-V outlet.

Think and Explain

1. Is this label on a household product cause for concern? "Caution: This product contains tiny electrically charged particles moving at speeds in excess of 10 000 000 kilometers per hour."

2. Do an *ampere* and a *volt* measure the same thing, or different things? What are those things, and which is a flow and which is the cause of the flow?

3. Why are thick wires rather than thin wires used to carry large currents?

4. Why is it important that the resistance of an extension cord be small when it is used to power an electric heater?

5. Why will an electric drill operating on a very long extension cord not rotate as fast as one operated on a short cord?

6. Will the current in a lightbulb connected to 220 V be more or less than when the same bulb is connected to 110 V? How much?

7. What is the effect on current if both the voltage and the resistance are doubled? If both are halved?

8. Would you expect to find dc or ac in the dome lamp in an automobile? In a lamp in your home?

9. In 60-Hz alternating current, how many times per second does an electron change its direction? (Don't say 60!)

10. Two lightbulbs designed for 120-V use are rated at 40 W and 60 W. Which lightbulb has the greater filament resistance? Why?

Think and Solve

1. How much voltage is required to make 2 amperes flow through a resistance of 8 ohms?

2. A battery does 18 joules of work on 3 coulombs of charge. What voltage does it supply?

3. Use the relationship power = current × voltage to find out how much current is drawn by a 1200-watt hair dryer when it operates on 120 volts. Then use Ohm's law to find the resistance of the hair dryer.

4. The wattage marked on a lightbulb is not an inherent property of the bulb but depends on the amount of voltage to which it is connected, usually 110 V or 120 V. Calculate the current through a 40-W bulb connected to 120 V.

5. Calculate the power dissipated in a toaster that has a resistance of 14 ohms plugged into a 120-V outlet.

6. Calculate the yearly cost of running a 5-W electric clock continuously in a location where electricity costs 10 cents per kW·h.

35 Electric Circuits

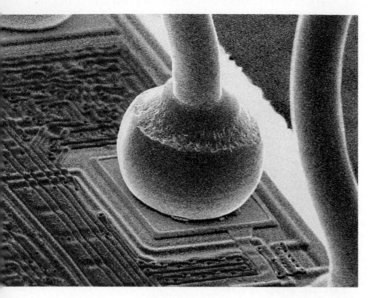

Connection to a microcircuit.

Mechanical things seem to be easier to figure out for most people than electrical things. Maybe this is because most people have had experience playing with blocks and mechanical toys when they were children. If you are among the many who have had far less direct experience with the inner workings of electrical devices than with mechanical gadgets, you are encouraged to put extra effort into the laboratory part of this course. You'll find hands-on laboratory experience aids your understanding of electric circuits. The experience can be a lot of fun, too!

35.1 A Battery and a Bulb

Take apart an ordinary flashlight like the one shown in Figure 35.1. If you don't have any spare pieces of wire around, cut some strips from some aluminum foil that you probably have in one of your kitchen drawers. Try to light up the bulb using a single battery* and a couple of pieces of wire or foil.

Figure 35.1 ▶
A flashlight taken apart.

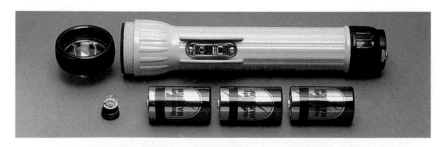

---

* Strictly speaking, a battery consists of two or more cells. What most people call a flashlight battery is more properly called a flashlight dry cell. To conform with popular usage, this chapter uses the term *battery* to mean either a single cell or series of cells.

Some of the ways you *can* light the bulb and some of the ways you *can't* light it are shown in Figure 35.2. The important thing to note is that there must be a complete path, or **circuit,** from the positive terminal at the top of the battery to the negative terminal, which is the bottom of the battery. Electrons flow from the negative part of the battery through the wire or foil to the side (or bottom) of the bulb, through the filament inside the bulb, and out the bottom (or side) and through the other piece of wire or foil to the positive part of the battery. The current then passes through the interior of the battery to complete the circuit.

Figure 35.2 ▲
(a) Unsuccessful ways to light a bulb. (b) Successful ways to light a bulb.

The flow of charge in a circuit is very much like the flow of water in a closed system of pipes. For the set up shown in Figure 35.2b, the battery is analogous to a pump, the wires are analogous to the pipes, and the bulb is analogous to any device that operates when the water is flowing. When a valve in the line is opened and the pump is operating, water already in the pipes starts to flow. Similarly, when a switch is turned on to complete an electric circuit, the mobile conduction electrons already in the wires and the filament begin to drift through the circuit. The water flows *through* the pump and electrons flow *through* the battery. Neither the water nor the electrons "squash up" and concentrate in certain places; they flow continuously around a loop, or circuit.

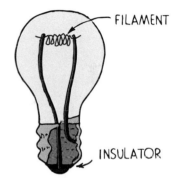

Figure 35.3 ▲
Electrons do not pile up inside a bulb, but instead flow through its filament.

35.2 Electric Circuits

Any path along which electrons can flow is a circuit. For a continuous flow of electrons, there must be a complete circuit with no gaps. A gap is usually provided by an electric switch that can be opened or closed to either cut off or allow electron flow.

The water analogy is quite useful for gaining a conceptual understanding of electric circuits, but it does have some limitations. An important one is that a break in a water pipe results in water spilling from the circuit, whereas a break in an electric circuit results in a complete stop in the flow of electricity. Another difference has to do with turning current off and on. When you *close* an electrical switch that connects the circuit, you allow current to flow in much the same way as you allow water to flow by *opening* a faucet. Opening a switch stops the flow of electricity. An electric circuit must be closed for electricity to flow. Opening a water faucet, on the other hand, starts the flow of water. Despite these and some other differences, thinking of electric current in terms of water current is a helpful way to study electric circuits.

Most circuits have more than one device that receives electrical energy. These devices are commonly connected in a circuit in one of two ways, *series* or *parallel*. When connected **in series,** they form a single pathway for electron flow between the terminals of the battery, generator, or wall socket (which is simply an extension of these terminals). When connected **in parallel,** they form branches, each of which is a separate path for the flow of electrons. Both series and parallel connections have their own distinctive characteristics. This chapter briefly treats circuits with these two types of connections.

35.3 Series Circuits

Figure 35.4 shows three lamps connected in series with a battery. This is an example of a simple **series circuit.** When the switch is closed, a current exists almost immediately in all three lamps. The current does not "pile up" in any lamp but flows *through* each lamp. Electrons in all parts of the circuit begin to move at once. Some electrons move away from the negative terminal of the battery, some move toward the positive terminal, and some move through the filament of each lamp. Eventually the electrons move all the way around the circuit. A break anywhere in the path results in an open circuit, and the flow of electrons ceases. Burning out of one of the lamp filaments or simply opening the switch could cause such a break.

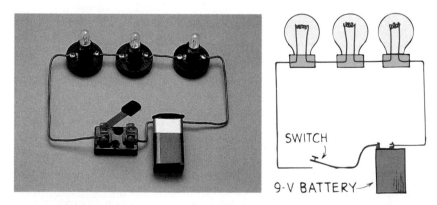

Figure 35.4 ▲
A simple series circuit. The 9-volt battery provides 3 volts across each lamp.

The circuit shown in Figure 35.4 illustrates the following important characteristics of series connections:

1. Electric current has but a single pathway through the circuit. This means that the current passing through each electrical device is the same.

2. This current is resisted by the resistance of the first device, the resistance of the second, and the third also, so that the total resistance to current in the circuit is the sum of the individual resistances along the circuit path.

3. The current in the circuit is numerically equal to the voltage supplied by the source divided by the total resistance of the circuit. This is Ohm's law.

4. Ohm's law also applies separately to each device. The *voltage drop,* or potential difference, across each device depends directly on its resistance. This follows from the fact that more energy is used to move a unit of charge through a large resistance than through a small resistance.

5. The total voltage impressed across a series circuit divides among the individual electrical devices in the circuit so that the sum of the voltage drops across each individual device is equal to the total voltage supplied by the source. This follows from the fact that the amount of energy used to move each unit of charge through the entire circuit equals the sum of the energies used to move that unit of charge through each of the electrical devices in the circuit.

■ Questions

1. What happens to current in other lamps if one lamp in a series circuit burns out?

2. What happens to the light intensity of each lamp in a series circuit when more lamps are added to the circuit?

It is easy to see the main disadvantage of a series circuit: If one device fails, current in the whole circuit ceases and none of the devices will work. Some cheap Christmas tree lights are connected in series. When one lamp burns out, it's "fun and games" (or frustration) trying to find which bulb to replace.

Most circuits are wired so that it is possible to operate electrical devices independently of each other. In your home, for example, a lamp can be turned on or off without affecting the operation of other lamps or electrical devices. This is because these devices are connected not in series but in parallel to one another.

■ Answers

1. If one of the lamp filaments burns out, the path connecting the terminals of the voltage source will break and current will cease. All lamps will go out.

2. The addition of more lamps in a series circuit results in a greater circuit resistance. This decreases the current in the circuit and therefore in each lamp, which causes dimming of the lamps. Energy is divided among more lamps so the voltage drop across each lamp will be less.

35.4 Parallel Circuits

Figure 35.5 shows three lamps connected to the same two points A and B. This is an example of a simple **parallel circuit.** Electrical devices connected in parallel are connected to the same two points of an electric circuit. Notice that each lamp has its own path from one terminal of the battery to the other. There are three separate pathways for current, one through each lamp. In contrast to a series circuit, the current in one lamp does not pass through the other lamps. Also, unlike lamps connected in series, the parallel circuit is completed whether all, two, or only one lamp is lit. A break in any one path does not interrupt the flow of charge in the other paths. Each device operates independently of the other devices.

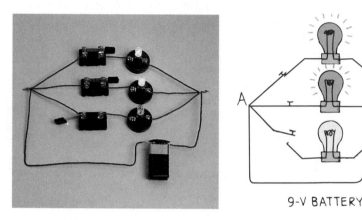

Figure 35.5 ▲
A simple parallel circuit. A 9-volt battery provides 9 volts across each lamp.

The circuit shown in Figure 35.5 illustrates the following major characteristics of parallel connections:

1. Each device connects the same two points A and B of the circuit. The voltage is therefore the same across each device.

2. The total current in the circuit divides among the parallel branches. Current passes more readily into devices of low resistance, so the amount of current in each branch is inversely proportional to the resistance of the branch. Ohm's law applies separately to each branch.

3. The total current in the circuit equals the sum of the currents in its parallel branches.

4. As the number of parallel branches is increased, the overall resistance of the circuit is *decreased*. Overall resistance is lowered with each added path between any two points of the circuit. This means the overall resistance of the circuit is less than the resistance of any one of the branches.

PARALLEL CIRCUIT

I'M GOING TO EXTEND THE TERMINALS OF THIS 12-VOLT BATTERY WITH THIS PAIR OF METAL RODS.

A LAMP WILL GLOW JUST AS BRIGHT WHEN CONNECTED HERE...

AS HERE!

OF COURSE... IT'S STILL ACROSS 12 VOLTS

NOW I FASTEN ANOTHER IDENTICAL LAMP -- WHAT HAPPENS TO THE BRIGHTNESS OF THE FIRST?

SAME SAME

SAME BRIGHTNESS BECAUSE 12 VOLTS IS ACROSS EACH!

THAT'S WHAT I SAID!

AND A THIRD LAMP ALSO --- SAME BRIGHTNESS!

WHY NOT? THERE'S STILL 12 VOLTS ACROSS EACH LAMP.

HOW DOES THE BATTERY "KNOW" TO SUPPLY 3 TIMES AS MUCH CURRENT TO 3 LAMPS??

PHYSICS OHM'S LAW PHYSICS PHYSICS

I KNOW! WITH 3 SEPARATE PATHS, THE BATTERY HAS ⅓ THE RESISTANCE BETWEEN ITS TERMINALS-- THAT'S WHY IT WILL DELIVER 3 TIMES AS MUCH CURRENT FOR ITS 12 VOLTS!

VERY GOOD! BUT DOESN'T THIS MEAN THE BATTERY IS PUTTING OUT 3 TIMES AS MUCH ENERGY AS WHEN LIGHTING ONLY ONE LAMP?

YES, AND YOU CAN EXPECT THE BATTERY TO WEAR DOWN 3 TIMES AS FAST!

SUCH EXCELLENCE ≈SIGH≈

1. What happens to the current in other lamps if one of the lamps in a parallel circuit burns out?

2. What happens to the light intensity of each lamp in a parallel circuit when more lamps are added in parallel to the circuit?

BATTERY

CONNECTING WIRE

OPEN SWITCH

CLOSED SWITCH

RESISTANCE

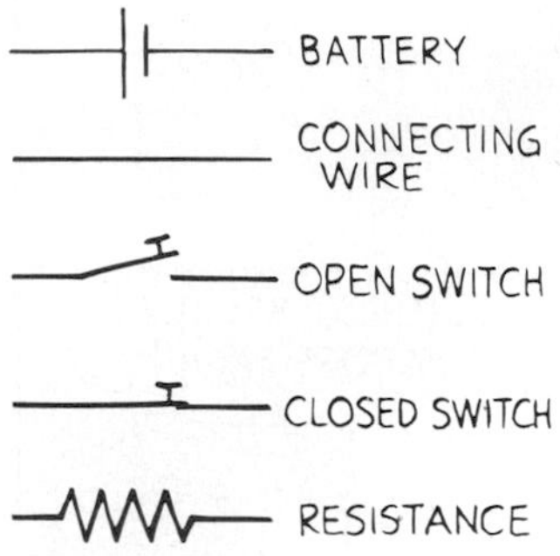

Figure 35.6 ▲
Symbols of some common circuit devices.

Figure 35.7 ▶
Schematic diagrams. (Left) The circuit of Figure 35.4, with three lamps in series. (Right) The circuit of Figure 35.5, with three lamps in parallel.

35.5 Schematic Diagrams

Electric circuits are frequently described by simple diagrams, called **schematic diagrams,** that are similar to those of the last two figures. Some of the symbols used to represent certain circuit elements are shown in Figure 35.6. Resistance is shown by a zigzag line, and ideal resistanceless wires are shown with solid straight lines. A battery is represented with a set of short and long parallel lines. The convention is to represent the positive terminal of the battery with a long line and the negative terminal with a short line. Sometimes a two-cell battery is represented with a pair of such lines, a three-cell with three, and so on. Figure 35.7 shows schematic diagrams for the circuits of Figures 35.4 and 35.5.

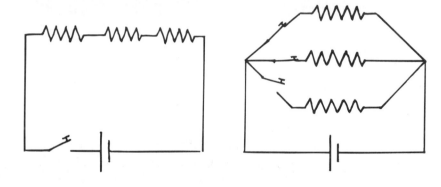

■ **Answers**

1. If one lamp burns out, the other lamps will be unaffected. The current in each branch, according to Ohm's law, is equal to (voltage)/(resistance), and since neither voltage nor resistance is affected in the branches, the current in those branches is unaffected. The total current in the overall circuit (the current through the battery), however, is decreased by an amount equal to the current drawn by the lamp in question before it burned out. But the current in any other single branch is unchanged.

2. The light intensity for each lamp is unchanged as other lamps are introduced (or removed). Only the total resistance and total current in the total circuit changes, which is to say, the current in the battery changes. (There is resistance in a battery also, which we assume is negligible here.) As lamps are introduced, more paths are available between the battery terminals, which effectively decreases total circuit resistance. This decreased resistance is accompanied by an increased current, the same increase that feeds energy to the lamps as they are introduced. Although changes of resistance and current occur for the circuit as a whole, no changes occur in any individual branch in the circuit.

35.6 Combining Resistors in a Compound Circuit

Sometimes it is useful to know the *equivalent resistance* of a circuit that has several resistors in its network. The equivalent resistance is the value of the single resistor that would comprise the same load to the battery or power source. The equivalent resistance can be found by the rules for adding resistors in series and parallel. For example, the equivalent resistance for a pair of 1-ohm resistors in series is simply 2 ohms.

The equivalent resistance for a pair of 1-ohm resistors in parallel is 0.5 ohm. (The equivalent resistance is *less* because the current has "twice the path width" when it takes the parallel path. In a similar way, the more doors that are open in an auditorium full of people trying to exit, the *less* will be the resistance to their departure.) The equivalent resistance for a pair of equal resistors in parallel is half the value of either resistor.

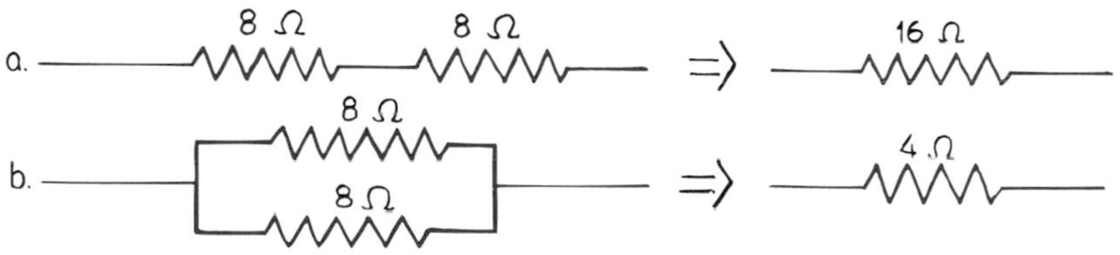

Figure 35.8 ▲
(a) The equivalent resistance of two 8-ohm resistors in series is 16 ohms.
(b) The equivalent resistance of two 8-ohm resistors in parallel is 4 ohms.

Figure 35.9 shows a combination of three 8-ohm resistors. The two resistors in parallel are equivalent to a single 4-ohm resistor, which is in series with an 8-ohm resistor and adds to produce an equivalent resistance of 12 ohms. If a 12-volt battery were connected to these resistors, can you see from Ohm's law that the current through the battery would be 1 ampere? (In practice it would be less, for there is resistance inside the battery as well, called the battery's *internal resistance.*)

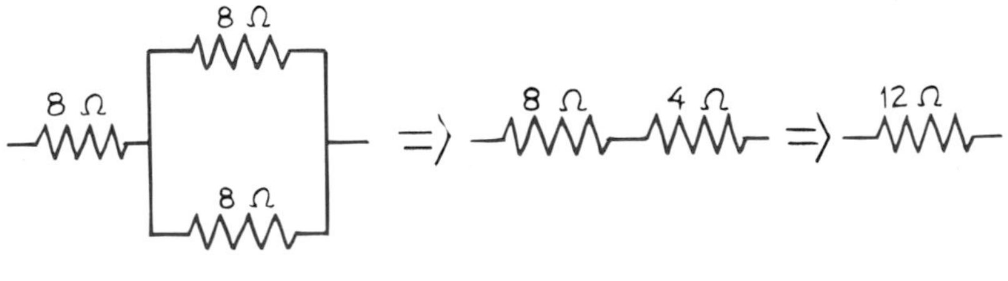

Figure 35.9 ▲
The equivalent resistance of the circuit is found by combining resistors in successive steps.

LINK TO TECHNOLOGY

Measuring with Current

A fuel gauge in an automobile uses variable resistance to measure the level in the gasoline tank. A float in the tank adjusts the resistance of a variable electric resistor. Maximum resistance occurs when the float bottoms out in the tank. Maximum resistance produces the minimum current, which barely deflects the pointer on the fuel gauge. When the tank is full, the variable resistor has its lowest resistance and the maximum current flows through the fuel gauge. For this current, the gauge is calibrated to read a full tank. Between empty and full, corresponding values of current produce appropriate deflections of the fuel gauge pointer.

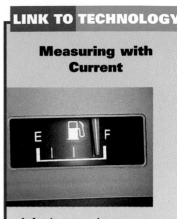

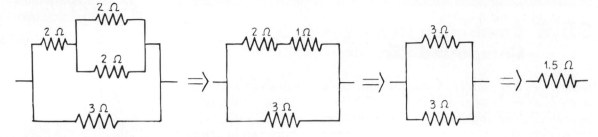

Figure 35.10 ▲
The equivalent resistance of the top branch is 3 ohms, which is in parallel with the 3-ohm resistance of the lower branch. The overall equivalent resistance is 1.5 ohms.

Two more complex combinations are broken down in successive equivalent combinations in Figures 35.10 and 35.11. It's like a game: Combine resistors in series by adding; combine a pair of equal resistors in parallel by halving.* The value of the single resistor left is the equivalent resistance of the combination.

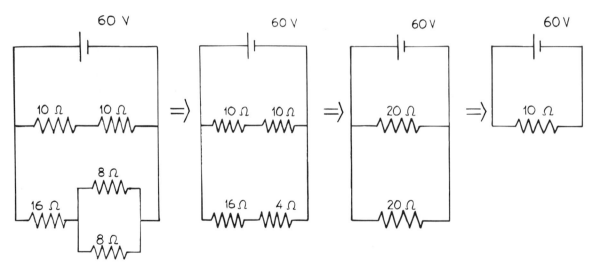

Figure 35.11 ▲
Schematic diagrams for an arrangement of various electric devices. The equivalent resistance of the circuit is 10 ohms. (The 60-V battery is for numerical convenience—most batteries are less than 60 V.)

---

* For a pair of non-equal resistors in parallel, the equivalent resistance is found by taking the product of the pair and dividing by the sum of the pair. That is

$$R_{equivalent} = \frac{R_1 R_2}{R_1 + R_2}$$

This rule of "product divided by sum" holds only for two resistors in parallel. For three or more parallel resistors, you can do a pair at a time (as is done in Figures 35.10 and 35.11), or use the more general formula

$$1/R_{equivalent} = 1/R_1 + 1/R_2 + 1/R_3 \text{ and so on}$$

Details can be found in other physics textbooks.

The following questions are based on the schematic diagrams in Figure 35.11.

1. What is the current in amperes through the battery? (Neglect the internal resistance of the battery.)

2. What is the current in amperes through the pair of 10-ohm resistors?

3. What is the current in amperes through each of the 8-ohm resistors?

4. How much power is provided by the battery?

35.7 Parallel Circuits and Overloading

Electricity is usually fed into a home by way of two lead wires called *lines*. These lines are very low in resistance and are connected to wall outlets in each room. About 110 to 120 volts are impressed on these lines by generators at the power utility. This voltage is applied to appliances and other devices that are connected in parallel by plugs to these lines.

As more devices are connected to the lines, more pathways are provided for current. What effect do the additional pathways produce? The answer is, a lowering of the combined resistance of the circuit. Therefore, a greater amount of current occurs in the lines. Lines that carry more than a safe amount of current are said to be *overloaded*. The resulting heat may be sufficient to melt the insulation and start a fire.

You can see how overloading occurs by considering the circuit in Figure 35.12. The supply line is connected to an electric toaster that draws 8 amperes, to an electric heater that draws 10 amperes, and to

■ **Answers**

1. The current in the battery (or total current in the circuit) is 6 A. You can get this from Ohm's law: current = (voltage)/(resistance) = (60 V)/(10 Ω) = 6 A. You know from the last step in the figure that the equivalent resistance of the circuit is 10 Ω.

2. Half the total circuit current, 3 A, will flow through the pair of 10 Ω resistors. You know this because you can see that both branches have equal resistances. This means that the total circuit current will divide equally between the upper and lower branches. (Can you think of another way to get the same answer?)

3. The current through the pair of 8 Ω resistors is 3 A, and the current through each is therefore 1.5 A. This is because the 3-A current divides equally through these equal resistances.

4. The battery supplies 360 watts. This is from the relationship

$$\text{power} = \text{current} \times \text{voltage} = (6 \text{ A}) \times (60 \text{ V}) = 360 \text{ watts}$$

This power will be dissipated among all the resistors in the circuit.

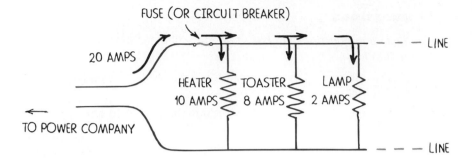

Figure 35.12 ▲
Circuit diagram for appliances connected to a household supply line.

an electric lamp that draws 2 amperes. When only the toaster is operating and drawing 8 amperes, the total line current is 8 amperes. When the heater is also operating, the total line current increases to 18 amperes (8 amperes to the toaster and 10 amperes to the heater). If you turn on the lamp, the line current increases to 20 amperes. Connecting any more devices increases the current still more.

To prevent overloading in circuits, *fuses* are connected in series along the supply line. In this way the entire line current must pass through the fuse. The safety fuse shown below is constructed with a wire ribbon that will heat up and melt at a given current. If the fuse is rated at 20 amperes, it will pass 20 amperes, but no more. A current above 20 amperes will melt the fuse, which "blows out" and breaks the circuit. Before a blown fuse is replaced, the cause of overloading should be determined and remedied. Often, insulation that separates the wires in a circuit wears away and allows the wires to touch. This effectively shortens the path of the circuit, and is called a *short circuit*. A short circuit draws a dangerously large current because it bypasses the normal circuit resistance.

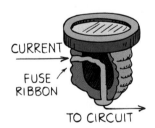

Circuits may also be protected by *circuit breakers,* which use magnets or bimetallic strips to open the switch. Utility companies use circuit breakers to protect their lines all the way back to the generators. Circuit breakers are used instead of fuses in modern buildings because they do not have to be replaced each time the circuit is opened. Instead, the switch can simply be moved back to the "on" position after the problem has been corrected.

35 Chapter Review

Concept Summary

Any path along which electric charge (usually electrons) can flow is a circuit.

- A complete circuit is needed to maintain a continuous flow of charge.

In a series circuit, electrical devices form a single pathway for electron flow.

- A break anywhere in the path stops the electron flow in the entire circuit.
- The total resistance is equal to the sum of individual resistances along the current path.
- The current is equal to the voltage divided by the total resistance.
- The voltage drop across each device is proportional to its resistance.
- The sum of voltage drops across the resistance of each individual device is equal to the total voltage.

In a parallel circuit, electrical devices form branches, each of which provides a separate path for the flow of electrons.

- Each device connects the same two points of the circuit; the voltage is the same across each device.
- The amount of current in each branch is inversely proportional to the resistance of the branch.
- The total current is equal to the sum of the currents in each branch.

Electric circuits are often described by schematic diagrams, in which each element of the circuit is represented by a symbol.

In a circuit with several resistors, the equivalent resistance is the value of the single resistor that would comprise the same load to the battery or power source.

- For resistors in series, the equivalent resistance is the sum of their values.
- For resistors in parallel, the equivalent resistance is less than the value of any individual resistor.

Lines carrying an unsafe amount of current are overloaded.

- To prevent overloading, fuses or circuit breakers are inserted in lines that provide power. Excessive current will "blow out" the fuse or "trip" the circuit breaker, stopping the current.
- A short circuit is often caused by faulty wire insulation.

Important Terms

circuit (35.1)
in parallel (35.2)
in series (35.2)
parallel circuit (35.4)
schematic diagram (35.5)
series circuit (35.3)

Review Questions

1. Are all the electrons flowing in a circuit provided by the battery? (35.1)

2. Why must there be no gaps in an electric circuit for it to carry current? (35.1)

3. Distinguish between a series circuit and a parallel circuit. (35.2)

4. If three lamps are connected in series to a 6-volt battery, how many volts are impressed across each lamp? (35.3)

5. If one of three lamps blows out when connected in series, what happens to the current in the other two? (35.3)

6. If three lamps are connected in parallel to a 6-volt battery, how many volts are impressed across each lamp? (35.4)

7. If one of three lamps blows out when connected in parallel, what happens to the current in the other two? (35.4)

8. a. In which case will there be more current in each of three lamps—if they are connected to the same battery in series or in parallel?

b. In which case will there be more voltage across each lamp? (35.4)

9. What happens to the total circuit resistance when more devices are added to a series circuit? To a parallel circuit? (35.6)

10. What is the equivalent resistance of a pair of 8-ohm resistors in series? In parallel? (35.6)

11. Why does the total circuit resistance decrease when more devices are added to a parallel circuit? (35.6)

12. What does it mean when you say that lines in a home are overloaded? (35.7)

13. What is the function of a fuse or circuit breaker in a circuit? (35.7)

14. Why will too many electrical devices operating at one time often blow a fuse or trip a circuit breaker? (35.7)

15. What is meant by a short circuit? (35.7)

Plug and Chug

1. Calculate the current in a 48-V battery that powers a pair of 30 Ω resistors connected in series.

2. Calculate the current in a 48-V battery that powers a pair of 30 Ω resistors connected in parallel.

Think and Explain

1. Sometimes you hear someone say that a particular appliance "uses up" electricity. What is it that the appliance actually "uses up," and what becomes of it?

2. Why are the wingspans of birds a consideration in determining the spacing between parallel wires in a power line?

3. Why are household appliances almost never connected in series?

4. As more and more lamps are connected in series to a flashlight battery, what happens to the brightness of each lamp?

5. As more and more lamps are connected in parallel to a battery, and if the current does not produce heating inside the battery, what happens to the brightness of each lamp?

6. In the circuit shown, how do the brightnesses of the identical bulbs compare? Which lightbulb draws the most current? What happens if bulb A is unscrewed? If bulb C is unscrewed?

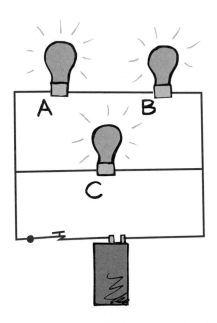

7. A number of lightbulbs are to be connected to a battery. Which will provide more overall brightness, connecting them in series or in parallel? Which will run the battery down faster, the bulbs connected in series or the bulbs connected in parallel?

8. A three-way bulb uses two filaments to produce three levels of illumination (50 W, 100 W, and 150 W) using a 120-V socket. When one of the filaments burns out, only one level, the 50 W or the 100 W, is available. Are the filaments connected in series or in parallel?

9. How does the line current compare with the total currents of all devices connected in parallel?

10. A 60-W bulb and a 100-W bulb are connected in series in a circuit. Which bulb has the greater current flowing in it? Which has the greater current when they are connected in parallel?

Think and Solve

1. A 16 Ω loudspeaker and an 8 Ω loudspeaker are connected in parallel across the terminals of an amplifier. Assuming the speakers behave as resistors, calculate the equivalent resistance of the two speakers.

2. Consider the combination series and parallel circuit shown here.

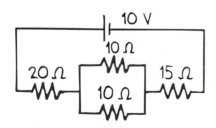

a. Identify the parallel part of the circuit. What is the equivalent resistance of this part? In other words, what single resistance could replace this part of the circuit and not change the total current from the battery?

b. What is the equivalent resistance of all the resistors? In other words, what single resistance could replace the whole circuit without changing the current from the battery?

3. How many 4 Ω resistors must be connected in parallel to create an equivalent resistance of 0.5 Ω?

4. What is the current in the battery of the circuit shown below? (What must you find before you can calculate the current?)

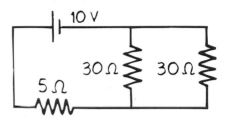

5. The rear window defrosters on automobiles are made up of several strips of heater wire connected in parallel. Consider the case of four wires, each of 6 Ω resistance, connected to 12 V.
a. What is the equivalent resistance of the four wires? (Consider the wires to be two groups of two.)

b. What is the total current drawn?

36 Magnetism

The earth is a magnet.

Magnets are fascinating. Bring a pair close together and they snap together and stick. Turn one of the magnets around and they repel each other. A magnet will stick to a refrigerator door, but it won't stick to an aluminum pan. Magnets come in all shapes and sizes. They are popular as toys, are utilized as compasses, and are essential elements in electric motors and generators. Magnetism is very common to everything you see, for it is an essential ingredient of light itself.

The term *magnetism* stems from certain rocks called *lodestones* found more than 2000 years ago in the region of Magnesia in Greece. In the twelfth century, the Chinese used them for navigating ships. In the eighteenth century, the French physicist Charles Coulomb studied the forces between lodestones. We now know that lodestones contain iron ore, which has been named magnetite.

Electricity and magnetism were regarded as unrelated phenomena into the early nineteenth century. This changed in 1820 when a Danish science professor, Hans Christian Oersted, discovered a relationship between the two while demonstrating electric currents in front of a class of students. When electric current was passed in a wire near a magnetic compass, both Oersted and his students noticed the deflection of the compass needle. This was the connecting link that had eluded investigators for decades.* Other discoveries soon followed. Magnets were found to exert forces on current-carrying wires, which led to electric meters and motors. The stage was set for a whole new technology, which would bring electric power, radio, and television.

---

* We can only speculate about how often such relationships become evident when they "aren't supposed to" and are dismissed as "something wrong with the apparatus." Oersted, however, had the insight characteristic of a good scientist to see that nature was revealing another of its secrets.

Figure 36.1 ▲
Which interaction has the greater strength—the gravitational attraction between the scrap iron and the earth, or the magnetic attraction between the magnet and the scrap iron?

36.1 Magnetic Poles

Magnets exert forces on one another. They are similar to electric charges, for they can both attract and repel without touching, depending on which end is held near the other. Also, like electric charges, the strength of their interaction depends on the distance of separation of the two magnets. Whereas electric charges produce electrical forces, regions called **magnetic poles** produce magnetic forces.

If you suspend a bar magnet from its center by a piece of string, it will act as a compass. The end that points northward is called the *north-seeking pole,* and the end that points southward is called the *south-seeking pole.* More simply, these are called the *north* and *south poles.* All magnets have both a north and a south pole. For a simple bar magnet these are located at the two ends. The common horseshoe magnet is a bar magnet that has been bent, so its poles are also at its two ends.

If the north pole of one magnet is brought near the north pole of another magnet, they repel. The same is true of a south pole near a south pole. If opposite poles are brought together, however, attraction occurs.*

Like poles repel; opposite poles attract.

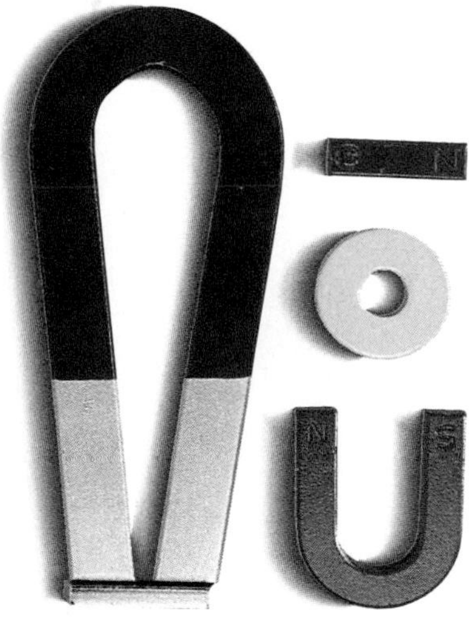

Figure 36.2 ▲
Common magnets.

---

* The force of interaction between magnetic poles is given by $F \sim pp'/d^2$, where p and p' represent magnetic pole strengths, and d represents the separation distance between them. Note the similarity of this relationship to Coulomb's law and Newton's law of gravity.

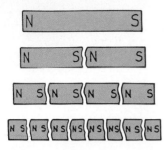

Figure 36.3 ▲
Break a magnet in half and you have two magnets. Break these in half and you have four magnets, each with a north and a south pole. Keep breaking the pieces further and further and you find the same results. Magnetic poles exist in pairs.

Magnetic poles behave similarly to electric charges in some ways, but there is a very important difference. Electric charges can be isolated, but magnetic poles cannot. Negatively charged electrons and positively charged protons are entities by themselves. A cluster of electrons need not be accompanied by a cluster of protons, and vice versa. But a north magnetic pole never exists without the presence of a south pole, and vice versa. The north and south poles of a magnet are like the head and tail of the same coin.

If you break a bar magnet in half, each half still behaves as a complete magnet. Break the pieces in half again, and you have four complete magnets. You can continue breaking the pieces in half and never isolate a single pole. Even when your piece is one atom thick, there are two poles. This suggests that atoms themselves are magnets.

■ Question

Does every magnet necessarily have a north and a south pole?

36.2 Magnetic Fields

Place a sheet of paper over a bar magnet and sprinkle iron filings on the paper. The filings will tend to trace out an orderly pattern of lines that surround the magnet. The space around a magnet, in which a magnetic force is exerted, is filled with a **magnetic field.** The shape of the field is revealed by *magnetic field lines.* Magnetic field lines spread out from one pole, curve around the magnet, and return to the other pole.

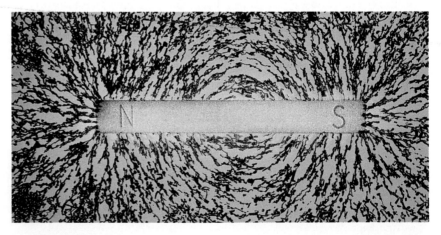

Figure 36.4 ▲
Iron filings trace out a pattern of magnetic field lines in the space surrounding the magnet.

■ Answer

Yes, just as every coin has two sides, a "head" and a "tail." (Some "trick" magnets have more than two poles.)

The direction of the field outside the magnet is from the north to the south pole. Where the lines are closer together, the field strength is greater. We see that the magnetic field strength is greater at the poles. If we place another magnet or a small compass anywhere in the field, its poles will tend to line up with the magnetic field.

Figure 36.5 ▲
Like the iron filings, the compasses line up with the magnetic field lines.

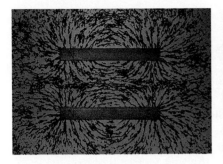 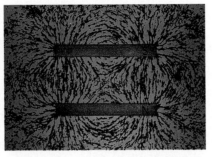

Figure 36.6 ▲
The magnetic field patterns for a pair of magnets when (left) opposite poles are near each other, and (right) like poles are near each other.

36.3 The Nature of a Magnetic Field

Magnetism is very much related to electricity. Just as an electric charge is surrounded by an electric field, the same charge is also surrounded by a magnetic field if it is moving. This is due to the "distortions" in the electric field caused by motion, and was explained by Albert Einstein in 1905 in his theory of special relativity. This text will not go into the details, except to acknowledge that a magnetic field is a relativistic by-product of the electric field. Charges in motion have associated with them both an electric and a magnetic field. A magnetic field is produced by the motion of electric charge.*

Where is the motion of electric charges in a common bar magnet? Although the magnet as a whole may be stationary, it is composed of atoms whose electrons are in constant motion about atomic nuclei. This moving charge constitutes a tiny current and produces a magnetic field. More important, electrons spin about their own axes like tops. A spinning electron constitutes a charge in motion and thus creates another magnetic field. In most materials, the field due to spinning predominates over the field due to orbital motion.

Every spinning electron is a tiny magnet. A pair of electrons spinning in the same direction makes up a stronger magnet. A pair of electrons spinning in opposite directions, however, work against each other. Their magnetic fields cancel. This is why most substances are not magnets. In most atoms, the various fields cancel each other

Figure 36.7 ▲
Both the orbital motion and the spinning motion of every electron in an atom produce magnetic fields. These fields combine constructively or destructively to produce the magnetic field of the atom. The resulting field is greatest for iron atoms.

---

* Interestingly enough, since motion is relative, the magnetic field is relative. For example, when a charge moves by you, there is a definite magnetic field associated with the moving charge. But if you move along with the charge so that there is no motion relative to you, there is no magnetic field associated with the charge. Magnetism is relativistic.

because the electrons spin in opposite directions. In materials such as iron, nickel, and cobalt, however, the fields do not cancel each other entirely. Each iron atom has four electrons whose spin magnetism is uncanceled. Each iron atom, then, is a tiny magnet. The same is true to a lesser degree for the atoms of nickel and cobalt.*

36.4 Magnetic Domains

The magnetic field of individual iron atoms is so strong that interactions among adjacent iron atoms cause large clusters of them to line up with each other. These clusters of aligned atoms are called **magnetic domains.** Each domain is perfectly magnetized, and is made up of billions of aligned atoms. The domains are microscopic (Figure 36.8), and there are many of them in a crystal of iron.

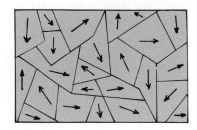

Figure 36.8 ▲
A microscopic view of magnetic domains in a crystal of iron. Each domain consists of billions of aligned iron atoms.

Figure 36.9 ▶
A piece of iron in successive stages of magnetism. The arrows represent domains, where the head is a north pole and the tail a south pole. Poles of neighboring domains neutralize each other's effects, except at the ends.

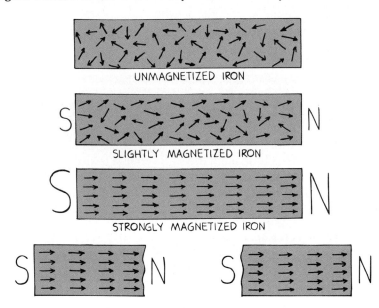

UNMAGNETIZED IRON

SLIGHTLY MAGNETIZED IRON

STRONGLY MAGNETIZED IRON

WHEN A MAGNET IS BROKEN
INTO TWO PIECES, EACH PIECE
RETAINS EQUALLY STRONG POLES

The difference between a piece of ordinary iron and an iron magnet is the alignment of domains. In a common iron nail, the domains are randomly oriented. When a strong magnet is brought nearby, two effects take place. One is a growth in size of domains that are oriented in the direction of the magnetic field. This growth is at the expense of domains that are not aligned. The other effect is a rotation of domains as they are brought into alignment. The

---

* Most common magnets are made from alloys containing iron, nickel, cobalt, and aluminum in various proportions. In these the electron spin contributes virtually all the magnetic properties. In the rare earth metals, such as gadolinium, the orbital motion is more significant.

BAR MAGNET

domains become aligned much as electric dipoles are aligned in the presence of a charged rod. When you remove the nail from the magnet, ordinary thermal motion causes most or all of the domains in the nail to return to a random arrangement.

Permanent magnets are made by simply placing pieces of iron or certain iron alloys in strong magnetic fields. Alloys of iron differ; soft iron is easier to magnetize than steel. It helps to tap the iron to nudge any stubborn domains into alignment. Another way of making a permanent magnet is to stroke a piece of iron with a magnet. The stroking motion aligns the domains in the iron. If a permanent magnet is dropped or heated, some of the domains are jostled out of alignment and the magnet becomes weaker.

■ Questions

1. How can a magnet attract a piece of iron that is not magnetized?

2. The iron filings sprinkled on the paper that covers the magnet in Figure 36.4 were not initially magnetized. Why, then, do they line up with the magnetic field of the magnet?

Figure 36.10 ▲
The iron nails become induced magnets.

■ Answers

1. Domains in the unmagnetized piece of iron are induced into alignment by the magnetic field of the nearby permanent magnet. See the similarity of this with Figure 32.12 back in Chapter 32. Like the pieces of paper jumping to the comb, pieces of iron will jump to a strong magnet when it is brought nearby. But unlike the paper, they are not repelled. Can you think of the reason why?

2. Domains align in the individual filings, causing them to act like tiny compasses. The poles of each "compass" are pulled in opposite directions, producing a torque that twists each filing into alignment with the external magnetic field.

36.5 Electric Currents and Magnetic Fields

A moving charge produces a magnetic field. Many charges in motion —an electric current—also produce a magnetic field. The magnetic field that surrounds a current-carrying conductor can be demonstrated by arranging an assortment of magnetic compasses around a wire (Figure 36.11) and passing a current through it. The compasses line up with the magnetic field produced by the current and show it to be a pattern of concentric circles about the wire. When the current reverses direction, the compasses turn completely around, showing that the direction of the magnetic field changes also. This is the effect that Oersted first demonstrated in the classroom.

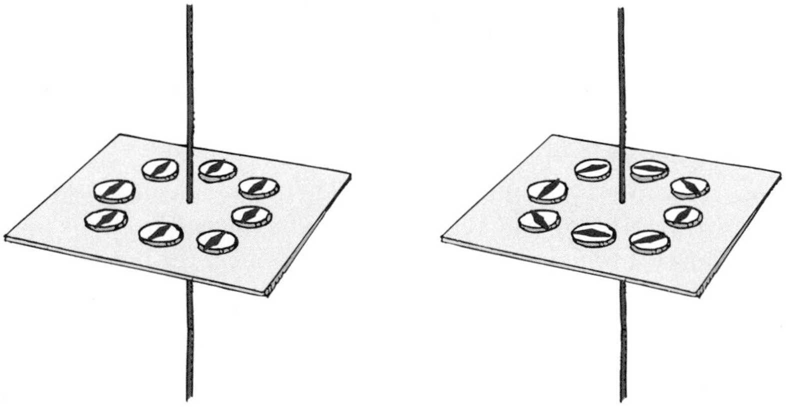

Figure 36.11 ▶
(Left) When there is no current in the wire, the compasses align with the earth's magnetic field. (Right) When there is a current in the wire, the compasses align with the stronger magnetic field near the wire. The magnetic field forms concentric circles about the wire.

If the wire is bent into a loop, the magnetic field lines become bunched up inside the loop (Figure 36.12). If the wire is bent into another loop, overlapping the first, the concentration of magnetic field lines inside the double loop is twice as much as in the single loop. It follows that the magnetic field intensity in this region is increased as the number of loops is increased. A current-carrying coil of wire with many loops is an **electromagnet.**

Figure 36.12 ▶
Magnetic field lines about a current-carrying wire crowd up when the wire is bent into a loop.

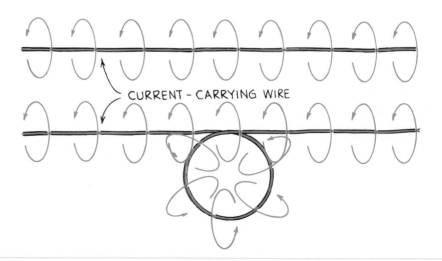

CURRENT - CARRYING WIRE

Maglev Transportation

An exciting application of superconducting electromagnets to watch for is magnetically levitated, or "maglev" transportation. Shown here is an experimental maglev train being developed in Japan. The vehicle carries superconducting coils on its underside. As it moves, these coils induce currents in wire coils set in the guideway that act as mirror-image magnets and levitate the vehicle. It floats several inches above the guideway, and its speed is limited only by air friction and passenger comfort. Someday in the future you may ride swiftly and smoothly from one city to another in a maglev train.

Sometimes a piece of iron is placed inside the coil of an electromagnet. The magnetic domains in the iron are induced into alignment, increasing the magnetic field intensity. Beyond a certain limit, the magnetic field in iron "saturates," so iron is not used in the cores of the strongest electromagnets, which are made of superconducting material (see Section 32.4).

A superconducting electromagnet can generate a powerful magnetic field indefinitely without using any power. At Fermilab near Chicago, superconducting electromagnets guide high-energy particles around the four-mile-circumference accelerator. Since the substitution of superconducting electromagnets for conventional ones in 1983, monthly electric bills are far less and the particles are accelerated to greater energy. Superconducting magnets can also be found in magnetic resonance imaging (MRI) devices in hospitals. They also hold much promise for high-speed transportation.

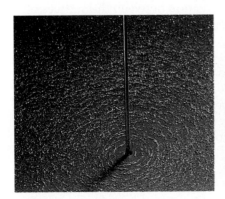

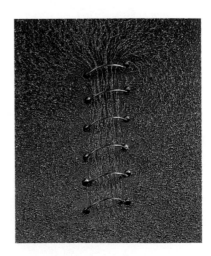

Figure 36.13 ▲
Iron filings sprinkled on paper reveal the magnetic field configurations about (left) a current-carrying wire, (center) a current-carrying loop, and (right) a coil of loops.

36.6 Magnetic Forces on Moving Charged Particles

A charged particle at rest will not interact with a static magnetic field. But if the charged particle *moves* in a magnetic field, the magnetic character of its motion becomes evident. The charged particle experiences a deflecting force.* The force is greatest when the particle moves in a direction perpendicular to the magnetic field lines. At other angles, the force is less; it becomes zero when the particle moves parallel to the field lines. In any case, the direction of the force is always perpendicular to both the magnetic field lines and the velocity of the charged particle (Figure 36.14). So a moving charge is deflected when it crosses magnetic field lines but not when it travels parallel to the field lines.

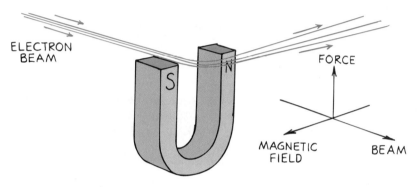

Figure 36.14 ▲
A beam of electrons is deflected by a magnetic field.

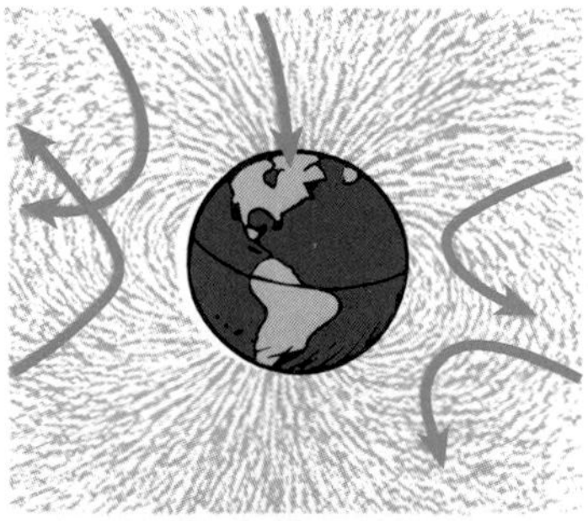

This sideways deflecting force is very different from the forces that occur in other interactions, such as the force of gravitation between masses, the electrostatic force between charges, and the force between magnetic poles. The force that acts on a moving charged particle does not act in a direction between the sources of interaction, but instead acts perpendicular to both the magnetic field and the electron velocity.

It's nice that charged particles are deflected by magnetic fields, for this fact is employed to spread electrons onto the inner surface of a TV tube and provide a picture. This effect of magnetic fields also works on a larger scale. Charged particles from outer space are deflected by the earth's magnetic field. The intensity of cosmic rays striking the earth's surface would be more intense otherwise (Figure 36.15).

Figure 36.15 ▲
The magnetic field of the earth deflects many charged particles that make up cosmic radiation.

---

* When particles of electric charge q and speed v move perpendicular to a magnetic field of strength B, the force F on each particle is simply the product of the three variables: $F = qvB$. For nonperpendicular angles, v in this relationship must be the component of velocity perpendicular to the field.

36.7 Magnetic Forces on Current-Carrying Wires

Simple logic tells you that if a charged particle moving through a magnetic field experiences a deflecting force, then a current of charged particles moving through a magnetic field also experiences a deflecting force. If the particles are trapped inside a wire when they respond to the deflecting force, the wire will also move (Figure 36.16).

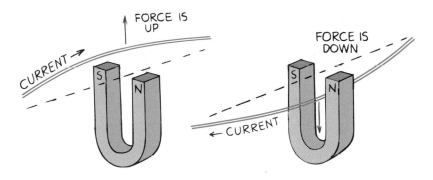

Figure 36.16 ▲
A current-carrying wire experiences a force in a magnetic field. (Can you see that this is a simple extension of Figure 36.14?)

If the direction of current in the wire is reversed, the deflecting force acts in the opposite direction. The force is maximum when the current is perpendicular to the magnetic field lines. The direction of force is along neither the magnetic field lines nor the direction of current. The force is perpendicular to both field lines and current, and it is a sideways force.

So just as a current-carrying wire will deflect a magnetic compass, a magnet will deflect a current-carrying wire. Both cases show different effects of the same phenomenon. The discovery that a magnet exerts a force on a current-carrying wire created much excitement, for almost immediately people began harnessing this force for useful purposes—with great sensitivity in electric meters, and with great force in electric motors.

■ Question

What law of physics tells you that if a current-carrying wire produces a force on a magnet, a magnet must produce a force on a current-carrying wire?

■ Answer

Newton's third law, which applies to *all* forces in nature.

Sound Reproduction

The loudspeakers of your radio and of other sound-producing systems change electrical signals into sound waves. The electrical signals pass through a coil wound around the neck of a paper cone. This coil acts as an electromagnet, which is located near a permanent magnet. When current flows one way, magnetic force pushes the electromagnet toward the permanent magnet, pulling the cone inward. When current flows the other way, the cone is pushed outward. Vibrations in the electric signal then cause the cone to vibrate. Vibrations of the cone produce sound waves in the air.

36.8 Meters to Motors

The simplest meter to detect electric current is shown in Figure 36.17. It consists of a magnetic needle on a pivot at the center of a number of loops of insulated wire. When an electric current passes through the coil, each loop produces its own effect on the needle so that a very small current can be detected. A sensitive current-indicating instrument is called a *galvanometer.**

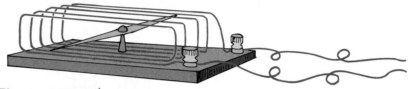

Figure 36.17 ▲
A very simple galvanometer.

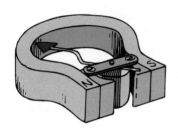

Figure 36.18 ▲
A common galvanometer design.

A more common design is shown in Figure 36.18. It employs more loops of wire and is therefore more sensitive. The coil is mounted for movement and the magnet is held stationary. The coil turns against a spring, so the greater the current in its loops, the greater its deflection.

A galvanometer may be calibrated to measure current (amperes), in which case it is called an *ammeter.* Or it may be calibrated to measure electric potential (volts), in which case it is called a *voltmeter.*

If the design of the galvanometer is slightly modified, you have an electric motor. The principal difference is that the current is made to change direction every time the coil makes a half revolution. After it has been forced to rotate one half revolution, it overshoots just in time for the current to reverse, whereupon it is forced to continue another half revolution, and so on in cyclic fashion to produce continuous rotation.

A simple dc motor is shown in bare outline in Figure 36.20. A permanent magnet is used to produce a magnetic field in a region where a rectangular loop of wire is mounted so that it can turn about an axis as shown. When a current passes through the loop, it flows in opposite directions in the upper and lower sides of the loop. (It has to do this because if charge flows into one end of the loop, it must flow out the other end.) If the upper portion of the loop is forced to the left, then the lower portion is forced to the right, as if it were a galvanometer. But unlike a galvanometer, the current is reversed during each half revolution by means of stationary contacts on the shaft. The parts of the wire that brush against these contacts are called *brushes.* In this way, the current in the loop alternates so that the forces in the upper and lower regions do not change directions as the loop rotates. The rotation is continuous as long as current is supplied.

---

* The galvanometer is named after Luigi Galvani (1737–1798), who discovered while dissecting a frog's leg that electric charge caused the leg to twitch. This chance discovery led to the invention of the chemical cell and the battery.

Larger motors, dc or ac, are usually made by replacing the permanent magnet by an electromagnet that is energized by the power source. Of course, more than a single loop is used. Many loops of wire are wound about an iron cylinder, called an *armature*, which then rotates when energized with electric current.

The advent of the motor made it possible to replace enormous human and animal toil by electric power in most parts of the world. Electric motors have greatly changed the way people live.

PHYSICS ON THE JOB

Oceanographer

Earth scientists who study the oceans are oceanographers. Using space-like submersibles, oceanographers study the composition and characteristics of the ocean floor. Some use sensitive instruments to identify and measure the magnetic fields found in deep ocean rocks. By studying these magnetic patterns, they have learned that the Atlantic Ocean gets a little bit wider each year. Job opportunities exist for oceanographers in academic, private, and government research laboratories.

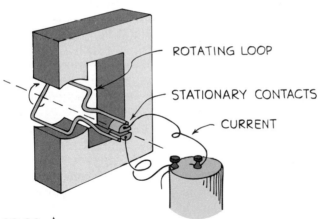

ROTATING LOOP

STATIONARY CONTACTS

CURRENT

Figure 36.20 ▲
A simplified dc motor.

■ Question

How is a galvanometer similar to a simple electric motor? How do they fundamentally differ?

■ Answer

Both a galvanometer and a motor are similar in that they both employ coils positioned in a magnetic field. When a current passes through the coils, forces on the wires rotate the coils. The fundamental difference is that the maximum rotation of the coil in a galvanometer is one half turn, whereas in a motor the coil (armature) rotates through many complete turns. In the armature of a motor, the current is made to alternate with each half turn of the armature.

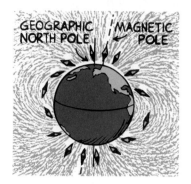

Figure 36.21 ▲
The earth is a magnet.

Figure 36.22 ▲
Convection currents in the molten parts of the earth's interior may produce the earth's magnetic field.

36.9 The Earth's Magnetic Field

A compass points northward because the earth itself is a huge magnet. The compass aligns with the magnetic field of the earth. The magnetic poles of the earth, however, do not coincide with the geographic poles—in fact, they aren't even close to the geographic poles. The magnetic pole in the Northern Hemisphere, for example, is located nearly 1800 kilometers from the geographic North Pole, somewhere in the Hudson Bay region of northern Canada. The other magnetic pole is located south of Australia (Figure 36.21). This means that compasses do not generally point to true north. The discrepancy between the orientation of a compass and true north is known as the *magnetic declination.*

It is not known exactly why the earth itself is a magnet. The configuration of the earth's magnetic field is like that of a strong bar magnet placed near the center of the earth. But the earth is not a magnetized chunk of iron like a bar magnet. It is simply too hot for individual atoms to remain aligned.

Currents in the molten part of the earth beneath the crust provide a better explanation for the earth's magnetic field. Most earth scientists think that moving charges looping around within the earth create its magnetic field. Because of the earth's great size, the speed of moving charges would have to be less than one millimeter per second to account for the field.

Convection currents from the rising heat of the earth's core are another possible cause for the earth's magnetic field (Figure 36.22). Earth's heat comes from the release of nuclear energy—radioactive decay. Perhaps such convection currents combined with the rotational effects of the earth produce the earth's magnetic field. A firmer explanation awaits more study.

Whatever the cause, the magnetic field of the earth is not stable; it has wandered throughout geologic time. Evidence of this comes from analysis of the magnetic properties of rock strata. Iron atoms in a molten state tend to align themselves with the earth's magnetic field. When the iron solidifies, the direction of the earth's field is recorded by the orientation of the domains in the rock. The slight magnetism that results can be measured with sensitive instruments. As samples of rock from different strata formed throughout geologic time are tested, the magnetic field of the earth for different periods can be charted. The evidence from the rock shows that there have been times when the magnetic field of the earth has diminished to zero and then reversed itself.

More than twenty reversals have taken place in the past 5 million years. The most recent occurred 700 000 years ago. Prior reversals happened 870 000 and 950 000 years ago. Studies of deep-sea sediments indicate that the field was virtually switched off for 10 000 to 20 000 years just over 1 million years ago. This was the time that modern humans emerged.

We cannot predict when the next reversal will occur because the reversal sequence is not regular. But there is a clue in recent measurements that show a decrease of over 5% of the earth's magnetic field strength in the last 100 years. If this change is maintained, we may well have another magnetic field reversal within 2000 years.

36 Chapter Review

Concept Summary

North and south magnetic poles produce magnetic forces.

- Like poles repel; opposite poles attract.
- North and south poles always occur in pairs.

A magnetic field is produced by the motion of electric charge.

- In magnetic substances such as iron, the magnetic fields created by spinning electrons do not cancel each other out; large clusters of magnetic atoms align to form magnetic domains.
- In nonmagnetic substances, electron pairs within the atoms spin in opposite directions; there is no net magnetic field.

An electric current produces a magnetic field.

- Bending a current-carrying wire into coils intensifies the magnetic field.
- Placing a piece of iron into a current-carrying coil creates an electromagnet.

A moving charged particle may be deflected by a magnetic field.

- Deflection is greatest for particles moving perpendicular to the magnetic field, and zero for particles moving parallel to the field.

An electric current is also deflected by a magnetic field.

- The force is maximum when the current is perpendicular to the field.
- Galvanometers, ammeters, voltmeters, and electric motors are based on this effect.

The earth itself is a magnet; its magnetic poles are almost 2000 km from its geographic poles.

Important Terms

electromagnet (36.5)
magnetic domain (36.4)
magnetic field (36.2)
magnetic pole (36.1)

Review Questions

1. What do electric charges have in common with magnetic poles? (36.1)

2. What is a major difference between electric charges and magnetic poles? (36.1)

3. What is a magnetic field, and what is its source? (36.2)

4. Every spinning electron is a tiny magnet. Since all atoms have spinning electrons, why are not all atoms tiny magnets? (36.3)

5. What is so special about iron that makes each iron atom a tiny magnet? (36.3)

6. What is a magnetic domain? (36.4)

7. Why do some pieces of iron behave as magnets, while other pieces of iron do not? (36.4)

8. How can a piece of iron be induced into becoming a magnet? For example, if you place a paper clip near a magnet, it will itself become a magnet. Why? (36.4)

9. Why will dropping or heating a magnet weaken it? (36.4)

10. What is the shape of the magnetic field that surrounds a current-carrying wire? (36.5)

11. If a current-carrying wire is bent into a loop, why is the magnetic field stronger inside the loop than outside? (36.5)

12. What must a charged particle be doing in order to experience a magnetic force? (36.6)

13. With respect to an electric and a magnetic field, how does the direction of a magnetic force on a charged particle differ from the direction of the electric force? (36.6)

14. What role does the earth's magnetic field play in cosmic ray bombardment? (36.6)

15. How does the direction in which a current-carrying wire is forced when in a magnetic field compare with the direction that moving charges are forced? (36.7)

16. How do the concepts of force, field, and current relate to a galvanometer? (36.8)

17. Why is it important that the current in the armature of a motor that uses a permanent magnet periodically change direction? (36.8)

18. What is meant by magnetic declination? (36.9)

19. According to most geophysicists, what is the probable cause of the earth's magnetic field? (36.9)

20. What are magnetic pole reversals, and what evidence is there that the earth's magnetic field has undergone pole reversals throughout its history? (36.9)

Think and Explain

1. What kind of field surrounds a stationary electric charge? A moving electric charge?

2. Why can iron be made to behave as a magnet while wood cannot?

3. Since iron filings are not themselves magnets, by what mechanism do they align themselves with a magnetic field as shown in Figure 36.6?

4. A strong magnet and a weak magnet attract each other. Which magnet exerts the stronger force—the strong one or the weak one? (Could you have answered this way back in Chapter 6?)

5. Why will the magnetic field strength be further increased inside a current-carrying coil if a piece of iron is placed in the coil?

6. A cyclotron is a device for accelerating charged particles in circular orbits of ever-increasing radius to high speeds. The charged particles are subjected to both an electric field and a magnetic field. One of these fields increases the speed of the particles, and the other field holds them in a circular path. Which field performs which function?

7. A magnetic field can deflect a beam of electrons, but it cannot do work on them to speed them up. Why? (*Hint:* Consider the direction of the force relative to the direction in which the electrons move.)

8. In what direction relative to a magnetic field does a charged particle travel in order to experience maximum magnetic force? Minimum magnetic force?

9. Pigeons have multiple-domain magnetite magnets within their skulls that are connected to a large number of nerves to the pigeon's brain. How does this aid the pigeon in navigation? (Magnetic material also exists in the abdomens of bees.)

10. What changes in cosmic ray intensity at the earth's surface would you expect during periods in which the earth's magnetic field passed through a zero phase while undergoing pole reversals? (A widely held theory, supported by fossil evidence, is that the periods of no protective magnetic field may have been as effective in changing life forms as X rays have been in the famous heredity studies of fruit flies.)

37 Electromagnetic Induction

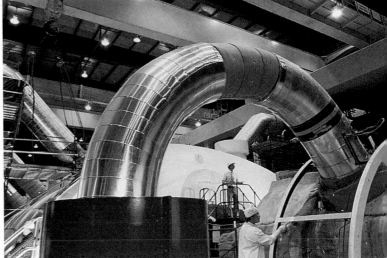

Nuclear power plant turbine room.

The discovery that an electric current in a wire produced magnetism was a turning point in physics and the technology that followed. The question arose as to whether magnetism could produce electric current in a wire. In 1831, two physicists, Michael Faraday in England and Joseph Henry in the United States, independently discovered that the answer is yes. Up until their discovery, the only current-producing devices were voltaic cells, which produced small currents by dissolving expensive metals in acids. These were the forerunners of our present-day batteries. The discovery of Faraday and Henry provided a major alternative to these crude devices. Their discovery was to change the world by making electricity so commonplace that it would power industries by day and light up cities by night.

37.1 Electromagnetic Induction

Faraday and Henry both discovered that electric current could be produced in a wire by simply moving a magnet in or out of a wire coil (Figure 37.1). No battery or other voltage source was needed—only the motion of a magnet in a coil or in a single wire loop. They discovered that voltage was induced by the relative motion between a wire and a magnetic field.

The production of voltage depends only on the relative motion between the conductor and the magnetic field. Voltage is induced whether the magnetic field of a magnet moves past a stationary conductor, or the conductor moves through a stationary magnetic field (Figure 37.2). The results are the same for the same *relative* motion.

The amount of voltage induced depends on how quickly the magnetic field lines are traversed by the wire. Very slow motion produces hardly any voltage at all. Quick motion induces a greater voltage.

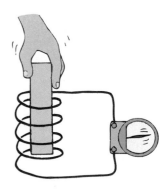

Figure 37.1 ▲
When the magnet is plunged into the coil, voltage is induced in the coil and charges in the coil are set in motion.

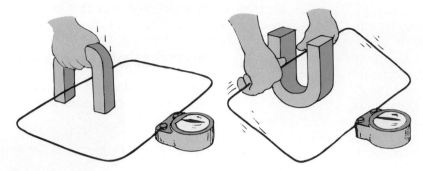

Figure 37.2 ▲

Voltage is induced in the wire loop whether the magnetic field moves past the wire or the wire moves through the magnetic field.

The greater the number of loops of wire that move in a magnetic field, the greater the induced voltage and the greater the current in the wire (Figure 37.3). Pushing a magnet into twice as many loops will induce twice as much voltage; pushing it into ten times as many loops will induce ten times as much voltage; and so on.

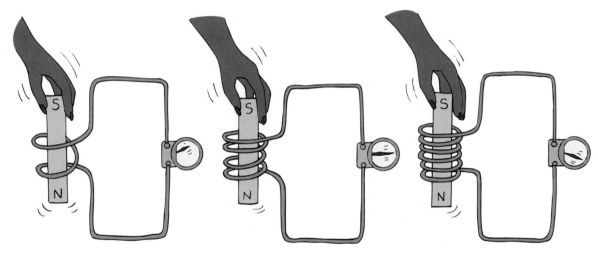

Figure 37.3 ▲

When a magnet is plunged into a coil of twice as many loops as another, twice as much voltage is induced. If the magnet is plunged into a coil with three times as many loops, then three times as much voltage is induced.

Does it seem that we get something (energy) for nothing by simply increasing the number of loops in a coil of wire? We don't. The law of energy conservation is not just a law of mechanics—it is a law of nature. It applies everywhere. In the experiments shown in Figures 37.3 and 37.4, the force that you exert on the magnet multiplied by the distance that you move the magnet is your input work. This work is equal to the energy expended (or possibly stored) in the circuit to which the coil is connected. If the coil is connected to a resistor, for example, more induced voltage in the coil means more current through the resistor, and that means more energy expenditure.

Another way to understand why it takes more force to move a magnet into a coil with more loops is to consider the magnetic effect of the coil acting back on you. When there are more loops, you cause more current to flow in the coil, and this makes the coil act as a more powerful electromagnet. It resists the motion of your magnet. So you must exert a force to move a magnet into the coil because the coil acts as an electromagnet that repels your magnet. A coil with more loops is a stronger magnet and pushes back harder.*

The amount of voltage induced depends on how quickly the magnetic field changes. Very slow movement of the magnet into the coil produces hardly any voltage at all. Quick motion induces a greater voltage.

It doesn't matter which moves—the magnet or the coil. It is the relative motion between the coil and the magnetic field that induces voltage. It so happens that any change in the magnetic field around a conductor induces a voltage. This phenomenon of inducing voltage by changing the magnetic field around a conductor is **electromagnetic induction.**

Figure 37.4 ▲
It is more difficult to push the magnet into the coil with more loops because more current flows and the coil generates a stronger magnetic field that resists the motion of the magnet.

37.2 Faraday's Law

Electromagnetic induction can be summarized in a statement that is called **Faraday's law:**

> The induced voltage in a coil is proportional to the product of the number of loops and the rate at which the magnetic field changes within those loops.

Voltage is one thing, current is another. The amount of current produced by electromagnetic induction depends not only on the induced voltage but also on the resistance of the coil and the circuit to which it is connected.** For example, you can plunge a magnet in and out of a closed rubber loop and in and out of a closed loop of copper. The voltage induced in each is the same, providing each intercepts the same number of magnetic field lines. But the current in each is quite different—a lot in the copper but almost none in the rubber. The electrons in the rubber sense the same electric field as those in the copper, but their bonding to the fixed atoms prevents the movement of charge that occurs so freely in the copper.

---

*   If the coil is not connected to anything, it takes no work to plunge the magnet into the coil, regardless of its number of turns. In this sense you do get something for nothing—an induced voltage, but it doesn't "do" anything. There is no current flow and no transfer of energy.

** Current also depends on the "reactance" of the coil. Reactance is similar to resistance and is important in ac circuits; it depends on the number of loops in the coil and on the frequency of the ac source, among other things. This complication will not be treated in this book.

37.3 Generators and Alternating Current

If one end of a magnet is plunged in and out of a coil of wire, the induced voltage alternates in direction. As the magnetic field strength inside the coil is increased (magnet entering), the induced voltage in the coil is directed one way. When the magnetic field strength diminishes (magnet leaving), the voltage is induced in the opposite direction. The greater the frequency of field change, the greater the induced voltage. The frequency of the induced alternating voltage equals the frequency of the changing magnetic field within the loop.

Rather than moving the magnet, it is more practical to move the coil. This is best accomplished by rotating the coil in a stationary magnetic field (Figure 37.5). This arrangement is called a **generator.** It is essentially the opposite of a motor. Whereas a motor converts electric energy into mechanical energy, a generator converts mechanical energy into electric energy.

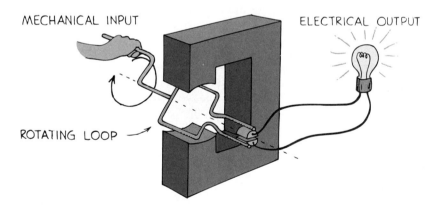

MECHANICAL INPUT

ELECTRICAL OUTPUT

ROTATING LOOP

Figure 37.5 ▲
A simple generator. Voltage is induced in the loop when it is rotated in the magnetic field.

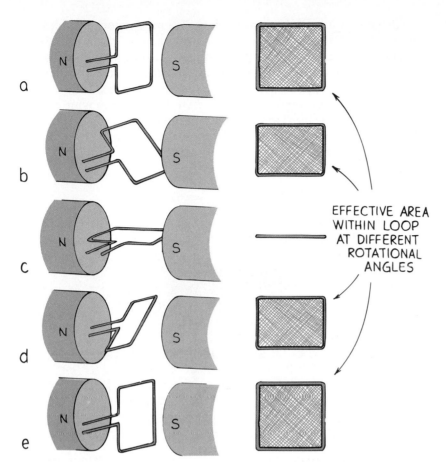

EFFECTIVE AREA WITHIN LOOP AT DIFFERENT ROTATIONAL ANGLES

When the loop of wire is rotated in the magnetic field, there is a change in the number of magnetic field lines within the loop as shown in Figure 37.6. In sketch (a) the loop has the largest number of lines inside it. As the loop rotates (b), it encircles fewer of the field lines until it lies along the field lines (c) and encloses none at all. As rotation continues, it encloses more field lines (d) and reaches a maximum when it has made a half revolution (e). As rotation continues, the magnetic field inside the loop changes in cyclic fashion.

The voltage induced by the generator alternates, and the current produced is alternating current (ac). The current changes magnitude and direction periodically (Figure 37.7). The standard alternating current in North America changes its magnitude and direction during 60 complete cycles per second—60 hertz.

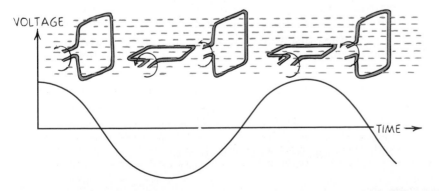

◀ **Figure 37.7**
As the loop rotates, the magnitude and direction of the induced voltage (and current) change. One complete rotation of the loop produces one complete cycle in voltage (and current).

The generators used in power plants are much more complex than the model discussed here. Huge coils made up of many loops of wire are wrapped on an iron core, to make an armature much like the armature of a motor. They rotate in the very strong magnetic fields of powerful electromagnets. The armature is connected externally to an assembly of paddle wheels called a *turbine*. Energy from wind or falling water can be used to produce rotation of the turbine, but most commercial generators are driven by moving steam. Usually a fossil fuel or nuclear fuel is used as the energy source for the steam.

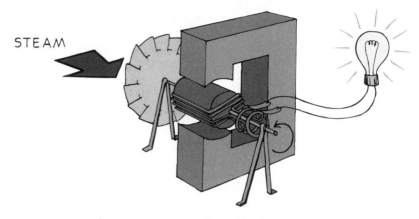

Figure 37.8 ▲
Steam drives the turbine, which is connected to the armature of the generator.

It is important to emphasize that an energy source of some kind is required to operate a generator. Some fraction of energy from the source, usually some type of fuel, is converted to mechanical energy to drive the turbine, and the generator converts most of this to electric energy. The electricity that is produced simply carries this energy to distant places. Some people think that electricity is a *source* of energy. It is not. It is a *form* of energy that must have a source.*

37.4 Motor and Generator Comparison

Chapter 36 discussed how an electric current is deflected in a magnetic field, which underlies the operation of the motor. This discovery occurred about 10 years before Faraday and Henry discovered electromagnetic induction, which underlies the operation of a generator. Both of these discoveries, however, stem from the same single fact: Moving charges experience a force that is perpendicular to both

---

* The conversion of mechanical energy to electric energy can be close to 100%, but because of the laws of thermodynamics, the conversion of heat to either mechanical energy or electric energy is much less efficient. In a typical thermal power plant, 35 to 40% of the fuel energy leaves the plant as electric energy.

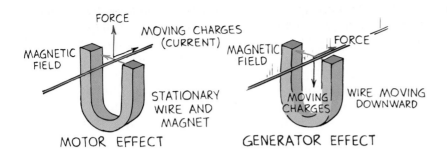

FORCE

MOVING CHARGES
(CURRENT)

MAGNETIC
FIELD

MAGNETIC
FIELD

FORCE

STATIONARY
WIRE AND
MAGNET

MOVING
CHARGES

WIRE MOVING
DOWNWARD

MOTOR EFFECT

GENERATOR EFFECT

◀ **Figure 37.9**
(Left) The motor effect. When a current moves to the right, there is a perpendicular upward force on the electrons. Since there is no conducting path upward, the wire is tugged upward along with the electrons. (Right) The generator effect. When a wire with no initial current is moved downward, the electrons in the wire experience a deflecting force perpendicular to their motion. There *is* a conducting path in this direction, and the electrons follow it, thereby constituting a current. (The current arrow is shown pointing conventionally, as positive charge would move.)

their motion and the magnetic field they traverse. See Figure 37.9. We will call the deflected wire the *motor effect* and the law of induction the *generator effect*. Each of these effects is summarized in the figure. Study them. Can you see that the two effects are related?

37.5 Transformers

Consider a pair of coils, side by side, as shown in Figure 37.10. One is connected to a battery and the other is connected to a galvanometer. It is customary to refer to the coil connected to the power source as the *primary* (input), and the other as the *secondary* (output). As soon as the switch is closed in the primary and current passes through its coil, a current occurs in the secondary also—even though there is no material connection between the two coils. Only a brief surge of current occurs in the secondary, however. Then when the primary switch is opened, a surge of current again registers in the secondary but in the opposite direction.

The explanation is that the magnetic field that builds up around the primary extends into the secondary coil. Changes in the magnetic field of the primary are sensed by the nearby secondary. These changes of magnetic field intensity at the secondary induce voltage in the secondary, in accord with Faraday's law.

If we place an iron core inside the primary and secondary coils of the arrangement shown in Figure 37.10, the magnetic field within the primary is intensified by the alignment of magnetic domains in

■ Question

When the switch of the primary in Figure 37.10 is opened or closed, the galvanometer in the secondary registers a current. But when the switch remains closed, no current is registered on the galvanometer of the secondary. Why?

■ Answer

When the switch remains in the closed position, there is a steady current in the primary, and a steady magnetic field about the coil. This field extends into the secondary, but unless there is a *change* in the field, electromagnetic induction does not occur.

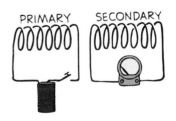

PRIMARY   SECONDARY

Figure 37.10 ▲
Whenever the primary switch is opened or closed, voltage is induced in the secondary circuit.

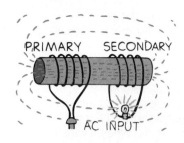

Figure 37.11 ▲
A simple transformer arrangement.

the iron. The magnetic field is also concentrated in the core, which extends into the secondary, so the secondary intercepts more of the field change. The galvanometer will show greater surges of current when the switch of the primary is opened or closed.

Instead of opening and closing a switch to produce the change of magnetic field, suppose that alternating current is used to power the primary. Then the rate at which the magnetic field changes in the primary (and hence in the secondary) is equal to the frequency of the alternating current. Now we have a **transformer** (Figure 37.11).

A more efficient arrangement is shown in Figure 37.12, where the iron core forms a complete loop to guide all the magnetic field lines through the secondary. All the magnetic field lines within the primary are intercepted by the secondary.

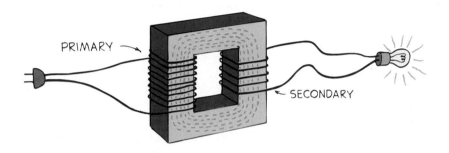

Figure 37.12 ▲
The iron core guides the changing magnetic field lines, which makes a more efficient transformer.

Voltages may be stepped up or stepped down with a transformer. To see how, consider the simple case shown in sketch (a) of Figure 37.13. Suppose the primary consists of one loop connected to a 1-V alternating source. Consider the symmetrical arrangement of a secondary of one loop that intercepts all the changing magnetic field lines of the primary. Then a voltage of 1 V is induced in the secondary.

Figure 37.13 ▶
(a) The voltage of 1 V induced in the secondary equals the voltage of the primary. (b) A voltage of 1 V is induced in the added secondary also because it intercepts the same magnetic field change from the primary. (c) The voltages of 1 V each induced in the two one-turn secondaries are equivalent to a voltage of 2 V induced in a single two-turn secondary.

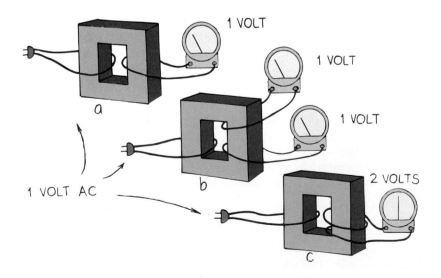

If another loop is wrapped around the core so the transformer has two secondaries (b), it intercepts the same magnetic field change. A voltage of 1 V is induced in it too. There is no need to keep both secondaries separate, for we could join them (c) and still have a total induced voltage of 1 V + 1 V, or 2 V. This is equivalent to saying that a voltage of 2 V will be induced in a single secondary that has twice the number of loops as the primary. So twice as much voltage will be induced in a secondary that has twice as many loops as the primary.

If the secondary is wound with three times as many loops, or *turns* as they are called, then three times as much voltage will be induced. If the secondary has a hundred times as many turns as the primary, then a hundred times as much voltage will be induced, and so on. This arrangement of a greater number of turns on the secondary than on the primary makes up a *step-up transformer*. Stepped-up voltage may light a neon sign or operate the picture tube in a television receiver.

If the secondary has fewer turns than the primary, the alternating voltage produced in the secondary will be *lower* than that in the primary. The voltage is said to be stepped down. If the secondary has half as many turns as the primary, then only half as much voltage is induced in the secondary. Stepped-down voltage may safely operate a toy electric train.

So electric energy can be fed into the primary at a given alternating voltage and taken from the secondary at a greater or lower alternating voltage, depending on the relative number of turns in the primary and secondary coil windings.

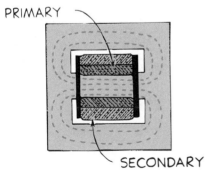

PRIMARY

SECONDARY

Figure 37.14 ▲
A practical transformer.

The relationship between primary and secondary voltages with respect to the relative number of turns is

$$\frac{\text{primary voltage}}{\text{number of primary turns}} = \frac{\text{secondary voltage}}{\text{number of secondary turns}}$$

It might seem that you get something for nothing with a transformer that steps up the voltage. Not so, for energy conservation always controls what can happen. The transformer actually transfers energy from one coil to the other. The rate at which energy is transferred is the power. The power used in the secondary is supplied by the primary. The primary gives no more power than the secondary uses, in accord with the conservation of energy. If the slight power losses due to heating of the core are neglected, then

$$\text{power into primary} = \text{power out of secondary}$$

Electric power is equal to the product of voltage and current, so we can say

$$(\text{voltage} \times \text{current})_{\text{primary}} = (\text{voltage} \times \text{current})_{\text{secondary}}$$

You can see that if the secondary has more voltage, it will have less current than the primary. Or vice versa; if the secondary has less voltage, it will have more current than the primary. The ease with which voltages can be stepped up or down with a transformer is the principal reason that most electric power is ac rather than dc.

■ Questions

The following questions refer to a transformer with 100 turns in the primary and 200 turns in the secondary.

1. If a voltage of 100 V is put across the primary, what will be the voltage output in the secondary?

2. The secondary is connected to a floodlight with a resistance of 50 ohms. Assuming the answer to the last question is 200 V, what will be the current in the secondary circuit?

3. What is the power in the secondary coil?

4. What is the power in the primary coil?

5. What is the current drawn by the primary coil?

6. The voltage has been stepped up, and the current has been stepped down. Ohm's law says that increased voltage will produce increased current. Is there a contradiction here, or does Ohm's law not apply to transformers?

■ Answers

1. From (100 V)/(100 primary turns) = (? V)/(200 secondary turns), you can see that the secondary puts out 200 V.

2. From Ohm's law, (200 V)/(50 Ω) = 4 A

3. Power = (200 V) × (4 A) = 800 W

4. By the law of conservation of energy, the power in the primary is the same, 800 watts.

5. Power = 800 W = (100 V) × (? A), so the primary must draw 8 A. (Note that the voltage is stepped up from primary to secondary and that the current is correspondingly stepped down.)

6. Ohm's law still holds and there is no contradiction. The voltage induced across the secondary circuit, divided by the load (resistance) of the secondary circuit, equals the current in the secondary circuit. The current is stepped down in comparison with the larger current that is drawn in the *primary* circuit.

37.6 Power Transmission

Almost all electric energy sold today is in the form of alternating current because of the ease with which it can be transformed from one voltage to another. Power is transmitted great distances at high voltages and correspondingly low currents, a process that otherwise would result in large energy losses owing to the heating of the wires. Power may be carried from power plants to cities at about 120 000 volts or more, stepped down to about 2200 volts in the city, and finally stepped down again to provide the 120 volts used in household circuits.

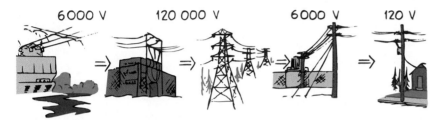

Figure 37.15 ▲
Power transmission.

Energy, then, is transformed from one system of conducting wires to another by electromagnetic induction. The same principles account for eliminating wires and sending energy from a radio-transmitter antenna to a radio receiver many kilometers away, and for the transformation of energy of vibrating electrons in the sun to life energy on Earth. The effects of electromagnetic induction are very far-reaching.

37.7 Induction of Electric and Magnetic Fields

Electromagnetic induction has thus far been discussed in terms of the production of voltages and currents. Actually, the more fundamental way to look at it is in terms of the induction of electric *fields*. The electric fields, in turn, give rise to voltages and currents. Induction takes place whether or not a conducting wire or any material medium is present. In this more general sense, Faraday's law states:

> An electric field is created in any region of space in which a magnetic field is changing with time. The magnitude of the created electric field is proportional to the rate at which the magnetic field changes. The direction of the created electric field is at right angles to the changing magnetic field.

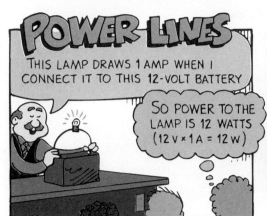

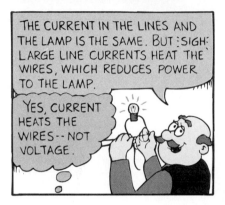

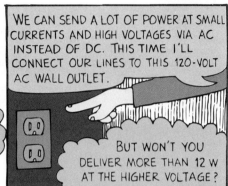

If electric charge happens to be present where the electric field is created, this charge will experience a force. For a charge in a wire, the force could cause it to flow as current, or to push the wire to one side. For a charge in an evacuated region, like in the chamber of a particle accelerator, the force can accelerate the charge to high speeds.

There is a second effect, which is the counterpart to Faraday's law. It is just like Faraday's law, except that the roles of electric and magnetic fields are interchanged. The symmetry between electric and magnetic fields revealed by this pair of laws is one of the many beautiful symmetries in nature. The companion to Faraday's law was advanced by the British physicist James Clerk Maxwell in the 1860s. According to Maxwell,

> A magnetic field is created in any region of space in which an electric field is changing with time. The magnitude of the created magnetic field is proportional to the rate at which the electric field changes. The direction of the created magnetic field is at right angles to the changing electric field.

These statements are two of the most important statements in physics. They underlie an understanding of electromagnetic waves.

37.8 Electromagnetic Waves

Shake the end of a stick back and forth in still water and you will produce waves on the water surface. Similarly shake a charged rod back and forth in empty space and you will produce electromagnetic waves in space. This is because the shaking charge can be considered an electric current. What surrounds an electric current? The answer is a magnetic field. What surrounds a *changing* electric current? The answer is, a changing magnetic field. What do we know about a changing magnetic field? The answer is, it will create a changing electric field, in accord with Faraday's law. What do we know about a changing electric field? The answer is, in accord with Maxwell's counterpart to Faraday's law, the changing electric field will create a changing magnetic field.

An electromagnetic wave is composed of vibrating electric and magnetic fields that regenerate each other. No medium is required. The vibrating fields emanate (move outward) from the vibrating charge. At any point on the wave, the electric field is perpendicular to the magnetic field, and both are perpendicular to the direction of motion of the wave (Figure 37.17).

How fast does the electromagnetic wave move? This is a very interesting question, and, in the history of physics, a very important one. If you ask how fast a baseball moves, or a car, or a spacecraft, or a planet, there is no single answer. It depends on how the motion got started and what forces are acting. But for electromagnetic radiation, there is only one speed—the speed of light—no matter what the frequency or wavelength or intensity of the radiation.

Figure 37.16 ▲
Shake a charged object back and forth and you produce electromagnetic waves.

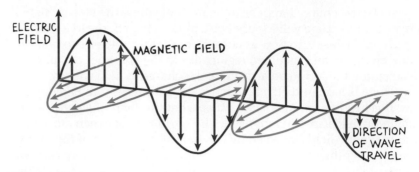

Figure 37.17 ▲
The electric and magnetic fields of an electromagnetic wave are perpendicular to each other.

This remarkable constancy of the speed of propagating electric and magnetic fields was discovered by Maxwell. The key to understanding it lies in the perfect balance between the two kinds of fields that must exist if they are to propagate as waves. The wave is continually self-reinforcing. The changing electric field induces a magnetic field. The changing magnetic field acts back to induce an electric field. Maxwell's equations showed that only one speed could preserve this harmonious balance of fields. If, hypothetically, the wave traveled at less than the speed of light, the fields would rapidly die out. The electric field would induce a weaker magnetic field, which would induce a still weaker electric field, and so on. If, still hypothetically, the wave traveled at more than the speed of light, the fields would build up in a crescendo of ever greater magnitudes—clearly a no-no with respect to energy conservation. At some critical speed, however, mutual induction continues indefinitely, with neither a loss nor a gain in energy.

From his equations of electromagnetic induction, Maxwell calculated the value of this critical speed and found it to be 300 000 kilometers per second. To do this calculation, he used only the constants in his equations determined by simple laboratory experiments with electric and magnetic fields. He didn't *use* the speed of light. He *found* the speed of light!

Maxwell quickly realized that he had discovered the solution to one of the greatest mysteries of the universe—the nature of light. If electric charges are set into vibration with frequencies in the range of 4.3×10^{14} to 7×10^{14} vibrations per second, the resulting electromagnetic wave will activate the "electrical antennae" in the retina of the eye. Light is simply electromagnetic waves in this range of frequencies! The lower end of this frequency range appears red, and the higher end appears violet. Maxwell realized that radiation of any frequency would propagate at the same speed as light.

On the evening of Maxwell's discovery, he had a date with a young woman he was later to marry. While walking in a garden, his date remarked about the beauty and wonder of the stars. Maxwell asked how she would feel to know that she was walking with the only person in the world who knew what the starlight really was. For it was true. At that time, James Clerk Maxwell was the only person in the world to know that light of any kind is energy carried in waves of electric and magnetic fields that continually regenerate each other.

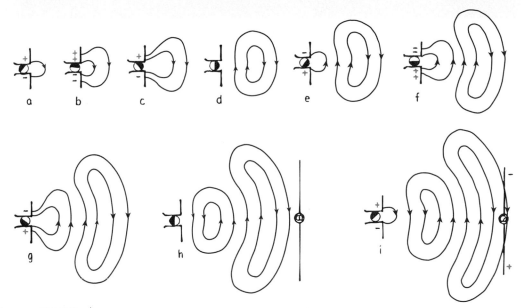

Figure 37.18 ▲

Electromagnetic wave emanation by a sending antenna, and reception by a receiving antenna. The rotating device alternately charges the upper and lower parts of the antenna positively and negatively. Successive views, (a) through (i), show how acceleration of the charges up and down the antenna transmit electromagnetic waves. Only sample electric field lines of the wave are shown—the magnetic field lines are perpendicular to the electric field lines and extend into and out of the page. When the waves are incident upon a receiving antenna, (h) and (i), electric charges in it vibrate in rhythm with the field variations.

SCIENCE, TECHNOLOGY, AND SOCIETY

Is ELF Radiation Dangerous?

We live in fear of the unsensed. Anything that exists, or is imagined to exist, yet escapes detection by our five senses, is often a source of fear.

Many people fear radiation. Some is hazardous, and some is not. No one doubts the hazards of radiation from some nuclear reactions, and no one seriously fears the low-frequency radiations of AM radio. But in recent years a series of books and magazine articles have fanned the flames of public fear by claiming that the *extremely low frequency* (ELF) *radiation* of common 60-Hz electric power causes certain forms of cancer.

Is this claim true? Some activists say yes, while most scientists rate it as one of many health scares that prove to be without basis when carefully studied. Bioscientists point out that the electric fields due to power lines at the location of a cell in the body are thousands of times smaller than those due to the normal electric activity of nearby cells. They also point out that cancer rates have remained constant or fallen over the last 50 years (with the exception of rising cancer rates due to smoking). Yet during this time, exposure to ELF radiation has increased tremendously. More detailed analysis of the studies that prompted the controversy show no link between ELF and cancer.

Suppose you're a scientist and you find uncertain evidence that some common food—tomatoes, for example—may be a serious health risk. What responsibility would you have to make your findings known to the general public? If you stress the uncertainty of your findings, perhaps no one will listen. Should you then make sensational claims, even unsupported, to get people's attention?

Concept Summary

Electromagnetic induction is the inducing of voltage by changing the magnetic field near a conductor.

- According to Faraday's law, the induced voltage in a coil is proportional to the product of the number of loops and the rate at which the magnetic field within the loops changes.

- A generator uses electromagnetic induction to convert mechanical energy to electric energy.

- A transformer uses electromagnetic induction to induce a voltage in the secondary that is different from that in the primary coil.

Electromagnetic induction may be described in terms of fields.

- A changing magnetic field induces an electric field.

- A changing electric field induces a magnetic field.

- Electromagnetic waves are composed of vibrating electric and magnetic fields that regenerate each other.

- All electromagnetic waves travel at the same speed, the speed of light.

Important Terms

electromagnetic induction (37.1)
Faraday's law (37.2)
generator (37.3)
transformer (37.5)

Review Questions

1. What did Michael Faraday and Joseph Henry discover? (37.1)

2. How can voltage be induced in a wire with the help of a magnet? (37.1)

3. A magnet moved into a coil of wire will induce voltage in the coil. What is the effect of moving a magnet into a coil with more loops? (37.1)

4. Why is it more difficult to move a magnet into a coil of more loops that is connected to a resistor? (37.1)

5. Current, as well as voltage, is induced in a wire by electromagnetic induction. Why is Faraday's law expressed in terms of induced voltage, and not induced current? (37.2)

6. How does the frequency of a changing magnetic field compare with the frequency of the alternating voltage that is induced? (37.3)

7. What is a generator, and how does it differ from a motor? (37.3)

8. Why is alternating voltage induced in the rotating armature of a generator? (37.3)

9. The armature of a generator must rotate in order to induce voltage and current. What causes the rotation? (37.3)

10. A motor is characterized by three main ingredients: magnetic field, moving charges, and magnetic force. What are the three main ingredients that characterize a generator? (37.4)

11. How can a change in voltage in a coil of wire (the primary) be transferred to a neighboring coil of wire (the secondary) without physical contact? (37.5)

12. Why does an iron core that extends inside and connects the primary and secondary coils intensify electromagnetic induction? (37.5)

13. What does a transformer actually transform—voltage, current, or energy? (37.5)

14. What does a step-up transformer step up— voltage, current, or energy? (37.5)

15. How does the relative number of turns on the primary and the secondary coil in a transformer affect the step-up or step-down voltage factor? (37.5)

16. If the number of secondary turns is 10 times the number of primary turns, and the input voltage to the primary is 6 volts, how many volts will be induced in the secondary coil? (37.5)

17. **a.** In a transformer, how does the power input to the primary coil compare with the power output of the secondary coil?

    b. How does the product of voltage and current in the primary compare with the product of voltage and current in the secondary? (37.5)

18. Why is it advantageous to transmit electrical power long distances at high voltages? (37.6)

19. What fundamental quantity underlies the concepts of voltages and currents? (37.7)

20. Distinguish between Faraday's law expressed in terms of fields and Maxwell's counterpart to Faraday's law. How are the two laws symmetrical? (37.7)

21. How do the wave speeds compare for high-frequency and low-frequency electromagnetic waves? (37.8)

22. What is light? (37.8)

Think and Explain

1. A common pickup for an electric guitar consists of a coil of wire around a permanent magnet. The permanent magnet induces magnetism in the nearby guitar string. When the string is plucked, it oscillates above the coil, thereby changing the magnetic field that passes through the coil. The rhythmic oscillations of the string produce the same rhythmic changes in the magnetic field in the coil, which in turn induce the same rhythmic voltages in the coil, which when amplified and sent to a speaker produce music! Why will this type pickup not work with nylon strings?

2. Why is a generator armature more difficult to rotate when it is connected to and supplying electric current to a circuit?

3. Some bicycles have electric generators that are made to turn when the bike wheel turns. These generators provide energy for the bike's lamp. Will a cyclist coast farther if the lamp connected to the generator is turned off? Explain.

4. An electric hair drier running at normal speed draws a relatively small current. But if somehow the motor shaft is prevented from turning, the current dramatically increases and the motor overheats. Why?

5. When a piece of plastic tape coated with iron oxide that is magnetized more in some parts than others is moved past a small coil of wire, what happens in the coil? What is a practical application of this?

6. When a strip of magnetic material, variably magnetized, is embedded in a plastic card that is moved past a small coil of wire, what happens in the coil. What is a practical application of this?

7. If a car made of iron and steel moves over a wide closed loop of wire embedded in a road surface, will the magnetic field of the earth in the loop be altered? Will this produce a current pulse? (Can you think of a practical application of this?)

8. How could you move a conducting loop of wire through a magnetic field without inducing a voltage in the loop?

9. Why does a transformer require alternating voltage?

10. Can an efficient transformer step up energy? Defend your answer.

11. What is wrong with this scheme? To generate electricity without fuel, arrange a motor to run a generator that will produce electricity that is stepped up with a transformer so that the generator can run the motor while furnishing electricity for other uses.

Think and Solve

1. A transformer has an input of 9 volts and an output of 36 volts. If the input is changed to 12 volts, what will the output be?

2. A portable tape deck requires 12 volts to operate correctly. A transformer nicely allows the device to be powered from a 120-volt outlet. If the primary has 500 turns, how many turns should the secondary have?

3. A model electric train requires a low voltage to operate. If the primary coil of its transformer has 400 turns, and the secondary has 40 turns, how many volts will power the train when the primary is connected to a 120-volt household circuit?

Atomic and Nuclear Physics

The natural heat of the earth that warms this natural hot spring, or that powers a geyser or powers a volcano, comes from nuclear power—the radioactivity of minerals in the earth's interior. Power from the atomic nucleus has been with us since the earth was formed and is not restricted to today's nuclear reactors, or "nukes" as they are called. Opinions about nuclear power are informed opinions if we first "know nukes"!

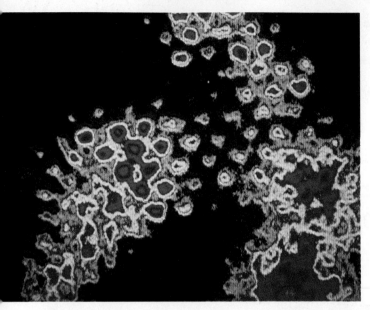

Uranium atoms as seen by a scanning tunneling microscope.

T he final unit of this book is about the realm of the unimaginably tiny atom. This chapter investigates atomic structure, which is revealed by analyzing light. Light has a dual nature, which in turn radically alters our understanding of the atomic world. The next chapter covers the structure of the atomic nucleus and radioactivity, and the concluding chapter is about the nuclear processes of fission and fusion.

38.1 Models

Nobody knows what an atom's internal structure looks like, for there is no way to see it with our eyes. To visualize the processes that occur in the subatomic realm, we construct models. In the planetary model—the one that most people think of when they picture an atom—the electrons orbit the nucleus like planets going around the sun. This was an early model of the atom suggested by the Danish physicist Niels Bohr in 1913 (Figure 38.1). We still tend to think in terms of this simple picture, even though it has been replaced by a more complex model in which the electrons are represented by clouds spreading over the interior of the atom (Figure 38.2). We will see that the planetary model of the atom is still useful for understanding the emission of light.

Models are assessed not in terms of their "truthfulness," but in terms of their "usefulness." Models help us to understand processes that are difficult to visualize. A useful model of the atom must be consistent with a model for light, for most of what we know about atoms we learn from the light and other radiations they emit. Most light has its source in the motion of electrons within the atom.

Down through the centuries there have been two primary models of light: the particle model and the wave model. Isaac Newton

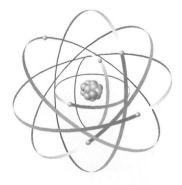

Figure 38.1 ▲
The old planetary model of an atom with electrons orbiting like little planets around a tiny sun.

believed in a particle model of light. He thought that light was composed of a hail of tiny particles. Christian Huygens believed that light was a wave phenomenon. The wave model was reinforced more than a century later when Thomas Young demonstrated interference. Later, James Clerk Maxwell proposed that light is electromagnetic radiation and a part of a broader electromagnetic spectrum. The wave model gained support when Heinrich Hertz produced radio waves that behaved as Maxwell had predicted. This seemed to verify the wave nature of light once and for all. But Maxwell's electromagnetic wave model was not the last word on the nature of light. In 1905 Albert Einstein resurrected the particle theory of light.

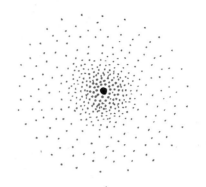

Figure 38.2 ▲
The current model of the atom, in which electrons are spread out in waves or clouds.

38.2 Light Quanta

Einstein visualized particles of light as concentrated bundles of electromagnetic energy. Einstein built on the idea of a German physicist, Max Planck, who a few years earlier had proposed that atoms do not emit and absorb light continuously, but do so in little chunks he called **quanta** (plural of *quantum*). Planck believed that light existed as continuous waves, just as Maxwell had asserted, but that emission and absorption occurred in quantum chunks. Einstein went further and proposed that light itself is composed of quanta. These quanta are now called **photons.**

A quantum is an elemental unit—a smallest amount of something. The idea that certain quantities are quantized—that they come in discrete (separate) units—was known in Einstein's time. Matter is quantized. The mass of a gold ring, for example, is equal to some whole-number multiple of the mass of a single gold atom. Electricity is quantized, as all electric charge is some whole-number multiple of the charge of a single electron.

Recent findings in physics tell us that there are other quantities that are also quantized—quantities such as energy and angular momentum. The energy in a light beam is quantized and comes in packets, or quanta; only a whole number of quanta can exist. The quanta of light, or of electromagnetic radiation in general, are the photons.

Photons have no rest energy. They move at one speed only—as fast as it is possible for anything to move—at the speed of light! The total energy of a photon is the same as its kinetic energy. This energy is directly proportional to the photon's frequency. When the energy E of a photon is divided by its frequency f, the quantity that results is always the same, no matter what the frequency. This quantity is a constant known as **Planck's constant,** h. The energy of every photon is therefore

$$E = hf$$

This equation gives the smallest amount of energy that can be converted to light of frequency f. Light is not emitted continuously, but as a stream of photons, each with an energy hf.

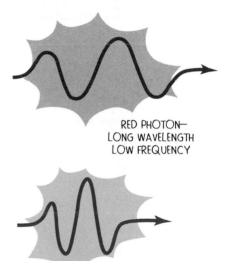

Figure 38.3 ▲
The energy of a photon of light is proportional to its vibrational frequency.

RED PHOTON— LONG WAVELENGTH LOW FREQUENCY

BLUE PHOTON— SHORT WAVELENGTH HIGH FREQUENCY

38.3 The Photoelectric Effect

Einstein found support for his quantum theory of light in the photo-electric effect. The **photoelectric effect** is the ejection of electrons from certain metals when light falls upon them. These metals are said to be *photosensitive* (that is, sensitive to light). This effect is used in electric eyes, in the photographer's light meter, and in picking up sound from the sound tracks of motion pictures.

Investigators discovered that high-frequency light, even from a dim source, was capable of ejecting electrons from a photosensitive metal surface; yet low-frequency light, even from a very bright source, could not dislodge electrons. Since bright light carries more energy than dim light, it was puzzling that dim blue or violet light could dislodge electrons from certain metals when bright red light could not.

Einstein explained the photoelectric effect by thinking of light in terms of photons. The absorption of a photon by an atom in the metal surface is an all-or-nothing process. One and only one photon is completely absorbed by each electron ejected from the metal. This means that the *number* of photons that hit the metal has nothing to do with whether a given electron will be ejected. If the energy per photon is too small, then the brightness or intensity of light does not matter. The critical factor is the frequency or color of the light. A few photons of blue or violet light can eject a few electrons, but hordes of red or orange photons cannot eject a single electron. Only high-frequency photons have the concentrated energy needed to pull loose an electron.

A wave has a broad front, and its energy is spread out along this front. For the energy of a light wave to be concentrated enough to eject a single electron from a metal surface is as unlikely as for an ocean wave to hit a beach and catapult a boulder far inland with an

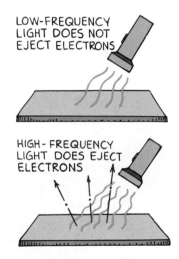

Figure 38.4 ▲
The photoelectric effect depends on the frequency of light.

■ Questions

1. Will brighter light eject more electrons from a photosensitive surface than dimmer light of the same frequency?

2. Will high-frequency light eject a greater number of electrons than low-frequency light?

Answers

1. Yes (if the frequency is great enough for any electrons to be ejected). The number of ejected electrons depends on the number of incident photons.

2. Not necessarily. The answer is yes if electrons are ejected by the high-frequency light but not by the low-frequency light, because its photons do not have enough energy. If the light of both frequencies can eject electrons, then the number of electrons ejected depends on the brightness of the light, not on its frequency.

energy equal to the energy of the whole wave. The photoelectric effect suggests that light interacts with matter as a stream of particle-like photons. The number of photons in a light beam controls the brightness of the *whole* beam, whereas the frequency of the light controls the energy of each *individual* photon. Light travels as a wave, and interacts with matter as a stream of particles.

Experimental verification of Einstein's explanation of the photoelectric effect was made 11 years later by the American physicist Robert Millikan. Every aspect of Einstein's interpretation was confirmed, including the direct proportionality of photon energy to frequency. It was for this (and not for his theory of relativity) that Einstein received the Nobel Prize.

38.4 Waves as Particles

Figure 38.5 is a striking example of the particle nature of light. The photograph was taken with exceedingly feeble light. Each frame shows the image progressing photon by photon. Note also that the photons seem to strike the film in an independent and random manner.

a.

b.

c.

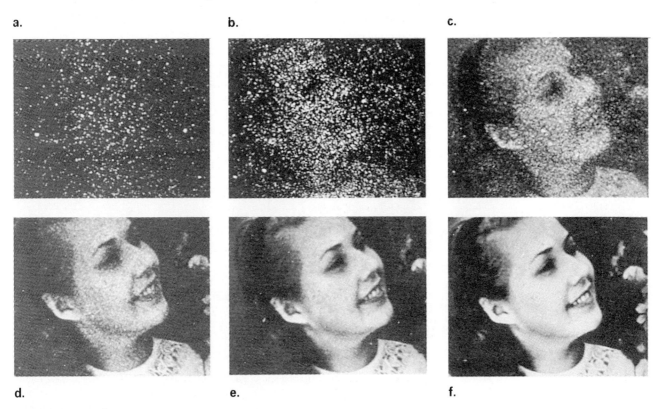

d.

e.

f.

Figure 38.5 ▲
Stages of exposure reveal the photon-by-photon production of a photograph. The approximate numbers of photons at each stage were (a) 3×10^3, (b) 1.2×10^4, (c) 9.3×10^4, (d) 7.6×10^5, (e) 3.6×10^6, (f) 2.8×10^7.

38.5 Particles as Waves

If waves can have particle properties, cannot particles have wave properties? This question was posed by the French physicist Louis de Broglie in 1924, while he was still a student. His answer to the question gave him a Ph.D. in physics, and later won him the Nobel Prize in physics.

De Broglie suggested that all matter could be viewed as having wave properties. All particles—electrons, protons, atoms, bullets, and even humans—have a wavelength that is related to the momentum of the particles by

$$\text{wavelength} = \frac{h}{\text{momentum}}$$

where h is, lo and behold, Planck's constant again. The wavelength of a particle is called the *de Broglie wavelength*. A particle of large mass and ordinary speed has too small a wavelength to be detected by conventional means. However, a tiny particle—such as an electron—moving at high speed has a detectable wavelength.* It is smaller than the wavelength of visible light but large enough for noticeable diffraction. A beam of electrons, interestingly enough, behaves like a beam of light. It can be diffracted and undergoes wave interference under the same conditions that light does (Figure 38.6).

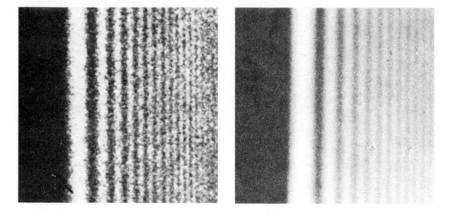

Figure 38.6 ▲
Fringes produced by the diffraction (left) of an electron beam and (right) of light.

---

* A bullet of mass 0.02 kg traveling at 330 m/s, for example, has a de Broglie wavelength of $h/mv = (6.6 \times 10^{-34} \text{ J·s})/[(0.02 \text{ kg}) \times (330 \text{ m/s})] = 10^{-34}$ m, an incredibly small size that is a million million million millionth the diameter of a hydrogen atom. An electron traveling at 2% the speed of light, on the other hand, has a wavelength of 10^{-10} m, which is equal to the diameter of the hydrogen atom. Diffraction effects for electrons are measurable, but diffraction effects for bullets are not.

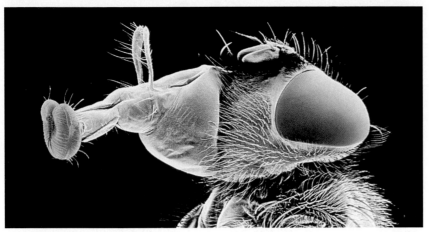

Figure 38.7
Detail of a fly's head as seen with
a scanning electron microscope.

An electron microscope makes practical use of the wave nature of electrons. The wavelength of electron beams is typically thousands of times shorter than the wavelength of visible light, so the electron microscope is able to distinguish detail not possible with optical microscopes (Figure 38.7).

38.6 Electron Waves

More far-reaching than the diffraction of electrons is de Broglie's model of matter waves in the atom. The planetary model of the atom developed by Niels Bohr was useful in explaining the atomic spectra of the elements. It explained why elements emitted only certain frequencies of light. An electron has different amounts of energy when it is in different orbits around a nucleus. From an energy point of view, an electron is said to be in different *energy levels* when it is in different orbits. The electrons in an atom normally occupy the lowest energy levels available.

An electron can be boosted by various means to a higher energy level. This occurs in gas discharge tubes, like those that make up neon signs. Electric current boosts the electrons of the gas to higher energy levels. As the electrons return to lower levels, photons are emitted. The energy of a photon is exactly equal to the difference in the energy levels in the atom. The characteristic pattern of lines in the spectrum of an element corresponds to electron transitions between the energy levels characteristic of the atoms of that element. By examining spectra, physicists were able to determine the various energy levels in the atom. This was a tremendous triumph for atomic physics.

One of the difficulties of this model of the atom, however, was reconciling why electrons occupied only certain energy levels in the atom—why, in effect, they were at discrete distances from the atomic nucleus. This was resolved by thinking of the electron not as a particle, a tiny BB whirling around the nucleus like a planet whirling around the sun, but as a wave. Investigators now consider the electron to be more *wave* than particle.

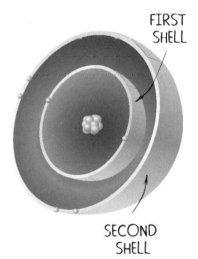

FIRST
SHELL

SECOND
SHELL

Figure 38.8 ▲
The Bohr model of the atom.

Figure 38.9 ▶

De Broglie's way of explaining discrete energy states in an atom. (a) Orbital electrons form standing waves only when the circumference of the orbit is equal to a whole-number multiple of the wavelength. (b) When the wave does not close in on itself in phase, it undergoes destructive interference.

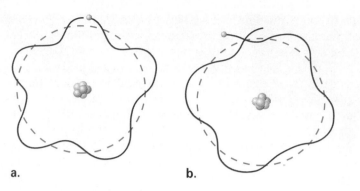

a.   b.

According to de Broglie's theory of matter waves, an orbit exists where an electron wave closes in on itself in phase. In this way it reinforces itself constructively in each cycle, just as the standing wave on a music string is constructively reinforced by its successive reflections. In this view, the electron is visualized not as a particle located at some point in the atom, but as though its mass and charge were spread out into a standing wave surrounding the nucleus. The wavelength of the electron wave must fit evenly into the circumferences of the orbits (Figure 38.9).

The circumference of the innermost orbit, according to this model, is equal to one wavelength of the electron wave. The second orbit has a circumference of two electron wavelengths, the third three, and so on (Figure 38.10). This is similar to a "chain necklace" made of paper clips. No matter what size necklace is made, its circumference is

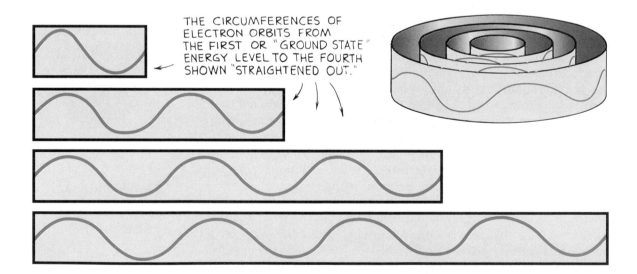

THE CIRCUMFERENCES OF ELECTRON ORBITS FROM THE FIRST OR "GROUND STATE" ENERGY LEVEL TO THE FOURTH SHOWN "STRAIGHTENED OUT."

Figure 38.10 ▲

A simplified version of the principle that the states of motion of electrons result from self-reinforcement of standing waves. In this illustration, the waves are shown only in circular paths around the nucleus. Orbit circumferences are whole-number multiples of the electron wavelengths, which differ for the various elements (and also for different orbits within the elements). This results in discrete energy levels, which characterize each element. In an actual atom, the standing waves make up spherical and ellipsoidal shells rather than flat, circular ones.

equal to some multiple of the length of a single paper clip.* Since the circumferences of electron orbits are discrete, it follows that the radii of these orbits, and hence the energy levels, are also discrete.

This view explains why electrons do not spiral closer and closer to the nucleus when photons are emitted. If each electron orbit is described by a standing wave, the circumference of the smallest orbit can be no smaller than one wavelength—no fraction of a wavelength is possible in a circular (or elliptical) standing wave.

In the still more modern wave model of the atom, electron waves move not only around the nucleus, but also in and out, toward and away from the nucleus. The electron wave is spread out in three dimensions. This leads to the picture of an electron "cloud," shown previously in Figure 38.2.

38.7 Relative Sizes of Atoms

The radii of the electron orbits in the Bohr model of the atom are determined by the amount of electric charge in the nucleus. For example, the single positively charged proton in the hydrogen atom holds one negatively charged electron in an orbit at a particular radius. If we double the positive charge in the nucleus, the orbiting electron will be pulled into a tighter orbit with half its former radius since the electrical attraction is doubled. This doesn't quite happen, however, because the double-positive charge in the nucleus attracts and holds a second electron, and the negative charge of the second electron diminishes the effect of the positive nucleus.

This added electron makes the atom electrically neutral. The atom is no longer hydrogen, but is helium. The two electrons assume an orbit characteristic of helium. An additional proton in the nucleus pulls the electrons into an even closer orbit and, furthermore, holds a third electron in a second orbit. This is the lithium atom, atomic number 3. We can continue with this process, increasing the positive charge of the nucleus and adding successively more electrons and more orbits all the way up to atomic numbers above 100, to the synthetic radioactive elements.**

As the nuclear charge increases and additional electrons are added in outer orbits, the inner orbits shrink in size because of the stronger electrical attraction to the nucleus. This means that the heavier elements

---

* Electron wavelengths are successively longer for orbits of increasing radii; so for a more accurate analogy, the construction of longer necklaces requires using not only *more* paper clips, but *larger* paper clips as well.

** There is a maximum number of electrons that each orbit may hold. A rule of quantum mechanics states that an orbit is filled when it contains a number of electrons given by $2n^2$, where n is 1 for the first orbit, 2 for the second orbit, 3 for the third orbit, and so on. For $n = 1$, there are 2 electrons; for $n = 2$, there are $2(2^2)$, or 8 electrons; for $n = 3$, there are a maximum of $2(3^2)$, or 18 electrons, and so on. The number n is called the *principal quantum number*. Because of complexities that arise in heavy atoms, the $2n^2$ rule is strictly valid only for the lighter atoms.

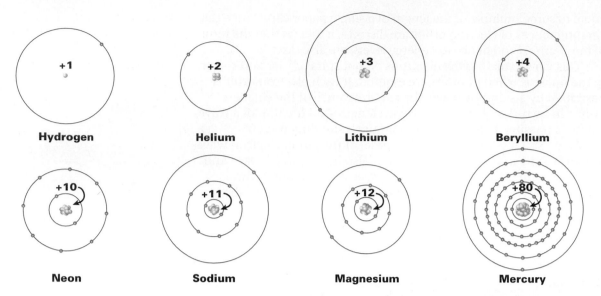

Figure 38.11 ▲

The orbital model illustrated for some light and heavy atoms drawn to approximate scale. Note that the heavier atoms are not appreciably larger than the lighter atoms.

are not much larger in diameter than the lighter elements. The diameter of the uranium atom, for example, is only about three hydrogen diameters even though it is 238 times more massive. The schematic diagrams in Figure 38.11 are drawn approximately to the same scale.

Each element has an arrangement of electron orbits unique to that element. For example, the radii of the orbits for the sodium atom are the same for all sodium atoms, but different from the radii of the orbits for other kinds of atoms. When we consider the 92 naturally occurring elements, we find that there are 92 distinct patterns or orbits. There is a different pattern for each element.

The Bohr model of the atom solved the mystery of the atomic spectra of the elements. It accounted for X rays that were emitted when electrons made transitions from outer orbits to innermost orbits. Bohr was able to predict X-ray frequencies that were later experimentally confirmed. He calculated the *ionization energy* of the hydrogen atom—the energy needed to knock the electron out of the atom completely. This also was verified by experiment. The Bohr model accounted for the general chemical properties of the elements and predicted properties of a missing element (hafnium), which led to its discovery.

The Bohr model was impressive. Nonetheless, Bohr was quick to point out that his model was to be interpreted as a crude beginning, and the picture of electrons whirling like planets about the sun was not to be taken literally (to which popularizers of science paid no heed). His discrete orbits were conceptual representations of an atom whose later description involved a wave description. Still, his planetary model of the atom with electrons occupying discrete energy levels underlies the more complex models of the atom today, which are built upon a completely different structure from that built by Newton and other physicists before the twentieth century. This is the structure called *quantum mechanics*.

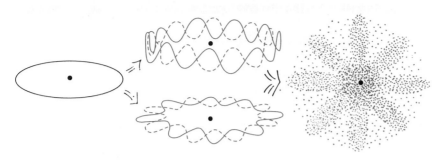

■ **Question**

What fundamental force dictates the size of an atom?

38.8 Quantum Physics

The more that physicists studied the atom, the more convinced they became that the Newtonian laws that work so well for large objects such as baseballs and planets (in the "macroworld") simply do not apply to events in the microworld of the atom. Whereas in the macroworld the study of motion is called *mechanics,* or sometimes classical mechanics, in the microworld of quanta it is called **quantum mechanics.** The more general study of quanta in the microworld is simply called **quantum physics.**

While one can be quite certain about careful measurements in the macroscopic world, there are fundamental uncertainties in the measurements of the atomic domain. For the measurement of macroscopic quantities, such as the temperature of materials, the vibrational frequencies of certain crystals, and the speeds of light and sound, there is no limit in practice to the accuracy with which the experimenter can measure. But subatomic measurements, such as the momentum and position of an electron or the mass of an extremely short-lived particle, are entirely different. In this domain, the uncertainties in many measurements are comparable to the magnitudes of the quantities themselves. The structure of quantum mechanics is based on probabilities, a notion that is difficult for many people to accept. Even Einstein did not accept this, which prompted his often-quoted statement, "I cannot believe that God plays dice with the universe."

If you continue to expose yourself to physics, you will likely study quantum mechanics in the future. Then you'll find that subatomic interactions are governed by laws of probability, not laws of certainty. It's fascinating material. Scientists and philosophers are still pondering the real meaning of quantum mechanics.

■ **Answer**

The electrical force.

38.9 Predictability and Chaos

When we know the initial conditions of an orderly system we can make predictions about it. For example, in the Newtonian macro-world, knowing with precision the initial conditions lets us state where a planet will be after a certain time, where a launched rocket will land, and when an eclipse will occur. Similarly, in the quantum microworld we can predict where an electron is likely to be in an atom, and the probability that a radioactive particle will decay in a given time interval. Predictability in orderly systems, both Newtonian and quantum, depends on knowledge of initial conditions.

Some systems however, whether Newtonian or quantum, are not orderly—they are inherently unpredictable. These are called "chaotic systems." Turbulent water flow is an example. No matter how precisely we know the initial conditions of a piece of floating wood as it flows downstream, we cannot predict its location later downstream. A feature of chaotic systems is that slight differences in initial conditions result in wildly different outcomes later. Two identical pieces of wood just slightly apart at one time are vastly far apart soon thereafter.

Weather is chaotic. Small changes in one day's weather can produce big (and largely unpredictable) changes a week later. Meteorologists try their best, but they are bucking the hard fact of chaos in nature. This barrier to good prediction first led the scientist Edward Lorenz to ask, "Does the flap of a butterfly's wings in Brazil set off a tornado in Texas?" Now we talk about the *butterfly effect* when we are dealing with situations where very small effects can amplify into very big effects.

PHYSICS OF SPORTS

Chaos on the Slopes

If you and a friend, sitting on snowboards at the top of a perfectly smooth ski slope, push off from positions close together with about the same velocity, you'll follow similar paths and end up near each other at the bottom of the slope. This is *orderly* behavior. Small differences in conditions at the beginning result in small differences in conditions at the end. But if the ski slope is full of hundreds of moguls (bumps) you'll likely find, after bouncing and jouncing down the slope, that you and your friend wind up many meters apart—no matter how close your initial conditions. This is *chaotic* behavior. Small differences in conditions at the beginning are likely to result in large differences in conditions at the end—so much so that you can't predict where you'll end up.

Interestingly, chaos is not all hopeless unpredictability. If you and your friend compare notes after your rides down the moguled slope, you might find that you had some very similar experiences. Maybe you skirted around the sides of many moguls in just the same way. Maybe you never went straight over the top of a mogul. So there is *order in chaos.* Scientists have learned how to treat chaos mathematically and how to find the parts of it that are orderly.

38 Chapter Review

Concept Summary

Models of atoms have involved particles and waves.

- In Bohr's planetary model of the atom, electrons orbit the nucleus.

- In de Broglie's early wave model, electron waves circle the nucleus only at radii where there are exact integral numbers of wavelengths.

- In the modern wave model, electron waves fill the three-dimensional space around the nucleus and can be visualized as clouds.

Models of light have involved particles and waves.

- Newton proposed a particle model of light.

- Huygens' wave model of light was reinforced by Maxwell's electromagnetic wave model.

- Einstein returned to a particle model, proposing that light is composed of discrete bundles, or quanta, of electromagnetic energy.

Einstein's light quantum is called a photon.

- The energy of a photon is proportional to its frequency.

- The constant of proportionality is Planck's constant, *h*.

The photoelectric effect, the ejection of electrons from certain metals that are struck by light, reinforced the particle theory of light.

- When the frequency of light is so low that the photons have insufficient energy to cause the effect, then increasing the intensity of the light does not matter.

Material particles and light have both wave properties and particle properties.

- Electrons, like photons, have a detectable wavelength, can be diffracted, and undergo interference.

Bohr's planetary model explained the atomic spectra of the elements.

- Spectral lines are due to transitions of electrons between energy levels that correspond to different electron orbits.

The radius of the atoms making up each element is unique to that element.

- As the nuclear charge increases, inner electron orbits shrink. As a result, heavier elements are not much larger in diameter than lighter elements.

Quantum mechanics is the study of motion in the microworld.

- Newtonian mechanics does not apply to events on the atomic scale.

- Quantum measurements, such as the momentum and position of an electron, involve fundamental uncertainties.

Important Terms

photoelectric effect (38.3)
photon (38.2)
Planck's constant (38.2)
quantum (pl. quanta) (38.2)
quantum mechanics (38.8)
quantum physics (38.8)

Review Questions

1. What is a model? Give two examples for the nature of light. (38.1)

2. What is a quantum? Give two examples. (38.2)

3. What is a quantum of light called? (38.2)

4. What is Planck's constant, and how does it relate to the frequency and energy of a quantum of light? (38.2)

5. Which has more energy per photon—red light or blue light? (38.2)

6. What is the photoelectric effect? (38.3)

7. Why does blue light eject electrons from a certain photosensitive surface, whereas red light has no effect on that surface? (38.3)

8. Will bright blue light eject more electrons than dim light of the same frequency? (38.3)

9. Does the photoelectric effect support the particle model or the wave model of light? (38.3)

10. **a.** Do particles of matter have wave properties?

    b. Who was the first physicist to give a convincing answer to this question? (38.5)

11. As the speed of a particle increases, does its associated wavelength increase or decrease? (38.5)

12. Does the diffraction of an electron beam support the particle model or the wave model of electrons? (38.5)

13. How does the energy of a photon compare with the difference in energy levels of the atom from which it is emitted? (38.6)

14. What does it mean to say that an electron occupies discrete energy levels in an atom? (38.6)

15. Does the particle view of an electron or the wave view of an electron better explain the discreteness of electron energy levels? Why? (38.6)

16. What does wave interference have to do with the electron energy levels in an atom? (38.6)

17. Why is a helium atom smaller than a hydrogen atom? (38.7)

18. Why are the heaviest elements not appreciably larger than the lightest elements? (38.7)

19. What is quantum mechanics? (38.8)

20. Can the momentum and position of electrons in an atom be measured with certainty? (38.8)

Think and Explain

1. What does it mean to say that a certain quantity is quantized?

2. What evidence can you cite for the wave nature of light? For the particle nature of light?

3. A very bright source of red light has much more energy than a dim source of blue light, but the red light has no effect in ejecting electrons from a certain photosensitive surface. Why is this so?

4. Which photon has the most energy—one from infrared, visible, or ultraviolet light?

5. If a beam of red light and a beam of green light have exactly the same energy, which beam contains the greater number of photons?

6. Suntanning produces cell damage in the skin. Why is ultraviolet light capable of producing this damage while infrared radiation is not?

7. Electrons in one electron beam have a greater speed than those in another. Which electrons have the longer de Broglie wavelength?

8. We do not notice the wavelength of moving matter in our ordinary experience. Is this because the wavelength is extraordinarily large or extraordinarily small?

9. The equation $E = hf$ describes the energy of each photon in a beam of light. If Planck's constant, h, were larger, would photons of light of the same frequency be more energetic or less energetic?

10. Why will helium rather than hydrogen more readily leak through an inflated rubber balloon?

39 The Atomic Nucleus and Radioactivity

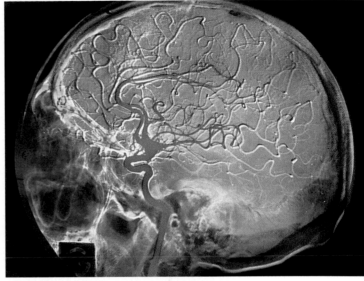

Radioactivity, a tool of medicine.

The configuration of electrons in an atom determines whether and how the atom bonds to form compounds. It also dictates the melting and freezing temperatures and the thermal and electrical conductivity, as well as the taste, texture, appearance, and color of substances. Changes in electron energies produce visible light. Larger changes produce X rays. This chapter goes even deeper into the atom—to the atomic nucleus.

39.1 The Atomic Nucleus

It would take 30 000 carbon nuclei to stretch across a single carbon atom. The nucleus within the atom is as inconspicuous as a cookie crumb in the middle of the Rose Bowl football stadium. Despite the small size of the nucleus, much has been learned about its structure. The nucleus is composed of particles called **nucleons,** which when electrically charged are protons, and when electrically neutral are neutrons.* Neutrons and protons have close to the same mass, with the neutron's being slightly greater. Nucleons have nearly 2000 times the mass of electrons, so the mass of an atom is practically equal to the mass of its nucleus alone.

The positively charged protons in the nucleus hold the negatively charged electrons in their orbits. Each proton has exactly the same magnitude of charge as the electron, but the opposite sign. So in an electrically neutral atom, there are as many protons in the

---

* Protons and neutrons are themselves composed of subnuclear particles called *quarks.* Are quarks themselves made of still smaller particles? They may be, but so far there is no evidence or theoretical reason for believing so. Theoretical physicists say today that quarks are the elementary particles of which all nucleons and other strongly interacting particles are made.

Figure 39.1 ▲
The number of electrons that surround the atomic nucleus is matched by the number of protons in the nucleus.

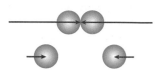

Figure 39.2 ▲
The nuclear strong force is a very short-range force. For nucleons very close or in contact, it is very strong (large force vectors). But a few nucleon diameters away (small force vectors) it is nearly zero.

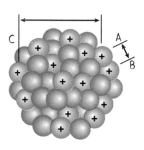

Figure 39.3 ▲
A strong attractive nuclear force acts between nearby protons A and B, but not significantly between A and C. The longer-range electric force repels protons A and C as well as A and B. The mutual repulsion of all the protons in a heavy nucleus tends to make such nuclei unstable.

nucleus as there are electrons outside. The number of protons in the nucleus therefore determines the chemical properties of that atom, because the positive nuclear charge determines the possible structures of electron orbits that can occur.

The number of neutrons in the nucleus has no direct effect on the electron structure, and hence does not affect the chemistry of the atom. The principal role of the neutrons is to act as a sort of nuclear cement to hold the nucleus together. Nucleons are bound together by an attractive nuclear force appropriately called the **strong force.**

The nuclear force of attraction is strong only over a very short distance (Figure 39.2). Whereas the electrical force between charges decreases as the inverse square of the distance, the nuclear force decreases far more rapidly. When two nucleons are just a few nucleon diameters apart, the nuclear force they exert on each other is nearly zero. This means that if nucleons are to be held together by the strong force, they must be held in a very small volume. Nuclei are tiny because the nuclear force is very short-range.

It is an interesting feature of quantum mechanics that particles held close together have large kinetic energy and tend to fly apart. So, although the nuclear force is strong, it is only barely strong enough to hold a pair of nucleons together. For a pair of protons, which repel each other electrically, the nuclear force is not quite strong enough to keep them together. When neutrons are present, however, the attractive strong force is increased relative to the repulsive electric force (since neutrons have no charge). Thus, the presence of neutrons adds to the nuclear attraction and keeps protons from flying apart.

The more protons there are in a nucleus, the more neutrons are needed to hold them together. For light elements, it is sufficient to have about as many neutrons as protons. For heavy elements, extra neutrons are required. The most common form of lead, for example, has 82 protons and 126 neutrons, or about one and a half times as many neutrons as protons. For elements with more than 83 protons, even the addition of extra neutrons cannot stabilize the nucleus.

39.2 Radioactive Decay

One of the factors that sets a limit on how many stable nuclei can exist is the instability of the neutron. A lone neutron will spontaneously decay into a proton plus an electron (and also an antineutrino, a tiny particle we will not treat here). Out of a bunch of lone neutrons, about half of them will decay in 11 minutes. Particles that decay in this or in similar ways are said to be **radioactive.** A lone neutron is radioactive.

The rules of radioactivity inside atomic nuclei are governed by the mass-energy equivalence. Particles decay only when their combined products have less mass after decay than before. The mass of a neutron is slightly greater than the total mass of a proton plus electron (and the antineutrino). So when a neutron decays, there is less mass after decay than before. Decay will not spontaneously occur for reactions where more mass results. The reverse reaction, a proton decaying into a neutron, can occur only with external energy input.

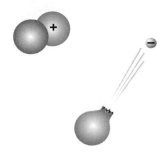

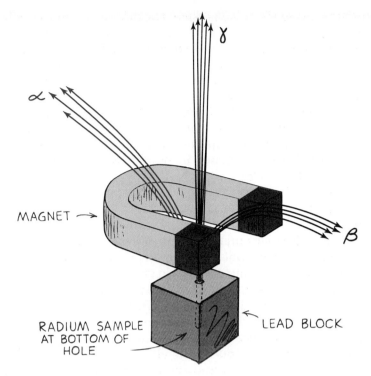

MAGNET →

γ

α

β

RADIUM SAMPLE AT BOTTOM OF HOLE

← LEAD BLOCK

Figure 39.5 ▲
The separation of alpha, beta, and gamma rays in a magnetic field. The rays come from a radioactive source placed at the bottom of a hole drilled in a lead block.

Figure 39.4 ▲
A neutron-proton combination is stable, but a neutron by itself is unstable and turns into a proton by emitting an electron (as well as an antineutrino—not shown).

All the elements heavier than bismuth (atomic number 83) decay in one way or another. Thus, these elements are radioactive. Their atoms emit three distinct types of rays. The rays are called by the first three letters of the Greek alphabet, α, β, and γ, named *alpha, beta*, and *gamma*, respectively. Alpha rays have a positive electric charge, beta rays are negative, and gamma rays are electrically neutral. The three rays can be separated by putting a magnetic field across their path (Figure 39.5).

An alpha ray is a stream of particles that are made of two protons and two neutrons and that are identical to the nuclei of helium atoms. These are called *alpha particles* (Figure 39.6).

A beta ray is simply a stream of electrons. An electron is ejected from the nucleus when a neutron is transformed into a proton. It may seem that the electrons are "buried" inside the neutron, but this is not true. An electron does not exist in a neutron any more than a spark exists inside a rock about to be scraped across a rough surface. The electron that pops out of the neutron, like the spark that pops out of the scraped rock, is produced during an interaction.

A gamma ray is massless energy. Like visible light, gamma rays are simply photons of electromagnetic radiation, but of much higher frequency and energy. Visible light is emitted when electrons jump from one orbit to another of lower energy. Gamma rays are emitted when nucleons do a similar sort of thing inside the nucleus. Sometimes there are great energy differences in nuclear energy levels, so the photons (gamma rays) emitted carry a large amount of energy.

Figure 39.6 ▲
An alpha particle contains two protons and two neutrons bound together and is identical to a helium nucleus.

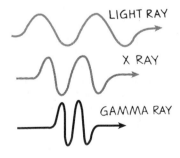

LIGHT RAY

X RAY

GAMMA RAY

Figure 39.7 ▲
A gamma ray is simply electromagnetic radiation, much higher in frequency and energy per photon than light and X rays.

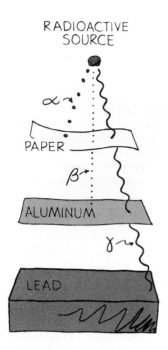

RADIOACTIVE
SOURCE

α

PAPER

β

ALUMINUM

γ

LEAD

Figure 39.8 ▲
Alpha particles penetrate least
and can be stopped by a few
sheets of paper; beta particles by
a sheet of aluminum; gamma
rays by a thick layer of lead.

39.3 Radiation Penetrating Power

There is a great difference in the penetrating power of the three types of rays emitted by radioactive elements. Alpha rays are the easiest to stop. They can be stopped by a reasonably heavy piece of paper or a few sheets of thin paper. Beta rays go right through paper but are stopped by several sheets of aluminum foil. Gamma rays are the most difficult to stop and require lead or other heavy shielding to block them.

An alpha particle is easy to stop because it is relatively slow and its double-positive charge interacts with the molecules it encounters along its path. It slows down as it shakes many of these molecules apart and leaves positive and negative ions in its wake. Even when traveling through nothing but air, an alpha particle will come to a stop after only a few centimeters. It soon grabs up a couple of stray electrons and becomes nothing more than a harmless helium atom.

A beta particle normally moves at a faster speed than an alpha particle, carries only a single negative charge, and is able to travel much farther through the air. Most beta particles lose their energy during the course of a large number of glancing collisions with atomic electrons. Except for rare direct hits, energy is lost in many small steps. Beta particles slow down until they reach the speeds of thermal motion, becoming a part of the material they are in, like any other electron.

Gamma rays are the most penetrating of the three because they have no charge. With no electric attraction or deflection, a gamma ray photon interacts with the absorbing material only via a direct hit with an atomic electron or a nucleus. Unlike charged particles, a gamma ray photon can be removed from its beam in a single encounter. Dense materials such as lead are good absorbers mainly because of their high electron density.

■ Question

Pretend you are given three radioactive cookies—one alpha, one beta, and the other gamma. Pretend that you must eat one, hold one in your hand, and put the other in your pocket. Which would you eat, hold, and pocket, if you are trying to minimize your exposure to radiation?

■ Answer

Ideally, of course, get as far from the cookies as possible. But if you must eat one, hold one, and put one in your pocket, then hold the alpha; the skin on your hand will shield you. Put the beta in your pocket; your clothing will likely shield you. Eat the gamma; it will penetrate your body in any of these cases, anyway. (In real life always use appropriate safeguards when near radioactive materials.)

39.4 Radioactive Isotopes

It has already been stated that the number of protons in an atomic nucleus determines the number of electrons surrounding the nucleus in a neutral atom. If there is a difference in the number of electrons and protons, the atom is charged and is called an *ion*. An ionized atom is one that has a different number of electrons than nuclear protons.

◀ **Figure 39.9**
Which of these diagrams shows an ion?

The number of neutrons in the nucleus, however, has no bearing on the number of electrons the atom may have. This means that the number of neutrons has no direct bearing on the chemistry of an atom. Let's consider a hydrogen atom. The common form of hydrogen has a bare proton as its nucleus. Any nuclear configuration that has only one proton in its nucleus *is* hydrogen—by definition. There can be different kinds, or **isotopes,** of hydrogen, however. In one isotope, the nucleus consists of only a single proton. In a second isotope, the proton is accompanied by a neutron. In a third isotope, there are two neutrons. All the isotopes of a particular element are chemically identical. The orbital electrons are affected only by the positive charge in the nucleus, not by its neutrons.

We distinguish between the different isotopes of hydrogen by $_1^1H$, $_1^2H$, and $_1^3H$, where the lower number is the **atomic number**—the number of protons—and the upper number is the **atomic mass number**—the total number of nucleons.

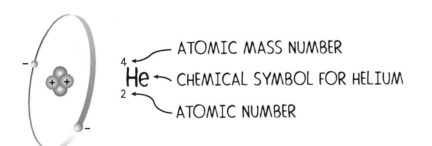

◀ **Figure 39.10**
The atomic number is equal to the number of protons in the nucleus, and the atomic mass number is equal to the number of nucleons in the nucleus (both protons and neutrons).

The common isotope of hydrogen, $_1^1H$, is a stable element. So is the isotope $_1^2H$, called *deuterium.* "Heavy water" is the name usually given to H_2O in which the H's are deuterium atoms. The triple-weight hydrogen isotope $_1^3H$, called *tritium,* however, is unstable and undergoes beta decay. This is the radioactive isotope of hydrogen. All elements have isotopes. Some are radioactive and some are not. All the isotopes of elements above atomic number 83, however, are radioactive.

Figure 39.11 ▶
Three isotopes of hydrogen.
Each nucleus has a single proton
that holds a single orbital electron, which, in turn, determines
the chemical properties of the
atom. The different number of
neutrons changes the mass of
the atom, but not its chemical
properties.

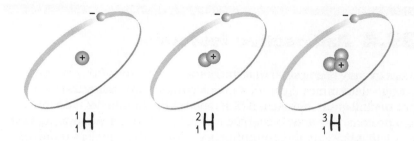

^1_1H ^2_1H ^3_1H

The common isotope of uranium is $^{238}_{92}\text{U}$, or U-238 for short. It is radioactive, but with a smaller decay rate than $^{235}_{92}\text{U}$, or U-235. Any nucleus with 92 protons is uranium, by definition. Nuclei with 92 protons but different numbers of neutrons are simply different isotopes of uranium.

Figure 39.12 ▶
All isotopes of uranium are
unstable and undergo radioactive decay.

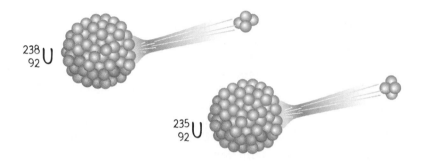

$^{238}_{92}\text{U}$

$^{235}_{92}\text{U}$

■ Questions

1. When the tritium nucleus, ^3_1H, undergoes beta decay, one of its neutrons is converted to a proton. What is the resulting nucleus?

2. The nucleus of beryllium-8, ^8_4Be, undergoes a special kind of radioactive decay: it splits into two equal halves. What nuclei are the products of this decay? Why is this a form of alpha decay?

3. The electric force of repulsion between the protons in a heavy nucleus acts over a greater distance than the attractive forces among the neutrons and protons in the nucleus. Given this fact, explain why all of the very heavy elements are radioactive.

■ Answers

1. The resulting nucleus contains two protons and one neutron. This is an isotope of helium, the second element in the periodic table. It is helium-3, or ^3_2He.

2. When beryllium-8, which contains 4 protons and 4 neutrons, splits into equal halves, a pair of nuclei with 2 protons and 2 neutrons are created. These are nuclei of helium-4, ^4_2He, also called alpha particles. So this reaction is a form of alpha decay.

3. Each proton in an atomic nucleus is repelled by every other proton in the nucleus, but it is attracted only by the nucleons closest to it. In a large nucleus, where protons such as those on opposite sides are far apart, electric repulsion can exceed nuclear attraction. This instability makes all the heaviest atoms radioactive.

39.5 Radioactive Half-Life

Radioactive isotopes decay at different rates. The radioactive decay rate is measured in terms of a characteristic time, the **half-life.** The half-life of a radioactive material is the time needed for half of the radioactive atoms to decay. Radium-226, for example, has a half-life of 1620 years. This means that half of any given specimen of Ra-226 will have undergone radioactive decay by the end of 1620 years. In the next 1620 years, half of the remaining radium will decay, leaving only one-fourth the original number of radium atoms. The other three-fourths are converted, by a succession of disintegrations, to lead. After 20 half-lives, an initial quantity of radioactive atoms will be diminished to about one-millionth of the original quantity.

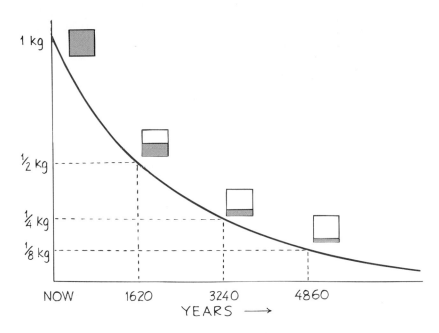

Figure 39.13 ▲
Every 1620 years the amount of radium decreases by half.

The isotopes of some elements have a half-life of less than a millionth of a second, while U-238, for example, has a half-life of 4.5 billion years. The isotopes of each radioactive element have their own characteristic half-lives.

Rates of radioactive decay appear to be absolutely constant, unaffected by any external conditions, however drastic. High or low pressures, high or low temperatures, strong magnetic or electric fields, and even violent chemical reactions have no detectable effect on the rate of decay of an element. Any of these stresses, however severe by ordinary standards, is far too mild to affect the nucleus deep in the interior of the atom.

How do physicists measure radioactive half-lives? They cannot always do it by observing a specimen and waiting until the quantity reduces to half. This is often much longer than a human life span! One can measure, however, the rate at which a substance decays. There are various radiation detectors for doing this (Figure 39.14). The half-life of an isotope is related to its rate of disintegration. In general, the shorter the half-life of a substance, the faster it disintegrates, and the more active is the substance. The half-life can be computed from the rate of disintegration, which can be measured in the laboratory.

Figure 39.14 ▶
Radiation detection. A Geiger counter detects incoming radiation by its ionizing effect on enclosed gas in the tube. A scintillation counter (not shown) detects incoming radiation by flashes of light that are produced when charged particles or gamma rays pass through it.

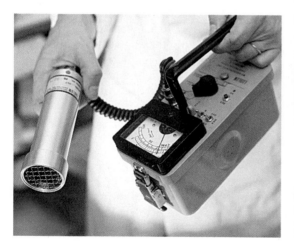

■ Questions

1. If a sample of a radioactive isotope has a half-life of 1 year, how much of the original sample will be left at the end of the second year?

2. If you have equal amounts of radioactive materials, one that has a short half-life and another that has a long half-life, which will give a higher reading on a radiation detector?

■ Answers

1. One-quarter of the original sample will be left. The three-quarters that underwent decay become one or more different elements altogether.

2. The material with the shorter half-life is more active and will give a higher reading on a radiation detector.

39.6 Natural Transmutation of Elements

When a nucleus emits an alpha or a beta particle, a different ele–ment is formed. The changing of one element to another is called **transmutation.** Consider common uranium, for example. Uranium has 92 protons. When an alpha particle is ejected, the nucleus is reduced by two protons and two neutrons. This is because they make up the alpha particle that leaves. The 90 protons and 144 neutrons left behind are then the nucleus of a new element. This element is *thorium.* This reaction is expressed as

$$^{238}_{92}\text{U} \rightarrow \, ^{234}_{90}\text{Th} + \, ^{4}_{2}\text{He}$$

An arrow is used here to show that the $^{238}_{92}\text{U}$ changes into the other elements. When this happens, energy is released in three forms: gamma radiation, the kinetic energy of the alpha particle ($^{4}_{2}\text{He}$), and the kinetic energy of the thorium atom. Be sure to notice in the nuclear equation that the mass numbers at the top balance (238 = 234 + 4) and that the atomic numbers at the bottom also balance (92 = 90 + 2).

Thorium-234, the product of this reaction, is also radioactive. When it decays, it emits a beta particle. Recall that a beta particle is an electron ejected from the nucleus. Once ejected, a beta particle is indistinguishable from an orbital electron or any other electron. When a beta particle is ejected, a neutron changes into a proton. In the case of thorium, which has 90 protons, beta emission leaves it with one fewer neutron and one more proton. The new nucleus then has 91 protons and is no longer thorium. It is the element *protactinium.* The reaction is*

$$^{234}_{90}\text{Th} \rightarrow \, ^{234}_{91}\text{Pa} + \, ^{0}_{-1}e$$

---

* Beta emission is accompanied by the ejection of an antineutrino, not shown here. We won't get into antineutrinos, the antiparticles of neutrinos, except to say they are extremely swift (moving at or close to the speed of light) and extremely plentiful. Whether they have mass is still questionable. If they do, it is thousands of times less than the mass of the electron. Neutrinos have no charge and seldom interact with matter. As you are reading this sentence, a thousand billion neutrinos emanating from the sun pierce through your body. This is true day or night, since at nighttime the solar neutrinos travel through the earth and pierce you from below. A cause for concern? No, that's just nature in action!

Note that although the atomic number has increased by 1 in this process, the mass number (number of nucleons) remains the same. Also note that the beta particle (electron) is written as $_{-1}^{0}e$. The −1 is the charge of the electron. The 0 indicates that its mass is insignificant when compared with the mass of the protons and neutrons that alone contribute to the mass number. Beta emission has hardly any effect on the mass of the nucleus; only the charge (atomic number) changes.

As the example of uranium-238 decay shows, when an atom ejects an alpha particle from its nucleus, the mass number of the resulting atom decreases by 4, and its atomic number decreases by 2. The resulting atom belongs to an element two spaces back in the periodic table.* When an atom ejects a beta particle from its nucleus, it loses no nucleons, so there is no change in mass number but its atomic number *increases* by 1.** The resulting atom belongs to an element one place forward in the periodic table. Thus, radioactive elements decay backward or forward in the periodic table. A radioactive nucleus may emit gamma radiation along with an alpha particle or a beta particle. Gamma emission has no effect on the mass number or the atomic number.

The radioactive decay of $_{92}^{238}$U to an isotope of lead, $_{82}^{206}$Pb, is shown on the next page in Figure 39.15. The steps in the decay process are shown in the diagram, where each nucleus that plays a part in the series is shown by a burst. The vertical column that contains the burst shows the atomic number of the nucleus, and the horizontal column shows its mass number. Each arrow that slants downward toward the left shows an alpha decay. Each arrow that points to the right shows a beta decay. Notice that some of the nuclei in the series can decay either way. This is one of several similar radioactive series that occur in nature.

■ Questions

1. Complete the following nuclear reactions.

   a. $_{88}^{228}$Ra → $_{?}^{?}$? + $_{-1}^{0}e$ b. $_{84}^{209}$Po → $_{82}^{205}$Pb + $_{?}^{?}$?

2. What finally becomes of all the uranium-238 that undergoes radioactive decay?

Answers

1. a. $_{88}^{228}$Ra → $_{89}^{228}$Ac + $_{-1}^{0}e$

   b. $_{84}^{209}$Po → $_{82}^{205}$Pb + $_{2}^{4}$He

2. All uranium-238 will ultimately become lead. On the way to becoming lead, it will exist as a series of other elements, as indicated in Figure 39.15.

* For a periodic table, see Figure 17.11. Look in the periodic table for the elements mentioned in this section.

** Sometimes a nucleus emits a *positron,* which is the antiparticle of an electron. A positron has a charge of +1 and the same mass as the electron. In this case, a proton in the nucleus becomes a neutron, and the atomic number is decreased by 1 with no change in mass number.

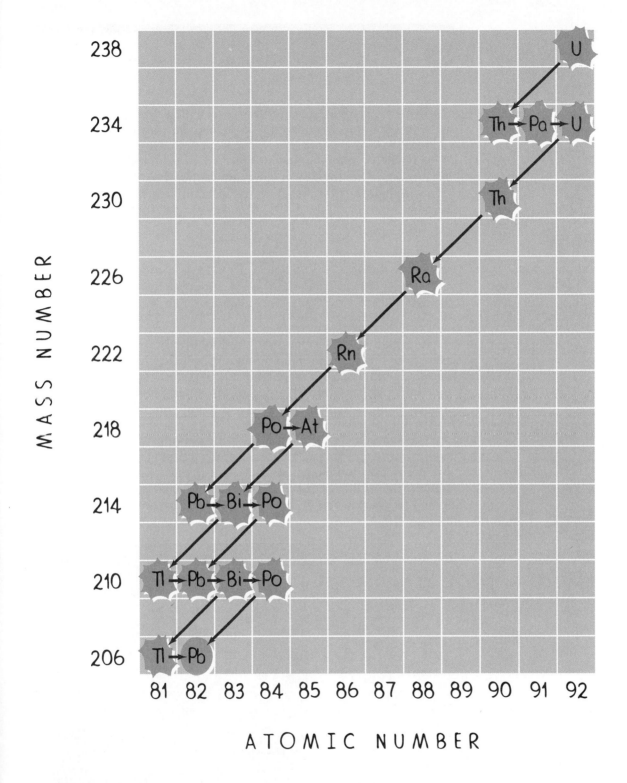

Figure 39.15 ▲

U-238 decays to Pb-206 through a series of alpha and beta decays.

39.7 Artificial Transmutation of Elements

The British physicist Ernest Rutherford, in 1919, was the first of many investigators to succeed in artificially transmuting a chemical element. In a sealed container he bombarded nitrogen nuclei with alpha particles from a radioactive piece of ore and then found traces of oxygen and hydrogen that were not there before. Rutherford accounted for the presence of the oxygen and hydrogen with the nuclear equation

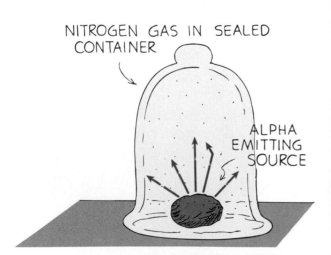

$$^{14}_{7}N + {}^{4}_{2}He \rightarrow {}^{17}_{8}O + {}^{1}_{1}H$$

After Rutherford's experiment there followed many such nuclear reactions—first with natural bombarding particles from radioactive elements, and then with more energetic particles (protons, deuterons, and alpha particles) hurled by giant atom-smashing particle accelerators. Artificial transmutation is an everyday fact of life to the researchers of today.

The elements beyond uranium in the periodic table—the *transuranic* elements—have been produced through artificial transmutation. All of these elements have half-lives that are much less than the age of the earth. Whatever transuranic elements might have existed naturally when the earth was formed have long since decayed.

NITROGEN GAS IN SEALED CONTAINER

ALPHA EMITTING SOURCE

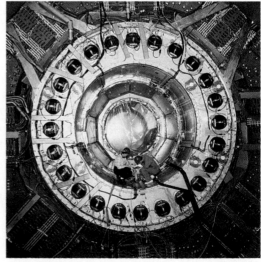

Figure 39.16 ▲
Artificial transmutation can be accomplished by simple means or by elaborate means.

39.8 Carbon Dating

The earth's atmosphere is continuously bombarded from above by *cosmic rays*—mainly high-energy protons—from beyond the earth. This results in the transmutation of many atoms in the upper atmosphere. Protons, neutrons, and other particles are scattered throughout the atmosphere. Most of the protons quickly capture stray electrons and become hydrogen atoms in the upper atmosphere, but the neutrons keep going for long distances because they have no charge and do not interact electrically with matter. Sooner or later many of them collide with the nuclei of atoms in the lower atmosphere. If they are captured by the nucleus of a nitrogen atom, the following reaction can take place:

$${}^{14}_{7}\text{N} + {}^{1}_{0}n \rightarrow {}^{14}_{6}\text{C} + {}^{1}_{1}\text{H}$$

In this reaction, when nitrogen-14 is hit by a neutron (${}^{1}_{0}n$), carbon-14 and hydrogen are produced.

Most of the carbon that exists on Earth is the stable ${}^{12}_{6}\text{C}$, carbon-12. In the air, it appears mainly in the compound carbon dioxide. Because of the cosmic bombardment, less than one-millionth of 1% of the carbon in the atmosphere is carbon-14. Like carbon-12, it joins with oxygen to form carbon dioxide, which is taken in by plants. This means that all plants have a tiny bit of radioactive carbon-14 in them. All animals eat plants (or eat plant-eating animals), and therefore have a little carbon-14 in them. All living things contain some carbon-14.

Carbon-14 is a beta emitter and decays back into nitrogen by the following reaction:*

$${}^{14}_{6}\text{C} \rightarrow {}^{14}_{7}\text{N} + {}^{0}_{-1}e$$

In a living plant, which continues to take in carbon dioxide, a radioactive equilibrium is reached where there is a fixed ratio of carbon-14 to carbon-12. When a plant or animal dies, replenishment of the radioactive isotope stops. Then the percentage of carbon-14 decreases—at a known rate. The longer an organism is dead, the less carbon-14 remains.

The half-life of carbon-14 is 5730 years. This means that half of the carbon-14 atoms that are now present in the remains of a body,

---

* The accompanying antineutrino is not shown.

plant, or tree will decay in the next 5730 years. Half the remaining carbon-14 atoms will then decay in the following 5730 years, and so forth. The radioactivity of once-living things therefore gradually decreases at a predictable rate (Figure 39.17).

20 924 B.C.    15 194 B.C.    9464 B.C.    3734 B.C.    1996 A.D.

Figure 39.17 ▲
The radioactive carbon isotopes in the skeleton diminish by one-half every 5730 years.

Archeologists use the carbon-14 dating technique to establish the dates of wooden artifacts and skeletons. Because of fluctuations in the production of carbon-14 through the centuries (due partly to changes in the earth's magnetic field and the consequent changes in the cosmic ray intensity), this technique gives an uncertainty of about 15%. This means, for example, that a mastodon bone that is dated to be 10 000 years old may really be only 8500 years old on the low side, or 11 500 years old on the high side. For many purposes this is an acceptable level of uncertainty. If greater accuracy is desired, then other techniques must be employed.

■ **Questions**

1. An archeologist extracts a gram of carbon from an ancient bone and measures between 7 and 8 beta emissions per minute from the sample. A gram of carbon extracted from a fresh piece of bone gives off 15 betas per minute. Estimate the age of the ancient bone.

2. Suppose the carbon sample from the ancient bone were found to be only one-fourth as radioactive as a gram of carbon from new bone. Estimate the age of the ancient bone.

■ **Answers**

1. Since beta emission for the old sample is one-half that of the fresh sample, about one half-life has passed, 5730 years.

2. The ancient bone is two half-lives of carbon-14, about 11 460 years.

39.9 Uranium Dating

The dating of older, but nonliving, things is accomplished with radioactive minerals, such as uranium. The naturally occurring isotopes U-238 and U-235 decay very slowly and ultimately become isotopes of lead—but not the common lead isotope Pb-208. For example, U-238 decays through several stages to finally become Pb-206, whereas U-235 finally becomes the isotope Pb-207. Most of the lead isotopes 206 and 207 that exist were at one time uranium. The older the uranium-bearing rock, the higher the percentage of these lead isotopes.

From the half-lives of the uranium isotopes and the percentage of lead isotopes in uranium-bearing rock, a calculation can be made of the date when the rock was formed. Rocks dated in this way have been found to be as much as 3.7 billion years old. Samples from the moon, where there has been less obliteration of early rocks than on Earth, have been dated at 4.2 billion years. This is "only" 400 million years short of the well-established 4.6-billion-year age of the earth and solar system.

39.10 Radioactive Tracers

Radioactive isotopes of all the elements have been produced by bombarding the element with neutrons and other particles. These isotopes are inexpensive, quite available, and very useful in scientific research and industry.

Agricultural researchers mix a small amount of radioactive isotopes with fertilizer before applying it to growing plants. Once the plants are growing, the amount of fertilizer taken up by the plant can be easily measured with radiation detectors. From such measurements, researchers can tell farmers the proper amount of fertilizer to use. When used in this way, radioactive isotopes are called *tracers* (Figure 39.18).

Tracers are used in medicine to study the process of digestion and the way in which chemicals move about in the body. Food containing a tiny amount of radioactive isotopes is fed to a patient. The paths of the tracers in the food are then followed through the body with a radiation detector. The same method is used to study the circulation of the blood.

Engineers study how parts of an automobile test engine wear away by making the cylinder walls in the engine radioactive. While the engine is running, the piston rings rub against the cylinder walls. The tiny particles of radioactive metal that are worn away fall into the lubricating oil, where they can be measured with a radiation detector. This test is repeated with different oils. In this way the engineer can determine which oil gives the least wear and longest life to the engine.

Figure 39.18 ▲
Radioactive isotopes are used to check the action of fertilizers in plants and the progress of food in digestion.

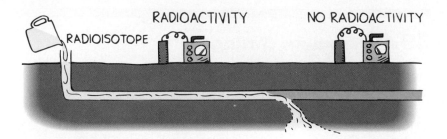

Figure 39.19 ▲
Tracking pipe leaks with radioactive isotopes.

There are hundreds more examples of the use of radioactive isotopes. The important thing is that this technique provides a way to detect and count atoms in quantities too small to be seen with a microscope and too small to be hazardous.*

39.11 Radiation and You

Radioactivity has been around longer than humans have. It is as much a part of our environment as the sun and rain. It is what warms the interior of the earth and makes it molten. In fact, radioactive decay inside the earth is what heats the water that spurts from a geyser or that wells up from a natural hot spring. Even the helium in a child's balloon is the result of radioactivity. Its nuclei are nothing more than alpha particles that were once shot out of radioactive nuclei.

As Figure 39.20 shows, most radiation you encounter originates in nature. It is in the ground you stand on, and in the bricks and stones of surrounding buildings. Even the cleanest air we breathe is slightly radioactive. This natural background radiation was present before humans emerged in the world. If our bodies couldn't tolerate it, we wouldn't be here.

Much of the radiation we are exposed to is cosmic radiation streaming down through the atmosphere. Most of the protons and other atomic nuclei that fly toward the earth from outer space are deflected away. The atmosphere, acting as a protective shield, stops most of the rest. But some cosmic rays penetrate the atmosphere, mostly in the form of secondary particles such as muons. At higher altitudes, radiation is more intense. In Denver, the "mile-high city," you receive more than twice the cosmic radiation you receive at sea level. A couple of round-trip flights between New York and San Francisco exposes you to as much radiation as in a normal chest X ray. The air time of airline personnel is limited because of this extra radiation.

---

* The use of intense radiation in treating cancer is different. The quantity of radioactive material is then far greater than in research using radioactive tracers, but the benefit is reckoned to outweigh the risk.

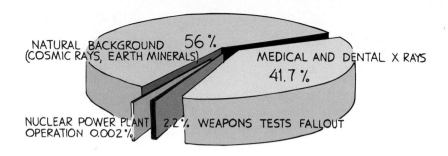

Figure 39.20 ▲
Origins of radiation exposure for an average individual in the United States.

We are bombarded most by what harms us least—neutrinos. Neutrinos are the most weakly interacting particles. They have near-zero mass, no charge, and are produced frequently in radioactive decays. They are the most common high-speed particles known, zapping the universe, and passing unhindered through our bodies by the billions every second. They pass completely through the earth with only occasional encounters. It would take a "piece" of lead 6 light-years in thickness to absorb half the neutrinos incident upon it. About once per year on the average, a neutrino triggers a nuclear reaction in your body. We don't hear much about neutrinos because they ignore us.

Of the types of radiation we have focused upon in this chapter, gamma radiation is by far the most dangerous. It emanates from radioactive materials and makes up a substantial part of the normal background radiation. Exposure to gamma radiation should be minimized. The cells of living tissue are composed of intricately structured molecules in a watery, ion-rich brine. When gamma radiation encounters this highly ordered soup, it produces damage on the atomic scale. Less damage is done by a beta particle because it does not penetrate as deeply into living matter. Regardless of whether damage is by gamma, beta, or some other kind of radiation, these altered molecules are often more harmful than useful to life processes. Altered DNA molecules, for example, can produce harmful genetic mutations.

Cells can repair most kinds of molecular damage if the radiation they are exposed to is not too intense. This is how we are able to tolerate small radiation doses. On the other hand, people who work around high concentrations of radioactive materials must be specially trained and protected to avoid an increased risk of cancer. This applies to medical people, workers in nuclear power plants, and personnel on nuclear-powered ships. People who receive high doses of radiation (on the order of 1000 times natural background or more) run a greater risk of cancer and have a shorter life expectancy than people who are not so exposed.

Whenever possible, exposure to radiation should be avoided. Unavoidable, however, is the natural radiation that all living beings have always absorbed.

Figure 39.21 ▲
This is the internationally used symbol to indicate an area where radioactive material is being handled or produced.

39 Chapter Review

Concept Summary

The atomic nucleus is composed of nucleons that consist of positively charged protons and electrically neutral neutrons.

- The number of protons determines the number of electrons and the chemical properties of an atom.
- Nucleons are bound together by the strong force.
- As the number of protons increases, more neutrons are needed to stabilize the nucleus.

Radioactive elements have unstable nuclei and emit various nuclear particles.

- Alpha particles consist of two protons and two neutrons and can be stopped by skin or heavy paper.
- Beta particles are electrons ejected from the nucleus. They can be stopped by clothing or aluminum foil.
- Gamma rays are high-energy photons. Heavy shielding, such as lead, is required for protection.

Isotopes of an element are chemically identical but differ in their number of neutrons.

- They have the same atomic number but different atomic mass numbers.
- Some isotopes are radioactive.
- Half-life is a measure of the decay rate of radioactive isotopes.

Transmutation is the changing of one element into another that occurs when a radioactive nucleus emits an alpha or beta particle.

- The transuranic elements were created by artificial transmutation.

Radioactive isotopes have several important uses.

- Carbon-14 is used to date wooden artifacts and the remains of plants and animals.

- Radioactive minerals, such as uranium-238 or uranium-235, are used to date older, nonliving materials.
- Radioactive tracers are used in agriculture and medicine.

Natural environmental radiation constantly bombards us.

- Additional exposure to radiation should be avoided whenever possible, because it is damaging to living molecules and cells.

Important Terms

atomic mass number (39.4)
atomic number (39.4)
half-life (39.5)
isotope (39.4)
nucleon (39.1)
radioactive (39.2)
strong force (39.1)
transmutation (39.6)

Review Questions

1. Which of the following are nucleons—protons, neutrons, or electrons? (39.1)

2. Do electrical forces tend to hold a nucleus together or push it apart? (39.1)

3. Between what kinds of particles does the nuclear strong force act? (39.1)

4. Which force has a longer range, the electric force or the strong force? (39.1)

5. When is a neutron unstable?

6. Distinguish between alpha, beta, and gamma rays. (39.2)

7. How do the penetrating powers of the three types of radiation compare? (39.3)

8. Distinguish between an ion and an isotope. (39.4)

9. How does the number of electrons in a normal atom compare with the number of protons in its nucleus? (39.4)

10. Which isotope has the greater number of neutrons, U-235 or U-238? (39.4)

11. What is meant by radioactive half-life? (39.5)

12. If the radioactive half-life of a certain isotope is 1620 years, how much of that substance will be left at the end of 1620 years? After 3240 years? (39.5)

13. When an atom undergoes radioactive decay, does it become a completely different element? (39.6)

14. a. What happens to the atomic number of an atom when it ejects an alpha particle?

   b. What happens to its atomic mass number? (39.6)

15. a. What happens to the atomic number of an atom when it ejects a beta particle?

   b. What happens to its atomic mass number? (39.6)

16. a. What element does thorium become if it emits an alpha particle?

   b. What if it emits a beta particle? (39.6)

17. a. What is a transuranic element?

   b. Why are there no ore deposits of transuranic elements in the earth? (39.7)

18. Which is radioactive, C-12 or C-14? (39.8)

19. Why is more C-14 found in new bones than in ancient bones of the same mass? (39.8)

20. Why would the carbon dating method be useless in dating old coins but not old pieces of adobe bricks? (39.8)

21. Why are there deposits of lead in all deposits of uranium ore? (39.9)

22. What isotopes accumulate in old, uranium-bearing rock? (39.9)

23. What is a radioactive tracer? (39.10)

24. From where does most of the radiation you encounter originate? (39.11)

25. Why is radiation more intense at high altitudes and near the earth's poles? (39.11

Think and Explain

1. What experimental evidence indicates that radioactivity is a process that occurs in the atomic nucleus?

2. Does your body contain more neutrons than protons? More protons than electrons? Discuss.

3. Why are the atomic masses of many elements in the periodic table not whole numbers?

4. How do the atomic number and atomic mass of an atom change when a proton is added to its nucleus? When a neutron is added? Which determines the chemical nature of the element?

5. What do different isotopes of a given element have in common? How are they different?

6. Why is a sample of radioactive material always a little warmer than its surroundings? Why is the center of the earth so hot?

7. A product of nuclear power plants is the isotope cesium-137, which has a half-life of 30 years. How long will it take for this isotope to decay to one-sixteenth its original amount?

8. Coal contains minute quantities of radioactive materials, yet there is more environmental radiation outside a coal-fired power plant than outside a fission power plant. What does this tell you about the shielding that typically surrounds these power plants?

9. When we speak of dangerous radiation exposure, are we generally speaking of alpha radiation, beta radiation, or gamma radiation? Discuss.

10. When the isotope bismuth-213 emits an alpha particle, it becomes a new element.

a. What are the atomic number and atomic mass number of the new element?

b. What element results if bismuth-213 emits a beta particle instead?

11. a. State the numbers of neutrons and protons in each of the following nuclei: $^{6}_{3}\text{Li}$, $^{14}_{6}\text{C}$, $^{56}_{26}\text{Fe}$, $^{201}_{80}\text{Hg}$, and $^{239}_{94}\text{Pu}$.

b. How many electrons will typically surround each of these nuclei?

12. Radiation from a point source follows an inverse square law. If a Geiger counter that is 1 m away from a small source reads 100 counts per minute, what will be its reading 2 m from the source? 3 m from it?

13. How is it possible for an element to decay "forward in the periodic table"—that is, to an element of higher atomic number?

14. People working around radioactivity wear film badges to monitor their radiation exposure. These badges are small pieces of photographic film enclosed in a lightproof wrapper. What kind of radiation do these devices monitor?

15. The age of the Dead Sea Scrolls was found by carbon dating. Could this technique work if they were instead stone tablets? Explain.

40 Nuclear Fission and Fusion

Chain reaction.

The discovery of radioactivity in 1896 sparked much interest among many kinds of people. Some people thought it was no more than a scientific curiosity, some thought it would be a cure for medical ailments, and a few thought it might turn out to be a source of plentiful energy to heat homes, power factories, and light up cities at night.

Radioactivity does release energy, but has not become a sub–stantial source of energy for humans. On a small scale it powers small energy sources in spacecraft, and makes a sample of radium warm. On a large scale it melts rocks and is the source of geothermal energy within the earth. In 1939, just before World War II, a nuclear reaction was discovered that released much more energy per atom than radioactivity, and had the potential to be used for both explosions and power production. This was the splitting of the atom, or *nuclear fission.*

A very different nuclear reaction, *nuclear fusion,* can also release huge amounts of energy. Both nuclear fission and nuclear fusion produce vastly more energy per kilogram of matter than any chemical reaction, and even more than most other nuclear reactions. The awesome release of this energy in atomic and hydrogen bombs ushered in the present "nuclear age." Out of the ashes of despair brought about by these bombs, hope grew that atoms could be used for peaceful purposes—that the awesome energy of nuclear reactions could be used for domestic power instead of for arsenals of war.

What exactly are nuclear fission and nuclear fusion? How do they differ? What is the physics that underlies why so much energy is released by these reactions? The answers to these questions are what this chapter is about.

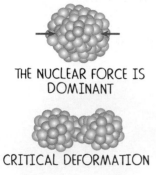

THE NUCLEAR FORCE IS
DOMINANT

CRITICAL DEFORMATION

THE ELECTRICAL FORCE IS
DOMINANT

Figure 40.1 ▲

Nuclear deformation leads to
fission when repelling electrical
forces dominate over attracting
nuclear forces.

40.1 Nuclear Fission

Biology students know that living tissue grows by the division of
cells. The splitting in half of living cells is called *fission*. In a similar
way, the splitting of atomic nuclei is called **nuclear fission.**

Nuclear fission involves the delicate balance between the attraction of nuclear strong forces and the repulsion of electrical forces
within the nucleus. In all known nuclei the nuclear strong forces
dominate. In uranium, however, this domination is tenuous. If the
uranium nucleus is stretched into an elongated shape (Figure 40.1),
the electrical forces may push it into an even more elongated shape.
If the elongation passes a critical point, electrical forces overwhelm
nuclear strong forces, and the nucleus splits. This is nuclear fission.

The absorption of a neutron by a uranium nucleus supplies
enough energy to cause such an elongation. The resulting fission
process may produce many different combinations of smaller nuclei.
A typical example is

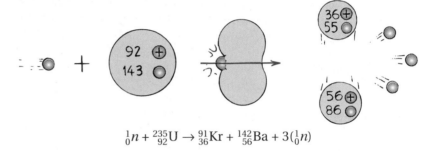

$$^{1}_{0}n + ^{235}_{92}U \rightarrow ^{91}_{36}Kr + ^{142}_{56}Ba + 3(^{1}_{0}n)$$

The energy that is released by the fission of one U-235 atom is
enormous—about seven million times the energy released by the
explosion of one TNT molecule. This energy is mainly in the form
of kinetic energy of the fission fragments, with some energy given
to ejected neutrons, and the rest to gamma radiation.

Note that one neutron starts the fission of the uranium atom,
and, in this example, three more neutrons are produced when the
uranium fissions. Between two and three neutrons are produced in
most nuclear fission reactions. These new neutrons can, in turn,
cause the fissioning of two or three other nuclei, releasing from four
to nine more neutrons. If each of these succeeds in splitting just one
atom, the next step in the reaction will produce between 8 and
27 neutrons, and so on. This makes a **chain reaction** (Figure 40.2).

Why do chain reactions not occur in naturally occurring uranium ore deposits? They would if all uranium atoms fissioned so easily. Fission occurs mainly for the rare isotope U-235, which makes up
only 0.7% of the uranium in pure uranium metal. When the prevalent isotope U-238 absorbs neutrons from fission, it does not
undergo fission. So a chain reaction can be snuffed out by the
neutron-absorbing U-238. It is rare for uranium deposits in nature
to spontaneously undergo a chain reaction.

If a chain reaction occurred in a chunk of pure U-235 the size of
a baseball, an enormous explosion would likely result. If the chain
reaction were started in a smaller chunk of pure U-235, however, no

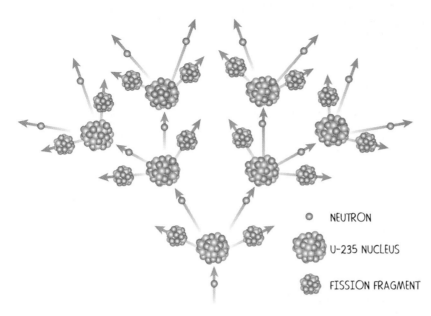

○ NEUTRON

U-235 NUCLEUS

FISSION FRAGMENT

explosion would occur. Why? Because a neutron ejected by a fission event travels a certain average distance through the material before it encounters another uranium nucleus and triggers another fission event. If the piece of uranium is too small, a neutron is likely to escape through the surface before it "finds" another nucleus. On the average, fewer than one neutron per fission will be available to trigger more fission, and the chain reaction will die out. In a bigger piece, a neutron can move farther through the material before reaching a surface. Then more than one neutron from each fission event, on the average, will be available to trigger more fission (Figure 40.4).The chain reaction will build up to enormous energy.*

◉ U 235 ○ U 238

Figure 40.3 ▲
Only 1 part in 140 of naturally occurring uranium is U-235.

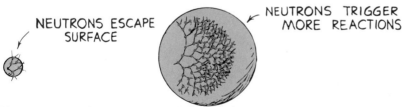

NEUTRONS ESCAPE SURFACE

NEUTRONS TRIGGER MORE REACTIONS

Figure 40.4 ▲
The exaggerated view shows that a chain reaction in a small piece of pure U-235 dies out, because neutrons leak from the surface too easily. In a larger piece, a chain reaction builds up because neutrons are more likely to trigger additional fission events than to escape through the surface.

The **critical mass** is the amount of mass for which each fission event produces, on the average, one additional fission event. It is just enough to "hold even." A *subcritical* mass is one in which the chain reaction dies out. A *supercritical* mass is one in which the chain reaction builds up explosively.

---

*   Another way to understand this is geometrically. Recall the concept of scaling in Chapter 18. Small pieces of material have more surface relative to volume than large pieces (there is more skin on a kilogram of small potatoes than on a single 1-kilogram large potato). The larger the piece of fission fuel, the less surface area it has relative to its volume.

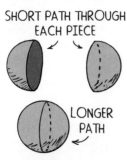

SHORT PATH THROUGH
EACH PIECE

LONGER
PATH

Figure 40.5 ▲

Each piece is subcritical. The average path of a neutron in each piece is short enough that a neutron is likely to escape. When the pieces are combined, the average path of a neutron is greater, and there is less chance that a neutron will escape. The combination may be supercritical.

In Figure 40.5 there are two pieces of pure U-235, each of them subcritical. Neutrons readily reach a surface and escape before a sizable chain reaction builds up. But if the pieces are joined together, there will be more distance available for neutron travel and a greater likelihood of their triggering fission before escaping through the surface. If the combined mass is supercritical, we have a nuclear fission bomb.

The construction of a uranium fission bomb is not a formidable task. The difficulty is separating enough U-235 from the more abundant U-238. It took Manhattan Project scientists and engineers more than two years to extract enough U-235 from uranium ore to make the bomb that was detonated over Hiroshima in 1945. Uranium isotope separation is still a difficult, expensive process today.

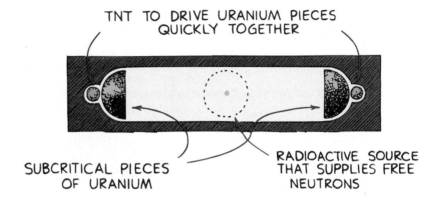

TNT TO DRIVE URANIUM PIECES
QUICKLY TOGETHER

SUBCRITICAL PIECES
OF URANIUM

RADIOACTIVE SOURCE
THAT SUPPLIES FREE
NEUTRONS

Figure 40.6 ▲

Simplified diagram of an idealized uranium fission bomb. (In an actual "gun-type" weapon, only one of the two pieces of uranium is fired toward the other one, which is the "target.")

■ Questions

1. What is nuclear fission?

2. What is a chain reaction?

3. Five kilograms of U-235 broken up into small separated chunks is subcritical, but if the chunks are put together in a ball shape, it is supercritical. Why?

■ Answers

1. Nuclear fission is the splitting of the atomic nucleus. When a heavy nucleus such as the U-235 nucleus splits into two main parts, there is a large release of energy.

2. A chain reaction is a self-sustaining reaction that, once started, continues because one reaction event triggers one or more additional reaction events.

3. Five kilograms of U-235 in small chunks will not support a sustained reaction because the path for a neutron in each chunk is so short that the neutron is likely to escape through the surface without causing fission. When the chunks are brought together, the average neutron path within the material is much longer and a neutron is likely to cause fission rather than escape.

40.2 The Nuclear Fission Reactor

A better use for uranium than for bombs is for power reactors. About 21% of electric energy in the United States is generated by nuclear fission reactors. These reactors are simply nuclear furnaces, which (like fossil fuel furnaces) do nothing more elegant than boil water to produce steam for a turbine (Figure 40.7). The greatest practical difference is the amount of fuel involved. One kilogram of uranium fuel, less than the size of a baseball, yields more energy than 30 freight-car loads of coal.

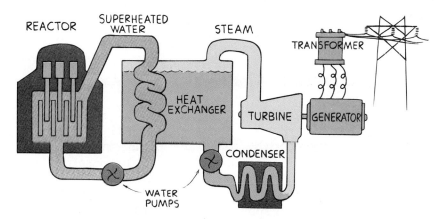

◄ **Figure 40.7**
Diagram of a nuclear fission power plant.

A reactor contains three main components: the nuclear fuel combined with a moderator to slow down neutrons, the control rods, and water used to transfer heat from the reactor to the generator. The nuclear fuel is uranium, with its fissionable isotope U-235 enriched to about 3%. The moderator may be graphite, a pure form of carbon, or it may be water. Because the U-235 is so highly diluted with U-238, an explosion like that of a nuclear bomb is not possible. Control rods that can be moved in and out of the reactor control the "multiplication" of neutrons, that is, how many neutrons from each fission event are available to trigger additional fission events. The control rods are made of a material, usually the metal cadmium or boron, that readily absorbs neutrons. Heated water around the nuclear fuel is kept under high pressure and thus brought to a high temperature without boiling. It transfers heat to a second, lower-pressure water system, which operates the electric generator in a

■ **Question**

What is the function of the control rods in a nuclear reactor?

■ **Answer**

Control rods absorb more neutrons when they are pushed into the reactor and fewer neutrons when they are pulled out of the reactor. They thereby control the number of neutrons that participate in a chain reaction.

conventional fashion. In this design two separate water systems are used so that no radioactivity can reach the turbine.

A major drawback to fission power is the waste products of fission. Recall that light atomic nuclei are most stable when composed of equal numbers of protons and neutrons, and that heavy nuclei need more neutrons than protons for stability. So there are more neutrons than protons in uranium—143 neutrons compared with 92 protons in U-235, for example. When uranium fissions into two medium-weight elements, the ratio of neutrons to protons in the product nuclei is greater than for medium-weight stable nuclei. These fission products are said to be "neutron rich." They are radioactive, most with very short half-lives, but some with half-lives of thousands of years. Safely disposing of these waste products requires special storage casks and procedures, and is subject to a developing technology that is less than ideal.

40.3 Plutonium

When a neutron is absorbed by a U-238 nucleus, no fission results. The nucleus that is created, U-239, emits a beta particle instead and becomes an isotope of the first synthetic element beyond uranium— the transuranic element called *neptunium* (named after the first planet discovered from the application of Newton's law of gravitation).*
This isotope, Np-239, in turn, very soon emits a beta particle and becomes an isotope of *plutonium* (named after Pluto, the second planet to be discovered via Newton's law). This isotope, Pu-239, like U-235, will undergo fission when it captures a neutron.

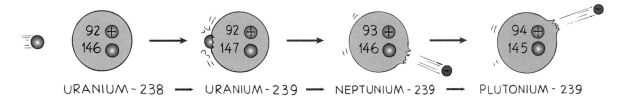

Figure 40.8 ▲
After U-238 absorbs a neutron, it emits a beta particle (and an antineutrino, not shown), which means that a neutron in the nucleus becomes a proton. The atom is no longer uranium, but neptunium. After the neptunium atom emits a beta particle it becomes plutonium.

The half-life of neptunium 239 is only 2.3 days, while the half-life of plutonium 239 is about 24 000 years. Since plutonium is an element distinct from uranium, it can be separated from uranium by ordinary chemical methods. Unlike the difficult process of separating U-235 from U-238, it is relatively easy to separate plutonium from uranium.

---

* At this writing, transuranic elements extend to atomic number 111. See the periodic table of the elements, Figure 17.11.

The element plutonium is chemically a poison in the same sense as are lead and arsenic. It attacks the nervous system and can cause paralysis. Death can follow if the dose is sufficiently large. Fortunately, plutonium does not remain in its elemental form for long because it rapidly combines with oxygen to form three compounds, PuO, PuO_2, and Pu_2O_3, all of which are chemically relatively benign. They will not dissolve in water or in biological systems. These plutonium compounds do not attack the nervous system and have been found to be biologically harmless.

Plutonium in any form, however, is radioactively toxic. It is more toxic than uranium, although less toxic than radium. Pu-239 emits high-energy alpha particles, which kill cells rather than simply disrupting them and leading to mutations. Interestingly enough, damaged cells rather than dead cells contribute to cancer, which is why plutonium ranks low as a cancer-producing substance. The greatest danger that plutonium presents to humans is its potential for use in nuclear fission bombs. Its usefulness is in breeder reactors.

40.4 The Breeder Reactor

When small amounts of Pu-239 are mixed with U-238 in a reactor, the fissioning of plutonium liberates neutrons that convert the abundant, nonfissionable U-238 into more of the fissionable Pu-239. This process not only produces useful energy, it also "breeds" more fission fuel. A reactor with this fuel is a **breeder reactor.** Using a breeder reactor is like filling a gas tank in a car with water, adding some gasoline, then driving the car and having more gasoline after the trip than at the beginning, at the expense of common water! After the initial high costs of building such a device, this is a very economical method of producing vast amounts of energy. After a few years of operation, breeder-reactor power utilities breed twice as much fuel as they start with.

Fission power has several benefits. First, it supplies plentiful electricity. Second, it conserves the many billions of tons of coal, oil, and natural gas that every year are literally turned to heat and smoke, and which in the long run may be far more precious as sources of organic molecules than as sources of heat. Third, it eliminates the megatons

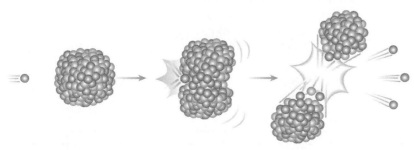

Figure 40.9 ▲
Pu-239, like U-235, undergoes fission when it captures a neutron.

of sulfur oxides and other poisons that are put into the air each year by the burning of these fuels.

The drawbacks include the problems of storing radioactive wastes, the production of plutonium and the danger of nuclear weapons proliferation, low-level release of radioactive materials into the air and groundwater, and the risk of an accidental release of large amounts of radioactivity.

Reasoned judgment is not made by considering only the benefits or the drawbacks of fission power. You must also compare its benefits and its drawbacks with those of alternate power sources. All power sources have drawbacks of some kind. The benefits versus the drawbacks of fission power is a subject of much debate.

■ Question

In a breeder reactor, what material is "bred"? What is this material bred from?

40.5 Mass-Energy Equivalence

Figure 40.10 ▲
Work is required to pull a nucleon from an atomic nucleus. This work goes into mass energy.

The key to understanding why a great deal of energy is released in nuclear reactions has to do with the equivalence of mass and energy. Recall from our study of special relativity in Chapter 16 that mass and energy are essentially the same—they are two sides of the same coin. Mass is like a super storage battery. It stores energy—vast quantities of energy—which can be released if and when the mass decreases.

If you stacked up 238 bricks, the mass of the stack would be equal to the sum of the masses of the bricks. Is the mass of a U-238 nucleus equal to the sum of the masses of the 238 nucleons that make it up? Like so much ruled by relativity, the answer isn't obvious. To find the answer, we consider the work that would be required to separate all the nucleons from a nucleus.

Recall that work, which transfers energy, is equal to the product of force and distance. Imagine that you can reach into a U-238 nucleus and, pulling with a force even greater than the attractive nuclear force, remove one nucleon. That would require considerable work. Then keep repeating the process until you end up with 238 nucleons, stationary and well separated. What happened to all the work done? You started with one stationary nucleus containing 238 particles and ended with 238 separate stationary particles. The work done shows up as *mass* energy. The separated nucleons have a total mass greater than the mass of the original nucleus. The extra mass, multiplied by the square of the speed of light, is exactly equal to your energy input: $\Delta E = \Delta mc^2$.

One way to interpret this mass change is to say that a nucleon inside a nucleus has less mass than its rest mass outside the nucleus. How much less depends on which nucleus. The mass difference is

■ Answer

Fissionable Pu-239 is bred from nonfissionable U-238.

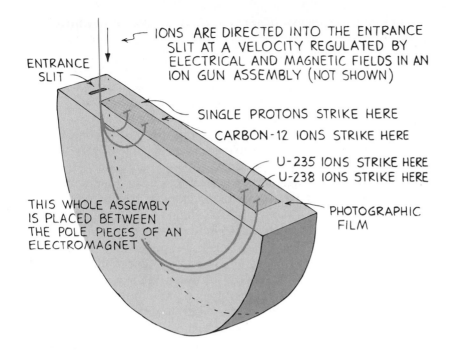

ENTRANCE
SLIT

IONS ARE DIRECTED INTO THE ENTRANCE
SLIT AT A VELOCITY REGULATED BY
ELECTRICAL AND MAGNETIC FIELDS IN AN
ION GUN ASSEMBLY (NOT SHOWN)

SINGLE PROTONS STRIKE HERE
CARBON-12 IONS STRIKE HERE

U-235 IONS STRIKE HERE
U-238 IONS STRIKE HERE

THIS WHOLE ASSEMBLY
IS PLACED BETWEEN
THE POLE PIECES OF AN
ELECTROMAGNET

PHOTOGRAPHIC
FILM

Figure 40.11 ▲
The mass spectrometer. Ions of a fixed speed are directed into the semicircular "drum," where they are swept into semicircular paths by a strong magnetic field. Because of inertia, heavier ions are swept into curves of larger
radii and lighter ions are swept into curves of smaller radii. The radius of the
curve is directly proportional to the mass of the ion.

related to the "binding energy" of the nucleus. For uranium, the mass
difference is about 0.7%, or 7 parts in a thousand. The 0.7% reduced
nucleon mass in uranium indicates the binding energy of the
nucleus, or how much work it would take to disassemble the atom.

The standard nucleus by which others are compared is carbon-
12, which has a mass of exactly 12.000 00 units.* In these units, a
proton outside the nucleus has a mass of 1.007 28, a neutron has a
mass of 1.008 66, and an electron has a mass of 0.000 55. The masses
of the pieces that make up the carbon atom—6 protons, 6 neutrons,
and 6 electrons—add up to 12.0989, about 0.8% more than the mass
of a C-12 atom. That difference indicates the binding energy of the
C-12 nucleus. We will see shortly that binding energy is greatest in
the nucleus of iron.

The masses of ions of isotopes of various elements can be accurately measured with a *mass spectrometer* (Figure 40.11). This important device uses a magnetic field to deflect ions into circular arcs.
The ions entering the device all have the same speed. The greater
the inertia (mass) of the ion, the more it resists deflection, and the
greater the radius of its curved path. In this way the nuclear masses
can be compared as the magnetic force sweeps heavier ions into
larger arcs and lighter ions into smaller arcs.

---

* These units are called *atomic mass units* and are the units for atomic mass used in
chemistry.

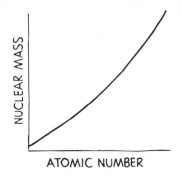

Figure 40.12 ▲
A graph that shows how nuclear mass increases with increasing atomic number. The curvature is somewhat exaggerated.

A graph of the nuclear masses for the elements from hydrogen through uranium is shown in Figure 40.12. The graph slopes upward with increasing atomic number as expected—elements are more massive as atomic number increases. The slope curves slightly because there are proportionally more neutrons in the more massive atoms.

A more important graph results from the plot of nuclear mass *per nucleon* from hydrogen through uranium (Figure 40.13). To obtain the nuclear mass per nucleon, simply divide the nuclear mass by the number of nucleons in the particular nucleus. (If you divided the mass of your whole class by the number of people in your class, you would get the average mass per person.) The graph indicates the different average effective masses of nucleons in atomic nuclei. A proton has the greatest mass when it is the nucleus of a hydrogen atom. None of the proton's mass is binding energy—it isn't bound to anything. Progressing beyond hydrogen, the masses of nucleons in heavier nuclei are effectively smaller. The low point of the graph occurs at the element iron. This means that pulling apart an iron nucleus would take more work per nucleon than pulling apart any other nucleus. Iron holds its nucleons more tightly than any other nucleus does. Beyond iron, the average effective mass of nucleons increases. For elements lighter than iron and heavier than iron, the binding energy per nucleon is less than it is in iron.

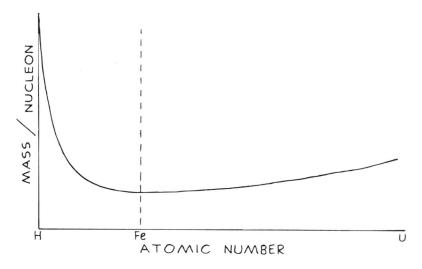

Figure 40.13 ▲
The graph shows that the mass per nucleon is not a constant for all nuclei. It is greatest for the lightest nuclei, the least for iron, and has an intermediate value for the heaviest nuclei. (The vertical scale covers only about 1% of the mass of a nucleon.)

From the graph you can see why energy is released when a uranium nucleus is split into nuclei of lower atomic number. If a uranium nucleus splits in two, the masses of the fission fragments lie about halfway between uranium and hydrogen on the horizontal scale of the graph. Most importantly, note that the mass per nucleon in the fission fragments is *less than* the mass per nucleon when the same set of nucleons are combined in the uranium nucleus. When this decrease in

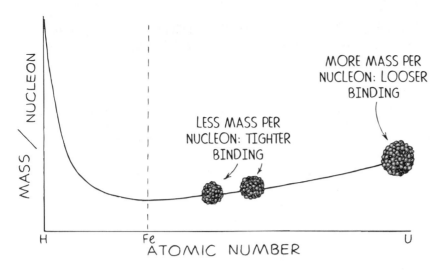

mass is multiplied by the speed of light squared, it is equal to the energy yielded by each uranium nucleus that undergoes fission.

You can think of the mass-per-nucleon graph as an energy valley that starts at hydrogen (the highest point) and drops steeply to the lowest point (iron), then rises gradually to uranium. Iron is at the bottom of the energy valley, which is the place with the greatest binding energy per nucleon. Any nuclear transformation that moves nuclei toward iron releases energy. Heavier nuclei move toward iron by dividing—nuclear fission. A drawback is the fission fragments, which are radioactive because of their greater-than-normal number of neutrons.

A more promising source of energy is to be found when lighter-than-iron nuclei move toward iron by *combining*—as indicated on the left side of the energy valley.

◄ **Figure 40.15**
The mass of a nucleus is *not* equal to the sum of the masses of its parts. Fission fragments of a heavy nucleus (including ejected neutrons) have less total mass than the nucleus. Why is there a difference in mass?

40.6 Nuclear Fusion

Inspection of the graph of Figure 40.13 will show that the steepest part of the energy hill is from hydrogen to iron. Energy is gained as light nuclei *fuse,* or combine, rather than split apart. This process is **nuclear fusion,** the opposite of nuclear fission. Whereas energy is released when heavy nuclei split apart in the fission process, energy is released when light nuclei fuse together. After fusion, the total mass of the light nuclei formed in the fusion process is less than the total mass of the nuclei that fused (Figure 40.16).

Atomic nuclei are positively charged. For fusion to occur, they normally must collide at very high speed in order to overcome electrical repulsion. The required speeds correspond to the extremely high

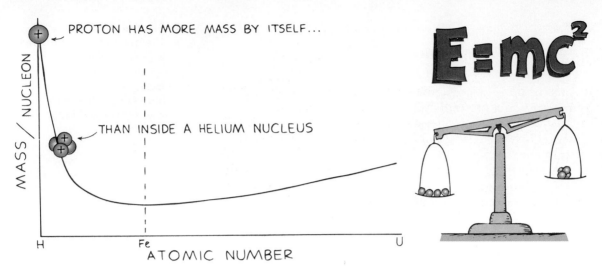

PROTON HAS MORE MASS BY ITSELF...

THAN INSIDE A HELIUM NUCLEUS

MASS / NUCLEON

ATOMIC NUMBER

H   Fe   U

$E = mc^2$

Figure 40.16 ▲
(Left) The mass of a single proton is more than the mass per nucleon in a helium-4 nucleus. When protons fuse to form helium, mass is reduced and energy is released. (Right) Two protons and two neutrons have more total mass when they are free than when they are combined in a helium nucleus.

temperatures found in the center of the sun and other stars. Fusion brought about by high temperatures is called **thermonuclear fusion**—that is, the welding together of atomic nuclei by high temperature. In the hot central part of the sun, approximately 657 million tons of hydrogen are converted into 653 million tons of helium each second. The missing 4 million tons of mass is discharged as radiant energy. Such reactions are, quite literally, nuclear burning.

Thermonuclear fusion is analogous to ordinary chemical combustion. In both chemical and nuclear burning, a high temperature starts the reaction; the release of energy by the reaction maintains a high enough temperature to spread the fire. The net result of the chemical reaction is a combination of atoms into more tightly bound molecules. In nuclear reactions, the net result is more tightly bound nuclei. The difference between chemical and nuclear burning is essentially one of scale.

■ Questions .

1. First it was stated that nuclear energy is released when atoms split apart. Now it is stated that nuclear energy is released when atoms combine. Is this a contradiction? How can energy be released by opposite processes?

2. To get energy from the element iron, should iron be fissioned or fused?

■ Answers

1. Energy is released only in a nuclear reaction in which the mass per nucleon decreases. Light nuclei, such as hydrogen, lose mass and release energy when they combine (fuse) to form heavier nuclei. Heavy nuclei, such as uranium, lose mass and release energy when they split to become lighter nuclei. For energy release, "Lose Mass" is the name of the game—any game.

2. Iron will release no energy at all, because it is at the very bottom of the energy valley. If fused with something else, it climbs the right side of the hill and gains mass. If fissioned, it climbs the left side of the hill and gains mass. In gaining mass, it absorbs energy instead of releasing energy.

40.7 Controlling Nuclear Fusion

Producing fusion reactions under controlled conditions requires temperatures of hundreds of millions of degrees. Producing and sustaining such high temperatures along with reasonable densities is the goal of much current research. There are a variety of techniques for attaining high temperatures. No matter how the temperature is produced, a problem is that all materials melt and vaporize at the temperatures required for fusion. The solution to this problem is to confine the reaction in a nonmaterial container.

A magnetic field is nonmaterial, can exist at any temperature, and can exert powerful forces on charged particles in motion. "Magnetic walls" of sufficient strength provide a kind of magnetic straitjacket for hot ionized gases called *plasmas.* Magnetic compression further heats the plasma to fusion temperatures.

Figure 40.17 ▲
A magnetic bottle used for containing plasmas for fusion research.

At a temperature of about a million degrees, some nuclei are moving fast enough to overcome electrical repulsion and slam together, but the energy output is much smaller than the energy used to heat the plasma. Even at 100 million degrees, more energy must be put into the plasma than will be given off by fusion. At about 350 million degrees, the fusion reactions will produce enough energy to be self-sustaining. At this *ignition temperature,* nuclear burning yields a sustained power output without further input of energy. A steady feeding of nuclei is all that is needed to produce continuous power.

Fusion has already been achieved in several devices, but instabilities in the plasma have thus far prevented a sustained reaction. A big problem is devising a field system that will hold the plasma in a stable and sustained position while an ample number of nuclei fuse. A variety of magnetic confinement devices are the subject of much present-day research.

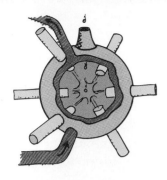

Figure 40.18 ▲
Fusion with multiple laser beams. Pellets of frozen deuterium are rhythmically dropped into synchronized laser cross fire. According to plan, the resulting heat will be carried off by molten lithium to produce steam.

Figure 40.19 ▶
Pellet chamber at Lawrence Livermore Laboratory. The laser source is Nova, the most powerful laser in the world, which directs 10 beams into the target region.

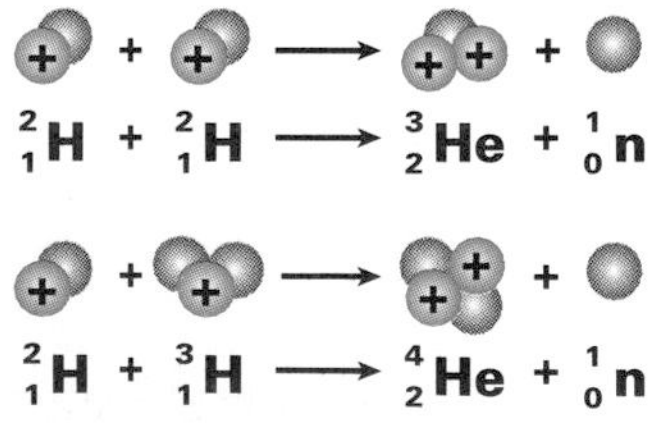

$${}^{2}_{1}\text{H} + {}^{2}_{1}\text{H} \longrightarrow {}^{3}_{2}\text{He} + {}^{1}_{0}\text{n}$$

$${}^{2}_{1}\text{H} + {}^{3}_{1}\text{H} \longrightarrow {}^{4}_{2}\text{He} + {}^{1}_{0}\text{n}$$

Figure 40.20 ▲
Fusion reactions of hydrogen isotopes that will be used for controlled thermonuclear power (the reactions in the sun are different). Most of the energy released is carried by the lighter-weight particles, protons and neutrons, which fly off at high speeds.

Another promising approach bypasses magnetic confinement altogether with high-energy lasers. One technique is to aim an array of laser beams at a common point and drop solid pellets composed of frozen hydrogen isotopes through the synchronous cross fire (Figure 40.18).

Other fusion schemes involve the bombardment of fuel pellets not by laser light but by beams of electrons, light ions, and heavy ions.

As this book goes to press we are still looking forward to the great "Break-Even Day" when one of the variety of fusion schemes will sustain a yield of at least as much energy as is required to initiate it.

Fusion power is nearly ideal. Fusion reactors cannot become "supercritical" and get out of control because fusion requires no critical mass. Furthermore, there is no air pollution because the only product of the thermonuclear combustion is helium (good for children's balloons). Except for some radioactivity in the inner chamber of the fusion device because of high-energy neutrons, the by-products of fusion are not radioactive. Disposal of radioactive waste is not a major problem.

The fuel for nuclear fusion is hydrogen—in particular, its heavier isotopes, deuterium (H-2) and tritium (H-3). Hydrogen is the most plentiful element in the universe. The thermonuclear reaction that occurs most readily at an achievable temperature is the so-called *dt* reaction, in which a deuterium nucleus and a tritium nucleus fuse. Both of these isotopes are found in ordinary water. For example, 30 liters of seawater contains 1 gram of deuterium, which when fused releases as much energy as 10 000 liters of gasoline or 80 tons of TNT. Natural tritium is much scarcer, but given enough to get started (it can be made in a fission reactor), a controlled thermonuclear reactor will breed it from deuterium in ample quantities. Because of the abundance of fusion fuel, the amount of energy that can be released in a controlled manner is virtually unlimited.

The development of fusion power has been slow and difficult, already extending over nearly fifty years. It is one of the biggest scientific and engineering challenges that we face. Yet there is every reason to believe that it will be achieved and will be a primary energy source for future generations.

Humans may one day travel to the stars in ships fueled by the same energy that makes the stars shine.

40 Chapter Review

Concept Summary

Nuclear fission, the splitting of atomic nuclei, occurs when the repelling electrical forces in the nucleus overpower the attracting nuclear strong forces.

- Fission is initiated by the absorption of a neutron by a nucleus. It can spread in a chain reaction in which neutrons ejected in one fission event trigger fission in other nuclei.

- A critical mass of a fissionable element is required for sustained fission to occur.

- In a subcritical mass, the chain reaction dies out. In a supercritical mass, the chain reaction builds up explosively.

- In a breeder reactor, extra neutrons from a small amount of fissionable plutonium-239 are absorbed by nonfissionable uranium-238, converting it into plutonium-239.

- Fission reactors are efficient energy producers but generate radioactive wastes.

Nuclei of the lightest elements have the most mass per nucleon, the nucleus of iron has the least mass per nucleon, and the heaviest elements have intermediate mass per nucleon.

- When fission occurs, the total mass of the fission fragments (including ejected neutrons) is less than the mass of the fissioning nucleus. The missing mass is equivalent to the great energy released.

In nuclear fusion, hydrogen nuclei fuse to form helium nuclei, releasing large amounts of energy.

- After fusion, the total mass of the products is less than the mass of the nuclei that fused.

- Thermonuclear fusion occurs at the high temperatures found in the center of the sun and other stars.

- Sustained fusion under controlled conditions is the goal of continuing research.

Important Terms

breeder reactor (40.4)
chain reaction (40.1)
critical mass (40.1)
nuclear fission (40.1)
nuclear fusion (40.6)
thermonuclear
    fusion (40.6)

Review Questions

1. What is the role of electrical forces in nuclear fission? (40.1)

2. What is the role of a neutron in nuclear fission? (40.1)

3. Of what use are the neutrons that are produced when a nucleus undergoes fission? (40.1)

4. Why does a chain reaction not occur in uranium ore? (40.1)

5. **a.** Which isotope of uranium is most common?

   b. Which isotope of uranium will fission? (40.1)

6. Jason sticks a dozen two-inch nails randomly into an apple. Jane sticks half a dozen of the same size nails randomly into each piece of an apple cut in half. Who will see more points of nails sticking out? (40.1)

7. Which has greater average path length inside the material—two separate pieces of uranium or the same pieces stuck together? (40.1)

8. Which will leak more neutrons—two separate pieces of uranium or the same pieces stuck together? (40.1)

9. Will a supercritical chain reaction be more likely in two separate pieces of U-235 or in the same pieces stuck together? (40.1)

10. What controls the chain reaction in a nuclear reactor? (40.2)

11. Are the fission fragments from a nuclear reactor light, medium, or heavy elements? (40.2)

12. Why are the fission-fragment elements radioactive? (40.2)

13. What happens when U-238 absorbs a neutron? (40.3)

14. How can plutonium be created? (40.3)

15. Is plutonium an isotope of uranium or is it a completely different element? (40.3)

16. What is the effect of putting a little Pu-239 with a lot of U-238 in a reactor? (40.4)

17. Is the mass per nucleon of a nucleus greater than, less than, or the same as the mass of a nucleon outside a nucleus? (40.5)

18. What device can be used to measure the relative masses of ions of isotopes? (40.5)

19. What is the primary difference in the graphs shown in Figures 40.12 and 40.13? (40.5)

20. What becomes of the loss in mass of nuclei when heavy atoms split? (40.5)

21. Why does helium not yield energy if fissioned? (40.5)

22. Why does uranium not yield energy if fused with something else? (40.6)

23. Why does iron not yield energy if fused with something else or fissioned? (40.6)

24. What becomes of the loss in mass when light atoms fuse to become heavier ones? (40.6)

25. Why are fusion reactors not a present-day reality like fission reactors? (40.7)

Think and Explain

1. Why does a neutron often make a better nuclear bullet than a proton?

2. Why does a chain reaction die out in small pieces of fissionable fuel, but not in large pieces?

3. If a piece of uranium is flattened into a pancake shape, will this make a supercritical chain reaction more or less likely? Why?

4. Why are there no appreciable deposits of plutonium in the earth's crust?

5. Your tutor says atomic nuclei are converted to energy in a nuclear reaction. Why should you seek a new tutor?

6. Is the mass of an atomic nucleus greater or less than the total mass of the nucleons that compose it?

7. The energy release of nuclear fission is tied to the fact that the mass per nucleon of medium-weight nuclei is about 0.1% less than the mass per nucleon of the heaviest nuclei. What would be the effect on energy release if the 0.1% figure were 1%?

8. To predict the approximate energy release of either a fission or a fusion reaction, explain how a physicist makes use of the curve of Figure 40.13, or a table of nuclear masses, and the equation $\Delta E = \Delta mc^2$.

9. Which process would release energy from gold—fission or fusion? From carbon? From iron?

10. If a uranium nucleus were to split into three pieces of approximately the same size instead of two, would more energy or less energy be released? Defend your answer in terms of Figure 40.13.

Appendix A: Units of Measurement

The units of measurement primarily used in this book are those used by scientists throughout the world—the International System of Units, or SI (after the French name, Système International). SI units are the outgrowth of the metric system of units. While familiar to scientists, many SI units are not generally familiar to students in high school. The SI units used in this book are the following.

Meter

The meter (m) is the SI unit of length. The standard of length for the metric system originally was defined in terms of the distance from the North Pole to the equator. This distance is close to 10 million meters. So one meter equals approximately one ten-millionth of the distance from the North Pole to the equator. A more exact definition is that one meter equals the length of the path traveled by light in a vacuum during a time interval of 1/299 792 458 of a second.

Common SI length units based on the meter are the *centimeter, millimeter,* and *kilometer.*

$$1 \text{ centimeter (cm)} = 1/100 \text{ meter}$$

$$1 \text{ millimeter (mm)} = 1/1000 \text{ meter}$$

$$1 \text{ kilometer (km)} = 1000 \text{ meters}$$

One meter is a little more than a yard. It is equal to 3.28 feet, or 39.37 inches. It takes 1.609 kilometers to make one mile.

Kilogram

The kilogram (kg), the SI unit of mass, is defined as the mass of a platinum-iridium cylinder preserved at the International Bureau of Weights and Measures in France. The kilogram originally was defined as the mass of one liter (1000 cubic centimeters) of water at the temperature at which it is most dense (now known to be 4° Celsius). Other common mass units are the *gram* and *milligram.*

$$1 \text{ gram (g)} = 1/1000 \text{ kilogram}$$

$$1 \text{ milligram (mg)} = 1/1000 \text{ gram}$$

$$1/1\,000\,000 \text{ kilogram}$$

The mass of a 1-pound object is 0.4536 kilogram. One kilogram weighs about 2.2 pounds at the earth's surface.

Second

The second (s) is the SI unit of time. Until 1956 the second was defined in terms of the mean solar day, which was divided into 24 hours. Each hour was divided into 60 minutes and each minute into 60 seconds. Thus there were 86 400 seconds per day, and the second was defined as 1/86 400 of the mean solar day. This was found to be unsatisfactory because the rate of rotation of the earth is gradually slowing. In 1956 the mean solar day of the year 1900 was chosen as the standard on which to base the second. Since 1964, the second has been officially defined as the time taken by a cesium-133 atom to make 9 192 631 770 vibrations.

Newton

The newton (N), the SI unit of force, is named after Sir Isaac Newton. One newton is the force required to give an object with a mass of one kilogram an acceleration of one meter per second squared.

One newton is a little less than a quarter of a pound—more accurately, 0.225 pound.

Joule

The joule (J), the SI unit of energy, is named after James Joule. One joule is equal to the amount of work done by a force of one newton acting over a distance of one meter.

The unit for power is derived from the unit for energy. Power is the rate at which energy is expended. Work done at the rate of one joule per second is equal to a power of one *watt* (W). The *kilowatt* (kW) equals 1000 watts. From the definition of power, it follows that energy can be expressed as the product of power and time. Electric energy is often expressed in units of *kilowatt-hours* (kW·h), where

$$1 \text{ kilowatt-hour (kW·h)} = 3.60 \times 10^6 \text{ joules}$$

The horsepower, a commonly used power unit for engines, is equal to 746 watts. A commonly used alternate energy unit for heat is the calorie. One calorie is equal to 4.184 joules.

Ampere

The ampere (A), the SI unit of electric current, is named after André-Marie Ampère. In this text the ampere is defined as the rate of flow of one coulomb of charge per second, where one coulomb is the charge of 6.24×10^{18} electrons. The official definition of the ampere is the intensity of constant electric current maintained in two parallel conductors of infinite length and negligible cross section that when placed one meter apart in a vacuum would produce between them a force of 2×10^{-7} newton per meter of length.

Kelvin

The kelvin (K), the SI unit of temperature, is named after the scientist Lord Kelvin. The kelvin is defined as 1/273.16 of the temperature change between absolute zero (the coldest possible temperature) and the triple point of water (the fixed temperature at which, for a certain pressure, ice, liquid water, and water vapor coexist in equilibrium). Temperatures are expressed in kelvins, and not in "degrees kelvin." On the Kelvin scale, absolute zero is 0 K. The temperature of melting ice at atmospheric pressure is 273.15 K, the triple point of water is 273.16 K, and the temperature of pure boiling water at atmospheric pressure is 373.15 K. There are 100 kelvins between the melting and boiling points of water, just as there are 100 Celsius degrees between these points. So a temperature change of 1 Celsius degree is the same as a temperature change of 1 kelvin.

A Fahrenheit degree measures a smaller temperature change. It takes a temperature change of 1.8 Fahrenheit degrees to equal a change of 1 kelvin or 1 Celsius degree. So the number of Fahrenheit degrees between the melting and boiling points of water is 180 (the difference between 212 and 32).

Measurements of Area and Volume

Area Area refers to the amount of surface. The unit of area is the surface of a square that has a standard unit of length as a side. In the SI system it is a square with sides one meter in length, which makes a unit of area of one square meter (1 m^2). A smaller unit area is represented by a square with sides one centimeter in length, which makes a unit of area of one square centimeter (1 cm^2).

The area of a rectangle equals the base times the height. The area of a circle is equal to πr^2, where $\pi = 3.14$, and r is the radius of the circle. Formulas for the surfaces of other shapes can be found in geometry textbooks.

One square meter is equal to 10.76 square feet.

Volume The volume of an object refers to the space it occupies. The unit volume is the space taken up by a cube that has a standard unit of length for its edge. In the SI system, it is the space occupied by a cube whose sides are one meter. This volume is one cubic meter (1 m^3) and is a relatively large volume by everyday standards. A smaller unit volume is the space occupied by a cube whose sides are one centimeter. Its volume is one cubic centimeter (1 cm^3), the space taken up by one gram of water at 4°C.

A liter (L) is equal to 1000 cm^3, and is a common measure of volume for liquids.

One liter, a little larger than a quart, is 1.057 quarts. It takes 3.785 liters to make one U.S. gallon.

Appendix B: Working with Units in Physics

A quantity in science is expressed by a number and a unit of measurement. A unit (singular) may be a combination of other units. The unit of acceleration, for instance, is m/s^2. Quantities may be actual measurements, or they may be obtained by performing calculations on measurements. Quantities may be added, subtracted, multiplied, or divided. There are rules for handling both the numbers and the units of measurement during these mathematical operations.

Addition

When you add quantities, all must have the *same* unit. Add up the numbers. The sum has the same unit as well.

Example:

(4 m) + (8 m) + (3 m) = 15 m

Subtraction

When you subtract one quantity from another, both must have the *same* unit. Subtract the numbers. The difference has the same unit.

Example:

(5.2 s) – (3.8 s) = 1.4 s

Multiplication

Quantities that are multiplied together need *not* have the same unit. Multiply the numbers. Multiply the units just as if they are algebraic variables.

When full names of units are used, use a hyphen between the units that are multiplied together.

Example:

(3 newtons) × (2 meters) = 6 newton-meters

When symbols are used, use a centered dot between the unit symbols that are multiplied together.

Example:

(3 N) × (2 m) = 6 N·m

When the units being multiplied are the same, the product is called the square (or cubic) unit. In symbols, a raised 2 after the unit symbol is used for the square. A raised 3 after the unit symbol is used for the cubic unit. These raised numerals are known as *exponents*.

Examples:

(3 meters) × (2 meters) = 6 meter-meters
$$= 6 \text{ square meters}$$
$$(3 \text{ m}) \times (2 \text{ m}) = 6 \text{ m·m} = 6 \text{ m}^2$$

(3 meters) × (2 meters) × (4 meters)
$$= 24 \text{ meter-meter-meters}$$
$$= 24 \text{ cubic meters}$$

(3 m) × (2 m) × (4 m) $= 24 \text{ m·m·m} = 24 \text{ m}^3$

Division

Quantities that are divided by each other need *not* have the same unit. Divide the numbers. Divide the units as though they are algebraic variables.

When the units are full names, use the word *per* after the unit that is being divided.

Example:

(100 kilometers) ÷ (2 hours)
$$= \frac{100 \text{ kilometers}}{2 \text{ hours}}$$
$$= 50 \text{ kilometers per hour}$$

When the units are symbols, use a slash after the unit symbol that is being divided.

Example:

$$(100 \text{ km}) \div (2 \text{ h}) = \frac{100 \text{ km}}{2 \text{ h}}$$
$$= 50 \text{ km/h}$$

When both units are the same, they "cancel" out and do not appear in the quotient.

Example:

$$(6 \text{ m}) \div (3 \text{ m}) = \frac{6 \,\cancel{\text{m}}}{3 \,\cancel{\text{m}}}$$
$$= 2$$

Complicated Multiplication and Division

In multiplication, when the quantities have units that are quotients of units, treat them as algebraic variables. Identical units in the numerator and denominator may be "canceled" out.

Example:

(25 meters per second) $\times$ (6 seconds)

$$= \left(25 \, \frac{\text{meters}}{\text{second}}\right) \times (6 \text{ seconds})$$

$$= 25 \times 6 \, \frac{\text{meters-}\cancel{\text{seconds}}}{\cancel{\text{second}}}$$

$$= 150 \text{ meters}$$

(25 m/s) $\times$ (6 s) $= (25 \, \frac{\text{m}}{\text{s}}) \times (6 \text{ s})$

$$= 25 \times 6 \, \frac{\text{m} \cdot \cancel{\text{s}}}{\cancel{\text{s}}}$$

$$= 150 \text{ m}$$

In division, when the quantities have units that are quotients of units, it is easiest to express the division in numerator and denominator form. That is, the number to be divided is the numerator (top value) and the divisor is the denominator (bottom value). Divide the numbers. Treat units as algebraic variables.

Examples:

(8.2 meters per second) $\div$ (2.0 seconds)

$$= \frac{8.2 \text{ meters per second}}{2.0 \text{ seconds}}$$

$$= \frac{8.2}{2.0} \, \frac{\text{meters}}{\text{second-second}}$$

$$= 4.1 \text{ meters per second squared}$$

(8.2 m/s) $\div$ (2.0 s) $= \frac{8.2 \text{ m/s}}{2.0 \text{ s}}$

$$= \frac{8.2}{2.0} \, \frac{\text{m}}{\text{s} \cdot \text{s}}$$

$$= 4.1 \text{ m/s}^2$$

Note that when *second* is multiplied by itself in the denominator, it is changed to *per second squared* (and not to *per square second*). Similarly, the symbols "m/s^2" are read as "meters per second squared."

Scientific Notation

It is convenient to use a mathematical abbreviation for large and small numbers. The number 40 000 000 can be obtained by multiplying 4 by 10, and again by 10, and again by 10, and so on until 10 has been used as a multiplier seven times. The short way of showing this is to write the number 40 000 000 as 4×10^7.

The number 0.0004 can be obtained from 4 by using 10 as a divisor four times. The short way of showing this is to write the number 0.0004 as 4×10^{-4}. Thus,

$$2 \times 10^5 = 2 \times 10 \times 10 \times 10 \times 10 \times 10 = 200\,000$$
$$5 \times 10^{-3} = 5/(10 \times 10 \times 10) = 0.005$$

Numbers expressed in this shorthand manner are said to be in *scientific notation.*

$1\,000\,000 = 10 \times 10 \times 10 \times 10 \times 10 \times 10$	$= 10^6$
$100\,000 = 10 \times 10 \times 10 \times 10 \times 10$	$= 10^5$
$10\,000 = 10 \times 10 \times 10 \times 10$	$= 10^4$
$1000 = 10 \times 10 \times 10$	$= 10^3$
$100 = 10 \times 10$	$= 10^2$
$10 = 10$	$= 10^1$
$1 = 1$	$= 10^0$
$0.1 = 1/10$	$= 10^{-1}$
$0.01 = 1/100$	$= 10^{-2}$
$0.001 = 1/1000$	$= 10^{-3}$
$0.0001 = 1/10\,000$	$= 10^{-4}$
$0.000\,01 = 1/100\,000$	$= 10^{-5}$
$0.000\,001 = 1/1\,000\,000$	$= 10^{-6}$

We can use scientific notation to express some of the physical data often used in physics.

Table B.1 Some Important Values in Physics

Speed of light in a vacuum $= 2.9979 \times 10^8$ m/s

Average earth-sun distance (1 astronomical unit (A.U.)) $= 1.50 \times 10^{11}$ m

Average earth-moon distance $= 3.84 \times 10^8$ m

Average radius of the sun $= 6.96 \times 10^8$ m

Average radius of Jupiter $= 6.99 \times 10^7$ m

Average radius of the earth $= 6.37 \times 10^6$ m

Average radius of the moon $= 1.74 \times 10^6$ m

Average radius of the hydrogen atom $\approx 5 \times 10^{-11}$ m

Mass of the sun $= 1.99 \times 10^{30}$ kg

Mass of Jupiter $= 1.90 \times 10^{27}$ kg

Mass of the earth $= 5.98 \times 10^{24}$ kg

Mass of the moon $= 7.35 \times 10^{22}$ kg

Proton mass $= 1.6726 \times 10^{-27}$ kg

Neutron mass $= 1.6749 \times 10^{-27}$ kg

Electron mass $= 9.11 \times 10^{-31}$ kg

Electron charge $= 1.602 \times 10^{-19}$ C

Appendix C: Graphing

Graphs—A Way to Express Quantitative Relationships

Graphs, like equations and tables, show how two or more quantities relate to each other. Since investigating relationships between quantities makes up much of the work of physics, equations, tables, and graphs are important physics tools.

Equations are the most concise way to describe quantitative relationships. For example, consider the equation $v = v_0 + gt$. It compactly describes how a freely falling object's velocity depends on its initial velocity, acceleration due to gravity, and time. Equations are nice shorthand expressions for relationships among quantities.

Tables give values of variables in list form. The dependence of v on t in $v = v_0 + gt$ can be shown by a table that lists various values v for corresponding times t. Table 2.2 on page 17 is an example. Tables are especially useful when the mathematical relationship between quantities is not known, or when numerical values must be given to a high degree of accuracy. Also, tables are handy for recording experimental data.

Graphs *visually* represent relationships between quantities. By looking at the shape of a graph, you can quickly tell a lot about how the variables are related. For this reason, graphs can help clarify the meaning of an equation or table of numbers. And, when the equation is not already known, a graph can help reveal the relationship between variables. Experimental data are often graphed for this reason.

Graphs are helpful in another way. If a graph contains enough plotted points, it can be used to estimate values between the points (interpolation), or following the points (extrapolation).

Cartesian Graphs

The most common and useful graph in science is the *Cartesian* graph. On a Cartesian graph, possible values of one variable are represented on the vertical axis (called the *y axis*) and possible values of the other variable are plotted on the horizontal axis (*x axis*).

Figure C-1 shows a graph of two variables, x and y, that are *directly proportional* to each other. A direct proportionality is a type of *linear* relationship. Linear relationships have straight-line graphs—the easiest kinds of graphs to interpret. On the graph shown in Figure C-1, the continuous straight-line rise from left to right tells you that as x increases, y increases. More specifically, it shows that y increases at a constant rate with respect to x. As x doubles, y doubles; as x triples, y triples, etc. The graph of a direct proportionality passes through the "origin"—the point at the lower left where $x = 0$ and $y = 0$.

Figure C-2 shows a graph of the equation $v = v_0 + gt$. Speed v is plotted along the y axis, and time t along the x axis. As you can see, there is a linear relationship between v and t.

Many physically significant relationships are more complicated than linear relationships, however. If you

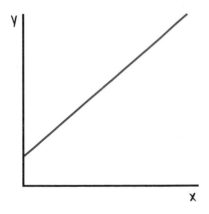

Figure C-1

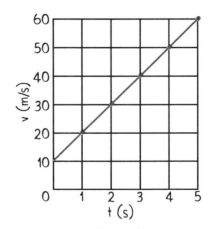

Figure C-2

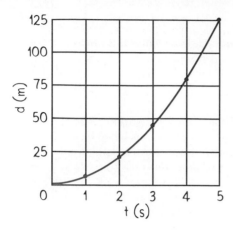

Figure C-3

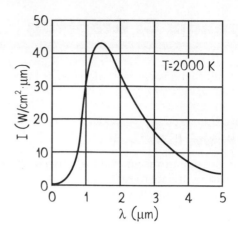

Figure C-4

double the size of a room, the area of the floor increases four times; tripling the size of the room increases the floor area nine times, and so on. This is one example of a *nonlinear* relationship. Figure C-3 shows a graph of another nonlinear relationship: distance vs. time in the equation of free fall from rest, $d = 1/2\ gt^2$.

Figure C-4 shows a *radiation curve*. The *curve* (or graph) shows the rather complex nonlinear relationship between intensity I and radiation wavelength λ for a glowing object at 2000 K. The graph shows that radiation is most intense when λ equals about 1.4 μm. Which is brighter, radiation at 0.5 μm or radiation at 2.0 μm? The graph can quickly tell you that radiation at 2.0 μm is appreciably more intense.

Slope and Area Under the Curve

Quantitative information can be obtained from a graph's *slope* and the *area under the curve*. The slope of

the graph in Figure C-2 represents the rate at which v increases relative to t. It can be calculated by dividing a segment Δv along the y axis by a corresponding segment Δt along the x axis. For example, dividing Δv of 30 m/s by Δt of 3 s gives $\Delta v / \Delta t = 10$ m/s·s $= 10$ m/s^2, the acceleration due to gravity. By contrast, consider the graph in Figure C-5, which is a horizontal straight line. Its slope of zero shows zero acceleration—that is, constant speed. The graph shows that the speed is 30 m/s, acting throughout the entire 5-second interval. The rate of change, or slope, of the speed with respect to time is zero—there is no change in speed at all.

The area under the curve is an important feature of a graph because it often has a physical interpretation. For example, consider the area under the graph of v versus t shown in Figure C-6. The shaded region is a rectangle with sides 30 m/s and 5 s. Its area is 30 m/s $\times$ 5 s = 150 m. In this example, the area is the distance covered by an object moving at constant speed of 30 m/s for 5 s ($d = vt$).

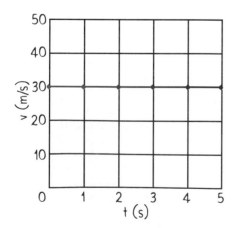

Figure C-5

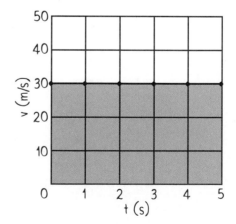

Figure C-6

The area need not be rectangular. The area beneath any curve of *v* versus *t* represents the distance traveled in a given time interval. Similarly, the area beneath a curve of acceleration versus time gives the change of velocity in a time interval. The area beneath a force-versus-time curve gives the change of momentum. (What does the area beneath a force-versus-distance curve give?) The nonrectangular area under various curves, including rather complicated ones, can be found by way of an important branch of mathematics—*integral calculus*.

Graphing with Conceptual Physics

You will develop basic graphing skills in the laboratory part of this course. The lab *Conceptual Graphing* introduces you to graphing concepts. It also gives you a chance to work with a computer and sonic-ranging device. The lab *Trial and Error* will show you the useful technique of converting a nonlinear graph to a linear one to discover a direct proportionality. The area under the curve is the basis of the lab activities *Impact Speed* and *Wrap Your Energy in a Bow*. You will learn about graphing by doing it in other labs as well.

You will also learn in the lab part of your *Conceptual Physics* course that computers can graph data for you. You are not being lazy when you graph your data

with a software program. Instead of investing time and energy scaling the axes and plotting points, you spend your time and energy investigating the meaning of the graph, a high level of thinking!

■ Questions

Figure C-7 is a graphical representation of a ball dropped into a mine shaft.
1. How long did the ball take to hit the bottom?
2. What was the ball's speed when it struck bottom?
3. What does the decreasing slope of the graph tell you about the acceleration of the ball with increasing speed?
4. Did the ball reach terminal speed before hitting the bottom of the shaft? If so, about how many seconds did it take to reach its terminal speed?
5. What is the approximate depth of the mine shaft?

■ Answers
1. 9 s
2. 25 m/s
3. Acceleration decreases as speed increases (due to air resistance)
4. Yes (since slope curves to zero), about 7 s
5. Depth is about 170 m. (The area under the curve is about 17 squares, each of which represents 10 m.)

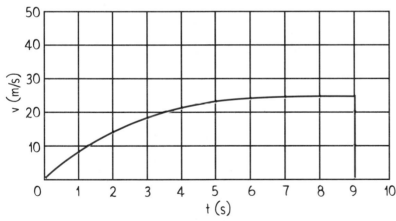

Figure C-7

Appendix D: Vector Applications

Appendix D expands the concept of vectors introduced in Chapters 3 and 5. Vectors are illustrated here by two fascinating cases: a sailboat sailing into the wind, and the passage of light through polarization filters, as discussed in Chapter 27. Vector explanations for both the sailboat and the transmission of light through polarizing filters involve a blend of geometry and physics.

The Sailboat

Sailors have always known that a sailboat can sail downwind (in the same direction as the wind). The ships of Columbus were designed to sail principally downwind. Not until modern times did sailors learn that a sailboat can sail upwind (against the wind). It turns out that many types of sailboats can sail faster "cutting" upwind than when sailing directly with the wind. The old-timers didn't know this, probably because they didn't understand vectors and vector components. Luckily, we do, and today's sailboats are far more maneuverable than the sailboats of the past.

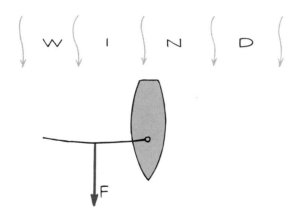

Figure D-1

To understand all this, first consider the relatively simple case of sailing downwind. Figure D-1 shows a force vector *F* due to the impact of the wind against the sail. This force tends to increase the speed of the boat. If it were not for resistive forces, mainly water drag, the speed of the boat would build up to nearly the speed of the wind. (It could be pushed no faster than wind speed because the wind would no longer have any speed relative to the sails. The sails would sag and the force *F* would shrink to zero.) It is important to note

that the faster the boat goes, the smaller will be the magnitude of *F*. So we see that a sailboat sailing directly with the wind can sail no faster than the wind.

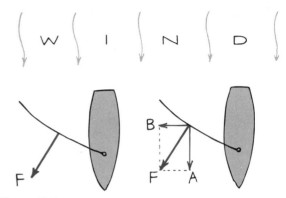

Figure D-2

If the sail is oriented as shown in Figure D-2 (left), the boat will move forward but with less speed for two reasons. First, the force *F* on the sail is less because the sail does not intercept as much wind at this angle. Second, the force on the sail is not in the direction of the boat's motion. It is instead perpendicular to the sail's surface. Generally speaking, whenever any fluid (liquid or gas) interacts with a smooth surface, the force of interaction is perpendicular to the smooth surface. In this case, the boat will not move in the direction of *F* because of its deep finlike keel, which knifes through the water and resists motion in sideways directions.

We can understand the motion of the boat by resolving *F* into perpendicular components, as shown in Figure D-2 (right). The important component is the one parallel to the keel and is labeled *A*. Component *A* propels the boat forward. The other component (*B*) is useless and tends to tip the boat over and move it sideways. The tendency to tip is offset by the heavy deep keel. Again, maximum speed can only approach wind speed.

When a sailboat's keel points in a direction other than exactly downwind and its sails are properly oriented, it can exceed wind speed. In the case of cutting across at an angular direction to the wind (Figure D-3, left), the wind continues to move relative to the sail even after the boat achieves wind speed. A surfer, in a similar way, exceeds the speed of the propelling wave

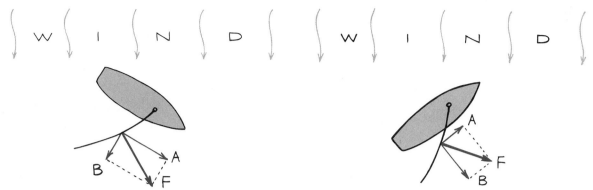

Figure D-3

by angling the surfboard across the wave. Greater angles to the propelling medium (wind for the boat, water wave for the surfboard) result in greater speeds. Can you see why a sailcraft can sail faster cutting across the wind than it can sailing downwind?

As strange as it may seem to people who do not understand vectors, maximum speed is attained by cutting into (against) the wind—that is, by angling the sailcraft in a direction upwind (Figure D-3, right)! Although a sailboat cannot sail *directly* upwind, it can reach a destination upwind by angling back and forth in zigzag fashion. This is called *tacking*. As the speed increases, the relative speed of the wind, rather than decreasing, actually *increases*. Wind impact increases. (If you run outdoors in a slanting rain, the drops will hit you harder if you run into the rain rather than away from the rain!) The faster the boat moves as it tacks upwind, the greater the magnitude of *F*. Thus, component *A* will continue pushing the boat along in the forward direction. The boat reaches its terminal speed when opposing forces, mainly water drag, balance the force of wind impact.

Icecraft, which are equipped with runners for sliding on ice, encounter no water drag. They can travel at several times wind speed when they tack upwind. Terminal speed is reached not so much because of resistive forces, but because the wind direction shifts relative to the moving craft. When this happens, the wind finally moves parallel to the sail rather than against it. This appendix will not go into detail about this complication; nor will it discuss the curvature of the sail, which also plays an important role.

The central concept underlying sailcraft is the vector. It ushered in the era of clipper ships and revolutionized the sailing industry. Sailing, like most things, is more enjoyable if you understand what is happening.

The Vector Nature of Light

Recall from Chapter 27 that light is electromagnetic energy that travels as a transverse wave. The wave is made up of an oscillating electric field vector and an oscillating magnetic field vector that is at right angles to the electric field vector (Figure D-4). It is the orientation of the electric field vector that defines the direction of polarization of light waves.

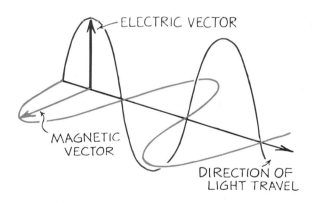

Figure D-4

The electric field vectors in waves of light from the sun or from a lamp are generated in all conceivable directions. Such light is nonpolarized. When the electric vectors of the waves are aligned parallel to each other, the light is considered to be *polarized*. Light can be polarized when it passes through polarizing filters. The most familiar are Polaroid sunglasses. Ordinary nonpolarized light incident upon a polarizing filter emerges as polarized light.

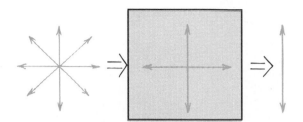

Figure D-5

Think of a beam of nonpolarized light coming straight toward you. Consider the electric vectors in that beam. Some of the possible directions of the vibrations are as shown in Figure D-5 (left). For this unpolarized light, there are as many vectors in any direction as in any other. These many directions of vibration can be replaced by just two directions, horizontal and vertical, since any electric vector can be resolved into horizontal and vertical components. The

center sketch shows the light, with its horizontally and vertically vibrating electric vectors, falling on a polarizing filter with its polarization axis vertically oriented. Only vertical components of light pass through the filter, and the light that emerges is vertically polarized, as shown on the right.

Figure D-6 shows that no light can pass through a pair of Polaroid filters when their axes are at a right angle to each other, but some light does pass through when their axes are at any other angle. This fact can be understood with vectors and vector components.

Recall from Chapter 3 that any vector can be resolved into two components at right angles to each other. The two components are often chosen to be in the horizontal and vertical directions, but they can be in *any* two perpendicular directions. In fact, the number of sets of perpendicular components possible for any vector is infinite. A few of them are shown for the vector *V* in Figure D-7. In every case, components *A* and *B* make up the sides of a rectangle that has *V* as its diagonal.

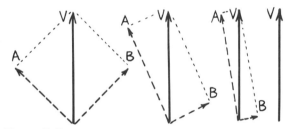

Figure D-7

You can see this somewhat differently by thinking of component *A* as always being vertical and *B* as being horizontal, and picturing vector *V* as rotating instead (Figure D-8). This time the different orientations of *V* are superimposed on a polarizing filter with its polarization axis vertical. In the first sketch on the left, when the electric field vector is vertical, all of *V* gets through. As *V* rotates, only the vertical component *A* passes through. This component *A* gets shorter and shorter until it is zero when *V* is completely horizontal.

Can you now understand how light gets through the second pair of Polaroid sunglasses in Figure D-6?

Figure D-6

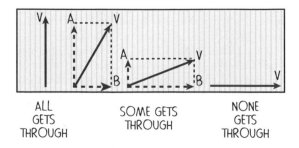

Figure D-8

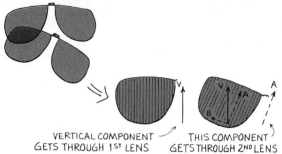

VERTICAL COMPONENT
GETS THROUGH 1ST LENS

THIS COMPONENT
GETS THROUGH 2ND LENS

Figure D-9

Look at Figure D-9, where for clarity the two crossed lenses of Figure D-6, which are one atop the other, are instead shown side by side. The vector V that emerges from the first lens is vertical. However, it has a component A in the direction of the polarization axis of the second lens. Component A passes through the second lens, while component B is absorbed.

To really appreciate this, you must toy around with a couple of polarizing filters, which you can do in a lab exercise. Rotate one above the other and see how you can regulate the amount of light that gets through. Can you think of practical uses for such a system?

■ **Question**

Consider a pair of polarizing filters crossed so that no light gets through. If you place a third filter *in front of* the pair, still no light gets through. The same is true if you place a third filter *in back of* the pair. But if you sandwich a third filter *between* the two with its polarization axis in a different direction from the other two, light *does* get through! (See Figure 27.19.) Magic? No, just physics. Can you explain why this happens?

■ **Answer**

Work on this one and doodle with vectors. If you solve it, help your classmates if they've tried without success and want help. As a last resort, ask your teacher for help.

Appendix E: Exponential Growth and Doubling Time*

You can't fold a piece of paper in half, then fold it again upon itself successively for 9 times. It gets too thick to keep folding. And if you could fold a fine piece of tissue paper upon itself 50 times, it would be more than 20 million kilometers thick! The continual doubling of a quantity builds up astronomically. Double one penny 30 times, so that you begin with one penny, then have two pennies, then four, and so on, and you'll accumulate a total of $10 737 418.23! One of the most important things we have trouble perceiving is the process of exponential growth, and why it proliferates out of control.

When a quantity such as money in the bank, population, or the rate of consumption of a resource steadily grows at a fixed percent per year, the growth is said to be *exponential.* Money in the bank may grow at 5 or 6 percent per year; world population is presently growing at about 2 percent per year; the electric power generating capacity in the United States grew at about 7 percent per year for the first three quarters of the twentieth century. The important thing about exponential growth is that the time required for the growing quantity to double in size (increase by 100 percent) is constant. For example, if the population of a growing city takes 10 years to double from 10 000 to 20 000 people and its growth remains steady, in the next 10 years the population will double to 40 000, and in the next 10 years to 80 000, and so on.

There is an important relationship between the percent growth rate and its *doubling time,* the time it takes to double a quantity:**

$$\text{doubling time} = \frac{69.2 \text{ percent}}{\text{percent growth per unit time}}$$

$$= \frac{70 \text{ percent}}{\text{percent growth rate}}$$

---

* This appendix is adapted from material written by University of Colorado physics professor Albert A. Bartlett, who strongly asserts, "The greatest shortcoming of the human race is man's inability to understand the exponential function." Look up Professor Bartlett's still timely and provocative article, "Forgotten Fundamentals in the Energy Crisis," in the September 1978 issue of the *American Journal of Physics,* or his revised version in the January 1980 issue of the *Journal of Geological Education.*

** For exponential decay we speak about *half-life,* the time for a quantity to reduce to half its value. An example of this case is radioactive decay, treated in Chapter 39.

This means that to estimate the doubling time for a steadily growing quantity, we simply divide 70 percent by the percentage growth rate. For example, when electric power generating capacity in the United States was growing at 7 percent per year, the capacity doubled every 10 years (since [70%]/[7%/year] = 10 years). If world population grew steadily at 2 percent per year, the world population would double every 35 years (since [70%]/[2%/year] = 35 years). A city planning commission that accepts what seems like a modest 3.5-percent-per-year growth rate may not realize that this means that doubling will occur in 20 years (since [70%]/[3.5%/year] = 20 years). That means double capacity for such things as water supply, sewage treatment plants, and other municipal services every 20 years.

Steady growth in a steadily expanding environment is one thing, but what happens when steady growth occurs in a finite environment? Consider the growth of bacteria that grow by division, so that one bacterium becomes two, the two divide to become four, the four divide to become eight, and so on. Suppose the division time for a certain kind of bacterium is one minute. This is then steady growth—the number of bacteria grows exponentially with a doubling time of one minute. Further, suppose that one bacterium is put in a bottle at 11:00 a.m. and that growth continues steadily until the bottle becomes full of bacteria at 12 noon.

It is startling to note that at 2 minutes before noon the bottle was only 1/4 full, and at 3 minutes before noon only 1/8 full. Table E-1 summarizes the amount of space left in the bottle in the last few minutes before noon. If bacteria could think, and if they were concerned about their future, at which time do you think they would sense they were running out of space? Do you think a serious problem would have been evident at, say, 11:55 a.m., when the bottle was only 3-percent full (1/32) and had 97 percent open space (just yearning for development)? The point here is that there isn't much time between the moment the effects of growth

---

■ Question

When was the bottle half full?

■ Answer

At 11:59 a.m., since the bacteria will double in number every minute!

Table E-1 The Last Minutes in the Bottle

Time	Portion Full		Portion Empty	
11:54 a.m.	1/64	(1.5%)	63/64	(98.5%)
11:55 a.m.	1/32	(3%)	31/32	(97%)
11:56 a.m.	1/16	(6%)	15/16	(94%)
11:57 a.m.	1/8	(12%)	7/8	(88%)
11:58 a.m.	1/4	(25%)	3/4	(75%)
11:59 a.m.	1/2	(50%)	1/2	(50%)
12:00 noon	Full	(100%)	None	(0%)

become noticeable and the time when they become overwhelming.

Suppose that at 11:58 a.m. some farsighted bacteria see that they are running out of space and launch a full-scale search for new bottles. And further suppose they consider themselves lucky, for they find three new empty bottles. This is three times as much space as they have ever known. It may seem to the bacteria that their problems are solved—and just in time.

■ Question

If the bacteria are able to migrate to the new bottles and their growth continues at the same rate, what time will it be when the three new bottles are filled to capacity?

Table E-2 illustrates that the discovery of the new bottles extends the resource by only two doubling times. In this example the resource is space—such as land area for a growing population. But it could be coal, oil, uranium, or any nonrenewable resource.

Table E-2 Effects of the Discovery of Three New Bottles

Time	Effect
11:58 a.m.	Bottle 1 is 1/4 full; bacteria divide into four bottles, each 1/16 full
11:59 a.m.	Bottles 1, 2, 3, and 4 are each 1/8 full
12:00 noon	Bottles 1, 2, 3, and 4 are each 1/4 full
12:01 p.m.	Bottles 1, 2, 3, and 4 are each 1/2 full
12:02 p.m.	Bottles 1, 2, 3, and 4 are each all full

Continued growth and continued doubling lead to enormous numbers. In two doubling times, a quantity will double twice ($2^2 = 4$), or quadruple in size; in

■ Answer

All four bottles will be filled to capacity at 12:02 p.m.!

■ Question

According to a French riddle, a lily pond starts with a single leaf. Each day the number of leaves doubles, until the pond is completely full on the thirtieth day. On what day was the pond half covered? One-quarter covered?

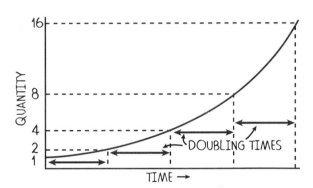

Figure E-1 Graph of a quantity that grows at an exponential rate. Notice that the quantity doubles during each of the successive equal time intervals marked on the horizontal scale. Each of these time intervals represents the doubling time.

three doubling times, its size will increase eightfold ($2^3 = 8$); in four doubling times, it will increase sixteenfold ($2^4 = 16$); and so on. This is best illustrated by the story of the court mathematician in India who years ago invented the game of chess for his king. The king was so pleased with the game that he offered to repay the mathematician, whose request seemed modest enough. The mathematician requested a single grain of wheat on the first square of the chessboard, two grains on the second square, four on the third square, and so on, doubling the number of grains on each succeeding square until all squares had been used. At this rate there would be 2^{63} grains of wheat on the sixty-fourth square alone. The king soon saw that he could not fill this "modest" request, which amounted to more wheat than had been harvested in the entire history of the earth!

As Table E-3 shows, the number of grains on any square is one grain more than the total of all grains on the preceding squares. This is true anywhere on the board. For example, when four grains are placed on the third square, that number of grains is one more than the total of three grains already on the board. The number of grains (eight) on the fourth square is one more than the total of seven grains already on the

■ Answer

The pond was half covered on the 29th day, and was one-quarter covered on the 28th day!

Figure E-2 A single grain of wheat placed on the first square of the chess board is doubled on the second square, and this number is doubled on the third square, and so on. There is not enough wheat in the world for this process to continue to the sixty-fourth square!

board. The same pattern occurs everywhere on the board. In any case of exponential growth, a greater quantity is represented in one doubling time than in all the preceding growth. This is important enough to be repeated in different words: Whenever steady growth occurs, the numerical count of a quantity that exists after a single doubling time is one greater than the total count of that quantity in the entire history of growth.

The consequences of unchecked exponential growth are staggering. It is very important to ask: Is growth really good? Is bigger really better? Is the motto of some businesses, "Grow or die," a good guide for humankind as a whole?

Table E-3 Filling the Squares on the Chessboard

Square Number	Grains on a Square	Total Grains Thus Far
1	1	1
2	2	3
3	$4 = 2^2$	7
4	$8 = 2^3$	15
5	$16 = 2^4$	31
6	$32 = 2^5$	63
7	$64 = 2^6$	127
•	•	•
•	•	•
•	•	•
64	2^{63}	$2^{64} - 1$

■ Questions

1. In 1984 the population growth rate of the world was 1.7 percent per year, and the total world population was 4.8 billion. At this rate, how long will it take for the world population to double?
2. What annual percentage increase in world population would be required to double the world population in 100 years?

■ Answers

1. The doubling time is 41 years, since (70%)/(1.7%/year) = 41 years.
2. 0.7 percent, since (70%)/(0.7%/year) = 100 years. You can rearrange the equation so it reads percent growth rate = (70%)/(doubling time). Using the rearranged equation gives (70%)/(100 years) = 0.7%/year.

Appendix F: Equations of Physics

The equations summarized here are shorthand for relations among concepts you have learned in this book. You can use the equations as guides to thinking, for they tell you at a glance which variables relate to which.

Chapter 2 Linear Motion

Average speed or velocity: $\overline{v} = d/t$

Acceleration (definition of): $a = \Delta v/\Delta t$

Velocity for constant acceleration from rest: $v = at$

Distance traveled for constant acceleration from rest: $d = \frac{1}{2}at^2$

Chapter 5 Newton's Second Law of Motion—Force and Acceleration

Acceleration (cause of): $a = F/m$

Pressure: $P = F/A$

Chapter 7 Momentum

Momentum: momentum = mv

Impulse: impulse = Ft

Impulse-momentum relationship: $Ft = \Delta(mv)$

Chapter 8 Energy

Work: $W = Fd$

Power: power = W/t

Gravitational potential energy: PE = mgh

Kinetic energy: KE = $\frac{1}{2}mv^2$

Chapter 9 Circular Motion

Angular speed: $\omega = \Delta\text{rotations}/\Delta\text{time}$

Linear speed: $v = r\omega$

Centripetal force: $F = mv^2/r = mr\omega^2$

Chapter 11 Rotational Mechanics

Torque: torque = force × lever arm = Fd

Rotational inertia for mass much smaller in size than radial distance: $I = mr^2$

Angular momentum: angular momentum = $mvr = I\omega$

Conservation of angular momentum: $mv_1r_1 = mv_2r_2$; or $I_1\omega_1 = I_2\omega_2$

Chapter 12 Universal Gravitation

Law of universal gravitation: $F = G\dfrac{m_1 m_2}{d^2}$

Chapter 13 Gravitational Interactions

Weight: weight = GmM/R^2

Acceleration due to gravity: $a = GM/R^2$

Chapter 14 Satellite Motion

Orbital speed: $v = \sqrt{GM/d}$

Orbital period: $T = 2\pi\sqrt{d^3/GM}$

Escape speed: $v = \sqrt{2GM/d}$

Chapter 15 Special Relativity— Space and Time

Time dilation: $t = \dfrac{t_0}{\sqrt{1 - (v^2/c^2)}}$

Relativistic addition of velocities: $V = \dfrac{v_1 + v_2}{1 + v_1 v_2/c^2}$

Chapter 16 Special Relativity— Length, Momentum, and Energy

Length contraction: $L = L_0\sqrt{1 - (v^2/c^2)}$

Momentum: $p = \dfrac{mv}{\sqrt{1 - (v^2/c^2)}}$

Kinetic energy: KE = $\dfrac{mc^2}{\sqrt{1 - (v^2/c^2)}} - mc^2$

Energy-mass relationship: $E_0 = mc^2$

Chapter 18 Solids

Density: density = m/V

Weight density: density = mg/V

Hooke's law: $F = k\Delta x$

Chapter 19 Liquids

Liquid pressure: pressure = weight density $\times$ depth

Chapter 20 Gases

Boyle's law: $P_1 V_1 = P_2 V_2$

General gas law: $\dfrac{P_1 V_1}{T_1} = \dfrac{P_2 V_2}{T_2}$

Chapter 21 Temperature, Heat, and Expansion

Quantity of heat: $Q = mc\Delta T$

Linear expansion: $L = \alpha L_0 \Delta T$

Chapter 24 Thermodynamics

First law of thermodynamics: $\Delta Q = \Delta E + W$

Carnot efficiency: Ideal efficiency $= (T_h \times T_c)/T_h$

Entropy: $\Delta S = \Delta Q/T$

Chapter 25 Vibrations and Waves

Period of a simple pendulum: $T = 2\pi\sqrt{L/g}$

Frequency: $f = 1/T$

Period: $T = 1/f$

Wave speed: $v = f\lambda$

Chapter 29 Reflection and Refraction

Index of refraction: $n = c/v$

Snell's law: $n\sin\theta = n'\sin\theta'$

Chapter 30 Lenses

Thin-lens equation: $1/o + 1/i = 1/f$

Chapter 32 Electrostatics

Coulomb's law: $F = k\dfrac{q_1 q_2}{d^2}$

Chapter 33 Electric Fields and Potential

Electric field strength: $E = F/q$

Electric potential: $V = E/q$

Capacitance: $C = q/V$

Chapter 34 Electric Current

Electric current: $I = Q/t$

Ohm's law: $I = V/R$

Electric power: $P = IV$

Chapter 35 Electric Circuits

Equivalent resistance for resistors in series:
$R_{eq} = R_1 + R_2 + \cdots + R_n$

Equivalent resistance for resistors in parallel:
$1/R_{eq} = 1/R_1 + 1/R_2 + \cdots + 1/R_n$

Chapter 36 Magnetism

Magnetic force between magnetic poles: $F \sim p_1 p_2/d^2$

Magnetic force on charge moving across magnetic field: $F = qvB$

Chapter 37 Electromagnetic Induction

Faraday's law: $V \approx -n\Delta B/\Delta t$

Chapter 38 The Atom and the Quantum

Energy of a photon: $E = hf$

De Broglie matter waves: wavelength $= h/\text{momentum}$

Appendix G: Preparing for a Career in Physics

Physics is the basis of varied scientific applications, ranging from the search for better energy sources to the drive for the fastest computer. For this reason, there are many career opportunities in applied physics. There are also career opportunities in pure physics research, where the goal is a better formulation of physical laws.

Few careers are more exciting, rewarding, and important to society than physics. A grasp of conceptual physics is an important first step toward such a career. Another important step is taking as much math as possible, for doing physics involves using its mathematical language.

One of the biggest myths about physics is that you have to be some sort of an Einstein to succeed in it. This is not true. Interest, motivation, and patience, not spectacular intelligence, are the key qualifications for success.

College and University Studies

Students working toward bachelor's degrees in physics typically spend about 30 percent of their time in physics courses and the rest in other areas such as math, chemistry, computer programming, and liberal arts. The first college physics course covers a range of topics similar to those in this book. The focus is on problem solving, using basic calculus. Later courses explore single areas of physics in greater depth, using more advanced mathematics. In advanced laboratory courses, students may use sophisticated electronic equipment and join a research team.

Graduate students concentrate fully on physics. The master's program typically takes two years, while two to four more years are usually needed to earn a Ph.D. An essential ingredient of a Ph.D. program is the *doctoral dissertation*, a major piece of research.

The Physics Workplace

Many physicists work in industrial, academic, and national laboratories. Most of these researchers have Ph.D. degrees. Other physicists, with Ph.D., master's, or bachelor's degrees, work in hospitals, power plants, the military, the astronaut corps, museums, patent law firms, business and government, and as teachers.

Glossary

This glossary gives meanings for terms printed in boldface in the text. The section reference at the end of each meaning is that of the section where the term is introduced.

A simple, phonetic spelling is given for terms that may be unfamiliar or hard to pronounce. CAPITAL LETTERS indicate the syllable that receives the heaviest stress. Accent marks are used when two syllables in a word are stressed; a lowercase syllable followed by an accent mark receives the secondary stress. The phonetic spellings are simple enough so that most can be interpreted without referring to the following key, which gives examples for the vowel sounds and for consonants that are commonly used for more than one sound.

Pronounciation Key

a	cat	ew	new	or	for
ah	father	g	grass	ow	now
ar	car	i,ih	him	oy	boy
aw	walk	$\bar{\text{i}}$	kite	s	so
ay	say	j	jam	sh	shine
ayr	air	ng	sing	th	thick
e,eh	hen	o	hot	u,uh	sun, forces
ee	meet	$\bar{\text{o}}$	hole	z	zebra
eer	deer	oo	moon	zh	pleasure
er	her	ooq	pull		

A The symbol for *ampere*. (34.2) Also, when in lower-case italic, the symbol for *acceleration*. (2.4)

aberration (ab-er-RAY-shun) Distortion in an image produced by a lens. (30.8)

absolute zero The temperature at which a substance has no kinetic energy per particle (thermal) to give up. This temperature corresponds to 0 K, or to –273°C. (21.1, 24.1)

acceleration (ak-sel´-er-RAY-shun) The rate at which velocity is changing. The change may be in magnitude, direction, or both. (2.4)

action force One of the pair of forces described in Newton's third law. (6.2)

additive primary colors Red, blue and green light. These colors when added together produce white light. (28.5)

adiabatic (ay-dee-ah-BAT-ik) Term applied to expansion or compression of a gas occurring without gain or loss of heat. (24.3)

air resistance Friction, or drag, that acts on something moving through air. (5.4)

alternating current (ac) Electric current that repeatedly reverses in direction, twice each cycle. Usually at 60 cycles per second, or hertz (Hz), in North America, or 50 hertz elsewhere. (34.7)

ampere (AM-peer) SI unit of electric current. A flow of one coulomb of charge per second is one ampere (symbol A). (34.2)

amplitude (AMP-lih-tewd) The distance from the midpoint to the maximum (crest) of a wave or, equivalently, from the midpoint to the minimum (trough). (25.2)

aneroid barometer (AN-er-oyd buh-ROM-uh-ter) An instrument used to measure atmospheric pressure; based on the movement of the lid of a metal box. (20.4)

angle of incidence (IN-sih-dens) Angle between an incident ray and the normal to a surface (see Figure 29.3). (29.2)

angle of reflection Angle between a reflected ray and the normal to a surface (see Figure 29.3). (29.2)

angular momentum (mo-MEN-tum) Product of rotational inertia and rotational velocity. (11.6)

antinodes The positions on a standing wave where the largest amplitudes occur. (25.8)

apogee (AP-uh-jee) The point in a satellite's elliptical orbit farthest from the center of the earth. (14.4)

Archimedes' principle (ark-uh-MEE-deez) The relationship between buoyancy and displaced fluid: An immersed object is buoyed up by a force equal to the weight of the fluid it displaces. (19.3)

astigmatism (uh-STIG-muh-tizm) A defect of the eye caused when the cornea is curved more in one direction than in another. (30.7)

atom The smallest particle of an element that can be identified with that element. Consists of protons and neutrons in a nucleus surrounded by electrons. (17.0)

atomic mass number Total number of nucleons (neutrons and protons) in the nucleus of an atom. (39.4)

atomic number Number of protons in the nucleus of an atom. (17.7, 39.4)

average speed Path distance divided by time interval. (2.2)

axis (AK-sis) (a) The straight line around which an object may rotate or revolve. (9.1) (b) A horizontal or vertical reference line in a graph. (Appendix C)

barometer An instrument used to measure the pressure of the atmosphere. (20.3)

beats A throbbing variation in the loudness of sound caused by interference when two tones of slightly different frequencies are sounded together. (26.10)

Bernoulli's principle (ber-NOO-leez) The statement that the pressure in a fluid decreases as the speed of the fluid increases. (20.7)

bimetallic strip (bi´-meh-TAL´-ik) Two strips of different metals, such as one of brass and one of iron, welded or riveted together into one strip. Because the two substances expand at different rates, when heated or cooled the strip bends. Used in thermostats. (21.8)

black hole A mass that has collapsed to so great a density that its enormous local gravitational field prevents light from escaping. (13.6)

blue shift An increase in the measured frequency of light from an approaching source; called the *blue shift* because the apparent increase is toward the high-frequency, or blue, end of the color spectrum.

Also occurs when an observer approaches a source. (25.9)

boiling   The change of phase from liquid to gas that occurs beneath the surface in the liquid. The gas forms bubbles that rise to the surface and escape. (23.4)

bow wave   The V-shaped wave produced by an object moving on a liquid surface faster than the wave speed. (25.10)

Boyle's law   For a constant number of molecules of gas at constant temperature, the product of pressure and volume is constant. (20.5)

breeder reactor   A nuclear fission reactor that not only produces power but produces more nuclear fuel than it consumes by converting a nonfissionable uranium isotope into a fissionable plutonium isotope. (40.4)

Brownian motion   Random movement observed among microscopic particles suspended in a fluid medium. (17.4)

buoyancy (BOY-un-see)   The apparent loss of weight of an object immersed or submerged in a fluid. (19.2)

buoyant force (BOY-unt)   The net upward force exerted by a fluid on a submerged or immersed object. (19.2)

C   (a) The symbol for *coulomb*. (32.3) (b) When preceded or followed by the degree symbol °, the symbol for *Celsius*. (21.1)

cal   The symbol for *calorie*. (21.5)

calorie (KAL-er-ee)   A unit of heat. One calorie (symbol cal) is the heat required to raise the temperature of one gram of water one Celsius degree. One Calorie (with a capital *C*) is equal to one thousand calories and is the unit used in describing the energy available from food. (1 cal = 4.184 J, or 1 J = 0.24 cal) (21.5)

capacitor (kuh-PAS-ih-ter)   A device used to store charge in a circuit. (33.6)

Carnot efficiency (KAR-no)   Ideal maximum percentage of input energy that can be converted to work in a heat engine. (24.5)

Celsius scale (SEL-see-us)   A temperature scale with 0 the melt-freeze temperature for water and 100 the boil-condense temperature of water at standard pressure (one atmosphere at sea level). (21.1)

center of gravity   Point at the center of an object's weight distribution, where the force of gravity can be considered to act. Abbreviated CG. (10.1)

center of mass   Point at the center of an object's mass distribution, where all its mass can be considered to be concentrated. For everyday conditions, it is the same as the center of gravity. (10.2)

centrifugal force (sen-TRIH-fuh-gul)   An apparent outward force on a rotating or revolving body. It is fictitious in the sense that it is not part of an interaction but is due to the tendency of a moving body to move in a straight-line path. (9.4)

centripetal force (sen-TRIH-peh-tul)   A center-directed force that causes an object to move in a curved (sometimes circular) path. (9.3)

chain reaction   A self-sustaining reaction in which one reaction event stimulates one or more additional reaction events to keep the process going. (40.1)

charge   The fundamental electrical property to which the mutual attractions or repulsions between electrons or protons is attributed. (32.1)

chemical formula   A description that uses numbers and symbols of elements to describe the proportions of elements in a compound or reaction. (17.6)

circuit (SER-kit)   Any complete path along which charge can flow. (35.1)

coherent (ko-HEER-ent)   As applied to light waves, having identical frequency and identical phase, and traveling in the same direction. Lasers produce coherent light. (31.7)

complementary colors (kom´-pluh-MENT´-uh-ree)   Two colors of light beams that when added together appear white. (28.6)

component (kom-PO-nent)   One of the vectors, often mutually perpendicular, whose sum is a resultant vector. Any resultant vector may be regarded as the combination of two or more components. (See *resultant*.) (3.3)

compound   A chemical substance made of atoms of two or more different elements combined in a fixed proportion. (17.6)

compression (kom-PRE-shun) (a) In mechanics, the act of squeezing material and reducing its volume. (18.4) (b) In sound, a pulse of compressed air (or other matter); opposite of rarefaction. (26.2)

condensation (kon´-den-SAY´-shun) The change of phase of a gas into a liquid; the opposite of evaporation. (23.2)

conduction (a) In heat, energy transfer from particle to particle within certain materials, or from one material to another when the two are in direct contact. (22.1) (b) In electricity, the flow of charge through a conductor. (32.4)

conductor (a) Material through which heat can be transferred. (22.1) (b) Material, usually a metal, through which electric charge can flow. Good conductors of heat are generally good charge conductors. (32.4)

conservation of charge The principle that net electric charge is neither created nor destroyed but is transferable from one material to another. (32.2)

conserved Term applied to a physical quantity, such as momentum, energy, or electric charge, that remains unchanged during interactions. (7.4)

constructive interference Addition of two or more waves when wave crests overlap to produce a resulting wave of increased amplitude. (25.7)

convection A means of heat transfer by movement of the heated substance itself, such as by currents in a fluid. (22.2)

converging lens A lens that is thickest in the middle, causing parallel rays of light to converge to a focus. (30.1)

cornea (KOR-nee-uh) The transparent covering over the eyeball. (30.6)

correspondence principle If a new theory is valid, it must account for the verified results of the old theory in the region where both theories apply. (16.5)

coulomb (KOO-lom) SI unit of charge. One coulomb (symbol C) is equal to the total charge of 6.24×10^{18} electrons. (32.3)

Coulomb's law The relationship among electrical force, charges, and distance: The electrical force between two charges varies directly as the product of the charges and inversely as the square of the distance between them. (32.3)

crest One of the places in a wave where the wave is highest or the disturbance is greatest. (25.2)

critical angle The minimum angle of incidence for which a light ray is totally reflected within a medium. (29.12)

critical mass The minimum mass of fissionable material in a nuclear reactor or nuclear bomb that will sustain a chain reaction. (40.1)

crystal (KRIS-tul) A regular geometric shape found in a solid in which the component particles are arranged in an orderly, three-dimensional, repeating pattern. (18.1)

current See *electric current*.

density (DEN-sih-tee) A property of a substance, equal to its mass per volume. (18.2)

destructive interference Combination of waves where crest parts of one wave overlap trough parts of another, resulting in a wave of decreased amplitude. (25.7)

diode (DI-od) An electronic device that restricts current to flow in a single direction in an electric circuit. (34.8)

diffraction (dih-FRAK-shun) The bending of a wave around a barrier, such as an obstacle or the edges of an opening. (31.2)

diffraction grating A series of closely spaced parallel slits or grooves that are used to separate colors of light by interference. (31.4)

diffuse reflection (dih-FYOOS) The reflection of waves in many directions from a rough surface (see Figure 29.7). (29.4)

direct current (dc) Electric current whose flow of charge is always in one direction. (34.7)

dispersion (dih-SPER-zhun) The separation of light into colors arranged according to their frequency, by interaction with a prism or diffraction grating, for example. (29.10)

displaced Term applied to the fluid that is moved out of the way when an object is placed in fluid. A submerged object always displaces a volume of fluid equal to its own volume. (19.2)

diverging lens A lens that is thinnest in the middle and that causes parallel rays of light to diverge. (30.1)

Doppler effect (DOP-ler) The change in frequency of a wave due to the motion of the source or of the receiver. (25.9)

eddy   Changing, curling paths in turbulent flow of a fluid. (20.7)

efficiency   In a machine, the ratio of useful energy output to total energy input, or the percentage of the work input that is converted to work output. (8.8)

elapsed time   The time that has passed since beginning of an event. (2.5)

elastic   Term applied to a material that returns to its original shape after it has been stretched or compressed. (18.3)

elastic collision   Collision in which colliding objects rebound without lasting deformation or heat generation. (7.5)

elasticity (ih-las-TIH-sih-tee)   The property of a solid wherein a change in shape is experienced when a deforming force acts on it, with a return to its original shape when the deforming force is removed. (18.3)

elastic limit   The distance of stretching or compressing beyond which an elastic material will not return to its original shape. (18.3)

electric charge   See *charge*.

electric current   The flow of electric charge; measured in amperes (coulombs per second). (34.2)

electric field   A force field that fills the space around every electric charge or group of charges. Measured by force per charge (N/C). (33.1)

electric potential   Electric potential energy per coulomb (J/C) at a location in an electric field; measured in volts and often called *voltage*. (33.5)

electric potential energy   Energy a charge has due to its location in an electric field. (33.4)

electric power   The rate at which electric energy is converted into another form, such as light, heat, or mechanical energy (or converted *from* another form into electric energy). (34.11)

electric resistance   The resistance of a material to the flow of electric current through it; measured in ohms (symbol Ω). (34.4)

electrical force   A force that one charge exerts on another. When the charges are the same sign, they repel; when the charges are opposite, they attract. (32.1)

electrically polarized   Term applied to an atom or molecule in which the charges are aligned so that one side is slightly more positive or negative than the opposite side. (32.7)

electromagnet   (ih-lek´-tro-MAG´-net) Magnet with a field produced by electric current; usually in the form of a wire coiled around a piece of iron. (36.5)

electromagnetic induction   (ih-lek´-tro-mag-NET´-ik in-DUK-shun) The phenomenon of inducing a voltage in a conductor by changing the magnetic field near the conductor. (37.1)

electromagnetic spectrum   The range of electromagnetic waves extending from radio waves to gamma rays. (27.3)

electromagnetic wave   A wave that is partly electric and partly magnetic and carries energy. Emitted by vibrating electric charges. (27.3)

electrostatics   (ih-lek´-tro-STAT´-iks) The study of electric charges at rest. (32.0)

element   A substance made of only one kind of atom. Examples of elements are carbon, hydrogen, oxygen, and nitrogen. (17.1)

ellipse (ih-LIPS)   An oval-shaped curve that is the path of a point that moves such that the sum of its distances from two fixed points (foci) is constant (see Figure 14.7). (14.3)

energy   That property of an object or a system which enables it to do work; measured in joules. (8.3)

entropy A measure of the amount of disorder in a system. (24.7)

equilibrium   (ee-kwih-LIH-bree-um) In general, a state of balance. Examples: The state of a body on which no net force acts. (4.7) The state of a body on which no net torque acts. (11.2) The state of a liquid in which the processes of evaporation and condensation are taking place at equal rates. (23.3)

escape speed   The minimum speed necessary for an object to escape permanently from a gravitational field that holds it. (14.5)

evaporation (ih-vap´-or-AY´-shun)   The change of phase from liquid to gas that takes place at the surface of a liquid. (23.1)

eyepiece   Lens of a telescope closest to the eye; enlarges the real image formed by the first lens. (30.5)

fact   A close agreement by competent observers of a series of observations of the same phenomena. (1.4)

Fahrenheit scale (FA-ren-hit)   The temperature scale in common use in the United States. The number 32 is assigned to the freezing point of water, and the number 212 to the boiling point of water (at standard atmospheric pressure). (21.1)

Faraday's law (FA-ruh-dayz)   Induced voltage in a coil is proportional to the product of the number of loops and the rate at which the magnetic field changes within those loops. (37.2) In general, an electric field is induced in any region of space in which a magnetic field is changing with time. The magnitude of the induced electric field is proportional to the rate at which the magnetic field changes. (37.7)

farsighted   Term applied to a person who has trouble focusing on nearby objects because the eyeball is so short that images form behind the retina. (30.7)

field   See *force field*.

first law of thermodynamics   Heat added to a system is transformed to an equal amount of some other form of energy; a version of the law of energy conservation. (24.2)

first postulate of special relativity   All the laws of nature are the same in all uniformly moving reference frames. (15.4)

fission   See *nuclear fission*.

fluid   Anything that flows; in particular, any liquid or gas. (5.4, 19.6)

focal length   The distance between the center of a lens and either focal point. (30.1)

focal plane   A plane passing through either focal point of a lens that is perpendicular to the principal axis. For a converging lens, any incident parallel beam of light converges to a point somewhere on a focal plane. For a diverging lens, such a beam appears to come from a point on a focal plane. (30.1)

focal point   For a converging lens, the point at which a beam of light parallel to the principal axis converges. For a diverging lens, the point from which such a beam appears to come. (30.1)

focus (FO-kus); **pl. foci** (FO-si)   (a) For an ellipse, one of the two points for which the sum of the distances to any point on the ellipse is a constant. A satellite orbiting the earth moves in an ellipse that has the earth at one focus. (14.3) (b) For optics, the point where parallel light rays converge. (30.1)

force   Any influence that tends to accelerate an object; a push or pull; measured in newtons. A vector quantity. (4.3)

forced vibration   The vibration of an object that is made to vibrate by another vibrating object that is nearby. The sounding board in a musical instrument amplifies the sound through forced vibration. (26.6)

force field   That which exists in the space surrounding a mass, electric charge, or magnet, so that another mass, electric charge, or magnet introduced to this region will experience a force. Examples of force fields are gravitational fields, electric fields, and magnetic fields. (13.1)

free fall   Motion under the influence of the gravitational force only. (2.5)

freezing   Change in phase from liquid to solid. (23.5)

frequency (FREE-kwen-see)   The number of events (cycles, vibrations, oscillations, or any repeated event) per time; measured in hertz (or events per time). Inverse of period. (25.2)

friction   The force that acts to resist the relative motion (or attempted motion) of objects or materials that are in contact. (4.3)

fulcrum   (FOOL-krum) The pivot point of a lever. (8.7)

fusion   See *nuclear fusion*.

g   (a) The symbol for gram. (4.5) (b) When in lowercase italic, the symbol for the acceleration due to gravity (at the earth's surface, 9.8 m/s^2). (2.5) (c) When in lowercase bold, the gravitational field vector (at the earth's surface, 9.8 N/kg). (13.1) (d) When in uppercase italic, the symbol for the universal constant of gravitation (6.67 × 10^{-11} Nm^2/kg^2). (12.4)

general theory of relativity   Einstein's generalization of special relativity, which features a geometric theory of gravitation. (15.0)

generator   A machine that produces electric current by rotating a coil within a stationary magnetic field. (37.3)

global warming   See *greenhouse effect*.

gravitational field   (grav´-ih-TAY´-shun-ul) A force field that exists in the space around every mass or group of masses. (13.1)

greenhouse effect   The warming effect whose cause is that short-wavelength radiant energy from the sun can enter the atmosphere and be absorbed by the earth more easily than long-wavelength energy from the earth can leave. (22.7)

grounding   Allowing charges to move freely along a connection between a conductor and the ground. (32.6)

group   Elements in the same column of the periodic table. (17.8)

h   (a) The symbol for *hour* (though hr is often used). (2.2) (b) When in italic, the symbol for *Planck's constant*. (38.2)

half-life   The time required for half the atoms of a radioactive isotope of an element to decay. Also used for decay processes in general. (39.5)

heat   Energy transfer via random molecular motions, resulting in gain or loss of internal energy. (21.2)

heat engine   A device that changes internal energy to mechanical work. (24.5)

hertz (HERTS)   The SI unit of frequency. One hertz (Hz) is one vibration per second. (25.2)

hologram (HOL-uh-gram)   A three-dimensional version of a photograph produced by interference patterns of laser beams. (31.8)

Hooke's law   The distance of stretch or squeeze (extension or compression) of an elastic material is directly proportional to the applied force. (18.3)

Huygens' principle (HI-gunz)   Every point on any wave front can be regarded as a new point source of secondary waves. (31.1)

hypothesis (hi-POTH-uh-sis)   An educated guess; a reasonable explanation of an observation or experimental result that is not fully accepted as factual until tested over and over again by experiment. (1.3)

Hz   The symbol for *hertz*. (25.2)

impulse (IM-puls)   Product of force and time interval during which the force acts. Impulse equals momentum change. (7.2)

incoherent (in´-ko-HEER´-ent)   As applied to light waves, having a jumbled mixture of frequency, phase, and possibly direction. (31.7)

induced (in-DEWSD)   (a) Term applied to electric charge that has been redistributed on an object because of the presence of a charged object nearby. (32.6) (b) Term applied to a voltage, electric field, or magnetic field that is created due to a change in or motion through a magnetic field or electric field. (37.1, 37.7)

induction (in-DUK-shun)   The charging of an object without direct contact. (32.6) See also *electromagnetic induction*.

inelastic   Term applied to a material that does not return to its original shape after it has been stretched or compressed. (Also called plastic.) (18.3)

inelastic collision   A collision in which the colliding objects become distorted and/or generate heat during the collision. (7.5)

inertia (ih-NER-shuh)   The reluctance of any body to change its state of motion. Mass is the measure of inertia. (4.3)

infrared   Electromagnetic waves of frequencies lower than the red of visible light. (27.3)

infrasonic (in´-fruh-SON´-ik)   Term applied to sound pitch too low to be heard by the human ear, that is, below 20 hertz. (26.1)

in parallel   Term applied to portions of an electric circuit that are connected at two points and provide alternative paths for the current between those two points. (35.2)

in phase (FAYZ)   Term applied to two or more waves whose crests (and troughs) arrive at a place at the same time, so that their effects reinforce each other. (25.7)

in series   Term applied to portions of an electric circuit that are connected in a row so that the current that goes through one must go through all of them. (35.2)

instantaneous speed (in-stan-TAY-nee-us)   Speed at any instant of time. (2.2)

insulator (IN-suh-lay-ter)   (a) A material that is a poor conductor of heat and that delays the transfer of heat. (22.1) (b) A material that is a poor conductor of electricity. (32.4)

interaction   A mutual action between objects where each object exerts an equal and opposite force on the other. (6.1)

interference pattern (in´-ter-FEER´-ens)   A pattern formed by the overlapping of two or more waves that arrive in a region at the same time. (25.7)

internal energy   The total energy stored in the atoms and molecules within a substance. (21.4)

inverse-square law   A physical quantity varies inversely as another quantity squared. Example: Illumination varies inversely as the square of the distance from the source. (12.5)

inversely   When two values change in opposite directions, so that if one is doubled the other is reduced to one half, they are said to be inversely proportional to each other. (5.2)

ion (I-un)   An atom (or group of atoms bound together) with a net electric charge, which is due to the loss or gain of electrons. (17.8)

iridescence (ih-rih-DES-ens)   The phenomenon whereby interference of light waves of mixed frequencies reflected from the top and bottom of thin films produces a spectrum of colors. (31.6)

iris (I-ris)   The colored part of the eye that surrounds the black opening through which light passes. The iris regulates the amount of light entering the eye. (30.6)

isotope (I-suh-top)   A form of an element having a particular number of neutrons in the nuclei of its atoms. Different isotopes of a particular element have the same atomic number but different atomic mass numbers. (17.7, 39.4)

J   The symbol for *joule*. (8.1)

joule (JOOL)   The SI unit of work and of all other forms of energy. One joule (symbol J) of work is done when a force of one newton is exerted on an object moved one meter in the direction of the force. (8.1)

K   (a) The symbol for *kelvin*. (21.1) (b) When in lowercase, the symbol for the prefix *kilo-*.

kcal   The symbol for *kilocalorie*. (21.5)

kelvin (KEL-vin)   The SI unit of temperature. A temperature measured in kelvins (symbol K) indicates the number of units above absolute zero. Since the divisions on the Kelvin scale and Celsius scale are the same size, a change in temperature of one kelvin equals a change in temperature of one Celsius degree. (21.1)

Kelvin scale   A temperature scale whose zero (called absolute zero) is the temperature at which it is impossible to extract any more internal energy from a material. 0 K = –273°C. There are no negative temperatures on the Kelvin scale. (21.1)

kg   The symbol for *kilogram*. (4.5)

kilocalorie (KIL-o-kal-er-ee)   A unit of heat. One kilocalorie equals 1000 calories, or the amount of heat required to raise the temperature of one kilogram of water by 1°C. (21.5)

kilogram (KIL-o-gram)   The fundamental SI unit of mass. One kilogram (symbol kg) is the amount of mass in one liter of water at 4°C. See Appendix A. (4.5)

kinetic energy (kih-NET-ik)   Energy of motion, equal (nonrelativistically) to half the mass multiplied by the speed squared. (8.5)

km   The symbol for *kilometer*. (2.2)

L   The symbol for *liter*. (19.3)

laser (LAY-zer)   An optical instrument that produces a beam of coherent light—that is, having the waves all the same frequency, phase, and direction. (31.7)

law   A general hypothesis or statement about the relationship of natural quantities that has been tested over and over again and has not been contradicted. Also known as a *principle*. (1.4)

law of conservation of angular momentum   An object or system of objects will maintain a constant angular momentum unless acted upon by an unbalanced external torque. (11.7)

law of conservation of energy   Energy cannot be created or destroyed. It may be transformed from one form into another, but the total amount of energy never changes. (8.6)

law of conservation of momentum   In the absence of a net external force, the momentum of an object or system of objects is unchanged. (7.4)

law of inertia Every body continues in its state of rest, or of motion in a straight line at constant speed, unless it is compelled to change that state by a net force exerted upon it. Also known as *Newton's first law.* (4.4)

law of reflection The angle of incidence for a wave that strikes a surface is equal to the angle of reflection. This is true for both partially and totally reflected waves. (29.2)

law of universal gravitation For any pair of objects, each object attracts the other object with a force that is directly proportional to the product of the masses of the objects, and inversely proportional to the square of the distance between their centers of mass. (12.4)

lens (LENZ) A piece of glass (or other transparent material) that can bend parallel rays of light so that they cross, or appear to cross, at a single point. (30.1)

lever (LEH-ver, LEE-ver) A simple machine, made of a bar that turns about a fixed point. (8.7)

lever arm The perpendicular distance between an axis and the line of action of a force that tends to produce rotation about that axis. (11.1)

lift In application of Bernoulli's principle, the net upward force produced by the difference between upward and downward pressures. When lift equals weight, horizontal flight is possible. (20.8)

light-year The distance light travels through a vacuum during one year. (27.2)

line spectrum Pattern of distinct lines of color, corresponding to particular wavelengths, that are seen in a spectroscope when a hot gas is viewed. (28.11)

linear momentum Product of the mass and the velocity of an object. Also called *momentum.* (This definition applies at speeds much less than the speed of light.) (11.6)

linear speed The path distance moved per unit of time. Also called simply *speed.* (9.2)

longitudinal wave (lon-jih-TEWD-ih-nul) A wave in which the vibration is in the same direction as that in which the wave is traveling, rather than at right angles to it. (25.6)

lunar eclipse The cutoff of light from the full moon when the earth is directly between the sun and the moon, so that the earth's shadow is cast on the moon. (13.4)

m (a) The symbol for *meter.* (2.2) (b) When in italic, the symbol for *mass.* (4.5)

machine A device for increasing (or decreasing) a force or simply changing the direction of a force. (8.7)

magnetic domain A microscopic cluster of atoms with their magnetic fields aligned. (36.4)

magnetic field A force field that fills the space around every magnet or current-carrying wire. Another magnet or current-carrying wire introduced into this region will experience a magnetic force. (36.2)

magnetic pole One of the regions on a magnet that produces magnetic forces. (36.1)

mass A measure of an object's inertia; also a measure of the amount of matter in an object. Depends only on the amount of and kind of particles that compose an object—not on its location (as weight does). (4.5)

mechanical advantage The ratio of output force to input force for a machine. (8.7)

mechanical energy The energy due to the position or the movement of something; potential or kinetic energy (or a combination of both). (8.3)

mirage (mih-RAHZH) A floating image that appears in the distance and is due to the refraction of light in the earth's atmosphere. (29.9)

molecule (MOL-uh-kyool) Two or more atoms of the same or different elements bonded to form a larger particle. (17.5)

momentum The product of the mass and the velocity of an object (provided the speed is much less than the speed of light). Has magnitude and direction (a vector quantity). Also called *linear momentum.* (7.1)

monochromatic (mon´-o-kro-MAT´-ik) Having a single color or frequency. (31.4)

N The symbol for *newton.*

natural frequency A frequency at which an elastic object, once energized, will vibrate. Minimum

energy is required to continue vibration at that frequency. Also called *resonant frequency*. (26.7)

neap tide A tide that occurs when the moon is halfway between a new moon and a full moon, in either direction. The tides due to the sun and the moon partly cancel, so that the high tides are lower than average and the low tides are not as low as average. (13.4)

nearsighted Term applied to a person who can clearly see nearby objects but not clearly see distant objects. Eyeball is elongated so that images focus in front rather than on the retina. (30.7)

net force The combination of all the forces that act on an object. (4.6)

neutral equilibrium The state of an object balanced so that any small movement neither raises nor lowers its center of gravity. (10.5)

neutron An electrically neutral particle that is one of the two kinds of particles that compose an atomic nucleus. (17.7)

newton SI unit of force. One newton (N) is the force applied to a one-kilogram mass that will produce an acceleration of one meter per second per second. (4.5)

Newton's first law See *law of inertia*.

Newton's law of cooling The rate of cooling of an object—whether by conduction, convection, or radiation—is approximately proportional to the temperature difference between the object and its surroundings. (22.6)

Newton's second law The acceleration produced by a net force on a body is directly proportional to the magnitude of the net force, is in the same direction as the net force, and is inversely proportional to the mass of the body. (5.3)

Newton's third law Whenever one body exerts a force on a second body, the second body exerts an equal and opposite force on the first. (6.2)

node Any part of a standing wave that remains stationary. (25.8)

normal A line perpendicular to a surface. (29.2)

normal force For an object resting on a horizontal surface, the upward force that balances the weight of the object; also called the *support force*. (4.7)

nuclear fission (FIH-shun) The splitting of an atomic nucleus, particularly that of a heavy element such as uranium-235, into two main parts accompanied by the release of much energy. (40.1)

nuclear fusion (Few-zhun) The combining of nuclei of light atoms, such as hydrogen, into heavier nuclei accompanied by the release of much energy. (40.6)

nucleon (NEW-klee-on) The principal building block of the nucleus; a neutron or a proton. (17.7, 39.1)

nucleus The positively charged center of an atom, which contains protons and neutrons and has almost all the mass of the entire atom but only a tiny fraction of the volume. (17.7)

objective lens In an optical device using compound lenses, the lens closest to the object observed. (30.5)

ohm (OM) The SI unit of electrical resistance. One ohm (symbol Ω) is the resistance of a device that draws a current of one ampere when a voltage of one volt is impressed across it. (34.4)

Ohm's law The statement that the current in a circuit is directly proportional to the voltage impressed across the circuit, and is inversely proportional to the resistance of the circuit. (34.5)

opaque Term applied to materials that absorb light without reemission, and consequently do not allow light through them. (27.5)

optical fiber A transparent fiber, usually of glass or plastic, that can transmit light down its length by means of total internal reflection. (29.12)

out of phase Term applied to two waves for which the crest of one wave arrives at a point at the same time that a trough of the second wave arrives. Their effects cancel each other. (25.7)

parallel circuit An electric circuit in which devices are connected to the same two points of the circuit, so that any single device completes the circuit independently of the others. (35.4)

pascal (pas-KAL) The SI unit of pressure. One pascal (symbol Pa) of pressure exerts a normal force of one newton per square meter. (5.5)

Pascal's principle Changes in pressure at any point

in an enclosed fluid at rest are transmitted undiminished to all points in the fluid and act in all directions. (19.6)

penumbra   A partial shadow that appears where light from part of the source is blocked and light from another part of the source is not blocked. (27.6)

perigee (PEH-rih-jee)   The point in a satellite's elliptical orbit where it is nearest the center of the earth. (14.4)

period   (a) The time required for a complete orbit. (14.2) (b) The time required for a pendulum to make one to-and-fro swing. In general, the time required to complete a single cycle. (25.1)

periodic table   A chart that lists elements by atomic number and by electron arrangements, so that elements with similar chemical properties are in the same column (Figure 17.11). (17.8)

perturbation   The deviation of an orbiting object from its path around a center of force caused by the action of an additional center of force. (12.6)

phase   One of the four possible forms of matter: solid, liquid, gas, and plasma. Often called *state*. (23.0)

photoelectric effect   The ejection of electrons from certain metals when exposed to certain frequencies of light. (38.3)

photon (FO-ton)   In the particle model of electromagnetic radiation, a particle that travels only at the speed of light and whose energy is related to the frequency of the radiation in the wave model. (27.1, 38.2)

pigment   A material that selectively absorbs colored light. (28.3)

pitch   Term that refers to how high or low sound frequencies appear to be. (26.1)

Planck's constant   A fundamental constant of quantum theory that determines the scale of the small-scale world. Planck's constant (symbol h) multiplied by the frequency of radiation gives the energy of a photon of that radiation. (38.2)

plasma (PLAZ-muh)   A fourth phase of matter, in addition to solid, liquid, and gas. In the plasma phase, which exists mainly at high temperature, matter consists of positively charged ions and free electrons. (17.9)

polarization (po-lcr-ih-ZΛY´-shun)   The aligning of vibrations in a transverse wave, usually by filtering out waves of other directions. (27.7)

postulate (POS-tyoo-lit)   A fundamental assumption. (15.3)

potential   See *electric potential*.

potential difference   The difference in electric potential (voltage) between two points. Free charge flows when there is a difference, and will continue until both points reach a common potential. (34.1)

potential energy   Energy of position, usually related to the relative position of two things, such as a stone and the earth, or an electron and a nucleus. (8.4)

power   Rate at which work is done or energy is transformed, equal to the work done or energy transformed divided by time; measured in watts. (8.2)

pressure   Force per surface area where the force is normal to the surface; measured in pascals. (5.5)

principal axis   The line joining the centers of curvature of the surfaces of a lens. (30.1)

principle   A general hypothesis or statement about the relationship of natural quantities that has been tested over and over again and has not been contradicted; also known as a *law*. (1.4)

principle of flotation   A floating object displaces a quantity of fluid of a weight equal to its own weight. (19.5)

projectile   Any object that moves through the air or through space, acted on only by gravity (and air resistance, if any). (3.4)

proton   A positively charged particle that is one of the two kinds of particles found in the nucleus of an atom. (17.7)

pulley   A type of lever that is a wheel with a groove in its rim, and one that is used to change the direction of a force. A pulley or system of pulleys can also multiply forces. (8.7)

pupil   The opening in the eyeball through which light passes. (30.6)

quantum (pl. quanta) (KWONT-um)   The fundamental "size" unit; the smallest amount of anything. One quantum of light energy is called a *photon*. (38.2)

quantum mechanics   The branch of physics that is the study of the motion of particles in the microworld of atoms and nuclei. (38.8)

quantum physics   The branch of physics that is the general study of the microworld of photons, atoms, and nuclei. (38.8)

quark (KWORK, KWARK)   One of the elementary particles of which nucleons (protons and neutrons) are made. (39.1)

radiant energy   Any energy, including heat, light, and X rays, that is transmitted by radiation. It occurs in the form of electromagnetic waves. (22.3)

radiation   (a) Energy transmitted by electromagnetic waves. (22.3) (b) The particles given off by radioactive atoms such as uranium. (39.2)

radioactive   Term applied to an atom with a nucleus that is unstable and that can spontaneously emit a particle and become the nucleus of another element. (39.2)

rarefaction (rayr-uh-FAK-shun)   A disturbance in air (or matter) in which the pressure is lowered. Opposite of compression. (26.2)

rate   In physics, how fast something happens, or how much something changes per unit of time; a change in a quantity divided by the time it takes for the change to occur. (2.0)

ray   A thin beam of light. (27.6)

ray diagram   A diagram showing rays that can be drawn to determine the size and location of an image formed by a mirror or lens. (30.3)

reaction force   The force that is equal in strength and opposite in direction to the action force, and one that acts simultaneously on whatever is exerting the action force. (6.2)

real image   An image that is formed by converging light rays and that can be displayed on a screen. (30.2)

red shift   A decrease in the measured frequency of light (or other radiation) from a receding source; called the *red shift* because the decrease is toward the low-frequency, or red, end of the color spectrum. (25.9)

reflection   The bouncing back of a particle or wave that strikes the boundary between two media. (29.1)

refraction   The change in direction of a wave as it crosses the boundary between two media in which the wave travels at different speeds. (29.6)

regelation   The phenomenon of ice melting under pressure and freezing again when the pressure is reduced. (23.7)

relative   Regarded in relation to something else. Depends on point of view, or frame of reference. Sometimes referred to as "with respect to." (2.1)

relative humidity   A ratio between how much water vapor is in the air and the maximum amount of water vapor that could be in the air at the same temperature. (23.2)

relativistic kinetic energy   Kinetic energy at very high speeds approaching the speed of light. (16.2)

relativistic momentum   Momentum at very high speeds approaching the speed of light. (16.4)

resolution (rez-uh-LOO-shun)   (a) The process of resolving a vector into components. (3.3) (b) In optics, a measure of how well closely adjacent optical images are distinguished.

resonance (REZ-uh-nuns)   A phenomenon that occurs when the frequency of forced vibrations on an object matches the object's natural frequency, and a dramatic increase in amplitude results. (26.8)

rest energy   The "energy of being" given the equation, $E_0 = mc^2$. (16.3)

rest mass   The intrinsic mass of an object, a fixed property independent of speed or energy. (16.2)

resultant (rih-ZUL-tunt)   The vector sum of two or more component vectors. (3.2)

retina (RET-ih-nuh)   The layer of light-sensitive tissue at the back of the eye. (30.6)

reverberation (rih-verb-er-AY-shun)   Persistence of a sound, as in an echo, due to multiple reflections. (29.5)

revolution   Motion of an object turning around an axis outside the object. (9.1)

rotation   The spinning motion that takes place when an object rotates about an axis located within the object (usually an axis through its center of mass). (9.1)

rotational inertia   The reluctance of an object to change its state of rotation, determined by the distribution of the mass of the object and the location of the axis of rotation or revolution. (11.4)

rotational speed   The number of rotations or revolutions per unit of time; often measured in rotations or revolutions per second or per minute (RPM). (9.2)

rotational velocity   Rotational speed together with a direction for the axis of rotation or revolution. (11.6)

s   The symbol for *second*. (2.2)

satellite   An object that falls around the earth or some other body rather than falling into it. (3.6, 14.1)

saturated   Term applied to a substance, such as air, that contains the maximum amount of another substance, such as water vapor, at a given temperature and pressure. (23.2)

scalar quantity   A quantity in physics, such as mass, volume, and time, that can be completely specified by its magnitude, and has no direction. (3.1)

scaling   The study of how size affects the relationship between weight, strength, and surface area. (18.5)

scatter   To absorb sound or light and reemit it in all directions. (28.8)

schematic diagram   Diagram that describes an electric circuit, using special symbols to represent different devices in the circuit. (35.5)

scientific method   An orderly method for gaining, organizing, and applying new knowledge. (1.3)

second law of thermodynamics   Heat will never of itself flow from one object to another of higher temperature. (24.4)

second postulate of special relativity   The speed of light in empty space always has the same value regardless of the motion of the source or the motion of the observer. (15.5)

semiconductor Material that can be made to behave as either a conductor or an insulator of electricity. (32.4)

series circuit   An electric circuit in which devices are arranged so that charge flows through each in turn. If one part of the circuit should stop the current, it will stop throughout the circuit. (35.3)

shadow   A shaded region that results when light falls on an object and thus cannot reach into the region on the far side of the object. (27.6)

shell model of the atom   A model in which the electrons of an atom are pictured as grouped in concentric shells around the nucleus. (17.8)

shock wave   A cone-shaped wave produced by an object moving at supersonic speed through a fluid. (25.11)

simple harmonic motion   Periodic motion in which acceleration is proportional to the distance from an equilibrium position and is directed toward that equilibrium position. (25.2)

sine curve   A curve whose shape represents the crests and troughs of a wave, as traced out by a swinging pendulum that drops a trail of sand over a moving conveyor belt. (25.2)

solar eclipse   The cutoff of light from the sun to an observer on the earth when the moon is directly between the sun and the earth. (13.4)

sonic boom   The sharp crack heard when the shock wave that sweeps behind a supersonic aircraft reaches the listener. (25.11)

space-time   A combination of space and time, which are viewed in special relativity as two parts of one whole. (15.1)

special theory of relativity   The theory, introduced in 1905 by Albert Einstein, that describes how time is affected by motion in space at a constant velocity, and how mass and energy are related. (15.1)

specific gravity   The ratio of the mass (or weight) of a substance to the mass (or weight) of an equal volume of water. (18.2)

specific heat capacity   The quantity of heat required to raise the temperature of a unit mass of a substance by one degree Celsius. Often simply called "specific heat," or "heat capacity." (21.6)

spectroscope   An instrument used to separate the light from a hot gas or other light source into its constituent frequencies. (28.11)

spectrum   For sunlight and other white light, the spread of colors seen when the light is passed through a prism or diffraction grating. In general, the spread of radiation by frequency, so that each frequency appears at a different position. (28.1)

speed   How fast something is moving; the path distance moved per time. The magnitude of the velocity vector. (2.2)

spring tide   A high or low tide that occurs when the sun, earth, and moon are all lined up so that the tides due to the sun and moon coincide, making the high tides higher than average and the low tides lower than average. (13.4)

stable equilibrium   The state of an object balanced so that any small displacement or rotation raises its center of gravity. (10.5)

standing wave   Wave in which parts of the wave remain stationary and the wave appears not to be traveling. The result of interference between an incident (original) wave and a reflected wave. (25.8)

streamline   The smooth path of a small region of fluid in steady flow. (20.7)

strong force   The force that attracts nucleons to each other within the nucleus; a force that is very strong at close distances but decreases rapidly as the distance increases. (39.1)

subtractive primary colors   The colors of magenta, yellow and cyan. These are the three colors most useful in color mixing by subtraction. (28.7)

superconductor   Material that has infinite conductivity at very low temperatures, so that charge flows through it without resistance. (32.4)

support force   Force that completely balances the weight of an object at rest. (4.7, 13.3)

tangential speed   The speed of an object moving along a circular path. (9.2)

tangential velocity   Component of velocity tangent to the trajectory of a projectile. (12.2)

telescope   Optical instrument that forms images of very distant objects. (30.5)

temperature   The property of a material that tells how warm or cold it is relative to some standard. In an ideal gas, the molecular kinetic energy per molecule. (21.1)

terminal speed   The speed at which the acceleration of a falling object is zero because friction balances the weight. (5.7)

terminal velocity   Terminal speed together with the direction of motion (down for falling objects). (5.7)

terrestrial radiation   Radiant energy emitted from the earth. (22.7)

theory   A synthesis of a large body of information that encompasses well-tested and verified hypotheses about aspects of the natural world. (1.4)

thermal contact   The state of two or more objects or substances in contact such that it is possible for heat to flow from one object or substance to another. (21.2)

thermal equilibrium   The state of two or more objects or substances in thermal contact when they have reached a common temperature. (21.3)

thermodynamics   The study of heat and its transformation to mechanical energy. (24.0)

thermonuclear fusion   Nuclear fusion brought about by extremely high temperatures. (40.6)

thermostat   A type of valve or switch that responds to changes in temperature and that is used to control the temperature of something. (21.8)

time dilation   An observable stretching, or slowing, of time in a frame of reference moving past the observer at a speed approaching the speed of light. (15.1)

torque (TORK)   The rotational analog of force; the product of force and the lever arm (measured in newton-meters). Torque tends to produce rotational acceleration. (11.1)

total internal reflection   The 100% reflection (with no transmission) of light that strikes the boundary between two media at an angle greater than the critical angle. (29.12)

transformer   A device for increasing or decreasing voltage through electromagnetic induction. (37.5)

transmutation   The conversion of an atomic nucleus of one element into an atomic nucleus of another element through a loss or gain in the number of protons. (39.6)

transparent   Term applied to materials that allow light to pass through them in straight lines. (27.4)

transverse wave   A wave with vibration at right angles to the direction the wave is traveling. (25.5)

trough (TRAWF)   One of the places in a wave where the wave is lowest or the disturbance is greatest in the opposite direction from a crest. (25.2)

ultrasonic   Term applied to sound frequencies above 20 000 hertz, the normal upper limit of human hearing. (26.1)

ultraviolet   Electromagnetic waves of frequencies higher than those of violet light. (27.3)

umbra   The darker part of a shadow where all the light is blocked. (27.6)

unstable equilibrium   The state of an object balanced so that any small displacement or rotation lowers its center of gravity. (10.5)

universal gravitational constant   The constant G in the equation for Newton's law of universal gravitation; measures the strength of gravity. (12.4)

V   (a) The symbol for *volt*. (33.5) (b) In lowercase italic, the symbol for *speed* or *velocity*. (2.2, 2.3) (c) In uppercase italic, the symbol for *voltage*. (33.5)

vector   An arrow whose length represents the magnitude of a quantity and whose direction represents the direction of the quantity. (3.1)

vector quantity   A quantity in physics, such as force, that has both magnitude and direction. (3.1)

velocity   Speed together with the direction of motion. (2.3)

vibration   An oscillation, a repeating, back-and-forth motion, about an equilibrium position. (25.0)

virtual image   An image formed through reflection or refraction that can be seen by an observer but cannot be projected on a screen because light from the object does not actually come to a focus. (29.3, 30.2)

volt   The SI unit of electric potential. One volt (symbol V) is the electric potential difference across which one coulomb of charge gains or loses one joule of energy. (33.5)

voltage (VOL-tij)   (a) Electric potential; measured in volts. (33.5) (b) Potential difference; measured in volts. (34.1)

voltage source   A device, such as a dry cell or generator, that provides a potential difference. (34.3)

W   (a) The symbol for *watt*. (8.2) (b) When in italic, the symbol for *work*. (8.1)

watt (WAT)   The SI unit of power. One watt is expended when one joule of work is done in one second. (8.2)

wave   A "wiggle in space and time"; a disturbance that repeats regularly in space and time and that is transmitted progressively from one place in a medium to the next with no actual transport of matter. (25.0)

wave front   The crest, trough, or any continuous portion of a two-dimensional or three-dimensional wave in which the vibrations are all the same way at the same time (see Figure 29.14). (29.6)

wavelength   The distance from the top of the crest of a wave to the top of the following crest, or equivalently, the distance between successive identical parts of the wave. (25.2)

weight   The force on a body due to the gravitational attraction of another body (commonly the earth). (4.5)

weightlessness   The condition of free fall toward or around the earth, in which an object experiences no support force (and exerts no force on a scale). (13.3)

weight density   Weight of a substance divided by its volume. (18.2)

white light   Light, such as sunlight, that is a combination of all the colors. Under white light, white objects appear white and colored objects appear in their individual colors. (28.1)

work   The product of the force on an object and the distance through which the object is moved (when force is constant and motion is in a straight line in the direction of the force); measured in joules. (8.1)

Index

Acknowledgments

For improvements in this third edition I am enormously grateful to Kenneth Ford (Figure 26.15) and consultant Charles Spiegel (Figure 30.25). These two worked very hard to make it squeaky clean. Ken, retired Executive Director and CEO of The American Institute of Physics, reviewed every equation and sentence for physics accuracy, while Charlie did the same for syntax.

For getting this book started in the first place, and being great resources for subsequent editions, I am thankful to my two friends Marshall Ellenstein (Figure 1.3) and Paul Robinson (Figure 5.8), who are pilot teachers. I miss my late son James (Figure 11.24) who cut the stencils for shading the artwork in the first edition—I'm thankful he left me a grandson, Manuel (Figure 6.15).

For critiquing sample chapters of the first edition, which carries over to this edition, I am thankful to Hal Eastin, Donald Gielow, Robin Gregg, Charlie Hibbard, Willa Ramsey, and especially to pilot teachers Nathan Unterman and Nancy Watson, and consultants Clarence Bakken, Art Farmer, and Sheron Snyder.

For input to the second edition, which carries over to this edition, I thank Robert Baruffaldi, Howard Brand, Ted Brattstrom, Paul Doherty, Peter Crooker, Jerry Hosken, Meidor Hu, Mario Iona, Chelcie Liu, John Learned, Dack Lee, Tenny Lim (Figure 8.4), Ron Lindemann, Marshall Mok, Lev Okun, Ernie Santner, John Suchocki, Vic Stenger, Chester Vause, Kellie Werkmeister, and my daughter Leslie (Figure 17.1).

For input to this third edition, I am grateful to Clarence Bakken, Jim Court, Alan Davis, Sumner Davis, Ken Ganezer, Bruce Gregory, Lonnie Grimes, Manuel Hewitt, Jerry Hosken, Marilyn Hromatko, Mei Tuck Hu, John Hubisz, Dan Johnson, Annie Kwak, Jose Larios, Dack Lee, Cherie Lehman, Graham Robertson, Les Ross, Ronald Shaw, and David Yee. I thank Meidor Hu (Figure 33.17) for manuscript assistance. I am grateful to Helen Yan (Figure 22.13) for hand lettering the illustrations.

San Francisco, Paul G. Hewitt

Conceptual Physics Photo Album

Conceptual Physics is a very personal book, and this is reflected in the many photographs of family and friends found throughout this edition. My wife Millie appears with our grandson Manuel in the "You cannot touch without being touched" photo (Figure 6.15). My late son James, Manuel's father, is shown illustrating gyroscopic motion in Figure 11.23, and again as a tot in the magnifying glass in Figure 30.8. My son Paul illustrates adiabatic compression in Figure 24.3, and his wife Ludmila holds the polarizing filters in Figure 27.19. My favorite photo of my daughter Leslie holding a molecular model is seen in Figure 17.1. My brother Dave Hewitt (no, not a twin) and wife Barbara operate the water pump in Figure 20.9. Other family members include "nephews" Marques Jones, who compares thermal conductivities to open Unit 3, and Ryan Paterson showing magnetic induction in Figure 36.10.

The pilot with the sound-canceling earphones in Figure 26.15 is physicist Ken Ford, whose savvy for physics embellishes this edition, and to whom this Third Edition is dedicated. The photo of washtub bass shown in Figure 26.8 comes from the album of his son Paul's wedding.

Dear friends include Charlie Spiegel (Figure 30.25), whose grammatical touch is added to every page of this edition, and Marshall Ellenstein (Figure 1.3). Lab manual author Paul (Pablo) Robinson lies between beds of sharp nails in Figure 5.8. Friend and mathematics teacher Gwen Roberts's granddaughter, Amor Yates, opens Unit 4, and CCSF colleague David Yee's daughter Sarah opens Unit 5.

Meidor Hu, formerly in Figure 1.1 of the second edition, now stands electrified in Figure 33.17. Other dear friends include my former student Tenny Lim, a design engineer who puts energy into her bow in Figure 8.4, and Helen Yan, an orbit analyst for an aerospace company and coauthor of the Next-Time Questions. Helen holds the same box in Figure 22.13 that she held as a physics student for the first edition some ten years ago. Her hand lettering adorns the illustrations in this book. In Figure 7.11, colleague Will Maynez operates the air track he designed for the CCSF physics department.

These photographs of people very dear to me, all the more make *Conceptual Physics* a labor of love.

Photo Credits

iv T Tom Brakefield/DRK Photo; iv B NASA; v T Bill & Sally Fletcher/Tom Stack & Associates; v B F. Rickard-Artdia, Agence Vandystadt/Photo Researchers; vi T NIBSC/SPL/Photo Researchers; vi B Michael Townsend/Tony Stone Images; vii T Jerry Jacka Photography. Courtesy, The Heard Museum, Phoenix, AZ; vii B Pat Crowe/Animals, Animals; viii T Milton Rand/Tom Stack & Associates; viii B Marc Romanelli/The Image Bank; ix T Ken Karp*; ix B Patti McConville/The Image Bank; x Nuridsany et Perennou/Photo Researchers; xi Anne Dowie*; 0 NASA; 2 The Bettmann Archive; 4 Paul Hewitt; 7 Roger Ressmeyer/Starlight; 9 Liza Loeffler*; 10 Flip Chalfant/The Image Bank; 11 Tom Brakefield/DRK Photo; 12 Cheryl Fenton*; 22 Tim Davis/Photo Researchers; 28 Kim Taylor/Bruce Coleman Inc.; 34 Richard Megna/Fundamental Photographs; 37 David Madison; 38 David Madison/Duomo; 39 NASA; 43 S. Barrow/Superstock, Inc.; 47 Ken Karp*; 59 T M. Taylor/Superstock, Inc.; 59 B Michael Layton/Duomo; 61 Simon Bruty/Tony Stone Images; 65 Colin Vinnard; 69 T F. Rickard-Artdia, Agence Vandystadt/Photo Researchers; 69 B Stephen Dalton/NHPA; 70 Fundamental Photographs; 74 Timothy Shonnard/Tony Stone Images; 76 David Madison/Bruce Coleman Inc.; 82 Paul Hewitt; 86 Miller Photography; 88 ©The Harold E. Edgerton 1992 Trust, courtesy Palm Press, Inc.; 91 T Debra P. Hershkowitz/Bruce Coleman Inc.; 91 B Shirley Burman*; 96 Paul Hewitt; 99 Dan McCoy/Rainbow; 103 Thomas Braise/The Stock Market; 104 William R. Sallaz/Duomo; 105 Roger Ressmeyer/Starlight; 108 Paul Hewitt; 109 David Madison/Bruce Coleman Inc.; 111 Richard Megna/Fundamental Photographs; 118 T Mark A. Leman/Tony Stone Images; 118 B Walter Hodges/Tony Stone Images; 122 T Geoffrey Nilsen Photography*; 122 B Jeff Persons/Stock, Boston; 126 Harvey Lloyd/Peter Arnold, Inc.; 132 NASA; 136 Renee Lynn*; 137 Richard Megna/Fundamental Photographs; 140 London Transport Museum; 141 Denise Tackett/Tom Stack & Associates; 144 Kunio Owaki/The Stock Market; 145 David Madison; 146 Larry Brownstein/Rainbow; 150 Tim Davis; 151 Craig Aurness/Westlight; 152 Ken Karp*; 156 Tim Davis*; 160 David Madison; 161 R Richard Megna/Fundamental Photographs; 161 L Paul Hewitt; 162 David Madison/Duomo; 163 Bill & Sally Fletcher/Tom Stack & Associates; 164 Gerald Lacz/NHPA; 168 Geoffrey Nilsen Photography*; 178 Bill & Sally Fletcher/Tom Stack & Associates; 179 Roger Ressmeyer/Starlight; 182 Phil Schermeister/Tony Stone Images; 187 NASA; 192 Robert Harding Picture Library; 193 NASA; 199 NASA; 204 Cheryl Fenton*; 208 NASA; 212 Cheryl Fenton*; 213 Roger Ressmeyer/Starlight; 229 NASA; 232 SAO/IBM Research/TASDO/Tom Stack & Associates; 237 Manfred Gottschalk/Tom Stack & Associates; 240 R The Bettmann Archive; 240 L UPI/Bettmann; 243 Liza Loeffler*; 244 Bob Martin/Allsport; 245 Paul Hewitt; 248 T Enrico Fermi Institute/University of Chicago; 248 B Dr. Mitsuo Ohtsuki/SPL/Photo Researchers; 249 NIBSC/SPL/Photo Researchers; 251 Cheryl Fenton*; 255 Ken Graham/Tony Stone Images; 258 Mark Kelley/Stock, Boston; 259 Cheryl Fenton*; 264 Gabe Palmer/The Stock Market; 266 Barrie Rokeach/The Image Bank; 270 Joe Bensen/Stock, Boston; 273 Bill Bachman/Photo Researchers; 275 Michael A. Keller/The Stock Market; 280 Michael Townsend/Tony Stone Images; 290 Geoffrey Nilsen Photography*; 292 Ted Russell/The Image Bank; 295 Paul Hewitt; 296 T Ken Karp*; 296 B Paul Silverman/Fundamental Photographs; 299 Francis Lepine/Earth Scenes; 301 Liza Loeffler*; 306 Liza Loeffler*; 307 Cheryl Fenton*; 310 Jose Fuste Raga/The Stock Market; 312 GHP Studio*; 316 Breck P. Kent/Earth Scenes; 317 Ken Karp*; 325 Marc Romanelli/The Image Bank; 326 John Coletti/Stock, Boston; 332 T Anne Dowie*; 332 B Cheryl Fenton*; 334 Tom Arnborn/Marine Mammal Images; 339 Rick McIntyre/Tony Stone Images; 340 T Liza Loeffler*; 340 CR Pat Crowe/Animals, Animals; 340 CL Lawrence Migdale/Stock, Boston; 341 Cheryl Fenton*; 350 Bill Stormont/The Stock Market; 354 Simon Fraser/SPL/Photo Researchers; 358 Paul Hewitt; 360 Dan McCoy/Rainbow; 363 Jerry Jacka Photography. Courtesy, The Heard Museum, Phoenix, AZ; 364 Michael J. Howell/Stock, Boston; 366 Cory Wolinsky/Stock, Boston; 367 Catalyst/The Stock Market; 371 Liza Loeffler*; 372 David Weintraup/Photo Researchers; 373 Paul Hewitt; 375 Runk-Schoenberger/Grant Heilman Photography; 376

Kim Taylor/Bruce Coleman Inc.; 380 Education Development Center, Inc., Newton, MA. PSSC PHYSICS, 2nd Edition, 1965; 390 Cheryl Fenton*; 391 Richard Megna/Fundamental Photographs; 393 P. Saada/Eurelios/SPL/Photo Researchers; 394 Paul Hewitt; 396 AP/Wide World Photos; 398 Paul Hewitt; 399 Cheryl Fenton*; 400 Disario/The Stock Market; 404 John R. MacGregor/ Peter Arnold, Inc.; 411 Cheryl Fenton*; 412 Cheryl Fenton*; 413 The Exploratorium; 414 Ken Karp*; 415 Anne Dowie*; 421 T Charles Krebs/The Stock Market; 421 B The Bettmann Archive; 422 R Runk-Schoenberger/Grant Heilman Photography; 422 L Cheryl Fenton*; 423 R James H. Carmichael/The Image Bank; 428 Fundamental Photographs; 431 Charles Krebs/Tony Stone Images; 432 Charlie Ott/Photo Researchers; 433 Kathleen Campbell/Tony Stone Images; 434 Guido Cozzi/Bruce Coleman Inc.; 436 Don & Liysa King/The Image Bank; 437 Courtesy Pascoe Scientific; 438 Kodansha Ltd/The Image Bank; 442 Simon Bruty/Tony Stone Images; 446 Edna Douthat/Photo Researchers; 447 Dr. Jeremy Burgess/SPL/Photo Researchers; 448 Jane Lidz; 449 Education Development Center, Inc., Newton, MA. PSSC PHYSICS, 2nd Edition, 1965; 453 John M. Dunay IV/Fundamental Photographs; 454 Marc Romanelli/The Image Bank; 459 T Stephen J. Krasemann/Photo Researchers; 459 B Bill Pierce/Rainbow; 462 Cheryl Fenton*; 463 Cheryl Fenton*; 466 R Paul Hewitt; 466 L Renee Lynn*; 475 Ken Karp*; 476 Paul Hewitt; 480 Dan McCoy/Rainbow; 481 Burndy Library; 483 Education Development Center, Inc., Newton, MA. PSSC PHYSICS, 2nd Edition, 1965; 484 Richard Megna/Fundamental Photographs; 485 Norbert Wu/Peter Arnold, Inc.; 487 Educational Development Center, Inc., Newton, MA. PSSC PHYSICS, 2nd Edition, 1965; 488 TR Burndy Library; 488 TL David Muench; 488 B Paul Hewitt; 489 Michael Dalton/Fundamental Photographs; 490 T Kristen Brochmann/Fundamental Photographs; 490 B Milton Rand/Tom Stack & Associates; 491 T Bausch and Lomb Ferson Optics Division; 491 B GHP Studio*; 494 Ken Karp*; 499 Liza Loeffler*; 500 Richard Kaylin/Tony Stone Images; 503 Mitch Kezar/Phototake NYC; 508 Patti McConville/The Image Bank; 517 Adam Hart-Davis/SPL/Photo Researchers; 520 Harold Waage/Princeton University; 521 copyright BBC; 523 Ken Karp*; 526 Ken Karp*; 528 Anne Dowie*; 531 Ben Simmons/The Stock Market; 534 Ken Karp*; 536 Ken Karp*; 540 Ken Karp*; 544 Cheryl Fenton*; 548 T Andrew Syred/SPL/Photo Researchers; 548 B Cheryl Fenton*; 550 Ken Karp*; 552 Ken Karp*; 555 Ken Karp*; 562 Cheryl Fenton*; 563 T Eunice Harris/Photo Researchers; 563 B Cheryl Fenton*; 564 Richard Megna/Fundamental Photographs; 567 Lori Patterson; 569 T Richard Megna/Fundamental Photographs; 569 B Railway Technical Research Institute, Tokyo, Japan; 571 Dave Wilhelm/The Stock Market; 573 TL Ken Karp*; 573 B Stephen Rose/Rainbow; 577 Flip Chalfant/The Image Bank; 582 Chuck Fishman/Woodfin Camp & Associates; 585 Cheryl Fenton*; 595 Liza Loeffler*; 596 Dr. Mitsuo Ohtsuki/SPL/Photo Researchers; 598 Albert Rose; 600 R J. Valasek, *Introduction to Theoretical and Experimental Optics* (N.Y., Wiley, 1959); 600 L H. Raether, "Elektroninterferenzen," *Handbuch der Physick*, Vol 32 (Berlin, Springer Verlag, 1957); 601 Nuridsany et Perennou/Photo Researchers; 606 Phil Schermeister/Tony Stone Images; 609 CNRI/SPL/Photo Researchers; 616 T Bob Daemmrich/Stock, Boston; 616 B Hank Morgan/Photo Researchers; 620 Stanford Linear Accelerator Center; 625 Jose Fernandez/Woodfin Camp & Associates; 629 Cheryl Fenton*; 641 Lawrence Livermore National Laboratory; 644 Ken Karp*

* Photographed expressly for Addison-Wesley Publishing Company, Inc.